Contributions to Finance and Accounting

The book series 'Contributions to Finance and Accounting' features the latest research from research areas like financial management, investment, capital markets, financial institutions, FinTech and financial innovation, accounting methods and standards, reporting, and corporate governance, among others. Books published in this series are primarily monographs and edited volumes that present new research results, both theoretical and empirical, on a clearly defined topic. All books are published in print and digital formats and disseminated globally. This book series is indexed in Scopus.

Ignazio Basile • Pierpaolo Ferrari
Editors

Asset Management and Institutional Investors

Methods and Tools for Asset Allocation, Portfolio Management and Performance Evaluation

Second Edition

Springer

Editors
Ignazio Basile
Department of Economics and Management
University of Brescia
Brescia, Italy

Pierpaolo Ferrari
SDA Bocconi School of Management
Bocconi University
Milan, Italy

ISSN 2730-6038 ISSN 2730-6046 (electronic)
Contributions to Finance and Accounting
ISBN 978-3-031-59818-0 ISBN 978-3-031-59819-7 (eBook)
https://doi.org/10.1007/978-3-031-59819-7

This Springer imprint is published by the registered company Springer Nature Switzerland AG
The registered company address is: Gewerbestrasse 11, 6330 Cham, Switzerland

Foreword

It is a pleasure for me to present this book devoted to the investment management policy of institutional investors, which a group of friends and colleagues have written with great dedication and expertise, providing a methodical and in-depth perspective of a multi-faceted topic that is in continuous evolution.

The book is composed of four parts. The first one analyses the different types of institutional investors, institutions which, with different objectives, professionally manage portfolios of financial and real assets on behalf of a wide variety of individuals. The development factors and benefits for the financial system generated by institutional investors are identified, and a comparison is made at the international level. The first part goes on with an in-depth analysis of the economic, technical, and regulatory characteristics of the different types of investment funds, assuming the perspective of a European Union investor. Management strategies, restrictions to investment policies, public documentation, and charges of investment funds are explored. This section also analyses other types of asset management products which have a high rate of substitutability with investment funds and represent their natural competitors.

The second part of the book identifies and investigates the stages of the investment portfolio management, dividing the investment process into the following stages: the identification of the objectives and the constraints of the investment policy; the formalisation of the investment strategy; the implementation of the financial strategy; the periodic rebalancing of the portfolio; the assessment of results and risk control. Given the importance of strategic asset allocation in explaining the *ex post* performance of any type of investment portfolio, this section provides an in-depth analysis of asset allocation methods, illustrating the different theoretical and operational solutions available to institutional investors. This section focuses on the concepts and applications of traditional approaches to asset allocation, based on mean-variance optimisation, but also deepens the new risk-based approaches for asset allocation, which eliminate estimation risks associated with the traditional approaches. Finally, this second part concludes with a presentation of the methods and instruments for portfolio selection available to institutional investors for a more aware identification of the "optimal portfolio", taking into consideration management objectives and constraints.

The third part describes performance assessment, its breakdown, and risk control. The first step in performance assessment is the calculation of return, with the identification of the most appropriate measure among the different methods of calculation available. Performance evaluation requires then the identification of the risk of an investment portfolio, in its different forms of absolute, asymmetric, and relative risk, and the calculation of the related risk-adjusted performance measures, which will enable to assess the efficiency of the asset manager with respect to the benchmark, the competitors, and the *ex ante* risk limits. When comparing the performance of competitors, it is essential that homogenous peer groups be created, made up of portfolios with the same management approach. To this end, the most common operational solution is to create a peer group based on the investment style, using a deductive approach founded on the so-called returns-based style analysis. This section provides alternative methods and utilisation rationales of style analysis. In the case of non-indexed portfolios, an in-depth *ex post* performance assessment requires also the evaluation of the asset manager in terms of the ability to realise effective stock picking and market timing activities. For this purpose, this section presents the most appropriate performance attribution model aimed to shed light on those management choices that generated the gap between the overall result of the portfolio and the benchmark, breaking down relative performance into its determinants and attributing it to the various factors that contributed to its generation.

Finally, the fourth part deals with the subject of diversification towards alternative asset classes, identifying the common characteristics and their possible role within the framework of investment management policies. This section analyses hedge funds, identifying their operational characteristics, management strategies, the regulatory framework, and the specific performance assessment techniques. The distribution of the hedge funds returns highlights some statistical anomalies that compromise the validity of normality assumption, which is a necessary condition for the application of classic performance measures. The section continues with a complete and exhaustive analysis of the other types of alternative investments: private equity; private debt; real estate; infrastructures; commodities; currency overlay techniques. A common characteristic of these alternative asset classes is to have returns that are often uncorrelated with those of traditional asset classes, although direction and intensity of the correlation varies greatly over time and according to the specific category of alternative investment.

I believe it is worthwhile to highlight three strengths that enrich this work and make it an easy reading and of interest to a wide audience. First, an abundance of up-to-date data and numerical and graphic illustrative tables. This is a very important aspect that facilitates the understanding of even the most complex concepts. Second, the ability to analyse frontier themes in relation to investment management, without underrating the fundamental concepts that lie at the basis of the most advanced research. This feature makes the book of value and of service to experts, who are interested in cutting-edge models and techniques, and also to those who are less experienced on the subject. Lastly, there is the meticulous work done by the editors, who, thanks to the logical coherence and sequence of the subjects, have succeeded in the difficult task of making a book written by different qualified authors comparable to a manual written by a single author.

To conclude, in its thorough and precise analysis of all these subjects, the book edited by Ignazio Basile and Pierpaolo Ferrari is a useful and effective tool for tackling the topic of the investment management policy of institutional investors.

Bocconi University of Milan,
Milan, Italy

Assicurazioni Generali S.p.A.,
Trieste, Italy

Andrea Sironi

Contents

Part I Institutional Investors: Typologies, Roles and Products

1 **Institutional Investors** 3
Ignazio Basile
1.1 Institutional Investor Features 3
1.2 Development Factors 10
1.3 Benefits to the Financial System 11
1.4 An International Comparison 18
References 28

2 **Collective Investment Vehicles and Other Asset Management Products** 31
Pierpaolo Ferrari
2.1 Asset Management Products 31
2.2 Collective Portfolio Management 31
2.3 Investment Funds 33
2.4 Unit Trust Schemes 78
2.5 Open-End Investment Companies 79
2.6 Other Asset Management Products 81
References 90

Part II Investment Management Policy

3 **Stages of Investment Management Policy** 95
Pierpaolo Ferrari
3.1 Investment Management Policy 95
3.2 Identification of Objectives and Constraints 95
3.3 Formalisation of the Investment Policy 97
3.4 Implementation of the Financial Strategy 98
3.5 Periodic Portfolio Rebalancing 108
3.6 Performance Assessment and Risk Control 113
References 114

4 Strategic Asset Allocation with Mean-Variance Optimisation 117
Maria Debora Braga
4.1 Strategic Asset Allocation with Markowitz's Mean-Variance Optimisation 117
4.2 The Assumptions of Mean-Variance Optimisation 125
4.3 Estimation Risk and Practical Problems with Mean-Variance Optimisation 131
4.4 The Application of the Markowitz Approach: An Example 139
4.5 Strategic Asset Allocation and Estimation Risk Management: An Initial Description 150
4.6 The Additional Weight Constraints Method 152
4.7 The Resampling Method 160
4.8 Bayesian Strategic Asset Allocation: The Black-Litterman Model 172
4.9 Out-of-Sample Analysis to Evaluate Investment Strategies 193
References 200

5 Methods and Tools for Portfolio Selection 203
Maria Debora Braga
5.1 Portfolio Selection 203
5.2 Level of Diversification of Optimal Portfolios 204
5.3 Analysis of Portfolio Risk Using Asymmetric Measures 217
5.4 Constraints on Risk-Taking 226
5.5 Analysis of Future Scenarios of the Cumulative Wealth 233
References 241

6 Alternative Approaches to Traditional Mean-Variance Optimisation 243
Maria Debora Braga
6.1 Alternative Asset Allocation Approaches to Traditional Mean-Variance Optimisation 243
6.2 The Global Minimum-Variance Strategy 245
6.3 The Optimal Risk Parity Strategy 248
6.4 Portfolio Selection Under Alternative Asset Allocation Approaches 251
References 254

7 Integrating Sustainability Criteria into the Investment Process 257
Ignazio Basile and Pierpaolo Ferrari
7.1 Sustainable Investing 257
7.2 The European Union Framework 258
7.3 Relation Between Investment Sustainability and Performance . . 268
7.4 ESG Rating 270
7.5 Approaches to Sustainable Investing 276
References 280

Part III Performance Evaluation for Traditional Investment Portfolios

8 Performance Evaluation 285
Pierpaolo Ferrari
8.1 Performance Evaluation Stages 285
8.2 Return on an Investment Portfolio 286
8.3 Investment Portfolio Risk Measures 305
8.4 Risk-Adjusted Performance Measures 312
8.5 Evaluation of Stock Picking Ability 329
8.6 Assessment of Market Timing Ability 332
8.7 Performance Persistence 341
References 342

9 Returns-Based Style Analysis 345
Maria Debora Braga
9.1 The Role of Returns-Based Style Analysis 345
9.2 Methodological Aspects of Returns-Based Style Analysis 346
9.3 An Empirical Application of Returns-Based Style Analysis 355
9.4 Rolling Returns-Based Style Analysis 363
9.5 An Empirical Application of Rolling Returns-Based Style Analysis 366
References 370

10 Performance Attribution 373
Maria Debora Braga
10.1 Performance Attribution: Definition, Objectives, and Assumptions 373
10.2 Performance Attribution Using the Brinson Model 377
10.3 Integration of the Currency Effect into Performance Attribution 387
10.4 Multi-Period Arithmetic Performance Attribution 393
References 395

Part IV Portfolio Diversification Towards Alternative Asset Classes

11 Portfolio Diversification Policies: Alternative Asset Classes 399
Ignazio Basile
11.1 Alternative Investment Features 399
11.2 Alternative Investments from a Portfolio Perspective 400
11.3 The New European Union Regulations 407
References 411

12 Hedge Funds 413
Ignazio Basile
12.1 Hedge Fund Features 413
12.2 Management Strategies 416
References 428

13 Hedge Fund Performance 431
Roberto Savona
13.1 The Complexity of Hedge Fund Performance Measurement 431
13.2 Classical Performance Measures 433
13.3 Advanced Performance Measures 439
13.4 Measuring Hedge Fund Performance: An Empirical Application 444
References 449

14 Private Equity 451
Guido Abate
14.1 Private Equity Features 451
14.2 Investment Vehicles 453
14.3 Management Strategies 460
14.4 Risk and Return Measurement 465
References 479

15 Private Debt 481
Guido Abate
15.1 Private Debt Features 481
15.2 Investment Vehicles 482
15.3 Management Strategies 485
15.4 Risk and Return Measurement 487
References 493

16 Real Assets 495
Guido Abate
16.1 Real Asset Features 495
16.2 Real Estate 495
16.3 Real Estate Investment Vehicles 499
16.4 Real Estate Management Strategies 507
16.5 Risk and Return Measurement in Real Estate 509
16.6 Infrastructures 517
16.7 Infrastructures Management Strategies 518
16.8 Risk and Return Measurement in Infrastructure Investments 520
References 522

17 Commodities 525
Guido Abate
17.1 Commodity Features 525
17.2 Risk and Return of Commodities 530
17.3 Commodities Within the Financial Assets Portfolio 536
17.4 Investment Vehicles in Commodities 539
References 544

18 Currency Overlay 547
Guido Abate
18.1 Passive Versus Active Currency Overlay 547
18.2 Passive Currency Overlay 549
18.3 Active Currency Overlay 557
References 561

Contributors

Guido Abate Department of Economics and Management, University of Brescia, Brescia, Italy

Ignazio Basile Department of Economics and Management, University of Brescia, Brescia, Italy

Maria Debora Braga Department of Economics and Political Science, University of "Valle d'Aosta", Aosta, Italy
SDA Bocconi School of Management, Bocconi University, Milan, Italy

Pierpaolo Ferrari Department of Economics and Management, University of Brescia, Brescia, Italy
SDA Bocconi School of Management, Bocconi University, Milan, Italy

Roberto Savona Department of Economics and Management, University of Brescia, Brescia, Italy

Part I
Institutional Investors: Typologies, Roles and Products

Chapter 1
Institutional Investors

Ignazio Basile

1.1 Institutional Investor Features

In the absence of a specific legal framework, the expression "institutional investors" identifies all the entities that manage portfolios of financial and real assets professionally on behalf of a plurality of investors with different goals.[1]

The need to focus on the real evolution of the structure of financial systems, rather than on the relevant regulatory framework of financial activities, derives from the heterogeneity of the institutional players involved as well as the complexity of this phenomenon, which cannot be easily delimited.

Focusing on the European Union, the Markets in Financial Instruments Directive II (MiFID II) does not provide any deterministic specifications regarding the definition of institutional investors, though setting the foundations for a comprehensive reorganisation of the asset management activities, as it refers to the logical category under investigation only when segmenting customer clusters.

Institutional investors are indeed viewed as a subset of "professional clients", identified as follows: "professional client is a client who possesses the experience, knowledge and expertise to make its own investment decisions and properly assess the risks that it incurs".[2] In fact, the existing categorisation is not a defined list itself, since a general reference to "other institutional investors" is also recalled together with explicit references to insurance companies, collective investment schemes and management companies of these schemes, pension funds and management

[1] Davis and Steil (2001).

[2] Directive 2014/65/EU, Annex 2.

I. Basile (✉)
Department of Economics and Management, University of Brescia, Brescia, Italy
e-mail: ignazio.basile@unibs.it

I. Basile, P. Ferrari (eds.), *Asset Management and Institutional Investors*, Contributions to Finance and Accounting,
https://doi.org/10.1007/978-3-031-59819-7_1

companies of these funds with the following specification: "whose main activity is to invest in financial instruments, including entities dedicated to the securitisation of assets or other financing transactions".[3]

Since the MiFID Directive puts its focus on the level of protection to be granted to the various categories of investors, it is obvious that an explicit reference to institutional investors was only incidentally made, as the lowest common denominator of professional clients is not identified with their functional specialisation, but with the level of financial expertise they have attained. This condition is also confirmed by the fact that these investors have the option to request non-professional treatment and benefit from a higher level of protection.

That being stated, the lowest common denominator among all the categories of institutional investors is represented by the professional management of assets on behalf of a plurality of individuals. Such management can be carried out on a collective basis, as in the case of investment funds, or on an individual basis, as in the case of discretionary mandates given on a client-by-client basis.

If we accept this assumption, we can define the following investors as institutional:

- Collective investment vehicles (CIVs).
- Individual portfolio management, based on mandates given by clients on an individual and discretionary basis.
- Insurance companies.
- Pension funds.
- Other institutions for occupational or personal retirement provisions.
- Foundations and endowments.[4]

Nevertheless, some clarifications are needed. In the case of collective and individual portfolio management, reference is commonly made to the activity rather than to the institution which provides the service, because, on the one hand the three types of existing CIVs (common contractual funds, trusts, and investment companies) have very different technical and legal characteristics from country to country, and on the other individual portfolio management can be provided both by specialised institutions (management companies, investment companies, and other specialised financial intermediaries) and directly by universal banks.

In turn, pension funds and other institutions for occupational and personal retirement provisions, albeit sharing similar social security purposes, are subject to very different regulatory frameworks in several countries, which require distinct patterns of analysis.

Finally, as for foundations and endowments, the institutional and operating context in which they work allows to propose the necessary generalisation only in

[3] Directive 2014/65/EU, Annex 2.

[4] Assuming a more extensive definition of institutional investor, we can include banks, given their need of managing securities in the proprietary book, and also sovereign funds, for their ability to manage diversified asset portfolios.

some cases, for instance with respect to grant-making foundations which should invest their assets to generate income for their institutional objectives characterised by general public interest and socially-oriented goals.

Under this premise, the institutional/functional profiles of the various types of institutional investors previously identified can be outlined here. In particular, the activities of each category are summarised in Table 1.1.[5]

In terms of regulation, we should highlight that each institutional investor category refers to a specific regulatory framework, at either primary or secondary level, in all European Union countries. In many cases, these regulations are not incorporated within a unified framework but derive from the layering of successive measures adopted in time (laws, decrees, regulations, supervisory authorities' instructions).

Moreover, the organisational structure of supervisory functions does not provide any distribution of responsibility that refers to a unique reference model. While at the European Union level an "objective-oriented model" is preferred, with the single authorities being assigned specific objectives (stability, transparency, etc.), in some countries the so-called "institutional model" still prevails, with a single authority concentrating all functions on a given category of institutions. Furthermore, we can often detect a slow rationalisation and convergence process of the two models, resulting in the concentration of all power into a single supervisory entity.[6]

Taking into consideration management profiles, this topic can be investigated from two different perspectives, comparing the management and the organisational approach.

Table 1.2 highlights how the following aspects should be considered when comparing the institutional investors' management strategies:

- Objectives.
- Risk attitude.
- Time horizon.
- Financial profile.
- Regulatory constraints.
- Management policies.[7]

In the case of individual and collective portfolio management, the frequent absence of a specific time horizon makes it difficult to define return targets in absolute terms; it also makes inevitable the comparison with market parameters, on the one hand, and with competitors, on the other. This leads to the widespread adoption of relative risk measures and *ex post* evaluation logics based on risk-adjusted performance indicators. From a financial perspective, for open-end CIVs and individual portfolio management, the absence of exit barriers imposes the need to maintain cash reserves and practically precludes any form of medium- and long-term financial planning.

---

[5]The different categories of CIVs will be discussed further in Chap. 2.

[6]Çelik and Isaksson (2013).

[7]Bushee (2001).

Table 1.1 Institutional investors: distinctive characteristics

Common contractual funds	Common funds are collective investment schemes (CIVs)—without legal status—that allow a number of separate and unrelated investors to pool their capital in order to access to professional management and to achieve a broader diversification. They are run by a management company and represented by units.
Trust funds	The trust form of CIVs is found in jurisdictions where the English common law system prevails other than the United States. The investment manager and trustee jointly sign the "trust deed", which determines how the trust is to be established. Investors are beneficiaries of the trust. The most recurring scheme within this category is that of unit trust, which is an open-end scheme. The pool of investments is divided into equal portions called units.
Investment companies	Another alternative legal structure for collective portfolio management is that of open-end investment companies or closed-end investment companies, where the investment company is a separate legal entity and the investors are shareholders. This legal form is known in many countries, when the scheme is open-end, with the French expression "société d'investissement à capital variable" (SICAV), equivalent to "investment company with variable capital" (ICVC), and, when the scheme is closed-end, as "société d'investissement à capital fixe" (SICAF).
Individual portfolio management	Investment services in which the investor's assets are managed on a discretionary and customised basis, by allocating them to one or more financial instruments, under a mandate given by the investor to authorised intermediaries.
Insurance companies	Companies that institutionally and systematically undertake and manage risks transferred by other companies and individuals, against insurance premiums whose amounts depend on the probability of the occurrence of the events to which risks are related.
Pension funds	Pool of assets formed with the contributions to a pension plan for the exclusive purpose of financing pension plan benefits. Fund members have a legal or beneficial right or some other contractual claim against the assets of the pension fund. Pension funds take the form of either a special purpose entity with legal status (such as a trust, foundation, or corporate entity) or a legally separated fund without legal status managed by a pension fund management company or other financial institution on behalf of the fund members. Open pension funds support at least one plan with no restriction on membership. Closed pension funds only support pension plans that are limited to certain employees.

(continued)

Table 1.1 (continued)

Other institutions for occupational or personal retirement provisions	Institutions that aim to provide social security service to their members. They may have a voluntary or mandatory membership, depending on the local regulations, and take different legal forms in the various countries.
Foundations and endowments	Not-for-profit organisations that direct the income generated by their asset management activity towards collective interest and socially-oriented goals.

The situation is quite different for those institutions that have shaped their investment policy, based on the undertaken financial and time commitments, in a deterministic way or with a sufficiently accurate assessment. This is the case of insurance companies, pension funds, and other institutions for occupational or personal retirement provisions, which benefit from a greater programmability of their cash flow and a medium-term to long-term reference time horizon for their allocation choices. In this particular context, the discriminating factor is the nature—defined or not—of the benefits provided to policy-holders/members. In the absence of defined benefits, the allocation policies are similar to those of CIVs and individual portfolio management, while in the case of specific commitments in terms of guarantees on capital or returns, it is necessary to implement asset-liability management policies.[8]

From an operational point of view, it is essential indeed to distinguish between institutional investors who can adopt asset-only management policies and those who must implement asset-liability management policies.

In the first case, the institutional investor has no predefined performance commitments. Thus, this investor can adopt an investment policy based on the merely optimal allocation of assets, taking into account the needs of liquidity required by the specific management model.

In the second case, the institutional investor needs to set an investment policy in compliance with the commitments towards final investors. In this case, it is necessary to carry out close monitoring of the surplus or deficit between the accrued value of investments on a certain date and the current value of outstanding liabilities and commitments on the same date.

Finally, the position of foundations and endowments is hybrid, as economic performance is not the ultimate management goal, but is functional to achieve social purposes, which are to be considered as a priority. Likewise, the need of any periodic flow has to be mediated with the benefit of operating, as a rule, on a long-term time horizon, preserving the value of their assets in real terms.[9]

The choice of the organisational model for the management activities is only partially independent from the pursued objectives and is strongly influenced by the

[8] Blake (2006).

[9] Acharya and Dimson (2007).

Table 1.2 Investment strategies of the different institutional investor categories

	Objectives	Risk attitude	Time horizon	Financial profile	Management constraints	Management approach
CIVs and individual portfolio management	Maximising returns against a benchmark (if any) and against competitors, respecting given risk constraints	Variable, depending on the risk profile	Variable time period	Daily cash flows which generally cannot be planned	Stringent regulatory constraints	Asset-only management
Insurance companies	Investment of the premiums received in advance with the aim of generating financial resources to meet future policy-holders benefits and claims	Medium-low	Variable time period, depending on the contract category	Predictable liquidity needs	Stringent regulatory constraints	Asset-liability management
Defined benefit pension funds	Ensure a predefined benefit to members (also through the variation of contributions required)	Variable, depending on the level of coverage of liabilities and commitments	Long- term period	Predictable liquidity needs based on members' age and working life	Stringent regulatory constraints	Asset-liability management
Defined contribution pension funds	Maximising returns for a given level of risk	Variable, depending on the fund category	Long-term period	Predictable liquidity needs based on members' age and working life	Stringent regulatory constraints	Asset-only management
Other institutions for occupational or personal retirement provisions	Ensure social security benefits to members	Medium-low	Long-term period	Predictable liquidity needs based on members' age and working life	Often less stringent regulatory constraints	Asset-only or asset-liability management, depending on institution characteristics
Foundations and endowments	Preserve the real value of investments in order to fulfil statutory obligations	Variable, depending on institution characteristics	Long-term period	Recurrent and extraordinary liquidity needs	Often less stringent regulatory constraints	Asset-only management

regulatory environment. The possible options, which are articulated depending on the type of institution considered, are essentially two:

- Internal management.
- External delegated management.

The choice between the two options is influenced by various factors, unless they are hindered by any regulation constraints and concern:

- The skills and resources required to achieve the institutional objectives.
- The existence of any minimum organisational requirements on behalf of the supervisory authorities.
- The degree of diversification by asset class and market assigned in the strategic asset allocation process.
- The size of assets under management, which is inversely related to the degree of operational risk associated to the implementation of direct management strategies.

In such a complex scenario, it is obviously difficult to generalise. Nevertheless, at an international level, there is a clear and irreversible preference towards a partial or total external delegation of the different stages concerning the portfolio management process and in particular:

- The strategic asset allocation planning, in accordance with given objectives and constraints.
- The selection of one or more managers for each asset class.
- The performance and risk monitoring for the selected managers and for the portfolio as a whole.

Moreover, the advisory activity which accompanies institutional investors' portfolio decisions is often externalised.[10]

Externalisation can regard not only front office activities but also back office and internal control functions (risk management, internal auditing, and compliance), notwithstanding the need to have adequate internal organisational structures to ensure full control over functions is no less crucial than allocation choices.

[10] Advisory and management functions are integrated into the fiduciary manager model, which is largely widespread in Anglo-Saxon countries, which manages the assets of an institution as a whole, sub-delegating the management of specific sectors to ad hoc selected managers. In this way, without stirring conflicts of interest, a unitary direction to all management activities and a more effective risk control are ensured. Refer to van Nunen (2007).

1.2 Development Factors

When investigating complex phenomena such as the interdependence between the morphological characteristics of a financial system and the internal role played by institutional investors, it is not easy to identify the cause-and-effect relationships in an unambiguous manner. The same factors that fostered the growth of the sector in the first stage very often promoted, at the same time, a more significant presence of institutional investors in the system. Although a dynamic mutual relationship arises between institutional investors and the financial system in a real scenario, these relationships will be investigated separately as if they were one-directional relations, for the sake of clarity and to be more schematic.[11]

As extensive literature reports, a driving force of particular importance for the development of institutional investors is attributed to factors related to both demand and supply of financial instruments and services, on the one hand, and to some factors which are exogenous to the financial system, on the other.[12]

The surplus units have indeed expressed investment needs that are increasingly articulated and sophisticated, showing growing difficulties to make fully aware investment decisions, in an increasingly complex and uncertain environment. The evolutionary stage reached by financial systems, with a particular emphasis on the presence of increasingly developed and efficient securities markets, has prompted final investors to delegate their investment decisions to entities with higher professional skills.

In contrast, institutional investors have been able to perform some of the primary functions assigned to the financial system more effectively, providing greater opportunity for diversification and more efficient risk-return combinations. Their success comes both with a more favourable competitive environment and also thanks to the gradual fall of barriers to entry, with globalisation phenomena across the asset management industry.

Socio-political exogenous variables can also play a decisive role in accelerating the above-mentioned virtuous process. In the context of an irreversible common ageing process of the population, due to lower birth rates and the lengthening of life expectancy, the global demographic changes have questioned the sustainability of the current social security systems and public pension schemes everywhere, leading to the comprehensive redesigning of the typical financial choices of the life cycle phases. There has been a widespread relative increase in the age classes characterised by a higher savings rate, these groups being more sensitive to social security issues.

---

[11] Vittas (1998).

[12] Davis and Steil (2001) and Ruiz (2018).

1.3 Benefits to the Financial System

When institutional investors achieve a significant relative weight compared to the total intermediation volumes of the financial system, albeit not automatically, the latter receives positive impulses in terms of competitive structure and efficiency.

Institutional investors can, in fact, have a tangible and significant impact concerning:

- Structural setting of financial systems.
- Strategies of financial intermediaries.
- Market organisation.
- Market efficiency.
- Governance model and performance of investee companies.
- Regulation and control of the financial activity.
- Asset management industry.

1.3.1 Structure of Financial Systems

In developed countries, the institutionalisation of savings is seen as a symptom and at the same time as a driving force for the evolution of financial systems. The latter are constantly seeking, in the performance of all their functions (payment system management, transfer of financial resources, risk management, pricing of financial assets), to respond better and more effectively at regulatory, organisational, and operational levels, against increasingly more complex and sophisticated needs expressed by all involved players.

As a confirmation, all major measurement parameters of the structural characteristics of a financial system are indeed positively influenced by the growing weight of the institutional investors.[13] For example, the Financial Interrelation Ratio (FIR), which measures the relative size of the financial system compared to that of the underlying economic system, is much higher in the countries where savers tend to move towards the massive delegation of their investment decisions to specialised operators.

A significant impact of the presence of institutional investors on the financial structure of the economy is also measurable with the Financial Intermediation Ratio, that is the ratio between the assets of financial institutions and all financial liabilities in the economy, and with the Banking Intermediation Ratio, which derives from the comparison between the assets of banks and those of all the financial institutions as a whole. Both ratios tend to decrease, leading to the reduction of the incidence of intermediation costs, in the case of a rebalancing between direct and indirect

[13] OECD (2022).

financial circuits and the gradual and increasing development of the role played by institutional investors.

With regard to the above, it should be noted that everywhere, both in market-based and in bank-based financial systems, an irreversible trend emerges in which flows managed by pure direct and indirect financial circuits are gradually passing over to the so-called direct circuits assisted by financial institutions, which enhance the typical functions of investment banking and asset management industry. In this perspective, an increasing spread of the processes of partial or total delegation of the investment decisions is more than physiological.[14]

1.3.2 Strategies of Financial Institutions

One of the most noticeable effects of the gradual and growing expansion of the role of institutional investors is a sort of counterbalancing of the dominant power held by banks over individual financial decisions.

In terms of funding activities, banks have witnessed the growing popularity of money market investment funds in recent decades. These instruments have been in some cases almost a perfect replacement of traditional bank instruments such as current accounts and fixed-term deposits. The effect of the substitution of bank liabilities by market instruments has only been partially offset by the investments made by CIVs, insurance companies, and pension funds on senior and subordinated bonds issued by banks and other financial institutions and by banks' active role in offering mutual funds in the retail market.[15]

In relation to the lending activity, corporate financing policies have slowly emancipated from the banking system and turned to the market increasingly in order to finance themselves through the placement of equity and corporate bonds.

The spread of securitisation processes too, which have involved institutional investors heavily, has been a threat as well as an opportunity for commercial banks at the same time. This business area, mainly controlled by investment banks, has allowed commercial banks to gain flexibility in the management of their assets and minimum capital requirements and to develop new financing techniques for their corporate customers.

This briefly summarised scenario has obviously had a significant impact on the structure of the income statement of banks, which have seen a significant narrowing of spreads and interest margins, and a shift of their revenues mix towards fee and commission income.

[14] De Hann et al. (2015).

[15] Bratton and McCahery (2015).

1.3.3 Market Organisation

The phenomenon of saving institutionalisation has also led to heavy consequences on the market structure and operations.[16] The relative increase in the weight of institutional investors demand has indeed stimulated a clearer segmentation of the capital market between retail and wholesale business.

As a matter of fact, wholesale markets primarily respond to the needs of liquidity and certainty of trading conditions typical of institutional investors, where liquidity is measured on the basis of best-known parameters of market operational efficiency: width (horizontal size of the trading activity), depth (vertical dimension of the same), price elasticity, certainty and speed of the execution process.[17]

For this reason, markets with an organisational structure centred on the figures of dealers and market makers have arisen and developed. This structure can ensure that the market carries out the function of liquidity, sometimes even at the expense of efficiency. Not surprisingly, such an organisational form was originally typical of the London Stock Exchange, which is traditionally the marketplace privileged by institutional investors at an international level, and has only recently been adopted also by other regulated markets where retail demand has historically prevailed.

The impact was therefore much more pervasive in the bank-based financial systems of continental Europe, compared to the Anglo-Saxon systems, which are mainly market-oriented and, as such, more effectively equipped to respond to institutional clients' needs.

Last but not least, there has been the emergence of growing investors' demand for innovative investment and risk management solutions. In particular, a driving force for the derivatives market has always been attributed to the operators of the insurance and retirement segment who undertake binding commitments towards their clients, paying particular attention to the issue of risk management in terms of immunisation over a long time period.

From this point of view, the comparison between the asset allocation made directly by the surplus units and that defined through the delegation to institutional investors is symptomatic. In fact, the portfolio of the latter is characterised by:

- Prevalence of market instruments over bank instruments.
- Greater diversification in favour of riskier asset classes (equities, corporate bonds, asset-backed securities, alternative investments), at the expense of investments in cash and government securities.
- Longer time horizon.
- Adoption of dynamic management techniques and, consequently, higher portfolio turnover.
- Use of derivatives for hedging as well as speculative and investment purposes.
- Structured and continuous risk monitoring activity.

---

[16] Vittas (1998) and Clark and Monk (2017).

[17] Piotroski (2004).

Current changes, which were previously described in their basic trends at the secondary market level, are fully reflected also at the primary market level, where there is an easier access to capital markets by a broader base of corporate issuers, with the gradual replacement of the traditional hierarchical placement patterns, which were adopted in the past by main investment banks. Of course, all the domestic public debt securities markets, which are in open competition in attracting large international institutional capital both inside and outside the Eurozone, have been deeply and positively affected.

Competitiveness and a strive towards efficiency have produced transaction cost savings, both in primary and secondary markets. In this respect, especially investment funds, but more generally all institutional investors who actively manage their portfolios, have provided a decisive contribution to increase the level of technological sophistication, transparency, safety, liquidity, and efficiency of the trading systems adopted worldwide. The specific needs of institutional investors (*ex-ante* knowledge of trading conditions and liquidity of investments) justify the existence of trading venues reserved to them with the presence of dealers/market makers.

1.3.4 Market Efficiency

Increasing the relative weight of institutional investors on the overall securities markets demand goes in parallel with improving the market efficiency standards. Nevertheless, this positive trend does not seem to be significant for all the issuers; instead, it influences them in a selective way.[18]

In general, institutional investors tend to manage very concentrated portfolios, focusing their monitoring activity on a few sectors and companies, where they hold a significant stake in both absolute and relative terms and can count on a concrete information advantage.[19] In these cases, they are able to obtain better results in aligning interests between shareholders and managers, with consequent positive impact on the information efficiency of the market.[20]

More specifically, they favour large caps, regardless of past performances, and more liquid securities with low price to book value, with the effect of penalising, also in terms of performance, small caps.[21] By concentrating their investments in benchmarks' constituents, they improve their absolute performance, negatively influencing its volatility at the same time.[22]

[18]It is emblematic, in this regard, what happened during the 2008 crisis, when large institutional investors significantly penalized the corporate segment (more efficient and more liquid) to the advantage of securitisation bond (less efficient and less liquid), leading to adverse selection phenomena and mispricing on the markets. On the topic refer to Manconi et al. (2012).

[19]Choi et al. (2017).

[20]Fich et al. (2015).

[21]Gompers and Metrick (2001).

[22]Basak and Pavlova (2013).

For securities that maintain a marginal weight in their portfolios—the so-called inertia stock—there is, instead, the risk that information asymmetries can accentuate and develop adverse selection phenomena.

It should also be considered that the portfolio choices of institutional investors and the resulting information and signalling role are significantly influenced by their time horizon.

Investors with a short-term and long-term orientation tend to manage portfolios with a significantly different risk/return profile.

The former, managing portfolios typically characterised by higher exposure to systematic risk, have a higher propensity to diversification than the latter, with the aim of mitigating exposure to non-systematic risk. With their own trading activity, they try to exploit market inefficiencies and their own superior market timing ability. The result is a lower average contribution to the market efficiency accompanied with a parallel superior ability to remunerate the risk.

Instead, thanks to more targeted market approach and continuous monitoring activities on the governance of the companies in which they hold, individually or jointly, significant stakes, long-term investors can offer a greater contribution to financial systems' information efficiency, favouring the stability of risk-adjusted performance compared to its absolute level. The final impact is not easy to evaluate, as both the direct effects and indirect ones must be weighted.

1.3.5 Governance Model and Performance of Companies

Everyone in literature agrees that institutional investors can significantly influence the governance model, on the one hand, and the company strategies and policies, on the other.[23]

The engagement activity of institutional investors produces its effects on target companies at various levels, such as:

- overall relationships between shareholders and management.
- management remuneration and incentive models.
- degree of disclosure.[24]
- superiority and stability of company performance.[25]

The engagement activity is carried out in various ways, through exercise of voting right, direct discussions with other shareholders and management, on-going participation to corporate life, securities trading, public demonstrations of dissent, etc.

---

[23] OECD (2011).

[24] Pucheta-Martínez and García-Meca (2014)

[25] Gillan and Starks (2000) and Lewellen and Lewellen (2022).

In this regard, not all institutional investors should be considered operating as "activists" and the time horizon of reference for their policies, combined with size, geographic location, and economic sector, cannot be overlooked as a discriminating factor.

Investors, such as open-end funds and, above all, hedge funds, which do not act as activists and are characterised by a shorter time horizon, tend to be less loyal as investors and to exploit their market timing ability as a priority, leveraging only some fundamentals and not having a positive impact on the performance of the investee companies.[26]

Vice versa, investors characterised by a longer horizon (typically insurance companies and pension funds) are more effective in influencing management and in preventing his/her opportunistic behaviour. The companies that benefit of their presence as shareholders are also more stimulated to invest in research and development and to innovate at a technological level.

It should also be noted that, despite operating on different time horizons and pursuing objectives that are not perfectly aligned, short-term investors and long-term investors often tend to focus their attention on the same targets, in order to exploit possible synergy effects.

1.3.6 Regulation and Control of Financial Activities

Wholesale and retail markets differ in terms of transparency and information standards that are guaranteed to investors by supervisory authorities. Because institutional investors have a greater ability to raise capital and process private and public information available in the markets, they have been less sensitive than retail clients to the protection issue. Such a difference in judgement has been widely accepted in current regulations, from the European Union directives to the primary and secondary level regulations of each Member State. The competent authorities have also been forced to regulate in a more specific manner the issues of conflicts of interest and corporate governance, to which institutional investors have always paid special attention.[27] In this particular perspective, the role of independent directors acquires greater importance and results in claiming the right to participate actively, individually, or in a concerted way, in the management of investee companies.[28] This power goes hand in hand with the increasing importance that institutional investors have as shareholders of investee companies.[29]

At a somewhat different analysis level, having the role of institutional investors strengthened in all financial systems albeit with different timing, it became necessary to define a stronger regulatory and supervisory framework in terms of constraints on

[26] Cremers and Sepe (2017).

[27] Ryan (2002).

[28] Rubach (2009).

[29] About this widely investigated topic, see Webb et al. (2003), Ingley and van der Walt (2004) and Picou (2008).

investment policies, organisational adequacy, transparency, and internal and external control mechanisms.[30]

1.3.7 Asset Management Industry

All the issues developed in the previous paragraphs bring to the conclusions that the asset management industry has been very seriously affected by the growing institutionalisation of savings. Partly for strategic choice and partly as a response to the pressing demand, the supply side has modified its market policies in different ways:

- Segmenting markets, which has resulted in product differentiation, where the retail component is more sensitive to the brand and the institutional component to the performance and to the efficiency of the managers.
- Causing an increasingly sharp distinction between global players and niche players.
- Dedicating "high touch" distribution channels to institutional client: these channels are characterised by higher level of sophistication combined with more targeted customisation and greater flexibility, and developing cross-selling policies towards retail customers.
- Strengthening their internal control structures (risk management, internal audit and compliance) to guarantee full compliance with the complex mix of statutory, regulatory, and management constraints that are typical of each counterparty.
- Ensuring higher standards in terms of reporting and accounting.

This has been accompanied with the gradual fall of protective barriers that were historically raised for the benefit of domestic players. To date, this process has been completed in the Eurozone with the UCITS and AIFMD Directives and the full implementation of the principles of freedom to provide services, freedom of establishment (single licence), and mutual recognition, but it is also in an advanced stage with respect to non-European Union asset managers. Consequently, cross-border portfolio management activities are growing at an increasing rate, although mainly referring to specific asset classes such as equity. On the contrary, the phenomenon remains more limited for some demand segments (i.e., insurance companies), burdened with more stringent investment constraints.

[30] OECD (2011).

1.4 An International Comparison

The key trends of institutional investors' total assets can be identified from regular publications by the OECD and other reputable sources.

All institutional investors in OECD countries, including investment funds, insurance companies, pension funds, public pension reserve funds (PPRFs), and other institutional investors, have increased their assets during the recent years.[31]

Institutional investors worldwide have been able to benefit over the years of a substantial increase in their total assets, which achieved about 131 trillion US dollars at the end of 2021, limiting the analysis to investment funds, insurance companies, and pension funds of OECD countries. This trend is justified both by positive average returns on total investments and by the constant inflow of new financial resources.

A specific analysis of the collective investment vehicles (CIVs) can be made using the data of the International Investment Funds Association (IIFA), which collects data on investment funds globally.

The last available report indicates that the assets under management of investment funds as a whole amounted to almost 60 trillion dollars at the end of 2022. Out of these, 44.50% was attributable to equity funds, 19.20% to bond funds, 11.80% to balanced funds, 7.15% to money market funds, and the residual share to other types of funds (Table 1.3).

The share of exchange-traded funds was 14.87% % globally, while the weight of investment funds dedicated to institutional clients amounted to 8.49% (Table 1.3).

Table 1.4 shows the ten countries with the most important weight of CIVs on a global scale: the first country is the United States of America with a weight of 48.88%, followed by Luxembourg (9.17%), which is the hub in Europe for the domicile of UCITS compliant investment funds to be distributed in Europe on a mutual recognition basis. There is a high concentration of investment funds in the first ten countries which altogether account for 91.75% of the total (Table 1.4). The same table shows a marked difference between the average size of investment funds in the USA and worldwide. In the US case, they have an average size of nearly 2.8 billion dollars, while in other countries the size is normally much smaller, with the exception of the UK, where the average size goes up to 576 million dollars.

In terms of geographical breakdown, Table 1.5 shows the distribution of the universe of investment funds by region: 52.88% of investment funds can be attributed to the Americas, 31.69% to Europe, 15.13% to Asia and the Pacific area, and the remaining 0.31% to Africa. There is also a considerable difference among the average size of investment funds, due to the different average size of US investment funds. In the same way, it is possible to analyse the allocation of funds in different geographical areas by type of investment (Table 1.6). The weight of equity funds is very different, depending on the region considered and goes from a high of 54.02%

[31] OECD (2022).

Table 1.3 Total net assets under management by category of investment fund. In USD billions and as a percentage at the end of 2022

	Net assets under management	
	Absolute value	%
All funds[a]	60,150	100.00
Equity	26,948	44.50
Bond	11,549	19.20
Balanced/mixed	7094	11.80
Guaranteed/protected	31	0.05
Real estate	1369	2.28
Other	8856	14.72
Money market	4303	7.15
Memo items included above:		
ETFs fund	8943	14.87
Institutional	5108	8.49

Source: International Investment Funds Association (2023)
[a]Funds of funds are excluded where possible

Table 1.4 Top ten countries in global investment funds industry. In USD billion at the end of 2022

Top ten countries	Total net assets	Market share (%)	Number of CIVs	Average size of CIVs
United States of America	28,587	48.88%	10,237	2.793
Luxembourg	5364	9.17%	14,322	0.375
Ireland	3900	6.67%	8689	0.449
China	3266	5.58%	9276	0.352
Germany	2455	4.20%	7186	0.342
Australia	2429	4.15%	n.a.	n.a.
France	2237	3.82%	10,997	0.203
Japan	2040	3.49%	14,297	0.143
United Kingdom	1800	3.08%	3125	0.576
Canada	1586	2.71%	3757	0.422

Source: International Investment Funds Association (2023)

Table 1.5 Worldwide investment funds distribution. In USD billion at the end of 2022

	Total net assets	Market share (%)	Number of CIVs	Average size of CIVs
Americas	29,819	52.88%	35,637	0.837
Europe	17,868	31.69%	60,156	0.297
Asia and Pacific	8532	15.13%	40,330	0.212
Africa	173	0.31%	1769	0.098
World	56,392	100.00%	137,892	0.409

Source: International Investment Funds Association (2023)

in the Americas, to 39.44% in Asia and the Pacific area, to 32.17% in Europe and down to 25.54% in Africa (Table 1.6).

In relation to the role of insurance companies as institutional investors, the starting point is the analysis of the relative size of market shares. For this purpose, Fig. 1.1 shows the market share of the insurance companies in terms of direct gross premiums in all OECD countries. It shows how the US companies have a weight on the global market amounting to 53.43%, followed by the UK (7.34%), Germany (6.54%), Japan (6.20%), and France (5.97%).

Table 1.6 Asset classes breakdown in global investment funds industry. As a percentage of total assets at the end of 2022

	Equity (%)	Bond (%)	Balanced/ mixed (%)	Money market (%)	Guaranteed/ protected (%)	Real estate (%)	Other funds (%)
Americas	54.02	20.91	7.99	15.87	0.00	0.14	1.06
Europe	32.17	20.67	18.79	8.88	0.16	4.84	14.49
Asia and Pacific	39.44	10.43	9.71	23.03	0.00	4.36	13.03
Africa	25.54	5.43	46.74	11.96	0.00	1.63	8.70
World	44.80	19.20	11.79	14.72	0.05	2.28	7.15

Source: International Investment Funds Association (2023)

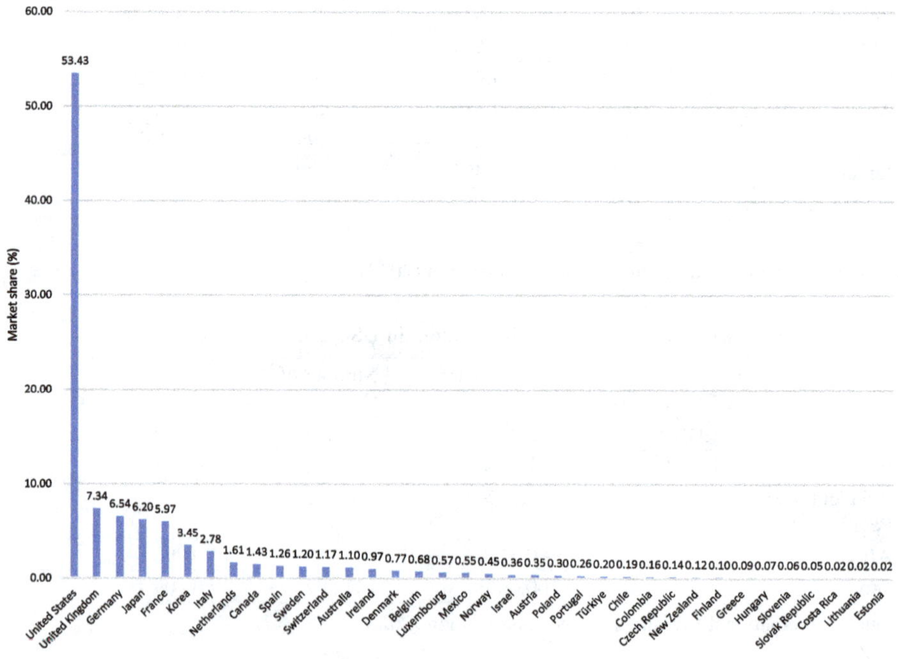

Fig. 1.1 Market share of insurance companies in OECD countries. Direct gross premium. Source: OECD (2021)

Regarding the investment policy, the investment portfolios of insurance companies in the majority of countries continue to be allocated towards bonds in a consistent way (Fig. 1.2).

Table 1.7 shows the total assets that are managed by pension funds in OECD countries and in selected non-OECD countries. The largest market for pension funds

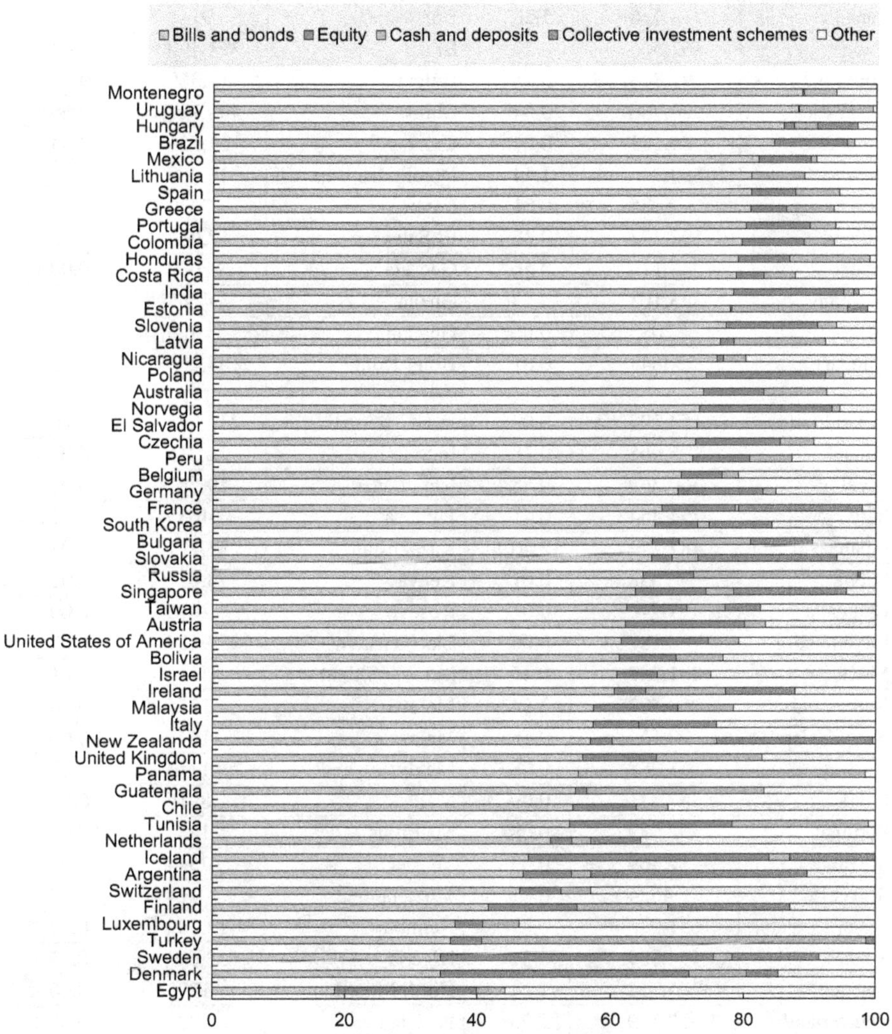

Fig. 1.2 Asset allocation of domestic insurance companies. As a percentage of total investment at the end of 2020. Source: OECD (2021)

Table 1.7 Total investments of pension funds in OECD and selected non-OECD countries. In millions of USD at the end of 2021

OECD countries	Total investments	Share (%)	Non-OECD countries	Total investments	Share (%)
Australia	2,291,929	3.70	Albania	43	0.00
Austria	30,550	0.05	Angola	861	0.00
Belgium	223,702	0.36	Armenia	1006	0.00
Canada	3,218,436	5.20	Botswana	9748	0.02
Chile	167,556	0.27	Brazil	393,490	0.64
Colombia	91,806	0.15	Bulgaria	11,360	0.02
Costa Rica	24,906	0.04	China	409,707	0.66
Czechia	26,183	0.04	Croatia	21,566	0.03
Denmark	889,945	1.44	Dominican Rep.	11,785	0.02
Estonia	5835	0.01	Egypt	7230	0.01
Finland	197,173	0.32	El Salvador	12,964	0.02
France	344,114	0.56	Georgia	639	0.00
Germany	325,103	0.53	Ghana	4665	0.01
Greece	2083	0.00	Guyana	426	0.00
Hungary	8844	0.01	Hong Kong (China)	198,208	0.32
Iceland	54,207	0.09	India	249,045	0.40
Ireland	175,066	0.28	Indonesia	22,215	0.04
Israel	360,569	0.58	Isle of Man	19,287	0.03
Italy	254,454	0.41	Jamaica	4635	0.01
Japan	1,487,779	2.40	Kazakhstan	31,291	0.05
Latvia	7638	0.01	Kenya	13,677	0.02
Lithuania	6944	0.01	Kosovo	2676	0.00
Luxembourg	2187	0.00	Liechtenstein	9439	0.02
Mexico	281,589	0.45	Macau (China)	4914	0.01
Netherlands	2,068,577	3.34	Malawi	1555	0.00
New Zealand	90,144	0.15	Malaysia	1348	0.00
Norway	51,788	0.08	Maldives	1143	0.00
Poland	57,321	0.09	Malta	8916	0.01
Portugal	52,539	0.08	Mauritius	1435	0.00
Slovakia	17,469	0.03	Morocco	7544	0.01
Slovenia	4624	0.01	Mozambique	224	0.00
South Korea	551,075	0.89	Namibia	13,302	0.02
Spain	194,509	0.31	Nigeria	32,644	0.05
Sweden	705,712	1.14	North Macedonia	1999	0.00
Switzerland	1,390,599	2.25	Pakistan	224	0.00
Turkey	18,430	0.03	Panama	697	0.00
United Kingdom	3,751,713	6.06	Peru	33,436	0.05
USA	40,027,760	64.67	Romania	21,168	0.03
			Russia	87,489	0.14
Total	**59,460,859**	**96.07**	Serbia	473	0.00

(continued)

Table 1.7 (continued)

OECD countries	Total investments	Share (%)	Non-OECD countries	Total investments	Share (%)
			Singapore	370,054	0.60
			South Africa	330,276	0.53
			Suriname	324	0.00
			Thailand	40,366	0.07
			Trinidad and Tobago	8140	0.01
			Uganda	5460	0.01
			Ukraine	123	0.00
			Uruguay	17,351	0.03
			Zambia	745	0.00
			Zimbabwe	2935	0.00
			Total	**2,430,251**	**3.93**

Source: OECD (2023)

is the United States of America, with a market share of 64.67%, followed by the United Kingdom (6.06%), Canada (5.20%), the Netherlands (3.34%), and Japan (2.40%).

One of the key indicators of pension funds' role is the market value of their total assets relative to the size of the economy measured by the Gross Domestic Product (GDP). The OECD considers a country's pension fund market "mature" if its assets-to-GDP ratio exceeds 20%.

In the last years, their importance to national economies grew considerably. Figures 1.3 and 1.4 illustrate the share of pension funds in national GDPs for OECD countries and selected non-OECD countries. The figures point out that in most OECD countries the pension fund assets-to-GDP ratio does not exceed 20%, as well as for many non-OECD countries, which suggests potential for growth. In the case of nine OECD countries (Denmark, Iceland, Netherlands, United States of America, Switzerland, Canada, Australia, United Kingdom, and Sweden), the pension fund assets-to-GDP ratio exceeds 100%, showing how pension funds represent an extremely important source of long-term financing and a significant growth factor for their economies.

The pension funds asset allocation depends on the development of the stock market, the distinctive features of pension systems, the structure of pension schemes, and the existing restrictions and requirements for investment policy established by national laws. Equities and bonds remain the predominant asset classes in pension funds' portfolios in both developed and emerging countries (Figs. 1.5 and 1.6). Seventeen OECD countries invested over 80% of their portfolios in these two asset

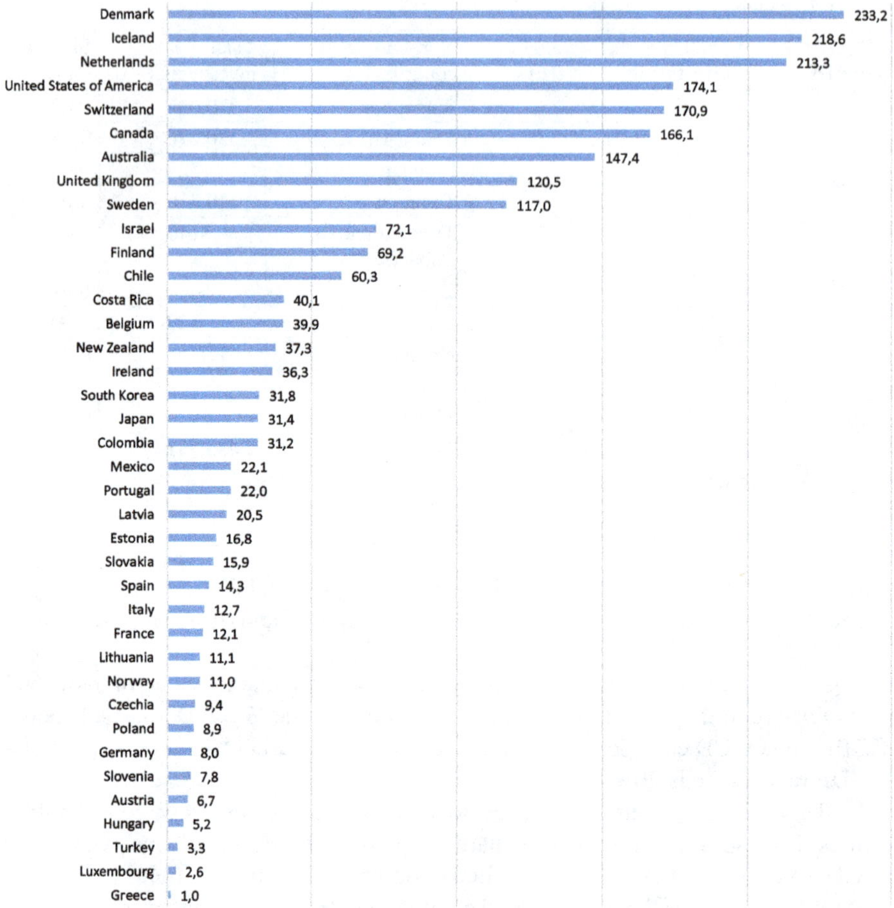

Fig. 1.3 Importance of pension funds relative to the size of the economy in OECD countries. As a percentage of GDP at the end of 2021. Source: OECD (2023)

classes in the last available year. Transition economies tend to prefer investments in government and corporate bonds, while developed countries prefer equity and CIVs units.

Pension funds become gradually interested in less volatile assets, including fixed income instruments, cash and deposits and alternative investments. In most countries under consideration, the government-to-corporate bonds ratio in the portfolios of pension funds is 60:40 or even higher on the government's side.

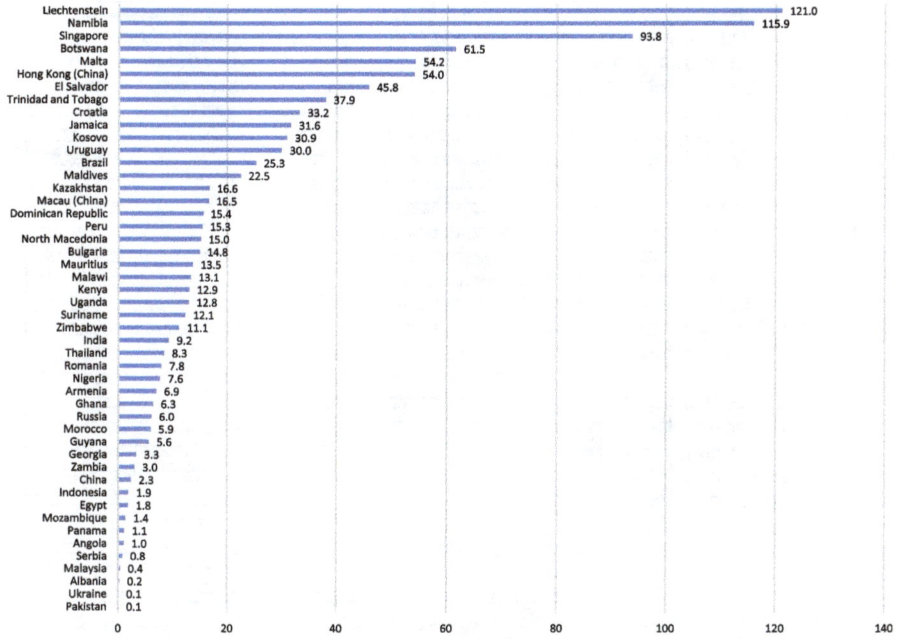

Fig. 1.4 Importance of pension funds relative to the size of the economy in selected non-OECD countries. As a percentage of GDP at the end of 2021. Source: OECD (2023)

OECD statistics of pension investments show that approximately a third of the countries changed their preferences from traditional investment instruments (equity, bonds, cash, and deposits) to other asset classes.[32] Pension funds in Japan, Germany, Denmark, Austria, Switzerland, United Kingdom, Finland, South Korea, France, Canada Ireland, Australia, Italy, and Netherlands have allocated over 20% of their portfolios in "other" assets. These include loans, real estate, hedge funds, private equity funds, and structured products.

Depending on how pension benefits are calculated and who bears the inherent risk, pension funds can either be defined contribution (DC) or defined benefit (DB) funds. In DC pension funds, participants bear the brunt of risk, while in traditional DB plans sponsoring employers assume most of the risks. In recent years, pension plan sponsors have shown a growing interest in DC plans, as demonstrated by the number of employers that, in many countries, have closed DB plans to new participants and encouraged employees to join DC plans, or even frozen benefit accruals for existing employees; in some cases. DB plans, however, still play an important role, largely due to their historical prominence in many countries.

[32] OECD (2023).

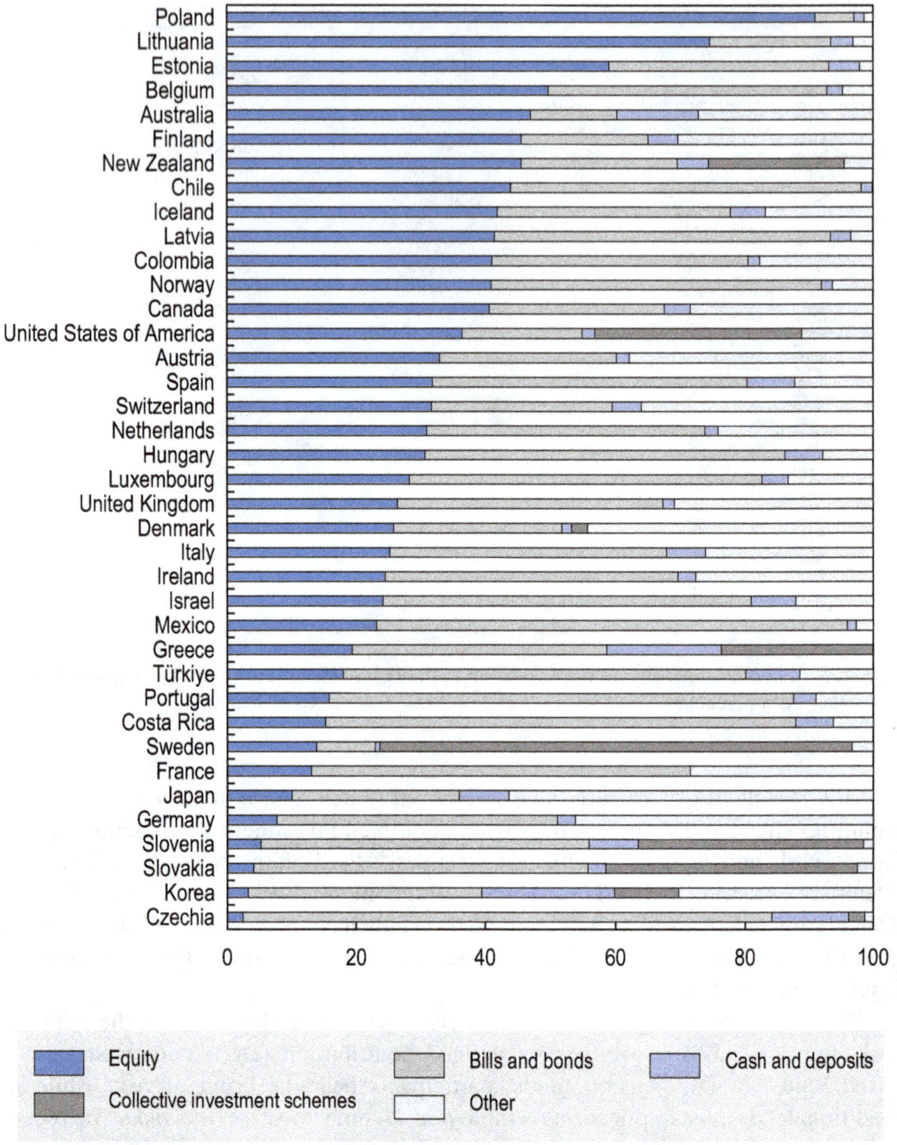

Fig. 1.5 Pension fund asset allocation for selected investment categories in selected OECD countries. As a percentage of total investments at the end of 2021. (1) The "Other" category includes loans, land and buildings, unallocated insurance contracts, hedge funds, private equity funds, structured products, other investment funds (i.e., not invested in cash, bills and bonds, shares or land and buildings) and other investments. Source: OECD (2023)

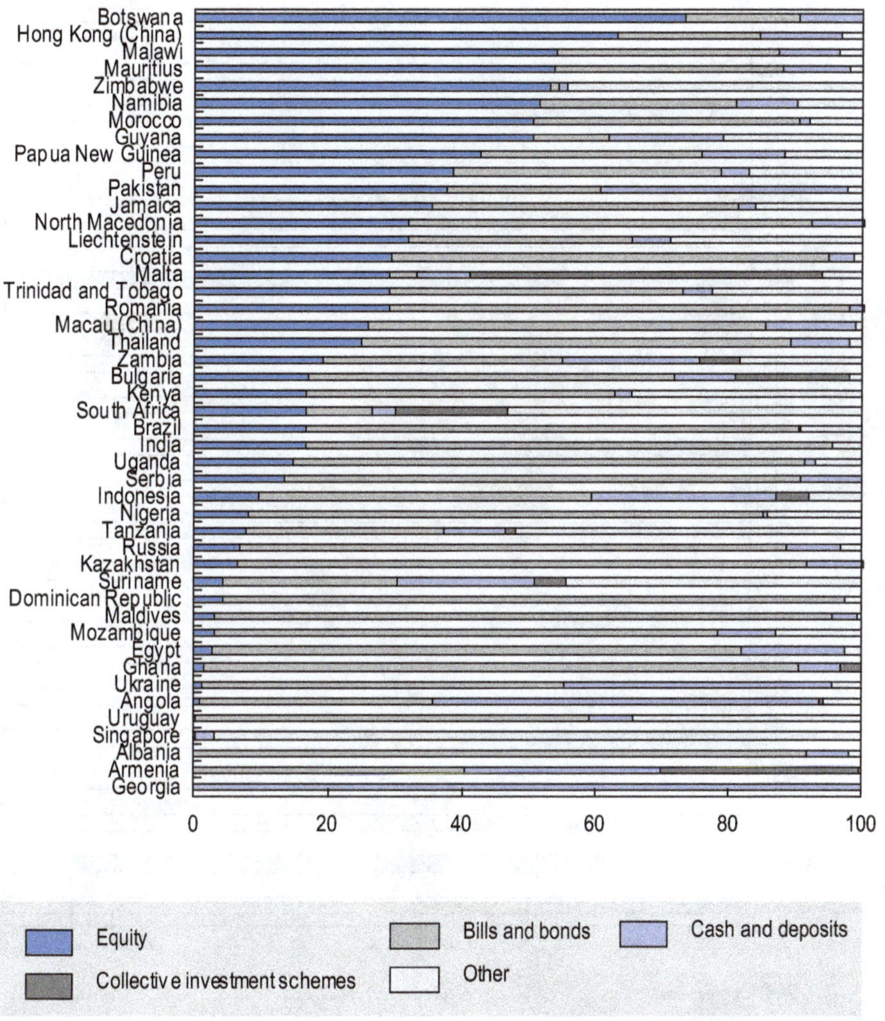

Fig. 1.6 Pension fund asset allocation for selected investment categories in selected non-OECD countries. As a percentage of total investment at the end of 2021. (1) The "Other" category includes loans, land and buildings, unallocated insurance contracts, hedge funds, private equity funds, structured products, other investment funds (i.e., not invested in cash, bills and bonds, equity or land and buildings), and other investments. Source: OECD (2023)

In 2021, DB plans accounted quite for the totality of pension fund assets in Finland and Switzerland; and, at the other extreme, all pension funds are classified as DC in Chile, the Czech Republic, Estonia, Lithuania, Hungary, the Slovak Republic, and Slovenia (Fig. 1.7).

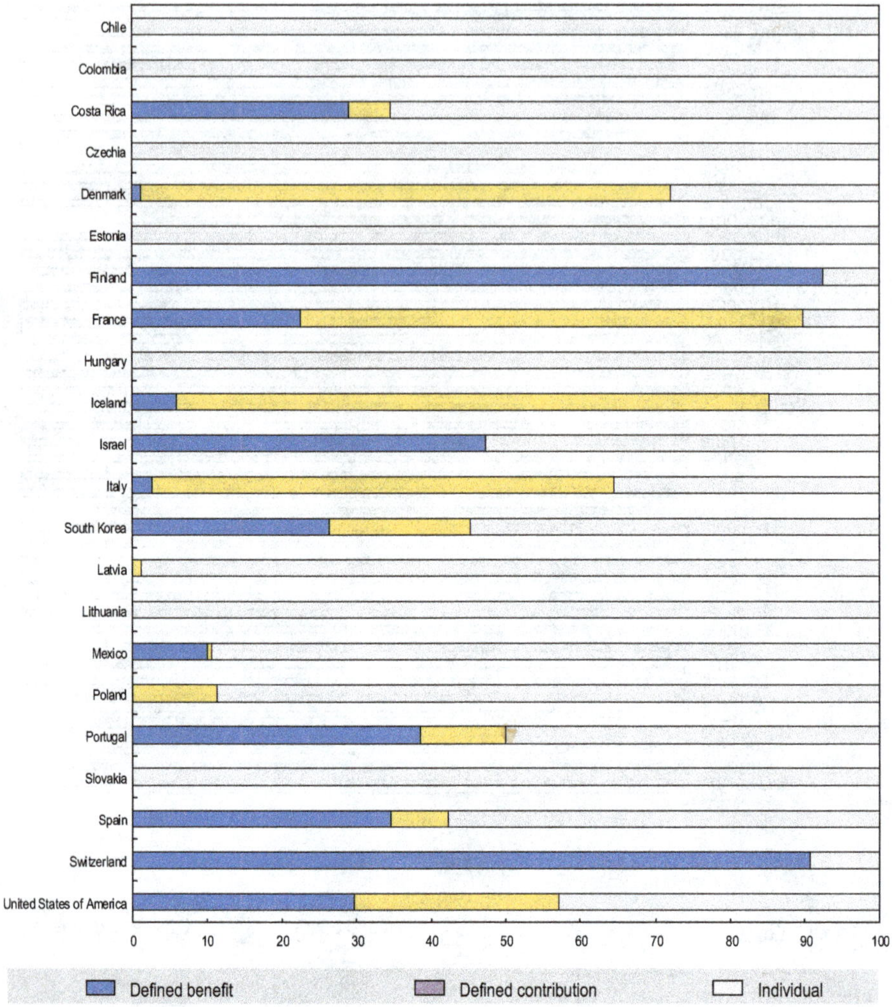

Fig. 1.7 Relative shares of DB and DC pension fund assets in selected OECD countries. As a percentage of total assets at the end of 2021. Source: OECD (2023)

References

Acharya S, Dimson E (2007) Endowment asset management. Oxford University Press, Oxford

Basak S, Pavlova A (2013) Asset prices and institutional investors. Am Econ Rev 103(5): 1728–1758

Blake D (2006) Pension finance. Wiley, Chichester

Bratton WW, McCahery JA (2015) Institutional investor activism. Oxford University Press, Oxford

Bushee B (2001) Do institutional investors prefer near-term earnings over long-run value? Contemp Account Res 18(2):207–246

Çelik S, Isaksson M (2013) Institutional investors and ownership engagement. Financ Market Trends 2(105):93–114
Choi N, Fedenia M, Skiba H, Sokolyk T (2017) Portfolio concentration and performance of institutional investors worldwide. J Financ Econ 123(1)
Clark GL, Monk AHB (2017) Institutional investors in global markets. Oxford University Press, Oxford
Cremers KJM, Sepe SM (2017) Institutional investors, corporate governance and firm value. Seattle Univ Law Rev 4:387–418
Davis EP, Steil B (2001) Institutional investors. MIT Press, Cambridge
De Hann J, Oosterloo S, Schoenmaker D (2015) Financial markets and institutions. A European perspective, 3rd edn. Cambridge University Press, Cambridge
Directive 2014/65/EU of the European Parliament and of the Council of 15 May 2014
Fich EM, Harford J, Tran A (2015) Motivated monitors: the importance of institutional investors' portfolio weights. J Financ Econ 118(1):21–48
Gillan SL, Starks LT (2000) Corporate governance proposals and shareholder activism: the role of institutional investors. J Financ Econ 5(2):275–305
Gompers PA, Metrick A (2001) Institutional investors and equity prices. Q J Econ 116(1):229–269
Ingley CB, van der Walt NT (2004) Corporate governance, institutional investors and conflicts of interest. Corp Gov 12(4):534–551
International Investment Funds Association (2023) Worldwide regulated open-end fund assets and flows. Fourth Quarter 2022, Toronto
Lewellen J, Lewellen K (2022) Institutional investors and corporate governance: the incentive to be engaged. J Financ LXXVII(1):213–264
Manconi A, Massa M, Yasuda A (2012) The role of institutional investors in propagating the crisis of 2007-2008. J Financ Econ 104(3):491–518
OECD (2011) The role of institutional investors in promoting good corporate governance. OECD, Paris
OECD (2021) Insurance markets in figures. OECD, Paris
OECD (2022) OECD institutional investors statistics. OECD, Paris
OECD (2023) Pension markets in focus. OECD, Paris
Picou A (2008) Does good governance matter to institutional investor? Evidence from the enactment of corporate governance guidelines. J Bus Ethics 65(1):55–67
Piotroski JP (2004) The influence of analyst, institutional investors and insiders on the incorporation of market, industry and firm-specific information into stock prices. Account Rev 79(4): 1119–1151
Pucheta-Martínez MC, García-Meca E (2014) Institutional investors on boards and audit committees and their effects on financial reporting quality. Corp Gov Int Rev 2(4):365–382
Rubach MJ (2009) Determinants of institutional investor activism: a test of the Ryan-Schneider Model (2002). J Manag Issues 21(2):245–261
Ruiz JL (2018) Financial development, institutional investors, and economic growth. Int Rev Econ Financ 54(C):218–224
Ryan LV (2002) The antecedents of institutional investor activism. Acad Manag Rev 27(4): 554–573
van Nunen A (2007) Fiduciary management. Blueprint for pension fund excellence. Wiley, Hoboken, NJ
Vittas D (1998) Institutional investors and securities markets. Which comes first? Policy research working paper. The World Bank
Webb R, Beck M, McKinnon R (2003) Problems and limitations of institutional investor participation in corporate governance. Corp Gov 11(1):65–73

Chapter 2
Collective Investment Vehicles and Other Asset Management Products

Pierpaolo Ferrari

2.1 Asset Management Products

This chapter looks at the economic, technical, and regulatory issues of asset management products, with particular attention paid to collective investment vehicles (CIVs), through an analysis of the different categories of CIVs, their technical features and management strategies, as well as the restrictions to their investment decisions, their public documentation, and charges.

The final part of the chapter examines the other types of asset management products which have a high rate of substitutability with CIVs, to which we may apply part of the economic and technical considerations made previously with regard to CIVs.

2.2 Collective Portfolio Management

Collective portfolio management is the activity accomplished through:

- The promotion, establishment, and organisation of CIVs and the administration of relations with the participants.
- The management of CIV assets by investing the assets under management in financial instruments, credits, and other financial or real assets.
- The distribution of units or shares of CIVs.

P. Ferrari (✉)
Department of Economics and Management, University of Brescia, Brescia, Italy

SDA Bocconi School of Management, Bocconi University, Milan, Italy
e-mail: pierpaolo.ferrari@unibs.it; pierpaolo.ferrari@unibocconi.it

I. Basile, P. Ferrari (eds.), *Asset Management and Institutional Investors*, Contributions to Finance and Accounting,
https://doi.org/10.1007/978-3-031-59819-7_2

In the various countries, the service of collective portfolio management is subject to legal restrictions and may only be practised by specific categories of financial intermediaries. In the member states of the European Union (EU), collective portfolio management can only be performed through:

- Regulated schemes that are recognised in each member state.
- Harmonised collective schemes which, according to EU rules, once authorised by the supervisory authority in their home member state, can offer their services in the various EU countries either by opening branches or thanks to the EU-wide freedom to provide services.

There are three basic legal structures for conducting collective portfolio management:[1]

- The contractual form, according to which the CIV is not a separate legal entity in its own right, but is instead the management company setting up the common fund. In this form, there is a contract under which the management company invests the money raised on behalf of final investors who own units of the common fund.
- The trust form, where the CIV is organised as a trust, a concept of Anglo-Saxon law according to which an identified group of assets is constituted and owned by a trustee for the benefit of another party (the beneficiary). Investors are beneficiaries of the trust and own units of the trust.
- The corporate form where the CIV is a separate corporate entity and investors are shareholders of this entity. This is the case of open-end investment companies (OEICs), known in many countries by the equivalent French expression or its acronym SICAV,[2] and is also the case of closed-end investment companies, also known using the short form of the French term SICAF.[3]

Some countries allow for only one legal form of collective investment vehicle, while others allow for more than one legal structure.[4]

Since 1985, the European Union has issued several directives on CIVs aimed at creating a homogeneous regulatory framework within member countries. These directives have progressively harmonised the regulations of member states, starting with the category of open-end funds that invest in transferable securities. If these CIVs implement investment policies in compliance with the constraints and

[1] In this chapter we will generically refer to investment funds and considerations, when possible, will be applicable to all three types of CIVs.

[2] Short for "société d'investissement à capital variable". In some countries, this legal structure is also known as an "investment company with variable capital" or its acronym ICVC and in the United States of America as a "mutual fund".

[3] Acronym of "société d'investissement à capital fixe".

[4] By value, corporate entity CIVs dominate the asset management sector worldwide, largely due to the scale of the US market, where open-end corporate CIVs, known as "mutual funds", predominate. In some European countries, on the other hand, the contractual form is predominant. OECD (2001), St Giles et al. (2003) and Loader (2007).

prohibitions imposed by the EU directives and are transposed into national legal systems, they benefit from the status of Undertakings for Collective Investment in Transferable Securities (UCITS). This status will enable CIVs to be freely marketed throughout the EU under a mutual recognition regime.

Subsequently, the EU legislature adopted a further fundamental element in the process of harmonising the European discipline of collective management: Directive 2011/61/EU on Alternative Investment Fund Managers (AIFMD). This is an EU directive on the financial regulation of hedge funds, private equity funds, private debt funds, real estate funds, and other alternative investment fund managers in the EU. These CIVs are officially defined as alternative investment funds (AIFs). The AIFMD provides harmonisation rules for managers and introduces a mutual recognition regime for professional clients only.

The provision of collective portfolio management services for both UCITS and AIF is subject to legal reserves in each country and is attributed exclusively to specific categories of financial intermediaries.

2.3 Investment Funds

An investment fund is a vehicle that allows a number of separate and unrelated investors, which may be a group of individuals or companies, to make investments together. By pooling their capital, investors can share costs and benefit from the advantages of investing larger amounts, including the possibility of achieving a broader diversification among a number of different assets and thus spreading their risks.

Depending on the amount invested, each participant in the fund holds a number of units, all of equal value, and participates in the profits and losses in proportion to the number of units held. The unit represents an equal fraction of the fund's net assets, whose value is calculated dividing the amount of net assets by the number of units outstanding.[5]

Each investment fund is a separate and independent pool of assets, distinct in all respects from the assets of the management company and of each participant, as well as from any other fund managed by the same company. An investment fund can also be organised as an umbrella scheme with separate sub-funds with segregated assets and liabilities.[6] The investment fund (or sub-fund) meets the obligations contracted

[5] In the case of investment companies, units of the CIV are shares.

[6] If an investment fund adopts the umbrella scheme, it comprises separate sub-funds, divided by type of asset class, specialised in different geographic areas or characterised by a different management style. The umbrella funds have traditionally benefited from the advantage of offering the investor the possibility to switch from one sub-fund to another with great simplifications from the regulatory, tax and charge point of view, although more recent regulations have significantly reduced the differences between single funds and umbrella funds. Both single funds and umbrella funds can adopt a multi-class scheme. All the classes of the same fund or sub-fund invest in the

exclusively with its assets, on which no actions are possible, either by creditors of the management company or by creditors of the depositary institution. The actions of creditors of individual investors are allowed only on the units held by them.

The operation of an investment fund is based on four types of essential activities:

- Promotion, establishment, and organisation of the fund and administration of the relationship with the participants.
- Investment of financial resources collected.
- Custody of financial instruments and liquidity in the fund's portfolio.
- Distribution of the fund units.

The first activity is carried out by a management company that promotes, establishes, and organises the fund and administers the relationships with unit-holders.[7]

The second activity is performed by a management company which may (and often does) coincide with the company that has promoted and established the fund, but may also be different. In the latter case, two different management companies are involved, one responsible for managing the administrative relationships with the participants and the other responsible for managing investments. In fact, on the basis of mandates, a management company may delegate specific tasks and functions to third parties so as to increase the efficiency of the conduct of its business.[8]

The third activity is entrusted to a depositary institution, which has custody of the CIV's financial instruments, holds the liquidity, and keeps a register of all movements. The depositary institution carries out the operational and accounting control, guaranteed by the subject involved in managing the funds being separate from the subject entrusted with custody. The depositary acts independently and in the interest of the CIV investors. More specifically, the depositary institution:

- Verifies the legitimacy of the operations of purchase, redemption, and switch of the CIV units, as well as the destination of incomes from the CIV.

same pool of securities and have the same investment objectives and policies, but each class may have different fees and expenses, may be reserved to special categories of investors, may have different income distribution arrangements, may be denominated in a different currency, and may hedge the currency risk or not. A multi-class scheme offers investors the possibility to select the most appropriate class for their investment goals.

[7] The present description is specifically referred to common funds, but the same considerations can be applied to both unit trusts and open-end investment companies (OEICs). Sections 2.4 and 2.5 contain a more detailed description of the specific features of unit trusts and OEICs.

[8] In order to guarantee a correct functioning of the home country control principle, EU member states permitting such delegations should ensure that the management company to which authorisation is granted does not delegate all its functions to one or more third parties, so as to become a "letter-box entity", and that the existence of mandates does not hinder effective supervision of the management company. However, the fact that the management company has delegated its functions should not affect the liabilities of that company, or of the depositary institution vis-à-vis the unit-holders and the competent authorities.

- Checks that the calculation of the net asset value of the units is correct or provides for that computation at the request of the management company.
- In all operations involving the CIV, verifies that the counter-obligation is fulfilled within the agreed deadlines.
- Carries out the management company's instructions, unless they conflict with the law, the requirements of the supervisory bodies or fund rules.
- Monitors the CIV's cash flows.

The depositary institution is liable to the CIV and to the investors for any loss they may suffer as a result of its failure to fulfil its obligations. Unless the depositary institution can prove that the loss was caused by accident or by force majeure, if the depositary institution should lose the financial instruments in its custody, it is obliged to replace them with financial instruments of the same kind, or with a sum to the corresponding amount, and shall be held liable for any other loss sustained by the CIV or investors consequent to the failure to respect its obligations, whether intentional or due to negligence.

The fourth activity is the result of recent EU legislation which has broadened the definition of collective investment management to include the distribution activities of investment funds so as to improve the efficiency of the distribution network, to promote the creation of a direct channel and to reduce the overall costs borne by investors.

From the investor's point of view, when compared to investing directly in individual shares, bonds, and money market instruments, investing through CIVs offers the following benefits:[9]

- Greater diversification and thus lower risk, regardless of the amount of capital invested.
- Access to a professional management service.
- Superior liquidity of the investment, at least for open-end funds.
- Greater transparency of information than other financial products, both in the underwriting phase and in the course of the investment, despite the considerable efforts of supervisors to bridge this gap between CIVs and many other financial instruments.
- Better protection, compared to other financial products, thanks to the fund assets being separate and segregated from the management company assets that runs it, and to the presence of a depositary institution which carries out accounting and operational control.

[9] Turner (2004) and Haslem (2010).

2.3.1 Investment Fund Classification

In order to fully understand the characteristics of investment funds and their specificity, it is useful to classify them on the basis of a set of alternative parameters. First of all, depending on the mode of entry and exit, it is possible to distinguish between:

- Open-end investment funds.
- Closed-end investment funds.

In open-end investment funds, purchases and redemptions may occur at any time. At any valuation date of the unit, the participant may subscribe new units of the fund, or request redemption of all or part of the units held. The total capital of the fund is therefore variable and depends on the combination of new subscriptions and redemptions produced daily or periodically. The company that manages the open-end fund must provide for reimbursement of the units within the maximum legal time-limit established by law and by the fund rules.

In the case of closed-end investment funds, the total capital of the fund to be subscribed is fixed when the fund is launched, and redemptions can only take place at predetermined maturities. Therefore, the number of units in these funds is fixed and does not vary as a result of new subscriptions and/or redemptions. Redemption before the deadline for repayment is impossible. Investors can only transfer their units to other investors through the sale of units on the secondary market, with liquidity problems and difficulty identifying a counterparty if the fund is not listed on a regulated market. This type of investment fund is particularly suitable when the object of the investment cannot be liquidated quickly without significant loss of value, as in the case of real properties (real estate funds), equity participation in unlisted companies (private equity and venture capital funds), and unlisted bonds and loans (private debt funds).

Depending on the object of the investment, it is possible to distinguish:

- Investment funds specialised in financial instruments.
- Investment funds specialised in real estate.

The former invest in transferable securities and other financial instruments, while the latter invest in real properties, property rights and real estate companies.

Depending on the mode of allocation of incomes, it is possible to distinguish:

- Investment funds that accumulate incomes.
- Investment funds that distribute incomes.[10]

[10]Fund rules indicate whether the management company reinvests periodical incomes received by the fund (accumulation fund) or whether it pays them to unit-holders every year, semester, quarter, or month (income fund). In the latter case, fund rules define the criteria and procedures for determining and allocating income to unit-holders.

In the first case, the income received periodically by the fund is reinvested in the fund's assets, while in the second case, the income is periodically distributed to unit-holders.

Depending on the country in which they are located, taking the perspective of the investor, and making specific reference to the EU member countries, it is possible to distinguish:

- National investment funds.
- Foreign investment funds.

National investment funds are established according to the law and rules in force in the investor's country of residence. In several countries, the most popular are open-end funds that invest in transferable securities.

Foreign investment funds are established abroad, in countries other than that of the investor's residence. Contrary to what might be expected, this second category does not only include the funds of foreign companies placed in one country in competition with national companies. In many cases, it is also about "national funds abroad" or round-trip funds, i.e. funds created abroad by banking, financial, and insurance groups from the different countries, but aimed largely at domestic participants from the same country as the banking, financial, and insurance group that established them abroad. In the recent past, one of the most important reasons for setting up management companies abroad has been the opportunity it provided to the same management companies to benefit from favourable tax effects on their income and a more favourable regulatory environment.

From the perspective of an EU investor, it is possible to distinguish the following investment funds based on compliance with the provisions of EU directives:

- Undertakings for collective investment in transferable securities.
- Alternative investment funds.

Undertakings for collective investment in transferable securities, known by the acronym UCITS, were the first category of investment funds harmonised in the EU. Therefore, they are often identified by market operators as harmonised funds. They must be based in one of the EU countries and meet the requirements laid down by the EU directives. Since 1985, many EU directives on investment funds have appeared with the aim of creating a uniform regulatory framework within all EU member states (Box 2.1). These directives have progressively harmonised the laws of member countries, with a focus on the category of open-end funds, which invest in transferable securities and raise capital by promoting the sale of their units to the public within the EU or any part of it.[11] These funds benefit from the status of UCITS

[11] The EU legislation currently covers six categories of investment fund with a common regulation: undertakings for collective investment in transferable securities (UCITS), the first category to have been harmonised at the Community level and by far the most important; money market funds (a category of UCITS funds), a denomination that, to be used within the EU, must comply with a series of requirements in terms of credit risk, interest rate risk, and currency risk; alternative investment funds (AIFs), including hedge funds, private equity funds, private debt funds, and real

funds if they implement investment policies within the limits and restrictions laid down in these directives and incorporated in each EU country. To comply with the EU directives, they must be open-end, invest in transferable securities, not be reserved to qualified investors, and adopt investment policies in line with the criteria for investment diversification and risk spreading provided by the same directives. UCITS funds benefit from EU-wide "passports", which means that, once authorised in one member state, they can be sold in any other member state without needing any additional authorisation. UCITS funds are now regarded globally as being efficiently regulated, with a strong emphasis on investor protection. As a result, UCITS status is recognised beyond the EU, and UCITS products are accepted for sale in many countries in Asia, the Middle East, and Latin America.

Non-UCITS funds are those funds which, despite being based in EU countries, do not comply with the requirements set by the UCITS EU directives and all funds based outside the EU, regardless of their characteristics. Within the EU legal framework, these funds are known as "alternative investment funds" (AIFs). In the case of EU-based AIFs, mutual recognition applies only to professional clients. The possible marketing of AIFs to retail customers, as well as any non-EU fund, is instead subject to prior authorisation by the supervisory authorities of the country where the product is intended to be placed and is valid only nationally, without automatic extension to other member countries.

Box 2.1 The Evolution of European Union Directives on UCITS Funds
The European directive UCITS I, issued at the end of 1985, represented the first effort towards the creation of a single market for investment funds in the European Union. This directive provided for the free movement of certain investment funds throughout the Union, under mutual recognition, on condition that they were open, invested at least 90% of net assets in transferable securities, were promoted to the public of the European Union or a part of it and respected many prohibitions and limitations in implementing their investment policy. In fact, all these restrictions and prohibitions affecting the investment policy of UCITS funds, together with the great discretion granted

(continued)

estate funds, whose managers are subject to the rules defined by a specific EU directive, known as "Alternative Investment Fund Managers Directive" (AIFMD); European venture capital funds (EuVECAs), a particular category of alternative investment funds focused on companies in the start-up phase; European social entrepreneurship funds (EuSEFs), a sub-category of alternative investment funds, open-ended or closed-ended, focused on qualifying investments for social entrepreneurship; and European long-term investment funds (ELTIFs), a particular category of closed-end alternative investment funds specialised in financing companies and infrastructure projects. For the money market funds, EuVECAs, EuSEFs, and ELTIFs, the objective of the EU harmonisation is to allow the investor to identify and choose the funds belonging to a given category, ensuring that the denomination corresponds to a fund with similar characteristics throughout the EU.

Box 2.1 (continued)
to individual member states in transposing the Directive, have come to represent the main shortcomings of the directive UCITS I as regards the creation of a real single market for investment funds.

The second European directive UCITS II, developed in the early 1990s with the aim of completing the path laid out by the UCITS I directive and therefore containing a series of proposed improvements, was never approved, since it was considered over-ambitious and too complex.

It was not until 2001 that two new directives were adopted which together constitute the measures known as UCITS III. While the first directive focused on the operational and managerial aspects of investment funds and the simplification of prospectuses, the second regarded the rules of investment diversification and risk spreading. These two directives introduced the so-called "European passport" for management companies according to which companies already authorised in their home country of origin were able to exercise the activities covered by the directives in any member state, either thanks to the freedom to provide services or through branches. UCITS III defined the new European standards for the realisation of the legal documentation and sales, thereby providing greater transparency, completeness, and simplicity: in particular, it made it obligatory to publish a simplified prospectus with standardised features, in addition to information documents. UCITS III expanded the range of investments that could be held by investment funds, while introducing the obligation to implement a rigorous and thorough process of managing and controlling risks associated with investing.

The UCITS IV directive of 2009 significantly innovated prior regulations in order to achieve multiple objectives. Firstly, it was aimed at transposing demands coming from the market, significantly simplifying cross-border distribution; secondly, it was designed to favour the reduction of management costs; finally, the aim was to provide investors with a simpler, more comprehensible disclosure. In this regard, the UCITS IV directive provided for the following changes:

- The introduction of a complete European passport.
- Simplification of the notification procedure between the various supervisory authorities.
- Facilitation of procedures for the merger of UCITS funds.
- Institution of the master-feeder structure.
- Introduction of the key investor information document (KIID).

With UCITS IV, the European passport which had already been introduced by UCITS III came to full maturity, being linked not only to distribution, but also to the management of funds. In fact, an asset management company established in a European Union country and authorised to practise collective

(continued)

Box 2.1 (continued)
portfolio management in that country is allowed both to market its own funds in all other member states and to manage funds set-up in another member state, without having to resort to any delegation management or set up a subsidiary in the host country.

Another novelty of UCITS IV is the initial phase of distribution in another member country: in the past, in every country of the European Union there was an ex-ante control by the competent national supervisory authority, under which foreign funds could be distributed in a country other than that of establishment only after a certain period of time, necessary for supervisory authorities to carry out controls. This delay between the time of the notice of the authorisation and the actual distribution of the funds represented a disadvantage when compared to other financial products offered to investors, often subject to less strict controls. UCITS IV directive intervened on this point in order to shorten waiting times and costs for the management companies. Now, in order to benefit the cross-border operation, there is a simple direct notification between supervisory authorities of the countries involved.

The great innovation introduced by UCITS IV concerns the possibility of mergers between UCITS funds, which makes it possible for two harmonised funds with the same investment objective to merge, regardless of their legal form and whatever their EU country of origin: in fact, there may be cross-border mergers (between two funds established in different member states) or domestic mergers (between funds of the same member state). These facilities at the merger processes are intended to contribute to the effective functioning of the complete European passport, since the fund affected by the merger can circulate within the European Union. The purpose of this change is to increase the assets under management and reduce costs for management companies, promoting economies of scale and making European funds more competitive, since they are currently at a disadvantage compared to US funds, which tend to be considerably larger in size.

The merger processes described above favour the centralised management of multiple funds by management companies with the so-called master-feeder structures, which are designed to encourage the growth in size of harmonised UCITS and allow them to gain significant economies of scale. They are characterised by the presence of a feeder fund which, despite the provisions on investment limits, is authorised to invest at least 85% of its assets in units of another investment fund, the so-called master. The remaining 15% of the assets of the feeder fund can be invested either in the same master fund or in cash and/or derivatives for hedging purposes. The master fund is characterised by numbering among its participants at least one feeder fund, by not being itself a feeder and by not holding units of a feeder. The purchase of master fund is reserved to feeder funds. The two types of funds may be managed by the

(continued)

Box 2.1 (continued)
same management company, as well as by different management companies. The master-feeder structures may have a purely national importance, if they are composed of investment funds set-up in the same member state of the European Union, or cross-border, in the event that master and feeder funds are established in different member states.

Another important innovation introduced by the UCITS IV directive for harmonised funds concerns the replacement of the simplified prospectus, introduced by the previous UCITS III directive, with another document—the key investor information document (KIID)—containing all relevant information for investors. The purpose of the KIID was to offer the investor a document easy to understand and able to illustrate the salient features of the investment in two pages, so as to make the participants fully aware of the fund's specific characteristics (including costs and risks). From January 2023, the KIID has been replaced by the key information document (KID), previously required for all the other packaged retail investment and insurance-based investment products (PRIIPs).

A further development of the European Union framework for UCITS funds came in 2014 with a directive that the market has renamed UCITS V, although, objectively speaking, its scope is much narrower than the previous guidelines. This directive has three principal objectives: strengthening the rules on the liability of the depositary institution; introducing requirements on remuneration policy within the management company; and harmonising administrative sanctions within the provisions of individual national supervisory authorities for violations of Community rules as introduced by the different UCITS Directives.[a]

[a]UCITS I is the directive 85/611/CEE. UCITS III rules are contained in the management company directive 2001/107/CE and in the product directive 2001/108/CE. UCITS IV is the directive 2009/65/CE of the European Parliament and the European Council of 13 July 2009. Finally, UCITS V is the directive 2014/91/EU of the European Parliament and the European Council of 23 July 2014. This last directive was introduced into the various member states by 18th March 2016.

Finally, from the perspective of an EU-based investor, foreign funds can be divided according to where they are marketed, as follows:

- Investment funds marketed in the investor's state (so-called "authorised funds").
- Investment funds not marketed in the investor's state (so-called "unauthorised funds").

In the former case, the fund is authorised for marketing in the investor's country of residence. In the case of UCITS funds, distribution is possible in all European Union member states without prior authorisation, simply on the basis of mutual recognition, following prior notification sent from the supervisory authority of the country of origin, in accordance with the provisions of the UCITS IV directive. In the case of AIFs, free circulation in the EU applies only to professional investors, while in the case of retail customers a specific authorisation regime is envisaged by the competent supervisory authority of the host country.

In the latter case, funds are not subject to distribution in the investor's country of residence, resulting in the prohibition of the solicitation of public savings, even if this does not exclude the possibility that the investor, in a regime of free movement of capital, can directly allocate part of his/her wealth to a fund not located in his/her own country and can buy it directly abroad, in compliance with the legal, fiscal, and financial monitoring rules established in each country.

Below, close attention will be paid to the case of open-end investment funds, since these funds play a predominant role in the industry. Table 2.1 contains a summary of the characteristics of certain specific types of investment funds. Specific considerations are made in Chaps. 14 and 15 for closed-end funds specialised in

Table 2.1 Some specific types of funds

Qualifying investor funds
Qualifying investor funds may be structured as open-end or closed-end. Fund rules provide that participation is reserved to "qualified" investors, specifying the characteristics they must hold to participate in the fund. Qualifying investor funds have greater flexibility in investment policy, since they may waive the prudential limits stated for UCITS funds.
Funds of funds
Funds of funds are investment funds that allocate the capital raised in other investment funds. If a fund of funds invests only in funds managed by the same management company, it is called a "fettered" fund of funds. If it invests in external funds run by other management companies, it is called "unfettered" and embraces a multi-management approach.
Guaranteed funds
Guaranteed funds assure investors that, at a specified time, they will be paid at least the value of the initial investment, even if the development in capital markets is unfavourable. If capital markets grow, the investor will participate in their development. The guarantee is usually based on an insurance policy taken out in favour of the fund. In guaranteed funds, the management company assumes an obligation of result.
Protected funds
Protected funds have an investment policy designed to protect the value of the investment, pursued through the application of quantitative techniques of investment management and the limitation of losses. There is no guarantee that the value of the investment will not fall below the level of protection. In this case, the management company assumes an obligation of means but not of result.
Ethical (or SRI or ESG) funds
Ethical funds or socially responsible investment (SRI) funds have an investment policy that bans the purchase of a set of securities that do not respect ethical standards and/or selects securities on the basis of criteria other than solely the maximisation of the expected return and/or adheres to an investment process that incorporates environmental, social, and governance (ESG) criteria.
Index funds
Index funds are investment funds with an investment policy aimed to replicate the risk-return profile of a market index calculated by third parties.

private equity, venture capital, and private debt, while Chap. 16 makes detailed reference to the case of real estate and infrastructure funds. A complete analysis of hedge funds is provided in Chap. 12.

2.3.2 Open-End Investment Funds

The key elements that characterise the participation in an open-end investment fund are:

- Purchase of units.
- Redemption of units.
- Switching.
- Valuation of units.
- Investment objective and policy.

Participation in the fund is achieved through the subscription of units of the fund, which can be done in two ways: by immediate payment of the full value of the units the investor has decided to buy (a capital investment plan) or by spreading the investment over time by subscribing to a capital accumulation plan (a savings plan).

In a capital investment plan, the purchase of units is carried out in a single instalment and in one operation. The investment must be at least equal to the minimum required by fund rules and also indicated in the prospectus. The minimum amount of the subscription varies depending on the management company and can also vary within the same management company, depending on the fund considered.

In a savings plan, the subscriber shares out the investment over time through a series of periodic payments, each of the same amount, the number of which may range, at the discretion of the subscriber, between a predetermined minimum and a maximum.[12] The accumulation plan may be suspended or terminated at any time. However, early redemption of the plan often involves higher subscription charges in relation to the amount invested.[13]

The number of units to be attributed to each participant is determined by dividing the amount of the capital invested, net of any entry charge and any fixed cost, by the net asset value on the reference date, i.e. the day when the management company received confirmation of the subscription or, if later, the day of the value date recognised to the means of payment used for the subscription. Conventionally, the application for subscription received within a preset time (cut-off time) is deemed to be received on a given day indicated in the prospectus of the fund.[14] In this regard,

---

[12] The contract of adhesion to a savings plan shows the total value of the investment, the number of payments, the amount and the timing of payments.

[13] This is tied to the procedures used for the calculation of the subscription fees in capital accumulation plans.

[14] The distributor must submit the application for subscription and means of payment no later than the day following the date of signing.

the management company is obliged to define procedures for subscription and redemption of units so as to avoid the risk of individual participants benefitting at the expense of the fund or other participants, as has happened in some financial systems in the past.[15]

On receipt of each subscription, the management company sends the subscriber a letter of confirmation containing the information regarding the date when the subscription request was received, the gross amount paid and the net amount invested, the date of entry, the unit value, and the number of units assigned.

Units of the investment fund can be registered or bearer, as provided in fund rules. If units are not intended for dematerialisation, the participant can always request issue of the representing certificate—both at purchase and afterwards.

In the case of open-end funds, investors can request the redemption of units held at any date of valuation. The request for reimbursement must be submitted or sent by the investor to the management company directly or through an entity in charge of the distribution which shall forward it to the management company no later than the first working day after receipt.[16]

In case of redemption, the amount to be repaid is determined starting from the net asset value on the day the management company receives the application for reimbursement, or, if that happens to be a day when the unit is not valued, on the first day of calculation thereafter. The fund rules define the criteria for identifying the day of receipt and also indicate the cut-off time within which the application must be received by the management company. The management company is obliged to settle the investor the amount due within the period stated in fund rules.

The participant may also request a planned redemption of units, indicating: the date from which he/she will effect the repayment plan; the periodic intervals of redemption transactions; the amount, or the number of units to be refunded; the way the amounts refunded are to be credited.

Fund rules must also specify cases of an exceptional nature, in which the redemption or issue of units may be suspended in the interest of the participants. In the case of redemptions, these events are generally related to situations where requests for reimbursement, because of their size and the existing market situation,

[15] For a discussion of possible forms of arbitrage (late trading and market timing) to the detriment of investment fund unit-holders, practised in the US financial system in the past, and for an analysis of the solutions adopted by the American Securities and Exchange Commission, one can refer to Haslem (2009) and McCabe (2009). In the case of late trading arbitrage, the subscription of units of the fund near the end of trading was settled, for some privileged customers, at the price of the current day or the next day, according to the market trend. In the case of market timing arbitrage (or time zone arbitrage), time differences were exploited, together with the related opportunities for arbitrage, especially as regards foreign shares and other foreign securities.

[16] The participant must be given the opportunity to: submit the redemption request directly to the management company, without going through distributors; disclose either the units or the amount to be repaid as the subject of divestment; request a partial refund; choose a method of payment among those indicated in fund rules.

would require disposals such as to potentially cause harm to the interests of the participants.

On the redemption of units of an umbrella fund, the investor is entitled to simultaneously subscribe units from other sub-funds managed by the same management company, thereby operating a switch.[17] This switch between sub-funds is performed as follows:

- The redemption value of the sub-fund of origin is determined on the day the management company receives the switch request.
- The day of subscription of the sub-fund of destination which the participant has chosen coincides with the payment date of the redemption.
- The number of units to be issued depends on the number of units of the original sub-fund to be switched, their redemption value and the subscription value of the new sub-fund.[18]

The unit value of an investment fund—usually called the fund's net asset value (NAV)—is derived from the sum of the current value of all assets in the portfolio (including cash), net of liabilities, divided by the total number of units of the fund. For example, a fund that, on a given day, has 50,000,000 units, owns 970 million euros in securities, 50 million euros in cash and debt for 20 million euros, has a unit value of 20 euros.

The calculation of the NAV per unit requires:

- The valuation of all investments and assets in the portfolio on the basis of the market price.
- The calculation of the cash belonging to the fund.
- The sum of these two components, which is the gross asset value of the fund.
- The calculation of the accrued expenses for management fees, performance fees, fees to the depositary institution and other expenses borne directly by the fund.
- The identification of any additional outstanding liability.

The difference between the total value of assets, on the one hand, and the debts, on the other, produces the overall net asset value, or NAV, which, divided by the number of units outstanding on that given day, gives us the NAV of the single unit, at which new subscriptions and redemptions received by the management company on that specific day are to be settled. A feature of investment funds is that, unlike what happens with the great majority of financial instruments traded continuously, for

[17] When the investor exchanges units held from one class in a sub-fund for units of another class in the same sub-fund, the operation is termed "conversion". On the other hand, when the investor exchanges units held from one class in a sub-fund for units in another sub-fund, the operation is termed "switch". As mentioned before, the switch operation has gradually lost many of the benefits in terms of regulations, taxes, and fees which it once enjoyed, since it is now treated as a redemption and a purchase.

[18] If the new sub-fund is denominated in a different currency from the original sub-fund, a conversion factor must be considered that depends on the exchange rate of the base currencies of the two different sub-funds.

each valuation date of the unit, there is a single NAV which coincides with the price of all subscriptions and all redemptions regulated on the same day, without a bid-ask spread between buying and selling prices.[19]

Depending on the investment policy and the subsequent adoption of a given benchmark index, it is possible to divide open-end investment funds into two categories:[20]

- Market funds.
- Non-market funds (also defined as "alternative" funds).

Depending on their investment strategy, market funds are designed to track or outperform a given benchmark index, representative of a segment of the financial market. In fact, such funds can be managed with an active management strategy or an indexed one. In the first case, in its role as active manager, the management company aims to beat the benchmark used as a reference. In the second case, the management company aims to be more faithful to the benchmark: a special case of indexed investment fund are exchange-traded funds (ETFs), passive funds traded continuously on one or more regulated markets, whose characteristics are described in Box 2.2.

Box 2.2 Characteristics of Exchange-Traded Funds

Exchange-traded funds (ETFs) belong to the family of passively managed investment funds, whose objective is to replicate the risk-return combination of a given benchmark index. The characteristic feature of ETFs compared to traditional index funds is the continuous marketability of the units (or shares, if the legal form is that of the investment company) on a regulated market at a price which is known in real-time, unlike what happens for traditional investment funds, characterised by a lag time between the purchase or redemption of units and the availability of the price at which the transaction is regulated.

Each ETF is linked to a particular market index and closely follows the trend of this index, being made up of a well-known portfolio of financial instruments which—physically or synthetically—replicates the performance of the underlying benchmark. Buying an ETF is therefore equivalent to investing in its benchmark index, despite imperfections in the replication due to periodic cash flows distributed or reinvested using different

(continued)

[19] An exception in this regard is represented by the unit trust—a collective investment scheme used in the UK and in other countries with common law systems other than the USA—where the trading can take place on the basis of a bid price and an ask price.

[20] CESR (2010a). The term "alternative funds" used here refers to strategy funds and to structured funds that adopt innovative investment strategies made possible by the new European Union rules but which are still, at least potentially, harmonised funds—in the sense of compliance with UCITS directives—and are not to be confused with hedge funds and other non-harmonised funds, despite sharing some of the objectives and strategies of these. On this point, see Chap. 12.

Box 2.2 (continued)
assumptions from those adopted by the index providers, as well as costs and charges that are levied on the assets of the ETF and do not exist on the benchmark.

ETFs are open-end funds, which are no-load funds (no entry or exit charges) with no performance fees, characterised by extremely low management fees. At the purchase and sale of ETF units (or shares), however, investors pay the costs of intermediation and bear the cost of the bid-ask spread related to the negotiation on the regulated market characterised by the presence of market makers.

The minimum investment is equal to one unit (or one share): the investor therefore has an opportunity to enter the market with a very diversified portfolio with small amounts.

To sum up, the advantages of ETFs are:

- Broad diversification with reduced capital.
- Very low management fees.
- Liquidity.
- Transparency.
- Buying and selling prices known in real-time.
- (Almost) perfect replication of the benchmark.
- Ability to be fully invested.

The disadvantages of ETFs are:

- Transaction costs of buying and selling.
- Bid-ask spread on the secondary market trading.

The creation of an ETF requires the existence of two main markets: a primary wholesale market, which involves only a few authorised participants, and a secondary market accessible to all investors.

The object of trades carried out on the primary market is represented only by wholesale creation units, which form a block trading of, i.e., 50,000 units. This market can only be accessed by authorised participants, who are the only ones who can apply for the issue of new units. The peculiarity is that the subscription of the units that make up the creation unit can be made "in kind", delivering the same financial instruments included in the benchmark index that is replicated by the ETF and any portion of liquidity, in order to make up the so-called "creation basket". Similarly, the redemption of the units on the primary market can be obtained only by authorised participants, which may also request redemption of units "in kind".

The division of the creation unit into individual units and their simultaneous distribution to investors is done by the same authorised participant through negotiation on the secondary market.

(continued)

Box 2.2 (continued)

Investors can then buy and sell ETF units on the secondary market, where the ETF is also traded through the intervention of one or more market makers who support the liquidity of the ETF trading book.

Specific rules on the wholesale primary market and its implications are the element that most distinguishes an ETF from a traditional open-end fund. The purchase or sale of ETF units by investors in the secondary market does not in fact generate the change in the number of outstanding units of the ETF but merely provides for the replacement of the previous investors (sellers) with new investors (buyers), thereby creating a scheme closer to that of closed-end funds. Unlike the latter, however, in the case of ETFs, the net asset value can be determined objectively and if a misalignment might arise between the theoretical NAV, based on the market prices of the basket of financial instruments, and the trading price on the secondary market, one or more authorised participants intervene immediately with an arbitrage transaction to restore parity between the theoretical NAV and the price on the secondary market. In the presence of a sufficiently liquid market, the option given to authorised participants to subscribe and redeem units (or shares) on the primary market through a mechanism of creation-redemption in kind makes sure that the trading price on the secondary market will converge with the ETF's theoretical NAV.

Like CIVs, ETFs are not affected by the management company's credit risk. However, they are not immune from counterparty risk if synthetic replication of the benchmark takes place, through swap contracts with no central counterparty, instead of through physical replication.

In recent times, increasingly complex ETFs have appeared on the market, such as ETFs based on new types of underlying indices (including real estate, private equity and hedge fund indices, volatility indices and indices that measure credit quality), structured ETFs (which allow the investor to take a prearranged investment strategy on a given index, based, for instance, on leverage) and actively managed ETFs (whose goal is not to replicate a benchmark, but to expose themselves to an "active" strategy developed and updated by the management company).

Sources: Gastineau (2010) and Abner (2010)

Non-market funds include various types of funds with investment policies and consequent risk-return combinations not summarised by a benchmark index.

These funds belong to one of these sub-categories:

– Absolute return funds.
– Total return funds.
– Life-cycle funds.
– Structured funds.

These new categories of funds are largely the result of new investment limits and additional European rules introduced by the EU directive UCITS III, which has made it possible for traditional investment funds to adopt investment strategies previously reserved to hedge funds, qualifying investor funds and other non-harmonised investment funds.[21]

Absolute return funds are managed according to investment policies or strategies which envisage a variable allocation of the portfolio of the fund across asset classes, under the constraint of a predetermined risk limit.[22]

Total return funds are engaged in investment strategies that pursue specified reward objectives by using flexible investments to participate in different financial asset classes (e.g. in both equity and fixed-income markets). Unlike absolute return funds, in which a target is set in terms of risk, in total return funds the target is set in terms of return.

Life-cycle funds are managed using strategies that entail a shifting of their portfolio allocation over time by reducing the market risk as the target maturity date approaches, in accordance with some predetermined rules.

Structured funds provide investors with algorithm-based payoffs at certain predetermined dates which are linked to price movements of financial instruments or indices. The final performance of the structured funds is then determined using established, objective, and unchangeable criteria, indicated in the predetermined algorithm.

Another classification system of investment fund investment policies in use in Europe has been developed by the European Fund and Asset Management Association (EFAMA) with the purpose of defining the parameters for a pan-European classification system.[23] The aim of classification is to clarify the main features of the investment fund and the main factors that can impact on its risk profile, thereby making possible a better perception of the risk-return combination of each classified fund.

This classification of funds is divided primarily into six macrocategories:

- Equity funds.
- Bond funds.
- Multi-asset funds.
- Money market funds.
- Absolute return innovative strategies funds.
- Other funds.

Each macro-category is characterised by minimum and maximum investment limits in the different categories of financial instruments, which provides the

[21] For this very reason, these funds are often called NEWCITS, as if they were a new category of UCITS. Stefanini et al. (2010).

[22] CESR (2010a).

[23] EFAMA (2008) and EFAMA (2023).

Table 2.2 EFAMA classification of investment funds

Macrocategories	Characteristics	Classification criteria
1. Equity funds	Funds with more than 85% exposure to equity.	Geographical exposure, sector exposure, market capitalisation, structural characteristics.[a]
2. Bond funds	Funds with at least 80% of their assets allocated to fixed-income securities. Investment in cash should not exceed 20%. Investment in other assets should not exceed 10%.	Geographical exposure, currency exposure, credit quality, interest rate exposure, emerging market exposure, structural characteristics.
3. Multi-asset funds	Funds that invest assets under management in both equity and fixed-income securities.	Geographical exposure, currency exposure, asset allocation, structural characteristics.
4. Money market funds	Funds that invest their asset in money market instruments of high quality or deposits with credit institutions.	Currency exposure, credit quality, interest rate exposure, structural characteristics.
5. Absolute return innovative strategies funds	Funds managed with the objective of generating a positive return over a cash benchmark, irrespective of market movements.	Directional, long/short, relative value, event driven, multistrategy, index trackers, funds of funds.
6. Other funds	Funds falling outside the five broad categories stated above.	Guaranteed, capital-protected, asset-backed securities, convertibles, life-cycle/target maturity, infrastructures, commodities, open-end real estate, closed-end real estate, REIT.

[a]Structural characteristics refer to features characterising funds, such as socially responsible investment funds, ETF, fund of funds, etc.
Sources: EFAMA (2008, 2023)

parameters for that fund's asset allocation, and is then divided into different categories defined on the basis of the risk factors that characterise each one (Table 2.2):

- Equity funds have an exposure to equity amounting to at least 85% of their assets and can be distinguished on the basis of geographical exposure, sector and market capitalisation.
- Bond funds have at least 80% of their assets invested in fixed-income securities. Investment in cash should not exceed 20%, including investment in other assets which should not exceed 10%. This category can be sub-divided according to credit quality, interest rate exposure, and currency exposure.
- Multi-asset funds are balanced funds that invest their assets under management both in equity and in fixed-income securities. They can be differentiated according to geographical exposure, asset allocation, and currency exposure.
- Money market funds can be classified, according to the European Securities and Markets Authority (ESMA) distinction,[24] in short-term money market funds and

[24]The CESR/ESMA guidelines on a common definition of European money market funds provide the following requirements for a fund in the European Union to be defined as a money market fund:

standard money market funds. Further sub-divisions of this category are based on currency exposure and, in the case of short-term funds, on the fund's valuation method.[25]

- Absolute return innovative strategies funds are managed with the aim of generating a positive return irrespective of market movements. Whilst equity, bond,

---

- The primary investment objective is to maintain the principal and provide a return in line with money market rates.
- The object of the investment policy is represented by high quality securities. A money market instrument is not considered to be high quality unless it has been awarded one of the two highest available short-term credit ratings by each recognised credit rating agency that has rated the instrument or, if unrated, it is of an equivalent quality as determined by the management company.
- Subscription, redemption, and NAV calculation are provided daily.
- Direct or indirect exposure to equity or commodities, including via derivatives, is not admitted.
- Investments must be limited to securities with a residual time to maturity of less than or equal to 2 years, provided that the time remaining until the next interest rate reset date is less than or equal to 397 days.
- Weighted average maturity (WAM) of the portfolio must be no more than 6 months.
- Weighted average life (WAL) of the portfolio must be no more than 12 months.

In the case of short-term money market funds, the last three requirements are replaced with the following:

- Investments must be limited to securities with a residual time to maturity of less than or equal to 397 days.
- Weighted average maturity (WAM) of the portfolio must be no more than 60 days.
- Weighted average life (WAL) of the portfolio must be no more than 120 days.

WAM is a measure of the average time to maturity of all of the securities in the fund, weighted to reflect the relative holdings in each instrument, assuming that the maturity of a floating rate instrument is the time remaining until the next interest rate reset to the money market rate, rather than the time remaining before the principal value of the security must be repaid.

WAL is the weighted average time to maturity of each security held in a fund, meaning the time until the principal is repaid in full. Unlike what happens in the calculation of the WAM, the calculation of the WAL for floating rate securities and structured financial instruments does not allow the use of interest-rate reset dates and uses instead only a security's stated final maturity.

[25] Short-term money market funds can have a constant net asset value or a variable net asset value, whereas money market funds, like any other kind of funds, must have a variable net asset value. A constant or stable NAV short-term money market fund seeks to maintain an unchanging face value NAV (e.g. $1 or €1 per unit or share). Income in the fund is accrued daily and can either be paid out to the investor or used to purchase more units in the fund. Assets are generally valued on an amortised cost basis which takes the acquisition cost of the security and adjusts this value for amortisation of premiums (or discounts) until maturity. According to CESR guidelines, the fund must ensure that such a method will not result in a material discrepancy between the market value of the instruments held by the fund and the value calculated according to the amortisation method, whether at the individual instrument level or at the fund level. The fund must periodically calculate both the market value of its portfolio and the amortised cost valuation and take action if any discrepancy becomes apparent. The constant NAV is not guaranteed, and whenever a discrepancy materialises between the market value and the amortised cost value of the portfolio, the short-term money market fund can no longer issue and redeem units at the stable NAV: this situation is often known as "breaking the buck". This may occur, for example, when there is a default by the issuer of an instrument in the portfolio. CESR (2010b).

multi-asset, and money market funds are long-only funds that typically aim to achieve a higher return than a benchmark index, innovative strategies funds make extensive use of derivatives to short/long the securities or the market as a whole.[26]

- "Other funds" is a residual category including different kinds of funds falling outside the five broad categories stated above. Examples of these are: asset-backed securities funds, guaranteed funds, capital-protected funds, commodity funds, infrastructure funds, and real estate funds.

2.3.3 The Benchmark and Investment Strategies

The benchmark is an objective reference parameter of common use, constructed by using financial indicators developed by third parties. The benchmark must be consistent with the risks of the CIV and is used to compare the performance of the fund with that of the benchmark itself.[27]

The benchmark is an *ex ante* tool able to signal risk characteristics of the fund and to permit a more informed investment choice by the subscriber. *Ex post*, the benchmark allows the results of the fund to be assessed, by comparing both the return and the absolute and relative risk of the fund in relation to those of the benchmark, over a given period.

It should be noted that, in general terms, the benchmark only reflects the risk-return profile of the fund. It is instead the kind of strategy adopted—active or indexed—which outlines the extent of the risk profile of the portfolio under management as compared to the market risk of the benchmark portfolio, considered "neutral". In the case of an active investment strategy, the benchmark does not have a compelling effect on the operations of the management company. In fact, active management increases the management company's room for manoeuvre and therefore adds an additional market risk, which can be positive, in case of an over-performance in relation to the benchmark, or negative, in case of an underperformance. Conversely, in the case of indexed investment strategy, the management company only replicates the composition of the benchmark, which therefore defines itself briefly the level of risk and return of the fund.[28]

---

[26]The EFAMA definition of "absolute return innovative strategies funds" does not correspond with the ESMA definition presented above, reflecting the existence of many different definitions of "absolute return" funds (as well as for "total return" funds) across the fund industry, which are not always similar in nature or easy to understand.

[27]Schoenfeld (2004), Gastineau et al. (2007), and Christopherson et al. (2009).

[28]Financial theory is the foundation of the development of indexed investments and this is a fairly isolated case in the history of the financial markets where, as a rule, product innovation has always preceded the formulation of the theories underlying it. The conceptual assumptions of the use of passive investment techniques, aimed at the replication of market indices, can be traced in the studies of Cowles (1933) regarding the limited predictive power of financial analysts in the USA from 1920 to 1940. These first empirical analyses were followed by the theoretical formalisation of

Even in the case of indexed investment strategies, the replication of the index and of its risk-return combination is not so obvious. In theory, the replication of a benchmark is elementary: it amounts to buying the same securities included in the benchmark in the same weights they have in the index. In practice, this replication is hindered by several factors, the most important of which are:[29]

- The need to recompose the portfolio periodically during each ordinary and extraordinary revision of the benchmark index which is being replicated.
- The presence, in some segments of the securities market, of minimum trading lots, which prevent perfect replication of the benchmark index.
- The existence of management fees and other expenses directly incurred by the fund, which affect the managed portfolio, but not the benchmark.
- The presence of transaction costs which affect the portfolio, but not the benchmark (trading charges and bid-ask spread).
- The occurrence of any tax burden on the performance of the managed portfolio and not on that of the benchmark.
- The impossibility for managed portfolios of open-end vehicles which face redemption requests to be fully invested, as a portion of assets must always be kept as cash so as to respond promptly to requests for reimbursement.
- The existence of regulatory constraints that prevent excessive concentration of the managed portfolio in a single financial instrument, in all those cases in which the replicated benchmark is composed of relatively few securities and one or more of these securities have an excessive weight compared to the concentration limits established for the fund.
- The presence of dividend and other income distributed by the securities included in the indexed portfolio but not incorporated into a price index.
- The reinvestment of all interest and other income earned on the managed portfolio on the basis of choices other than those assumed by the index provider that calculates the benchmark.

The choice between active management and indexed management is essentially linked to two factors:

---

the Capital Asset Pricing Model and, in particular, the introduction of the market portfolio and have come to fruition with the theory of efficient markets. After this theoretical evolution, Malkiel felt it was time to make an open appeal to the community of asset managers to provide “a new investment instrument: a no-load, minimum management-fee mutual fund that simply buys the hundreds of stocks making up the market averages and does no trading” (Malkiel 1973, p. 226). From this proposal emerged the first indexed investment fund targeted at retail investors: the Vanguard 500 Index Fund, established in the USA in 1976, which still exists today. Since this milestone, decades have passed and investments for replication of market indices have grown considerably and diversified by asset class and by different technical solutions. Meanwhile, the scientific literature has continued its studies along two main lines: an empirical line aimed at investigating the ability of actively managed portfolios to achieve a higher-than-average performance compared to indexed portfolios; and a theoretical line aimed at highlighting which are the methodologies best suited to assessing the efficiency of a stock index.

[29] Grinold and Kahn (2000), Siegel (2003), and IOSCO (2004).

- Costs.
- Market efficiency.

In relation to the first aspect, other conditions being equal, the cost of active management is certainly higher than that of an indexed portfolio, because of the differing role of activities of research and analysis and the increased portfolio turnover. Starting from this assumption, the theorem of active management developed by Sharpe showed that:[30]

- Ignoring management costs, the average gross return of actively managed portfolios must be equal to the average gross return of indexed portfolios.
- Including management costs, the average net return of actively managed portfolios is necessarily lower than the average net return of indexed portfolios.

Contrary to what is often said, this does not mean, as shown in Box 2.3, that, net of management costs, the average net return of actively managed investment funds must necessarily be lower than the average net return of indexed investment funds, because for "active manager" Sharpe meant not only active funds but any investor on a given market with a portfolio that does not exactly match the representative benchmark of the market itself.

Box 2.3 The Theorem of Active Management
In a famous paper of 1991, then partially revised in 2002, William Sharpe developed the so-called "law of active management", based on two different postulates:

1. Before management costs, the return on actively managed portfolios matches, on average, the return of passively managed portfolios.
2. Net of management costs, active portfolios as a whole must necessarily provide, on average, a lower return than indexed portfolios.

According to Sharpe, the two postulates are true on any time horizon and are demonstrable, under certain assumptions, making use of only the four elementary operations.

The starting point for the proof of the theorem is the following. A passive manager will include in his/her portfolio the same securities included in the benchmark, respecting their proportions. An active manager, on the other hand, will implement active choices that lead him/her to exclude, to underweight or to overweight some securities and include others.

If, in a given period, the market generates a return of +10%, for example, the passive managers will earn, by definition, a return which, before management costs, is identical to that of the market. But it follows that active

(continued)

[30] Sharpe (1991, 2002).

Box 2.3 (continued)
managers, too, will earn an average return, before management costs, of +10%. This is because in a market where there are only passive and active managers, if the average performance of the market is +10%, the gross return of indexed managers is certainly of 10%, but the logical consequence is that the average return of active managers before costs must necessarily be equal to 10%. Practically:

$$R_{\text{m}} = R_{\text{pm}} \cdot x_{\text{pm}} + R_{\text{am}} \cdot x_{\text{am}}$$

where

R_{m} = market return, as measured by a capitalisation-weighted benchmark index.
R_{pm} = average return before costs of passive management strategy with a full replication of the benchmark index that measures changes in the market.
x_{pm} = weight of passive portfolios in the market.
R_{am} = average return before costs of active managers investing in the benchmark index that measures changes in the market.
x_{am} = weight of active managers in the market.

It is evident that the reasoning is based on the assumption that the sum of the percentage weights of the assets managed by active managers and those held by indexed managers is equal to 100%. In other words, the term "active manager" is used by Sharpe to identify not only active asset managers but also any holder of financial instruments who has a discrepancy in his/her financial portfolio with respect to the composition of the benchmark used to measure market return. The figure of the "active manager" described by Sharpe therefore includes individual investors, institutional investors, and any other investor who holds instruments in his/her portfolio listed on the reference market in proportions different to those in the specified benchmark used to calculate the market return.

Having clarified the meaning of "active manager", it is undeniable that the return before costs of indexed portfolios has to coincide with the average return before costs of actively managed portfolios.

Once the first proposition has been proven, it is even easier to prove the second. The cost of active management is obviously higher than that of passive management, because of the higher costs of analysis, stock selection, and trading. By way of example, if the average cost of active management in a given market is 1.5% per year, and that of the passive management 0.5% per year, this proves the second part of the thesis. The average return of an actively managed portfolio, net of costs, must necessarily be lower than that of an indexed portfolio. While both start at gross returns of 10%, after management

(continued)

Box 2.3 (continued)
costs are deducted, it is easy to understand what Sharpe was saying, provided that this does not mean—as is often pointed out—that, on average, the net return of actively managed investment funds is by definition lower than the net return of passively managed investment funds.

The theme of the ability of active management to produce a greater return compared to indexed management has been the subject of debate in the scientific literature for decades, with mixed results.[31] If, on the one hand, the ability of the indexed manager to beat the reference benchmark is essentially non-existent, on the other, active managers, faced with higher management costs, have at least the potential ability to achieve outperformance compared to the market, though there may also be the risk of generating an underperformance. In fact, active managers add to the risk of the benchmark an additional market risk, the presence of which may be positive or negative: that is, so-called "active risk".

At this point, the second key variable comes into play: the degree of market efficiency. The fund manager may choose to follow an active strategy, since he/she is convinced that he/she possesses the specific skills to beat the benchmark, given the limited market efficiency. On the other hand, the fund manager may decide to follow an indexed investment strategy so as to replicate a predefined benchmark, because he/she believes that the market is so efficient as to make it impossible to obtain outperformances continuously and systematically, or because he/she believes the market to be so efficient that even if systematically positive excess returns were obtainable, these would be too low to justify the higher costs of analysis, research, and transaction involved in implementing an active investment strategy.

It is therefore left to the investor and, in a nutshell, to his/her capacity of evaluation and appetite for risk to decide whether to rely on an active manager or a passive manager, taking into account the difference in costs and evaluating the ability of the asset manager to beat the market return, especially as a function of the level of efficiency of the same market.

2.3.4 Bond Portfolio Strategies

Bond portfolio strategies can be classified according to the objectives pursued by the asset manager. This distinction leads to the identification of three strategies:

[31] Elton et al. (1993), Goetzmann and Ibbotson (1994), Malkiel (1995), Gruber (1996), and Haslem (2010).

- Active bond strategies.
- Indexed bond strategies.
- Bond strategies independent of a benchmark.

In using active bond strategies, the asset manager's objective is to beat the reference benchmark. These strategies offer the manager a wider degree of freedom in implementing the investment policy, with margins of discretion in the choice of securities, the timing of a reduced or increased exposure to risk, the choice of sectors, and the characteristics of the securities considered most interesting. In active strategies, the analysis of individual securities, expected market trends, sector evaluation, and characteristics of the best securities play key roles, entailing higher costs than those of indexed management. In the face of these higher costs, however, the investor gains an opportunity to achieve *ex post* a better risk-return combination than that provided by indexed investing if active management has been efficient.

In indexed bond strategies, the manager intends to replicate the benchmark using a series of alternative techniques. These strategies offer asset managers limited discretion, as they can only choose the technique of index replication. In indexed management, the analysis of individual securities, expected market trends, sector evaluation, and characteristics of the best securities play no role, involving much lower costs than those involved in active management.

Recently, a third category of bond investment strategies has emerged, known as "benchmark-free" bond strategies. These heterogeneous strategies all lack an index to beat or replicate. The three main strategies in this category are total return, absolute return, and target date. Implementing these strategies may involve the use of financial instruments other than bonds.

2.3.4.1 The Levers of Active Bond Strategies

In the case of bond portfolios, the added value of active management relative to indexed management can be derived from five types of choices, which constitute the so-called "levers of active management". These levers include:

- An exposure to interest rate risk different from the benchmark.
- A different breakdown by duration time bands through positioning strategies on the yield curve.
- An exposure to credit risk different from the benchmark.
- A sector rotation with respect to the benchmark.
- For portfolios fully or partially denominated in a foreign currency, an exposure to exchange rate risk different from the benchmark.

The first lever available to the bond manager who intends to beat the benchmark is the portfolio duration, used to amplify the impact of favourable changes in interest rates and mitigate the impact of unfavourable changes. This strategy aims to outperform the benchmark by varying a portfolio's degree of exposure to changes in market interest rates. In the phases in which interest rates are expected to fall, active bond managers take up an average portfolio duration that is greater than that of the benchmark. In phases in which interest rates are expected to rise, active bond managers take up an average portfolio duration shorter than the benchmark.

The second lever of active management, still referring to expected movements in interest rates, is based on positioning along the yield curve. Active managers can adjust the maturity structure of a bond portfolio based on expected changes in the relationship between bonds with different maturities. Assuming that yields do not increase or decrease to the same extent at different maturities but, rather, the possibility of a flattening of the yield curve by maturity or a steepening of the curve, the asset manager has the opportunity to position the portfolio in the segment of the yield curve that is likely to perform best in a given economic scenario.

The third lever available to the bond manager assumes a credit risk different from that of the benchmark. Active bond managers tend to deviate from the composition of the benchmark and seek better relative performance by investing in securities with credit risks different from the securities included in the benchmark. The use of this active management lever involves overweighting the bonds of issuers who could face a future upgrade in the market's credit risk assessment, with a consequent increase in price, and underweighting the bonds of issuers who could experience a future downgrade in the market's assessment of credit risk. An active manager can also select bonds not included in the benchmark if they are considered a good investment opportunity.

The fourth lever of active management for bond portfolios is sector rotation, which is an attempt to beat the market through the dynamic rotation of the weight of the production sectors according to the expectations of the evolution of the economic cycle. The sector rotation lever originates from the assumption that various production sectors react differently to the alternating phases of the economic cycle, and active managers can dynamically vary their weight within the portfolio to prevent the consequences of negative economic phases in some sectors or anticipate the effects of positive economic phases in other production sectors.

The fifth lever in the case of benchmarks comprising (at least partially) bonds denominated in foreign currencies is the degree of exposure to the exchange rate risk. In this case, the active manager can decide to adopt currency risk-hedging choices with a logic different from that adopted by the index provider to induce over-performance compared to the benchmark.

It is common practice to divide bond portfolio management strategies into nine categories derived from the combination of:

- The sensitivity of the portfolio to changes in market interest rates; here, portfolios are classified into short, intermediate, and long based on the average portfolio duration.
- The portfolio's credit quality; here, they are classified into high, medium, and low credit quality portfolios.

By crossing these variables, nine possible bond investment strategies are derived, characterised by different risk-return combinations, which can be represented through a matrix called the "fixed-income style box" (Fig. 2.1).

Portfolio duration				
Short	Intermediate	Long		
1	2	3	High	
4	5	6	Medium	Credit quality
7	8	9	Low	

Fig. 2.1 Fixed-income style box

Bond investment strategies with high sensitivity to changes in interest rates, low credit quality, and related strategies (quadrants 9, 6, and 8) are characterised by the greatest risk. Bond investment strategies with low sensitivity to changes in interest rates and high credit quality and similar strategies (quadrants 1, 2, and 4) are characterised by lower risk. Strategies with intermediate sensitivity to interest rate changes and average portfolio quality and similar strategies (quadrants 5, 3, and 7) are intermediate in terms of the risk-return combination.

2.3.4.2 Indexed Bond Strategies

From a technical standpoint, an indexed bond manager can track a benchmark index in three ways:

- Full physical replication.
- Stratified sampling.
- Synthetic replication.

Full physical replication involves inserting the same securities included in the benchmark in the same proportion. Therefore, this strategy is based on the full physical replication of the benchmark index.

Stratified sampling assumes the identification, based on alternative criteria, of a sample of securities representative of the entire benchmark capable of minimising the tracking error and ensuring significant savings in transaction costs compared to the purchase of the entire set of securities included in the benchmark.

Finally, synthetic replication involves replicating an index's performance through derivative contracts (typically futures or total return swaps). The synthetic replication strategy has the advantage of minimising transaction costs compared to physical replication; however, it adds a possible counterparty risk to the market risk associated with the performance of the benchmark index. If the replication is based on an over-the-counter derivative, such as a total return swap, and the counterparty ends up in default, a potential counterparty risk can arise if the derivative lacked a central clearing system.

2.3.4.3 Benchmark-Free Bond Strategies

A third category of bond investment strategy comprises those not tied to a pre-established benchmark index. This constitutes a residual category that includes all strategies that are not focused either on beating a benchmark taken as a reference or on replicating its risk-return combination.

Of all the possible strategies independent of a benchmark index, three deserve particular attention:

- Total return strategies.
- Absolute return strategies.
- Target date strategies.[32]

The first two strategies are derived from the EU UCITS III directive, which has expanded the range of operations of open-end investment funds UCITS-compliant, transforming a set of strategies that were previously reserved for hedge funds and other alternative investment vehicles into strategies available to retail investors. Through these strategies, any investor can now access liquid alternative investments UCITS-compliant, that is, alternative investments with a daily price and the possibility of subscription and redemption on any day.

The common objective of these two strategies is to achieve a relatively constant return over time, which does not correlate with the performance of the equity and bond markets. This result is made possible by: the opening of short positions, which makes it possible to hedge the risk on long positions and/or to profit from the fall in the price of the financial instruments sold; the possible (limited) use of financial leverage; and rigorous risk control intended to minimise fluctuations in the portfolio's market value by allowing managers broad freedom of operation and flexibility by which they can vary the allocation of investments over time, ranging from cash, bonds, derivatives, and even shares.

[32] The terminology introduced in the European Union by the ESMA is assumed here, even if, on a global scale, the expressions 'total return' and 'absolute return' are not always used with the meaning attributed to them by the ESMA.

At a global level, the expressions "total return" and "absolute return" are often used interchangeably or with nuances that vary according to the country or asset manager. At the EU level, however, the two terms have been codified by the ESMA to differentiate between them.

Total return strategies aim to achieve a threshold return above an absolute level (e.g. 2%) over a certain period as a combination of periodic income and capital appreciation. This target return does not guarantee a minimum return but rather the specification of a target that the investment fund aims to achieve. Therefore, if the performance is not consistent with the threshold return the manager intends to achieve, the investor is not entitled to that minimum return but will have to settle for the return actually obtained, even if negative. Among these strategies, bond investments are most prevalent, with the possibility of opening short positions and using derivative instruments. The investment approach is flexible and allows managers to select bonds of different issuer categories, with different technical characteristics, and denominated in different currencies.

Absolute return strategies provide wide freedom in the selection of financial instruments and markets, resulting in the variable allocation of the portfolio in different asset classes subject to compliance with a target risk. Absolute return strategies set the target risk, unlike total return strategies, which set a target return. Furthermore, in total return strategies, the assets under management are mainly allocated to bonds, albeit with a certain dynamism, whereas, in absolute return strategies, the assets can also be allocated to other financial instruments, thus giving rise to investment strategies that are not exclusively bond related.

In practice, an investment fund with a return objective may also have constraints in terms of risk, and a fund with a risk objective may also have a target return to aim for, making any clear distinction between the two strategies more complex.

The target date strategies (or maturity dates) envisage the construction of bond portfolios characterised by a maturity date coinciding with the end date of the management mandate or the fund's expiry date. Thus, the asset manager has the opportunity to implement an investment strategy bound to a pre-established final date with a level of risk that decreases as maturity approaches.[33] This strategy can be implemented in several ways, focusing on specific geographical areas or currencies or on certain categories of issuers or securities. If maturity coincides with the relevant holding period for the investor, this strategy creates an investment comparable to the purchase of a single bond but with the benefits of diversification, owing to the opportunity to invest in a bond portfolio. In this strategy, as in the previous

[33]We refer here to bond investment strategies with maturity, it being understood that the same terminology is used for balanced portfolios that have a predominantly equity exposure at the beginning, which decreases as time passes in favour of the bond component. This balanced strategy is widespread for pension funds and has the objective of making the asset allocation of the pension fund evolve in connection with the age of the members, thus providing, in a life cycle logic, a greater exposure to equity investments when members are young and a gradually decreasing exposure to equity markets as retirement approaches.

two, there is no capital guarantee but only the certainty that the market risk of the portfolio decreases as the scheduled maturity date approaches.

2.3.5 *Equity Portfolio Strategies*

Equity investment strategies can be classified according to the objectives pursued. This distinction leads to the identification of three strategies:

- Active equity strategies.
- Indexed equity strategies.
- Smart beta strategies.

In active management, the manager's objective is to beat the benchmark taken as a reference by using the levers of active management.

In indexed management, the manager intends to passively replicate the benchmark through a series of alternative techniques.

More recently, a third category of equity investment strategies has emerged, often referred to as "smart beta" strategies. These strategies, although indexed or governed by predefined rules that leave no room for manager discretion, aim to beat the risk-return combination of traditional value-weighted benchmarks. These strategies combine the characteristics of active and indexed management. Similar to active management, smart beta strategies aim to beat the benchmark's risk-reward combination by generating higher returns and/or lower risk. Like indexed management, they have lower management costs—although they are higher than those of pure indexed management—, predefined security selection rules, and zero (or limited) margins of manager discretion.

2.3.5.1 The Levers of Active Equity Strategies

In the case of equity strategies, the added value of active management compared to indexed management can be derived from four types of choices that constitute the levers of active management in equity portfolios. These levers are represented by:

- Market timing.
- Stock picking.
- Market rotation.
- Factor tilt.

Market timing represents the first lever of active management. This strategy aims to obtain results above the benchmark by varying the degree of portfolio exposure to systematic risk according to the expected market evolution. Therefore, the market timing lever assumes that the portfolio beta varies with respect to the benchmark. In phases of expected market growth (bull markets), managers should increase the portfolio beta, seeking an average beta greater than one, to maximise the profits resulting from the expected increase. In the phases of an expected reduction in market prices (bear market), the manager should reduce the portfolio beta, seeking an average beta lower than one to mitigate the impact of expected losses on the value of the portfolio. Practical implementation of a market timing strategy is based on the use of derivative instruments, debt, or changes in the proportion of portfolios allocated to cash.

The second level of active management is stock picking, also known as security selection. This strategy is based on an effective selection of shares to be included in the portfolio. In this case, the active manager has two options: to select the shares included in the benchmark to beat, assigning them a different weight from those they have within the reference index; or to select stocks not included in the benchmark but which, in the manager's opinion, represent attractive investment opportunities.

The third lever of active management consists of the use of the so-called market rotation and an attempt to beat the market through the dynamic rotation of the weight of the production sectors according to the expectations of the evolution of the economic cycle. As already seen in bond strategies, the market rotation lever originates from the assumption that different production sectors react differently to the alternating phases of the economic cycle. Therefore, the manager can exploit the different performance of the shares belonging to different sectors by dynamically varying their weight within the portfolio to amplify positive changes or attenuate negative changes.

A fourth lever derives from a refinement of the third lever and is represented by the different exposure to the factors that can explain the return and the risk of individual securities. This lever, known as factor tilt, is an investment methodology that is becoming increasingly popular for the management of equity portfolios, although the underlying theories are far from recent. In fact, these methodologies originate from multifactor models formulated many decades ago; they have been gradually refined to identify the most significant and rational factors. This lever foresees concentrating the portfolio on groups of securities distinguished by a series of alternative factors, the most frequent of which are shown in Fig. 2.2. Exposure to single factors (or to a plurality of factors simultaneously) has helped fuel a new category of equity investment strategies, defined as smart betas, which, unlike active strategies, define an investment rule based on one or more factors, and this rule remains stable over time. By contrast, the active manager has a margin of discretion in changing exposure to the factors considered most relevant according to the expected evolution of the market over time.

Two factors that have become particularly widespread on a global scale over time are:

- Investment style.
- Company size.

Regarding the first, the investment strategy alternately focuses on two share categories:

Fig. 2.2 Summary of the most relevant factors

- Value stocks.
- Growth stocks.

Value stocks are the equity shares of companies operating in traditional sectors with a reduced level of expected growth but good balance sheet fundamentals. The goal of a manager inspired by value philosophy is to look for undervalued companies with low price/earnings (P/E) and/or low price/book value (P/BV), with the aim of paying them as little as possible. Fundamental to the success of this strategy is the ability to distinguish between stocks with low P/E and low P/BV that are undervalued from those that are worse off in the sector.

Growth stocks are the equity shares of companies operating in innovative sectors with strong growth potential. These securities usually have high volatility, as high as their respective P/E and P/BV. Growth managers focus on stocks that have experienced significant price growth and are expected to grow further. Therefore, shares are paid more (in terms of P/E and P/BV) than those purchased, on average, by a value manager. Fundamental to the success of this strategy is the ability to distinguish between overvalued stocks with high P/E and high P/BV from those stocks that are better than the sector average.

The inability to allocate a stock to the value category or, alternatively, to the growth category can lead to the identification of a third intermediate category, blend (or core) stocks, which include securities with mixed characteristics that cannot be allocated with certainty to the former or latter category.

Regarding the size factor, the investment strategy focuses on stocks belonging to a specific segment:

- Stocks of companies with large capitalization (large cap).
- Stocks of companies with medium capitalization (medium cap or mid-cap).
- Stocks of companies with small capitalization (small cap).

Value	Blend	Growth	
1	2	3	Large cap
4	5	6	Medium cap
7	8	9	Small cap

Fig. 2.3 Equity-style box

In the first case, the manager focuses on the stocks of companies with a higher market capitalisation, the so-called "blue chips". In the second case, the manager focuses on mid-cap stocks. In the third case, the manager focuses on smaller companies' stocks.

Nine possible equity investment strategies are derived from the intersection of value, blend, and growth strategies, on the one hand, and large, mid, and small caps, on the other; these are characterised by different risk-return combinations and representable through a matrix called the "equity-style box". The interpretative logic of the nine strategies shown in Fig. 2.3 is the following (with some contradictions related to the global financial crisis that began in the summer of 2007 and exploded in 2008 and the consequent sharp fall in the price of bank stocks, until then considered value stocks). Value/large cap equity investment strategies and similar (quadrants 1, 2, and 4) are characterised by lower volatility (and lower returns). Growth/small caps and similar equity investment strategies (quadrants 9, 6, and 8) are characterised by the greatest volatility (and the highest returns). The medium cap/blend and similar strategies (quadrants 5, 3, and 7) are intermediate in terms of the risk-return combination.

2.3.5.2 Indexed Equity Strategies

For equity portfolios, an indexed asset manager can replicate the benchmark used as a reference in three ways:

- Full physical replication.
- Stratified sampling.
- Synthetic replication.

Full physical replication involves inserting in the portfolio the same stocks into the benchmark in the same proportions. Therefore, this strategy is based on the physical replication of the reference benchmark. The physical replication strategy is applied effectively when the index to be replicated contains relatively few highly liquid securities, when it is relatively stable in composition, and when the size of the replicating portfolio is sufficiently large. If the index is made up of many securities, is not highly liquid, and has frequent turnover, full physical replication results in higher transaction costs than the two other technical methods described below.

The stratified sampling technique assumes the identification, based on alternative criteria, of a sample of securities of the reference benchmark that can minimise the tracking error and ensure considerable savings in transaction costs. Stratified sampling requires that the composition of the portfolio be sufficiently similar to that of the benchmark to be replicated in terms of the risk-return combination. The most widespread techniques use multifactor models, which make it possible to identify the weights of the explanatory factors of portfolio performance. For these models to be effective, the relationship between the performance of the index and the single explanatory factors must be sufficiently solid and stable over time to avoid the so-called "basis risk", that is, the risk that the replicated portfolio and replicant may have a divergent trend over time.

Finally, the synthetic replication technique involves replicating the performance of the index through equity swap contracts under which parties exchange the performance measured by a given underlying index applied to predefined notional capital. The synthetic replication strategy has the advantage of minimising transaction costs relative to physical replication but adds a counterparty risk to the market risk associated with the benchmark return.

For the investor who buys an indexed product, the full physical replication technique is undoubtedly the safest method, as it is exempt from the additional risks or counterparty risks of the other two techniques. However, this increased safety could result in higher management fees because of the asset manager's higher cost for the index replication service.

2.3.5.3 Smart Beta Strategies

Smart beta strategies, sometimes also referred to as alternative or structured beta strategies, comprise a new, diversified category of approaches to equity investing that combine active and indexed management.

Unlike active management, investment selection logic is pre-established, and the manager's margin of discretion is practically nil. Unlike indexed strategies, smart beta strategies do not replicate a traditional value-weighted index, weighted on market capitalisation, or the monetary value of companies' free float, but identify objective rules for investing in market segments characterised by the presence of homogeneous factors. The methodologies behind smart beta portfolios are often designed to select an investment universe of stocks with specific characteristics that are believed to offer the opportunity for better returns and/or lower risk than a traditional value-weighted index.

The objective of smart beta strategies is to achieve better risk-return combinations than indexed strategies but at lower costs than active management strategies. Smart beta investment strategies aim to offer investors elements of both active and indexed management: the possibility to outperform a traditional market index, on the one hand, and the low cost and transparency of an investment approach based on predefined rules that do not generate analysis, research, or transaction costs due to repeated portfolio turnover, on the other.

Smart beta equity investing strategies fall into one of these categories:

- Return-oriented strategies aim to outperform the benchmark by selecting a group of stocks using a range of alternative factors, considered individually or jointly. The most common selection rule is to passively focus on *ad hoc* indices, including:
  - Stocks with a dividend yield above a certain threshold.
  - Stocks of exclusively value companies.
  - Stocks of exclusively growth companies.
  - Stocks of companies belonging to a single size range (small cap, mid-cap, or large cap).
  - Stocks with a favourable momentum.
  - Stocks of companies selected based on earnings stability and overcoming certain financial indicators (quality).
  - Stocks with certain liquidity characteristics.
  - Stocks selected based on a mix of the previous approaches.
- Risk-oriented strategies, which aim to achieve a lower risk than the traditional benchmark index, selecting a group of stocks using a range of alternative factors, such as:
  - The creation of minimum volatility portfolios, characterised by the lowest possible level of total risk, or other risk-based portfolios, characterised by precise rules defined ex-ante in terms of risk.
  - The creation of equity portfolios with betas falling within certain pre-defined limits.
- Equal weighting strategies attempt to dominate the risk-return combination of the traditional benchmark by replicating unweighted indices. In the case of equally weighted indices, the weight of each stock within the index is identical, and the smart beta investment strategy aims to replicate this index rather than the traditional value-weighted index.

2.3.6 Restrictions on the Investment Policy

Investments in the fund are determined by the management company in compliance with the provisions of applicable laws, regulations, and fund rules.

In the EU, specific restrictions on investment fund management policy have been adopted to create a single market for collective management schemes within the Union. When a fund matches the requirements established by EU regulations, it has a European-wide passport, which allows asset managers, once authorised, to trade and run funds throughout the Union regardless of their country of domiciliation, without the need for additional authorisation.

This paragraph presents the restrictions to investment policy on UCITS funds. The rules for alternative investment fund managers will be analysed in Chap. 11, with a focus on single categories of alternative funds (hedge funds, real estate funds, commodities funds) in the following chapters.

In the case of UCITS funds, there are specific limits for risk spreading and diversification of investments, the most important of which are:

- Constraints on the overall composition of the portfolio.
- General prohibitions.
- Limits on risk concentration.
- Limits on investments in other open-end funds.
- Limits on derivatives.
- Limits on borrowing.
- Limits to the holding of voting rights.

With regard to restrictions on the overall composition of the portfolio, a UCITS fund invests the capital raised in assets that are consistent with its investment policy, sufficiently liquid so as to ensure the ability to redeem units, where the risks are adequately controlled through internal risk management and with a maximum potential loss limited to the amount paid, or, in the case of derivatives, with an exposure not exceeding the fund's net asset value.

In terms of general prohibitions, UCITS funds may not:

- Grant loans in forms other than those provided in respect of forward transactions in financial instruments.
- Carry out uncovered short sales.
- Buy precious metals or certificates representing them.

The limits to the concentration of the risks of a harmonised fund configure a set of rules to be observed with continuity, the most important of which are the following.

- Limits on investments in securities of the same issuer which, in the case of transferable securities (shares, bonds and similar securities) and money market instruments, are fixed at 5% of the total assets of the fund.[34] Member states of the European Union may raise the 5% limit to a maximum of 10%. If they do so, however, the total value of the fund's investments of more than 5% with a single issuer may not exceed more than 40% of the whole portfolio. This limit is known in the jargon as the "10/40 rule".[35]

- Limits for investments in deposits pursuant to which a UCITS fund cannot allocate more than 20% of total assets in deposits with a single bank, a limit reduced to 10% in the case of deposits with its own depositary institution.
- Limits on over-the-counter derivatives transaction, which provide a cap to the risk exposure of a harmonised fund based on 10% of total assets, if the counterparty is a credit institution, and on a 5% of total assets in all other cases.

The limits to investments in other open-end funds mean that a harmonised fund cannot invest more than 10% of its total assets in another UCITS fund or open-end CIV, a limit that can be raised by individual member countries to 20%.[36] It is prohibited to invest in investment funds that invest in another CIV more than 10% of their assets.

In the case of derivatives, these can be used by UCITS funds not only for hedging purposes but also for more efficient portfolio management, aiming to benefit from cost savings or increased liquidity, and also for investment purposes. The total exposure in derivatives may not exceed the total net value of the fund. This implies a maximum leverage potentially equal to 2.[37] The fund also has the option to use derivatives for taking net short positions.[38]

As regards investment management activity, it is up to each member country of the European Union to authorise a harmonised fund to borrow money aimed at facing temporary mismatches in treasury management for a period not exceeding 6 months, within the limit of 10% of the total net assets of the fund. In relation to voting rights on securities representing investments in equity capital, a management company may not hold, for all the funds and sub-funds managed, the voting rights in

[34] This limit is raised to 35% when securities are issued or guaranteed by a European Union country, by its local authorities, by OECD countries or by supranational organisations in which at least one EU country participates. The limit is set to 5% for instruments other than transferable securities or money market instruments. In the case of index funds aimed at replicating the composition of a given financial index, the concentration limit in any single issuer is raised to 20% of the fund's total assets.

[35] There are also indices in the version 10/40 in order to let the asset manager to replicate the index or, in the case of active management, to make a meaningful benchmark available to the manager for returns comparison.

[36] Investments made in CIVs other than UCITS shall not exceed, in aggregate, 30% of the assets of the UCITS. Investments in closed-end CIVs are not allowed.

[37] In the fund's prospectus, the management company must state the purposes of the use of derivatives and the maximum level of leverage to be adopted in implementing the investment policy.

[38] This implies that the financial leverage of a UCITS fund could potentially range from −2 to +2.

the same company for an amount such as to exert a dominant influence on the same.[39]

As we said before, UCITS regulation is now regarded internationally as an efficient regulation scheme, able to ensure adequate protection to investors. As a result, the UCITS status is recognised beyond the EU and UCITS products are accepted for sale in many countries outside the EU.

2.3.7 *Mandatory Public Documents*

According to the regulations, investment funds are characterised in each country by extensive documentation which accompanies their institution and distribution. Within the EU, there have been a number of innovations in recent decades regarding the required documentation and its contents, subsequent to the appearance of the EU directives on UCITS and the inherent possibility of free movement in the various member countries.

The main documents required are:

- The prospectus.
- Fund rules.
- Key information document.
- Periodical reports.

The prospectus contains a clear and easily understandable explanation of the fund's risk profile, by presenting to investors all the information necessary to make an informed judgement of the investment proposed to them, and, in particular, of the related risks.

The fund rules govern the relationship of participation in the investment fund, defining the characteristics of the fund, regulating the operations, presenting the fund's essential features, and providing subscribers with the principal information about the fund, its management company, and the depositary institution. Fund rules also detail the specific elements of the fund with particular regard to investment policy, the procedures for participation, the terms and procedures for the issue and redemption of units, the subjects responsible for investment choices, the type of financial instruments and other assets in which the capital of the fund may be invested, the criteria for determination of income, the expenses borne by the fund and those borne by the management company, the criteria for determining the commissions due to the management company, and the costs charged to investors.

Fund rules can be an integral part of the prospectus and, in this case, are attached to it. If not contained in the prospectus, fund rules can be sent to investors, upon request, or retrieved from the place indicated in the statement itself. A company that

[39] Discretion is left to each member state to define the thresholds of the voting rights acquired by a management or investment company.

manages a number of funds of the same type may establish single fund rules, which shall contain the same provisions only once for all funds and specific parts dedicated to the unique features of each fund.

Starting from January 2023 in the EU it is mandatory also for UCITS funds to prepare the so-called key information document (KID), previously required for all the other packaged retail investment and insurance-based investment products (PRIIPs), including AIFs. The KID provides the fundamental characteristics of the fund in summary form and includes the following mandatory information: purpose of the document; name and identification data of the product; the characteristics of the product; risk and return scenarios; costs; the recommended holding period; procedures for submitting any complaints; other relevant information.

The updated KID is delivered free of charge to the investor prior to subscription of the fund and its information must be consistent with the relevant parts of the prospectus and fund rules.[40]

Within a maximum of 4 months from the end of each financial year, or within the shortest period in which the fund proceeds to distribute income, the management company must prepare the annual report for each fund managed, as well as, within a maximum period of 2 months from the end of the semester, preparing an interim report (the semi-annual report) in which it illustrates the fund's mid-year financial position.

The investment fund's annual report includes:

- The statement of net assets as of end of the present financial year, illustrating the composition of the fund's portfolio in terms of total assets, total liabilities, and net asset value.
- The number of units in circulation and the consequent net asset value per unit, with a comparative table covering the last three financial years, showing total net asset value and net asset value per unit at the end of each financial year.
- The statement of operations and changes in net assets, which explain the factors that have led to an increase or decrease of net asset value per unit in the course of the financial year.

[40] The most important information of the KID is contained in the sections "Product characteristics", "Risks and return scenarios", and "Costs". The "Product Characteristics" section indicates: the type of product and the category it belongs to; the investment policy undertaken and the benchmark index adopted; the distribution or accumulation of income; the recommended holding period; the disclosure on sustainability; the depositary institution. The "Risks and Return Scenarios" section presents both the product's risk level, based on a scale ranging from 1 (low risk) to 7 (high risk), and the performance scenarios, in which the possible return on exit after 1 year and after the recommended holding period are illustrated under 4 different scenarios: stress; unfavourable; moderate; favourable. The "costs" section of the KID divides costs into three categories: one-off costs (entry or exit fees); recurring costs (charges directly borne by the fund, excluding any incentive fee); incidental costs due to specific conditions (any incentive fee). In the European Union the KID has replaced from January 2023 the previous KIID required for UCITS funds and introduced by European legislation UCITS IV.

- The portfolio of investments as of the end of the last financial year, which indicates the composition of the asset portfolio and analyses it in accordance with the most appropriate criteria in the light of the investment policy of the UCITS (e.g. in accordance with economic, geographical, or currency criteria).
- A summary of significant accounting policies and other explanatory notes.
- An audit report by an independent auditor.

It is also accompanied by the directors' report, which aims to provide information on both the investment policy undertaken and the prospects for market growth in the fund's areas of interest. The semi-annual report only presents the assets and liabilities resulting on the last day of the semester, represented by the same criteria used for the preparation of the annual report, subject to the right of the management company to broaden its information content.

2.3.8 Investment Fund Charges

The costs affecting the service of collective portfolio management fall into two categories:

- Costs borne directly by investors.
- Costs charged directly to the assets of the fund and, therefore, borne indirectly by investors.

Costs incurred directly by investors include the entry charge, the switch charge, the exit charge, and fixed fees. These burdens are aimed mostly at rewarding the activities of those who distribute such products, in the sense that these fees are usually paid entirely to the distributor. More specifically, the fees incurred directly by the investor can be broken down as follows:[41]

- The entry charge (or subscription fee) is paid at the time of purchase of fund units and therefore reduces the capital actually invested. It can be of a fixed amount for each subscription or, more frequently, it may vary. In the latter case, it is indicated as a percentage and its amount can vary according to a mechanism normally given in brackets according to which its incidence decreases as the amount invested increases. The entry charge varies depending on the type of fund and is higher for products with a higher risk or a greater specialisation.
- The switch charge is payable in the event of switching between sub-funds of the same umbrella fund, managed by the same company. It can be of a fixed amount or, more frequently, may be indicated as a percentage. It is generally configured so as to avoid arbitrage opportunities for investors and to provide for the recovery

[41] In the US fund market, the terminology used to describe fees is slightly different, and terms that are synonymous here have a different meaning.

of any entry charge established in the sub-fund of destination not paid in the sub-fund of origin.
- The exit charge (or redemption fee) is paid by the investor when he/she requests the redemption of units and therefore involves the reduction of the capital actually repaid. The redemption fee, also known as a deferred sales charge, may be incurred as an alternative to the subscription fee. Sometimes it only applies if the investors ask for redemption before a certain period, within which the commission usually decreases as the number of years spent in the fund passes. Some funds allow investors to choose between either an entry charge or an exit charge.
- The fixed fees are paid as the management company directly withdraws a fixed amount for each transaction performed. Additional sums are deducted by the management company to cover expenses if the subscriber requests, when possible, the delivery of physical certificates representing the units.[42]

The charges described above can exist, but there are investment funds—called no-load funds—totally free of charges directly borne by the investor, for which fees are charged solely to the fund's assets. Indeed, a fund characterised by charges borne directly by the investor can virtually turn into a no-load fund, due to possible concessions. It is common knowledge that certain categories of subscribers are recognised benefits in the form of reduced fees for entry, switch, or exit. In other cases, it is the distributors who enjoy a certain margin of discretion in granting these concessions, even to investors not covered by any particular category. This allows for more flexibility in negotiations and enables investors to obtain more favourable treatment than would normally be the case. Such benefits can even result in the complete elimination of entry, switch, or exit charges.

Costs charged directly on the funds and supported indirectly by the subscribers are taken from the fund's assets. This category includes: management fees, performance fees (or incentive fees), brokerage fees, and other costs incurred by the fund. As we shall see, these charges are not always connected only to the remuneration of the management service but can also be paid in part to the distributors. In detail, the costs borne by an investment fund may consist of the following:

- The management fee has the aim of remunerating the management company for the service of collective portfolio management. It is expressed as a percentage and is calculated and charged daily (or at least with the same frequency of calculation of the NAV) on net asset value of the fund by the imputation of accrued expenses, although the actual drawing takes place periodically.[43] Despite being taken by the

[42] Some interesting facts and statistics regarding the costs of investment funds are provided by the European Fund and Asset Management Association (EFAMA) in Europe, by the Investment Company Institute (ICI) in the United States, as well as by the various analyses of specialised companies.

[43] The fund's prospectus indicates the average size of the portion of the management fee that is paid to distributors, although in practice this percentage is differentiated according to the existing agreements with each distributor and is proportionate to the nature of all the pre- and post-sale services that each distributor has undertaken to provide to investors.

management company, it is also possible that the management fee is partly transferred to distributors. The management fee is generally higher as the level of risk or the specialisation of the fund increases.

- The cost of calculating the value of the fund unit includes the charges incurred by the management company to calculate the NAV. This activity can be conducted by both the management company and a third party (including the depositary) under an outsourcing regime.
- The performance fee (or incentive fee) is levied by the management company if and when the fund reaches, in a given period, a higher return than a predetermined parameter. Therefore, it allows the management company to increase its remuneration in the case of over-performance in relation to a preset parameter. Not every fund has a performance fee. If it has, in this case too, as for the management fee, the imputation is done daily (or with the same frequency of calculation of the NAV) on the net asset value of the fund by charging the accrued portion. For a detailed description of the method of calculation of performance fees, see Box 2.4.

Box 2.4 Performance Fees on Investment Funds

Performance fees have two functions: to link the asset manager's remuneration to the return achieved by the investor and to share the result of the delegated management between the asset manager and investors.

The payment of performance fees is subject to the achievement of a performance threshold, identified in accordance with predefined criteria. Therefore, incentive fees have the effect of making the total fees collected by the asset manager uncertain, depending on the results of the collective management activity.

During 2020, the ESMA issued a series of "Guidelines on performance fees for UCITS and certain types of AIFs", aimed at ensuring that: the performance fee model used is consistent with the objectives and policy fund investment; the interests of the manager and those of the investors are aligned on an on-going basis; the calculation of the performance fee is verifiable and not susceptible to manipulation; performance fees are proportionate to the actual investment performance of the fund.

The key elements relating to incentive fees are:

- Reference index for assessing extra returns.
- Symmetrical or asymmetrical structures.
- Time horizon for evaluating extra returns.
- Basis for the calculation of performance fee.

The reference index to which the return achieved by the investor is compared must be consistent with the objectives pursued through the introduction of an incentive fee to reward the management company's ability and align the

(continued)

Box 2.4 (continued)
interests of the manager and investors. Consequently, for market funds, this index should normally be represented by the benchmark adopted for regulatory purposes; for non-market funds, with no benchmark, the reference index should be an absolute high-watermark, that is, the highest NAV level recorded in a predefined past time frame, or a hurdle rate, that is, a predefined return, consistent with the fund's absolute return objective, if any.

Incentive fees may have symmetrical or asymmetrical structures. A symmetrical performance fee awards an additional fee to the management company in the event that the fund is able to beat the reference index, but leads to a reduction in management fees in the event that the fund is beaten by the reference index. Conversely, an asymmetrical performance fee rewards the management company in the event of a positive extra return, but simultaneously protects the management fees from reduction if the fund is beaten by the reference index. Both structures are allowed in the EU, unlike the USA, where only symmetrical structures have been allowed since 1970. Therefore, the structure generally widespread in the European market is the asymmetric one, albeit mitigated by the need to recover any previous underperformance compared to the reference index for at least five years before the performance fee becomes payable.

The time horizon considered for the payment of performance fees should be broad enough to enable investors to accurately assess a management company's ability. The payment of performance fees over a too short-time horizon would expose investors to the risk of paying incentive fees on non-existent extra returns if more appropriately measured over a sufficiently long-time horizon. To this end, European rules set a periodicity of at least one year, unless management companies adopt a calculation mechanism based on high-watermark or high-on-high systems. In the case of a high-watermark (HWM), the performance fee may be charged only if the fund's NAV reaches a new high point (the so-called high-watermark) during the performance reference period. In the case of high-on-high (HoH), the performance fee may only be charged if the NAV exceeds the level at which the performance fee has become payable last time. These incentive fee calculation systems, originally applied only to hedge funds, guarantee the collection of incentive fees for real and effective performance, making the period over which performance fees are calculated essentially irrelevant.

The basis for calculating the performance fee is the net asset value of the fund on the calculation date, adjusted for any income distribution since the last performance fee was paid, or, if stated in the fund rules, the average net asset value of the fund during the evaluation period adjusted for any income distribution in the period itself.

Sources: Elton et al. (2003), Scherer and Lee (2012), and ESMA (2020)

- Portfolio-trade related costs include expenses that the fund sustains for the purchase and the sale of securities in which to invest assets under management and are therefore related to portfolio turnover.[44]
- The depositary institution fee is paid for the activity carried out by the custodian institution.
- Other costs are a residual category which includes the costs of publishing the net asset value of the units, the periodic statements and public documents, expenses for notices to investors required by the law, costs for audit accounting and reporting by an independent auditor, the supervisory contribution, and so on.

Distinct classes of fund units, characterised by different commissions, income allocation or participants, may be offered to investors. In some cases, this gives the investor the opportunity to choose the solution that best suits his/her preferences. For example, a single fund may provide two different unit classes: class A with an entry charge and class B with a redemption charge and/or a management fee higher than class A. In this way, the subscriber has the opportunity to choose when and how to pay the fees, whether at purchase or at redemption, or whether to bear higher management fees in exchange for the absence of subscription fees. In other cases, the classes with lower fees are reserved for institutional investors or investment with higher minimum subscription thresholds. Unit classes can be differentiated according to currency of denomination or, as an alternative, according to currency risk management. Moreover, unit class can be distinguished depending on the allocation of the income earned, creating a class that accumulates income and a class that distributes it periodically.

Several indicators allow a synthetic evaluation of the costs of investment funds and homogeneous comparisons between different funds. The three most important indicators are as follows:

- Total expense ratio.
- On-going costs.
- Reduction in yield.

These synthetic cost indicators can be drawn from both the mandatory public documents accompanying the distribution of investment funds and reports and databases of specialised companies that collect these cost indicators to allow a time series evaluation for the same fund and a cross-sectional comparison for several funds belonging to the same peer group.

The total expense ratio (TER) is the ratio between the total costs borne directly by the investment fund in a given period and the average assets under management of the fund in the same period. The total costs used for its calculation include: the

[44] The portfolio turnover rate is the ratio between the total purchases and sales of securities that have occurred in a given period, net of subscriptions and redemptions, and the average net assets of the fund over the same period, although alternative methods may be used to calculate this indicator.

management fee, costs of calculating the value of the unit, any performance fee, transaction costs related to portfolio turnover, remuneration paid to the depositary institution, and other costs borne directly by the fund. Therefore, the calculation of the TER does not consider any costs charged directly to the investor at the time of subscription, switch, or redemption, or any tax charge levied on the fund.[45]

On-going costs (OGC) coincide with the TER net for any performance fee. Like the TER, the OGC indicator does not consider any cost directly charged to the investor or tax charges.[46]

Starting from January 2023, for all funds in the EU (including UCITS), it is mandatory to indicate in the new KID the synthetic cost indicator identified by the PRIIPs regulation and known as the "annual cost impact" or "reduction in yield" (RIY), which measures the impact of total costs in terms of reducing investment return. The total costs used to calculate the RIY consider the costs directly borne by the investor and those directly borne by the investment fund (and indirectly borne by the investor). In contrast to the OGC indicator, the RIY indicator also includes incentive fees and one-off costs paid by the investor, such as entry costs. To allow the inclusion of one-off costs directly borne by the investor, it is necessary to assume one or more holding periods to amortise these costs over the investment time horizon and determine their annual incidence. Current EU regulations require the assumption of two time horizons: the first refers to the hypothesis of exit one year after investment, and the second refers to the hypothesis of exit at the end of the recommended holding period for the fund. Therefore, two different RIY are calculated according to the assumed holding period.

The prospectus and KID of investment funds pay special attention to costs. In fact, a section of the prospectus is devoted to "charges and expenses" and contains all relevant information regarding costs, detailing any charge borne by the subscriber as well as charges borne by the fund and the related concessions. This section analytically illustrates the methods used to calculate and charge the various fees and expenses. This information is consistent with that reported in the KID, which separately shows the one-off charges (entry and exit charges borne by the investor),

[45] In comparing the TER of investment funds belonging to different countries, it is essential to verify which charges directly borne by the fund are included or excluded. In the EU, for example, the UCITS III directive provided for the exclusion of the costs of buying and selling assets in the portfolio, now included instead. Once the inclusion or exclusion from the TER of the transaction related to portfolio turnover has been assessed, it should in any case be considered that homogeneous comparisons between funds is objectively impossible due to the presence of both explicit trading costs, which have an autonomous highlighting and increase the TER, and of trading costs implicit in the bid-ask spreads, which cannot be highlighted independently and thus do not affect the TER. This aspect is dealt with in the EU via a request to indicate, in the offer documentation, for which part of the portfolio-trade related costs it was not possible to determine explicitly the amount.

[46] In this case as well, it is essential to verify in international comparisons which charges are included or excluded among the on-going costs: while it is a worldwide practice to exclude performance fees, transaction costs related to the portfolio turnover can be included or excluded, depending on the country's regulation. In the case of the EU, the portfolio-trade related costs are included in the KID adopted from January 2023 also for UCITS funds.

on-going charges recorded in a year for the fund, charges taken from the fund under specific conditions (any incentive fee), and the RIY indicator.

2.4 Unit Trust Schemes

Collective portfolio management can be done by using alternative legal structures. The structure most used in some European countries is the common fund set-up in the form of a contract between the management company and the investors. In this case, the common fund is not a separate legal entity in its own right, but the legal entity is the management company setting up the fund. Another legal structure for organising a CIV is the trust form, in which the CIV is usually called "unit trust". The unit trust form of CIV is found in jurisdictions where the English common law system prevails, excluding the USA.[47]

A unit trust pools the money of many individual investors to create a portfolio of financial securities with a specific investment objective. The pool of investments is divided into equal portions called units. Like the common fund, the unit trust can also be organised as an umbrella unit trust scheme, comprising separate sub-funds with segregated assets and liabilities. The unit trust is a CIV structure constituted by a deed of trust between a trustee and a management company (manager). The deed of trust—which is the equivalent to the fund rules of common funds—is the primary legal document constituting the trust and setting out the various rights and obligations of the trustee, the management company, and the unit-holders who are beneficiaries of the trust.[48] The unit trust, like the common fund, is not a separate legal entity and therefore the trustee acts as legal owner of the fund's assets on behalf of the investors.

There are three parties to a unit trust: the trustee, the manager, and the unit-holders.

The trustee is usually a financial institution, which must be unconnected to the manager. The trustee is the legal owner of the trust property and its duties are:

- To protect the interest of unit-holders by ensuring that the fund is run in accordance with the law, regulations, and the trust deed.
- To act as custodian for the trust's assets, or to delegate this function to a trusted third party. The trust's assets are registered in the name of the trustees, who also hold the trust income.
- To create and maintain the register of unit-holders or to delegate this function to the managers.

---

[47] The legal structure of unit trust should not be confused with that of the unit investment trust, used mainly in the USA, which is a closed-end structure. Loader (2007).

[48] St Giles et al. (2003).

- To replace the manager if he/she is deemed not to be acting in investors' interest, goes insolvent or if the majority of investors vote to remove him/her.
- To report any significant irregularity to the supervisory authority.

A separate management company is always required and managerial responsibility rests with the board of directors of the management company. This separate management company can also manage other CIVs' assets. The manager's main business is the provision of fund management and administration services to the unit trust. Under the trust deed, the manager is responsible for:

- Managing the investment of each fund or sub-fund with the aim of achieving the investment objectives and policies of such funds or sub-funds and of carrying out the duties of a manager of a unit trust in accordance with the law and the regulations set by the supervisory authority.
- Carrying on the general administration of the fund.

The manager can delegate the execution of the investment management functions to the investment manager and the administrative functions to the administrator.

The investors in the unit trust are beneficially entitled to an undivided share of the investments subject to the trust and are referred to as unit-holders.

The price of units is determined by the manager of the trust at the current market value of the investments held in each fund or sub-fund. Units are created for new investors and cancelled for redeeming investors in the unit trust. Unit trusts have historically been dual priced with a difference between creation price and cancellation price, thus being the sole CIV with a bid-ask spread and not with a single price. The trust deed gives the manager the right to vary the bid-offer spread to reflect market conditions, with the purpose of allowing the manager to control liquidity. It is also possible now to zero the bid-ask spread.

The unit trust form was the traditional form of collective portfolio management in the United Kingdom and Ireland but both countries have been increasingly integrated in the European CIV market and the corporate structure is gaining in significance at the expense of the unit trust structure. The trust system is also found in Australia, Canada, China, Hong Kong, Malaysia, New Zealand, and Singapore. However variations among trust systems are significant in the various countries.[49]

2.5 Open-End Investment Companies

Another alternative legal structure for CIVs which is the most commonly used worldwide is the open-end investment company (OEIC), where the OEIC is a separate legal entity and the investors are shareholders. In many countries, this

[49] OECD (2001).

legal entity is known by the French expression "société d'investissement à capital variable" or by its English translation "investment company with variable capital".[50]

The distinguishing characteristic of OEICs as compared to common contractual funds and unit trusts is the fact that the investor takes the role of shareholder and not of simple unit-holder. In theory, therefore, because the investor is eligible to vote, he/she can influence the investment strategies of the OEIC, although in practice, active investor participation in the shareholders' meetings is fairly uncommon.[51]

In legal terms, an OEIC is a company, incorporated under special regulations, with a board of directors and shareholders. These special regulations provide that the OEIC must:

- Have an instrument of incorporation (roughly equivalent to the fund rules of a common fund and to the trust deed of a unit trust) which governs the OEIC's activities and contains details of the share classes which the OIEC can issue.
- Have a prospectus providing details of its objectives and operations, such as its investment policy and arrangements for valuing the investment property.
- Prepare annual and half-yearly reports, which must include the OEIC's audited accounts, and which have to be approved by shareholders at the annual general meeting.

Despite their corporate structure, OEICs are also subject to special rules. Unlike what happens in a normal corporation, the capital of an OIEC continues to fluctuate as a result of the issue and/or redemption of shares in connection with new subscriptions and redemptions.

The two bodies responsible for the operation of an OEIC are the authorised corporate director and the depository, with roles very similar to those of the manager and the trustee of a unit trust.

The OEIC's instrument of incorporation normally provides for the existence of several sub-funds, each one characterised by a different investment strategy and a different risk-return combination. In an umbrella OEIC, each sub-fund is a separate compartment, segregated from every other sub-fund. A special class of shares is issued for each of these, replicating the scheme and the benefits of umbrella common funds.

Within the EU, OEICs may delegate the management of their assets exclusively to management companies. Unlike the management companies of common funds and unit trusts, an OEIC may therefore only be a promoter, or promoter and manager, at the same time, but cannot be manager of the assets of other investment funds.

---

[50] This legal form is also known with the two acronyms SICAV or ICVC. In the United States of America the official name of this form is "mutual fund". This legal entity is predominant worldwide, mainly because of the size of the US market, and also in some European countries, like Luxembourg, it is the most widespread structure.

[51] St Giles et al. (2003), Turner (2004), and Haslem (2010).

To sum up, despite their different legal structure, in economic terms, OEICs are equivalent to open-end common funds. In fact, if they wish to benefit from UCITS status, OEICs are subject to the same rules and restrictions on their investment policy as harmonised common funds. The custody of the securities and cash must be entrusted to a depositary institution, which carries out the same functions described in the case of common funds. The public documentation required is the same as common funds, with the OEIC's instrument of incorporation replacing the fund rules of the common fund. Moreover, the costs of OEICs have the same structure as those of common funds.

2.6 Other Asset Management Products

The following paragraphs describe the characteristics of certain products which, from the investor's point of view, have a high rate of replaceability with CIVs and must be seen as their natural competitors. Strictly speaking, in some cases, they are not even real asset management products, as in the case of capital redemption policies, index-linked policies, and structured products, but the competition and, in some cases, the crowding-out effect that these products may have on CIVs make it necessary to take a close look at them. In the following order, we will be examining:

- Individual portfolio management, in accordance with mandates given by clients on a discretionary and single basis.
- Capital redemption policies.
- Linked policies.
- Structured products.

2.6.1 Individual Portfolio Management

In the investment service of individual portfolio management, the portfolio of the investor is managed on a discretionary basis by allocating it in one or more financial instruments, in accordance with a mandate given by the client on a discretionary and individual basis.[52]

The discretionary basis of the individual portfolio management gives the asset manager, in the execution of the investment service, full autonomy in every decision, in accordance with the rules of conduct laid down by regulations and limits established in the management contract.

[52] Within the European Union, the regulation of individual portfolio management has been developed separately from that of CIVs as an offshoot of the wider harmonised investment services regime in 1993 and is now included in the Markets in financial instruments directive II (MiFID II)/ Markets in financial instruments regulation (MiFIR) regime.

Individual portfolio management also needs to be carried out on a client-by-client basis. Unlike collective portfolio management, in which the money raised from participants is invested in a group of assets and the investment management service is identical for each investor in the same fund, the investment service of individual portfolio management is characterised by a closer and more personalised relationship between the asset manager and investors, which translates (or, at least, should translate) into a greater personalisation of the service on the basis of each investor's individual characteristics and needs.[53]

For this reason, individual portfolio management based on a discretionary mandate is an investment service with a very high threshold and reserved for high net worth investors. In fact, the construction of a truly individual portfolio for each investor is possible only above a certain minimum threshold, both for objective and for business reasons. The objective reasons are connected to the limit below which portfolio diversification is inefficient, while the business reasons refer to the marginal cost of production of personalised service to the individual investor.

In some cases, this has led those asset managers who offer the service of individual portfolio management to standardise the service, offering investors more management lines differentiated by type of investment and level of risk but with standardised features within each line. In this way, the asset managers can offer the "individual" service for relatively lower costs, by applying the same investment strategy for all those who subscribed the same line, with obvious cost savings and benefits in terms of diversification, although at the expense of actual customisation of the service.

In addition to this, the search for a further reduction of the threshold required for the service of individual portfolio management has led asset managers to use CIVs, instead of shares, bonds, or money market instruments. It is therefore possible to construct many combinations of portfolios adaptable to the specific needs of investors. Although, in theory, the possible combinations are endless, in practice, in this case too, each asset manager reduces the actual combinations to a fairly limited number of investment profiles, and individual portfolio management ends up meaning that each investor is assigned to the profile that best matches his/her characteristics, with the advantage of having many potential investors in each profile, due to the much lower threshold.

In the current offer of asset managers, therefore, there are two types of individual portfolio management, with the same discipline:

- Individual portfolio management allocated mainly in transferable securities and money market instruments, with higher entry thresholds;
- Individual portfolio management allocated mainly in CIVs, characterised by lower entry thresholds.

For both types of individual portfolio management, the management contract:

[53] Moloney (2014).

- Must be made in writing.
- May give the customer the power to issue binding instructions.
- Must give the customer the option to terminate the contract at any time, without being charged any penalty.
- Must give the customer the right to vote on the financial instruments representing investments in equity capital, while being allowed the opportunity to give the intermediary the power to exercise voting rights.

The agreement also gives the asset manager the possibility:

- To invest in unlisted securities, derivatives, illiquid or highly volatile securities.
- To carry out short sales, to use financial leverage, to lend securities and do any transaction involving the payment of margins, deposit of collateral or foreign exchange risk.

The investor may terminate the discretionary mandate at any time and consequently request repayment of the capital assigned for individual management or, alternatively, the transfer of financial instruments. The complexity of this service of individual portfolio management requires the asset manager to send periodic full statements, which enable the customer to be fully informed about the transactions carried out at the discretion of the manager and the composition of the managed portfolio, also in order to evaluate investment performance.[54]

The transferable securities, money market instruments, and cash held by the manager of individual clients constitute separate assets from that of the manager. The separation of the assets of each customer from that of the asset manager is particularly important in the event of insolvency of the same asset manager as it makes possible the restitution of the securities to individual customers.

In relation to costs, individual portfolio management includes:

- A possible initial charge.
- A management fee.
- Often a performance fee.

2.6.2 *Investment-Oriented Insurance Policies*

In the last few decades, investment-oriented insurance products have become more and more important as a tool for investment of private client resources in the financial systems of EU countries.

Investment-oriented policies are insurance contracts in which the financial dimension of the contract (the financial return obtained from the investment of premiums) is predominant or even exclusive, when compared to the size of the actuarial component (coverage of a pure risk).

[54] Maude (2006).

In the legal systems of many countries, such products often benefit from a range of regulatory and tax advantages that facilitate their distribution. The widespread use of such products in the financial systems of different countries was also due to the wide gap between insurance products and investment funds in terms of transparency of information, mandatory documentation and rules, a gap which led financial intermediaries to make greater use of insurance products so as to take advantage of simpler procedures, more efficient operating practises and, in some cases, even greater opacity of the contractual conditions.

2.6.2.1 Capital Redemption Policies

Capital redemption policies are contracts whereby the investor-contracting party pays a single premium or, more rarely, a recurring premium to an insurance company that agrees to repay, to a pre-established date, the amounts received increased by the interest accrued during the contract term and provide a minimum guaranteed return, without placing any restriction or reference to the life of the insured person. This particular insurance contract has names and technical characteristics that vary according to the rules of the different countries. The typical feature of capital redemption policies is that their content is exclusively financial, owing to the fact that the commitment of the insurance company does not depend on the occurrence of events relating to human life.

To face the commitments made by the capital redemption contracts, the insurance company establishes an internal management of investments, separate from other assets of the insurance company and governed by specific rules. The premiums paid, net of fees, are invested in this segregated pool, usually allocated predominantly in bonds and other fixed-income securities.

The capital so invested is revalued annually based on the financial performance achieved on the separate management, with a minimum guaranteed return. If the annual return of the internal separate management exceeds the minimum guaranteed return, the revaluation recognised annually is added to the returns already accrued and shall be forfeited definitively by the policyholders, on the basis of a principle known as consolidation of the results.[55]

In addition to the initial charge, the costs for the investor are represented by annual fees for the internal separate management,[56] the costs of security transactions in the separate management, and charges in the event of surrender during the contract period.[57] At the end of the contract, as an alternative to a revalued capital,

[55] As an alternative, it is possible for the accumulated income to be distributed periodically.

[56] In the case of annual fees, the contract often identifies a pre-established retrocession rate, which recognises the investor a percentage of the return on the internal separate management (e.g. 98.5%), while the remainder is held by the insurance company.

[57] As a rule, early surrender is possible one year after the contract is signed.

it is also possible in some cases to convert the accumulated capital into a periodical or life annuity contract.[58]

2.6.2.2 Linked Insurance Policies

The linked policies are forms of insurance in which the commitment of the insurance company is tied to the value of a reference entity. Theoretically, any amount can be taken as a reference. In fact, the identification of the reference base is constrained by regulation and by the market possibilities to make investments able to cover the corresponding technical provisions.[59]

Linked policies can be divided into:

- Unit-linked insurance policies, with performance tied to the value of one or more investment funds, internal or external to the insurance company.
- Index-linked insurance policies, characterised by performance whose size is calculated according to the value of a reference index, also composite.

Both in unit-linked policies and in index-linked policies, the financial component of the contract prevails over the actuarial component, which in both contracts takes the form of the premature death of the insured person when the contract is still valid.

Unit-linked insurance policies are contracts whereby the investor-policyholder pays a single premium or, more rarely, a recurring premium to an insurance company, which invests the capital raised in internal or external investment funds. At the end of the contract, if provided, or upon surrender, the principal repaid depends on the value of the internal or external investment funds in which the insurance company invested the premiums.

In addition to that, if the insured person should die when the contract is still in existence, the beneficiaries of the policy are entitled to the reimbursement of the principal accrued to that date, plus a percentage that will vary depending on the age of the insured person at time of death. In any case, the benefit in case of death is very limited, given the absolute prevalence in such insurance contracts of the financial component compared to the actuarial one.

In practice, in such contracts, the premium (net of fees and of the insurance component in case of death) is invested by the insurance company in internal investment funds (proprietary funds) or in external investment funds (external CIVs). The performance of the unit-linked policy at maturity depends on the performance of these investment funds. It follows that the commitments of the

---

[58] The agreement outlined here can be created with different variants. The most common is to put a very limited coverage in case of death. In this variation, the insurance contract, despite having a content almost exclusively financial, is no longer a capital redemption contract, but it becomes a life insurance contract. The limited insurance component makes it possible not to assign a deadline to the contract, which lasts as long as the insured person remains alive, and especially to benefit from the favorable legislation usually reserved to life insurance policies in the various jurisdictions.

[59] Jones (2002).

insurance company in unit-linked products are in fact related to the corresponding portion of the assets that constitute a separate pool, resulting in a complete transfer of investment risk from the insurance company to the insured.[60]

At the time of subscription, the investor is often given the opportunity to choose among alternative risk profiles, provided that, at a later date, switches are allowed from one profile to another. In all cases, at the time of subscription, each investor is recognised a number of units of internal or external investment funds underlying the unit-linked policy, depending on the premium paid and the initial commissions agreed. Because during the contract period the insurance company must update and disseminate the value of this unit (based on the same logic as NAV calculation for CIVs), every investor can evaluate his/her own position and assess trends over time.

The contract may have a specified maturity or may last the entire life of the insured person. In both cases, the right exists to surrender the contract and, in this case, the surrender value is linked to the value of the internal or external funds to which the unit-linked policy is tied. As an alternative to liquidating the entire capital accrued at maturity, it is possible to convert the accumulated capital into a periodical or life annuity. In the event of the death of the insured person, the beneficiaries collect the accumulated capital at the date of settlement, plus a percentage linked to the age of the insured person at the time of death.

As well as the possible initial charge, costs for the investor are represented by the annual management fee and all additional charges on internal or external investment funds,[61] charges for surrender during the contract period and the cost of the insurance component in case of death, but this component represents a marginal portion of the premium paid by the investor.[62]

The technical characteristics of unit-linked policies can vary from country to country depending on the different laws and regulations applied.

Index-linked insurance policies are contracts in which the investor-policyholder pays a single premium to an insurance company in order to receive a return linked to the development of an underlying index in the form of a participation, a recurring payment or a one-off payment at the expiry, with a guarantee of a minimum return (at least equal to zero), regardless of the index trend.[63]

In addition to that, if the insured person should die during the contract, beneficiaries of the policy are entitled to reimbursement of the principal accrued to that

---

[60] The most common version of unit-linked insurance policy does not provide any minimum guaranteed return, although this does not exclude the possibility that in some contracts such a guarantee may be provided.

[61] In the case of annual fees, the contract often identifies a retrocession rate, which recognises the investor a percentage of the return on the internal separate management (for example, 98%), while the remainder is held by the insurance company.

[62] As a rule, early surrender is possible one year after the contract is signed.

[63] The most common version of index-linked policy provides a minimum guaranteed return from the insurance company, but this does not mean that in some contracts this guarantee may not be provided.

date, plus a percentage that varies according to the age of the insured person at the time of death. As in the case of unit-linked policies, the benefit in case of death in index-linked policies is rather small, given the absolute prevalence in such insurance contracts of the financial component over the actuarial one.

In practice, in such contracts the premium received, net of fees and of the insurance component in case of death, is invested by the insurance company into structured bonds (or into an equivalent position in bonds and derivatives) on which the value of the index-linked policy depends once it matures. It follows that the insurance company's commitments in the index-linked products are in fact related to the corresponding portion of the assets that constitute a separate pool.[64]

At the time of subscription, each investor is recognised a number of units of the structured bond underlying the index-linked policy, depending on the premium paid and the initial commissions agreed. Because during the contract period the insurer must update and spread the value of the structured security, in index-linked policies too, every investor is able to evaluate his/her position and to assess trends over time.

By necessity, the maturity of the contract is specified, although the surrender option is allowed, with the insured person in this case receiving a sum tied to the current value of the structured bond to which the index-linked insurance policy is tied. Instead of liquidating the entire capital accrued at maturity, it is possible to convert the accumulated capital into a periodical or life annuity. In the event of the death of the insured person, beneficiaries collect the accumulated capital on that date, plus a percentage linked to the age of the insured person at the time of death.

As well as the possible initial charge, the costs for the investor are represented by the annual charges, the implicit cost emerging from the possible misalignment between the fair value of the derivative component included in the structured bond and the effective price paid by the policyholder, charges for surrender during the contract period and the cost of the insurance component in case of death, although this component represents a marginal portion of the premium paid by the policyholder. The technical features of index-linked policies may vary according to the specific rules in force in the different countries.

2.6.3 Structured Products

Structured products are financial instruments issued by banks or other financial institutions with varying terms, payouts, and risk profiles, providing a return linked to the performance of one or more reference entities. There are no standardised structured products: the terms, payouts, and risk profiles of each instrument are bespoke and determined at the time of issue. These products can be bought "off the

[64] The risk of default of the issuer of the structured security related to the index-linked policy can be taken or not by the insurance, depending on the country concerned and the applicable rules.

shelf' or individually tailored to the specific market needs or views of the investor, in order to match his/her desired risk-return combination and expectations.

One of the key characteristics of structured products is that the return is not determined by active investment but by a pre-specified formula that sets out how the product will perform in any possible future scenario. In any case, structured products are a combination of a traditional investment and one or more derivatives, which are bundled into a securitised instrument. Derivatives embedded in structured products may be plain vanilla or exotic and are linked to a variety of asset classes, mainly equity but also interest rates, currency rates, commodities, and other types of assets.

The main strengths inherent in investing in structured products are the following:

- Structured products are instruments that can be used as an alternative to traditional investment categories while providing additional attractive features, such as capital protection, yield enhancement, participation, leverage, or a combination of these.
- They offer a simple and cost-effective access to a wide range of markets.
- When listed, they can be traded in the same way as shares.
- They can offer geared or ungeared performance.
- Depending on their specific features, they can generate a positive return whether the performance of the underlying asset is static, bullish, or bearish.
- They can have built-in currency risk-management features.[65]
- The maximum potential loss on structured products is known.
- They do not require margin calls or dedicated accounts.
- They can be originated in order to be tax efficient, respecting the fiscal rules of each country.[66]

On the other hand, the main weaknesses and drawbacks of investing in structured products are:

- Unlike investment funds, structured products bear an issuer risk. In the event of the issuer's insolvency, repayment may not be made at maturity, which would mean the total loss of the capital invested. If the issuer's financial standing deteriorates during the life of the instrument, the price of the product on the secondary market may fall and a sale before maturity could lead to partial or even total loss of the capital invested. Even products with capital protection are exposed to issuer risk. The issuer's financial standing is thus extremely important, unlike in the case of investment funds.
- In most cases, the valuation and risk analysis of structured products is not straightforward and requires suitable pricing tools and financial knowledge of the different pricing models. As a consequence, these instruments can be affected,

[65] Where an underlying asset is denominated in a foreign currency, some structured products incorporate a "quanto" feature which ensures that the product is constantly fully hedged for currency risk. The cost of hedging varies and is built into the price of the instrument. Other structured products can be unhedged and are therefore exposed to currency risk.

[66] Tolle et al. (2008) and London Stock Exchange Group (2013).

at least in theory, by a potential mispricing between the fair value of the derivative embedded in the structured product and the price actually paid by the investor at the date of purchase.
- An investment in a structured product does not give the investor any of the rights associated with the ownership of the underlying, such as voting rights, subscription rights, dividends, or interest.
- When unlisted, they bear a strong liquidity risk, i.e. the possibility that the investor may not be able to dispose of a structured product at any given time or at a reasonable market price. The payout profile of a structured product defined in advance is always valid only at the end of the term. Before the end of the term, it may not be possible to sell the product at an acceptable price, for example because no binding prices are available for it.

Structured products can be split into different categories depending on their pay-off structure, as this has a direct impact on their potential risk and return. In particular, the European Structured Investment Products Association (EUSIPA) classifies structured products into four main categories:

- Capital-protection products.
- Yield enhancement products.
- Participation products.
- Leverage products.[67]

Capital-protection products offer partial or total protection of the par value (between 90% and 100%) at maturity, subject to credit risk of the issuer. In addition, these instruments offer a possible return linked to the development of an underlying, in the form of participation, a recurring payment, or a one-off payment at expiry. Products belonging to this category are suitable for investors with low risk appetite. Capital-protection products can be structured to perform positively in rising or falling markets and should be chosen in accordance with market expectations over the lifetime of the product.

Yield enhancement products offer a limited (capped) upside return, usually in the form of a fixed coupon or, as an alternative, in the form of a discount. Investors forgo an unlimited participation in favour of a recurring or one-off payment. Products belonging to this category may offer a conditional capital protection, in which case the protection is granted only if a predefined "condition" is met (i.e. a barrier has not been touched). Also these instruments are subject to the credit risk of the issuer. Yield enhancement products are suitable for investors with a moderate to increased appetite for risk and the expectation of markets moving sideways over the lifetime of the product.

Participation products usually offer unleveraged participation in the performance of one or multiple underlyings. These products may offer a conditional capital protection, in which case the protection is granted only if a predefined "condition"

[67] Swiss Bankers Association (2008) and EUSIPA (2019).

is met (i.e. a barrier has not been touched). Participation products are subject to credit risk. These products are suitable for investors with a moderate to increased appetite for risk. Market expectation should be a directional move (up or down) over the lifetime of the product.

Leverage products offer investors the opportunity to track an underlying instrument with a geared performance. For example, a three-to-one leverage tracker would increase in value at a rate that is three times the growth rate of the underlying asset. Gearing reflects how much the price of the structured product will change in reaction to a given change in the price or value of the underlying asset. Some instruments have distinct structures, resulting in an inverse relationship with the underlying asset, and are called inverse trackers. If the price of the underlying asset decreases, the value of the reverse tracker increases. These products may be suited to investors with a strong bearish view of the underlying asset. Leverage products, therefore, are only suitable for investors with a high risk appetite. Leverage products can also be used to hedge risks, in which case they reduce the risk of a portfolio. Market expectation should be a strong up- or downwards move, and products are best used as short-term speculative investments requiring daily supervision, or as a hedge.

It is generally considered that capital-protection products carry the least risk of the four categories, followed by yield-enhanced products, participation products, and finally leverage products, which carry the most risk. The reality is that any instrument, regardless of product class, can be structured with various features, and each instrument must be assessed according to its specific terms and not by general product grouping.

References

Abner D (2010) The ETF handbook. How to value and trade exchange-traded funds. Wiley, Hoboken, NJ

CESR (2010a) Guidelines on the methodology for the calculation of the synthetic risk and reward indicator in the key investor information document. CESR, Paris

CESR (2010b) Guidelines on a common definition of European money market funds. CESR, Paris

Christopherson JA, Carino DR, Ferson WE (2009) Portfolio performance measurement and benchmarking. McGraw Hill, New York

Cowles A (1933) Can stock market forecasters forecast? Econometrica 1(3):309–324

EFAMA (2008) The European fund classification. An EFAMA project to facilitate the comparison of investment funds: a pan-European approach. EFAMA, Bruxelles

EFAMA (2023) The European fund classification, 4th edn. EFC Categories, Bruxelles

Elton EJ, Gruber MJ, Das S, Hlavka M (1993) Efficiency with costly information. A reinterpretation of evidence from managed portfolios. Rev Financ Stud 6(1):1–22

Elton EJ, Gruber MJ, Blake CR (2003) Incentive fees and mutual funds. J Financ 58(2):779–804

ESMA (2020) Guidelines on performance fees for UCITS and certain types of AIFs| ESMA34-39-992

EUSIPA (2019) Structured products categorization map. EUSIPA, Bruxelles

Gastineau GL (2010) The exchange-traded funds manual, 2nd edn. Wiley, Hoboken, NJ

Gastineau GL, Olma AR, Zielinski RG (2007) Equity portfolio management. In: Maginn L, Tuttle DL, McLeavey DW, Pinto JE (eds) Managing investment portfolios. A dynamic process, 3rd edn. Wiley, Hoboken, NJ

Goetzmann WN, Ibbotson RG (1994) Do winners repeat? J Portf Manag 20(2):9–18

Grinold RC, Kahn RN (2000) Active portfolio management, 2nd edn. McGraw Hill, New York

Gruber MJ (1996) Another puzzle. The growth in actively managed mutual funds. J Financ 51(3): 783–810

Haslem JA (2009) Mutual fund arbitrage and transaction costs. J Invest 18(2):57–61

Haslem JA (2010) Mutual funds. Portfolio structures, analysis, management, and stewardship. Wiley, Hoboken, NJ

IOSCO (2004) Index funds and the use of indices by the asset management industry. Technical Committee of the International Organization of Securities Commissions. IOSCO, Madrid

Jones FJ (2002) Investment-oriented life insurance. In: Fabozzi FJ (ed) The handbook of financial instruments. Wiley, Hoboken, NJ

Loader D (2007) Fundamentals of fund administration: a guide. Elsevier, Amsterdam

London Stock Exchange Group (2013) Structured products. Precision tools. A guide for private investors. London Stock Exchange Group, London

Malkiel BS (1973) A random walk down Wall Street. W.W. Norton, New York

Malkiel BS (1995) Returns from investing in equity mutual funds 1971 to 1991. J Financ 50(2): 549–570

Maude D (2006) Global private banking and wealth management: the new realities. Wiley, Chichester

McCabe PE (2009) The economics of the mutual fund trading scandal. Finance and Economics Discussion Series, no 6. Federal Reserve Board

Moloney N (2014) EU securities and financial markets regulation, 3rd edn. Oxford University Press, Oxford

OECD (2001) Governance systems for collective investment schemes in OECD countries. Financial Affairs Division, Occasional Paper, no. 1

Scherer B, Lee X (2012) Performance based fees, incentives, and dynamic tracking error choice. In: Scherer B, Winston K (eds) The Oxford handbook of quantitative asset management. Oxford University Press, Oxford

Schoenfeld SA (ed) (2004) Active index investing. Maximizing portfolio performance and minimizing risk through global index strategies. Wiley, Hoboken, NJ

Sharpe WF (1991) The arithmetic of active management. Financ Anal J 47(1):7–9

Sharpe WF (2002) Indexed investing. A prosaic way to beat the average investor. Spring President's Forum. Monterey Institute of International Studies

Siegel LB (2003) Benchmarks and investment management. CFA Institute, Research Foundation Publication, Charlottesville

St Giles M, Alexeeva E, Buxton S (2003) Managing collective investment funds, 2nd edn. Wiley, New York

Stefanini F, Vismara S, Meoli M, Derossi T (2010) Newcits: investing in UCITS compliant hedge funds. Wiley, Chichester

Swiss Bankers Association (2008) Special risks in securities trading. Swiss Bankers Association, Basel

Tolle S, Hutter B, Ruthemann P, Wohlwend H (2008) Structured products in wealth management. Wiley, Singapore

Turner C (2004) International funds. A practical guide to their establishment and operations. Elsevier, Oxford

Part II
Investment Management Policy

Chapter 3
Stages of Investment Management Policy

Pierpaolo Ferrari

3.1 Investment Management Policy

Investment management policy for the various categories of institutional investors can be divided into the following stages:

- Identification of objectives and constraints.
- Formalisation of the investment strategy.
- Implementation of the financial strategy.
- Periodic rebalancing of the portfolio.
- Performance assessment and risk control.

The level of detail and structure of each stage of the investment management policy varies depending on the category of the institutional investor, size of the portfolio, complexity of the management model adopted, and specific regulation.

3.2 Identification of Objectives and Constraints

The starting point for setting up an investment management policy is the identification of the objectives and obligations of the investor. In general terms, the objectives of such a policy for an institutional investor are those of achieving efficient risk-return combinations on a given time horizon, bearing in mind any constraint or return

P. Ferrari (✉)
Department of Economics and Management, University of Brescia, Brescia, Italy

SDA Bocconi School of Management, Bocconi University, Milan, Italy
e-mail: pierpaolo.ferrari@unibs.it; pierpaolo.ferrari@unibocconi.it

I. Basile, P. Ferrari (eds.), *Asset Management and Institutional Investors*, Contributions to Finance and Accounting,
https://doi.org/10.1007/978-3-031-59819-7_3

guarantee to final investors. More specifically, there is a fundamental difference between institutional investors that need to implement investment policies based on asset-liability management and those that can adopt asset-only investment policies.[1]

In the first case, the investment policy must consider and assess the liabilities that correspond to the performance obligations guaranteed to final investors. In other words, the same investment policy serves to guarantee performance to the final investors within the specified contractual terms and conditions. This obliges the institutional investor to perform a strict and dynamic control of the surplus or deficit between the total amount of investments at a certain date and the current value of the liabilities and obligations in existence at that date, in order to promptly intervene in case of excessive deviation between the two total values.

In the second case, lacking any predefined performance obligation, the institutional investor can adopt an investment policy that exclusively focuses on optimal asset allocation, while maintaining the necessary caution with regard to the liquidity needs required by the adopted management model. Compared to the former case, the latter entails a less strict risk control, at least in relative terms, and in effect, for a given risk level, the investment policy has only an indirect impact on the performances provided to the final investors, in the form of failed performance maximisation.

Investment policy constraints include a series of factors that exercise a direct or indirect influence on the definition of the objectives of the investment policy. In the case of an institutional investor, these constraints can be traced back to the following factors:

- Liquidity requirements.
- Reference time horizon.
- Fiscal treatment of financial instruments.
- Legislative-regulatory obligations.
- Specific factors related to individual institutional investors.

Each of these factors has an impact on the investment policy, but the first two, in particular, influence the ability of the institutional investor to support risks and, therefore, have a direct influence on the identification of the most appropriate risk-return combination. The first significant factor is the need for the institutional investor to have access to cash flows during the life of the investment, in order to honour requests for reimbursements or performance obligations, whether expected or unexpected.[2] This requirement is connected to liquidity risk, which has a

[1] Sharpe et al. (2007).

[2] Liquidity risk implies funding and market liquidity risks for institutional investors. Funding liquidity risk entails the risk in which the institutional investor is not able to cope economically and in a timely manner with the expected and unexpected cash flows, linked to the reimbursement of liabilities or to compliance with performance obligations. Market liquidity risk implies the risk in which the institutional investor is unable to convert a position on a given financial instrument into cash, or is able to liquidate it, but incurs a reduction in price due to the lack of liquidity of the market in which this instrument is negotiated or due to the temporary malfunctioning of the market itself. While institutional investors with an investment policy that has been set according to the asset-liability management approach can deal with both risks, those with an asset-only approach encounter only the latter risk. See also Hull (2023).

significant impact on the definition of the investment policy and varies depending on the type of institutional investor and, above all, depending on the reimbursement, more or less discretionary, of the capital provided by final investors.[3]

A second factor that has a significant impact on the definition of the investment policy is the reference time horizon adopted by the investor. This factor assumes significance under two different aspects: the effect that it has on the ability of the investor to bear risks and the impact it has on the actual construction of the portfolio. In this regard, the traditional principle that is widely accepted by operators, albeit not without criticism, consists of the so-called time diversification, on the basis of which, as the time horizon adopted by the investor increases, so does the weight potentially assignable to assets with a higher risk.[4]

The third factor able to influence investment policy, in particular the security selection stage, is the tax treatment of the different financial instruments. In this case, tax legislation for each category of institutional investor must be analysed in relation to the income provided by the different financial instruments, with the objective of establishing an investment policy that can achieve an optimisation of the tax burden, while complying with other conditions.

Moreover, the fourth factor affecting the investment policy of institutional investors involves legislative-regulatory constraints, which may be more or less stringent depending, in particular, on the type of final investors.[5] These constraints may include restrictions to the overall composition of the portfolio, in terms of the categories of financial instruments and/or homogenous asset classes of financial instruments; restrictions to risk concentration; restrictions to the holding of voting rights, resulting in maximum thresholds for participation in the companies' equity capital; and general prohibitions that identify the type of operations that cannot be made by a particular category of institutional investors.

Lastly, the final factor is represented by non-regulatory restrictions that are set by the institutional investor. One example is the constraint, even when not prescribed by law, of implementing socially responsible investment policies, where not only the risk-return combination of the individual investment is assessed, but also the environmental, social, and governance impact that it generates. Therefore, the constraints individually set by each institutional investor have a significant effect on the concrete realisation of the investment policy.[6]

3.3 Formalisation of the Investment Policy

Once the objectives and constraints of the investment policy have been defined, institutional investors need to identify the financial strategy they intend to implement, in order to achieve efficient risk-return combinations from the resources

[3]Tschampion et al. (2007).

[4]For a critical analysis of the traditional principle of time diversification, see Bodie (1995).

[5]There is a clear distinction between final investors that are retail clients and those that are professional. In fact, legislation provides a greater degree of protection for the former than for the latter.

[6]Reilly and Brown (2012).

invested during a given period of time. This stage—beyond any regulatory requirement—demands the formalisation of the following:

- Objectives to achieve using the investment policy.
- Criteria to follow in its implementation.
- Tasks and responsibilities of the subjects involved in the process.
- Control and assessment system of the results obtained.

In order to reach these objectives, the investor must identify:

- The approach underlying the investment policy.
- Financial instruments in which to invest and the associated risks.[7]
- *Ex ante* constraints to risk exposure.[8]
- Management style adopted.
- Management method, whether internal or delegated, to adopt when such a choice is possible.
- Organisational division of tasks and responsibilities of the various subjects involved in the investment process.

The financial strategy must be subjected to periodical review, in order to assess its effective congruence with the objectives and requirements of the investor.

3.4 Implementation of the Financial Strategy

The next stage of investment management policy involves the implementation of the financial strategy, which depends on specific factors regarding the investor, as well as on economic and market factors. The approaches underlying the investment decisions are based on two alternative methods:

- The bottom-up approach.
- Top-down approach.[9]

In the bottom-up approach, the investment process starts directly with the selection of single securities in which to invest, on the basis of a sector analysis, an analysis of the company fundamentals, a quantitative analysis, or any other selection criterion deemed reliable. The resulting overall investment portfolio is, therefore, given by the sum of the single securities chosen during the initial stage.

In contrast, a top-down approach focuses, first, on the strategic allocation of the investor's portfolio, identifying the division between macro-classes of financial assets, consistent with the objectives of medium-long term investments, and only

[7] Quantitative limits (minimum and maximum) and qualitative limits (issuer reliability, issuer nationality, negotiation markets, etc.) for each financial instrument and/or homogenous asset class can be established for an even more rigorous control of the risks.

[8] See Chap. 5.

[9] Farrell (1997) and Kinlaw et al. (2021).

later concentrates on the selection of individual financial instruments. In the case of institutional investors, it is usual and, in some cases, compulsory to follow the top-down approach, in which the investor starts by dividing the portfolio into different investment classes and, in the final stage, selects individual financial instruments. This preference is linked to three factors.

The first factor is the hypothesis that the forecast of the expected return of the homogenous asset classes of financial instruments may be subject, on average, to more limited margins of error, if compared to the forecast of the return expected from each single security. If the specific risks of the single security included in a homogenous asset class are mutually offset, it would undoubtedly seem more rational and less complex to concentrate only on the forecast of the variables that are the source of total risk which cannot be diversified. The second factor is an improvement of the diversification effect, due to an easier forecast of the risks, at least in relative terms, and, above all, of the correlations between homogenous asset classes, instead of between single securities. Finally, the third factor is a better adaptation of the top-down approach to hierarchical decision-making structures and allocation of tasks, with the identification of the subjects responsible for each stage of the investment policy.[10]

The most well-known model for realising a top-down investment approach is Markowitz's mean-variance optimisation, which, after decades, still constitutes one of the most applicable methods for reconciling methodological rigour with practical feasibility.[11] Chap. 4 is dedicated to the analytical presentation of the model in the logic of strategic asset allocation, its theoretical assumptions, the problems associated with its practical application, and the relative solutions.[12] Based on a top-down approach, the management of an investment portfolio may be broken down into three consecutive stages:

- Strategic asset allocation.
- Tactical asset allocation (or market timing).
- Stock picking.

3.4.1 Strategic Asset Allocation

Strategic asset allocation consists of the identification of the weights that the different asset classes must keep within the portfolio in the medium to long-term.

[10] Gibson (2013).

[11] Markowitz (1952).

[12] A top-down alternative to Markowitz's mean-variance optimisation is the creation of a naïve portfolio. This consists in putting together equally weighted portfolios that are extremely diversified, in terms of asset classes, without making forecasts of the individual market sector trends. Although this strategy, also known as 1/N strategy, does have the undoubted advantage of highly diversifying the portfolio, it does not pursue any objective of maximising return for a given risk level. Other two alternatives to Markowitz's mean-variance optimisation are global minimum-variance strategy and optimal risk-parity strategies, which are presented in Chap. 6.

It is derived from the forecasts of the real and financial trends of each market sector, and from the consequent assessment of comparative convenience, bearing in mind the investor's objectives and constraints. Therefore, the target weights obtained with strategic asset allocation define the structure that the portfolio must maintain, within the time period assumed by the investor.

At the first level, strategic asset allocation envisages the identification of different asset classes in which to divide the investment universe, where each asset class represents a group of financial assets with a certain degree of homogeneity, in terms of risk-return combination. The identification of asset classes takes on particular importance as the prediction of market variables occurs through the formulation of expectations on the evolution of the general economic scenario and then obtaining forecasts on the future performance of the individual market segments, specifically represented by as many assets classes.

As mentioned, the creation of a connection between macroeconomic forecasts and the return of a single share or bond is a difficult task, given that the return of each security is significantly influenced by specific factors associated with each company. However, if we consider an entire asset class consisting of a range of securities with homogenous characteristics, the specific factors tend to mutually compensate and disappear, thereby tightening the connection between formulated economic forecasts and asset classes trends.[13]

The composition criteria vary according to whether we consider equity, bond, or money-market asset classes; however, in all three cases, it is essential that the selected asset classes satisfy the following three requirements:

- Completeness.
- Internal consistency.
- External differentiation.

The first requirement entails that the selected asset classes be able to completely represent the investment universe. The second requirement imposes that each asset class consists of financial instruments that are as homogenous as possible, and similarly exposed to systematic risk factors. Lastly, the third requirement posits that the different asset classes have different exposure levels to the various macroeconomic and political factors, or sources of systematic risk.

Box 3.1 provides the possible segmentation criteria for equity, bond, and money-market asset classes.

Box 3.1 Identification of Asset Classes

In the case of the equity market, asset classes are usually identified using:

- Geographical criteria, dividing the markets according to the country or economic area from which the shares originate, or
- Sector criteria, distinguishing the industry group of the companies.

(continued)

[13] Kaplan (2012).

Box 3.1 (continued)

The geographical criteria are justified in the fact that securities from the same market tend to move in a similar manner, since companies:

- Operate with the same currency and are subject to the effects of the same economic policy.
- Have the same basic interest rates.
- Are subject to the same country risk.

The sector criteria assume that securities within the same industry move in a similar way because the company's industry group determines the sensitivity of the securities *vis-à-vis* different macroeconomic and political factors, or sources of systematic risk.

The table below presents the distribution of the global equity MSCI All Country World Index (ACWI) in the different geographical markets and their different industry groups.

MSCI All Country World Index: Geographical breakdown					
MSCI World			MSCI Emerging Markets		
Developed markets			*Emerging markets*		
America	**Europe and Middle East**	**Pacific**	**America**	**Europe, Middle East, and Africa**	**Asia**
Canada USA	Austria Belgium Denmark Finland France Germany Ireland Israel Italy Netherlands Norway Portugal Spain Sweden Switzerland United Kingdom	Australia Hong Kong Japan New Zealand Singapore	Brazil Chile Colombia Mexico Peru	Czech Republic Egypt Greece Hungary Kuwait Poland Qatar Saudi Arabia South Africa Turkey United Arab Emirates	China India Indonesia Korea Malaysia Philippines Taiwan Thailand
MSCI All Country World Index: Industry breakdown					
Communication Services	Consumer Discretionary	Consumer Staples	Energy	Financials	
Health Care	Industrials	Information Technology	Materials	Real Estate	Utilities

(continued)

Box 3.1 (continued)

More recently, two further criteria for the segmentation of the equity market and the consequent identification of equity asset classes have been gaining ground, based on:

- The economic characteristics of companies and, in particular, on the price/earnings (P/E) and price/book value (P/BV).
- The size of companies in terms of market capitalisation or, as an alternative, the value of the free-float capital.

A study by Fama and French showed that the different share trends can be explained, to a large extent, by[a]:

- Sensitivity to the market trend as a whole, measured by the security beta.
- Size, represented by the market capitalisation of the listed company.
- Economic characteristics of the companies (growth or value shares), measured by the ratio between the price of the share and net earnings per share or, alternatively, by the ratio between the price and book value per share.

In addition to their betas, the different trends of the shares may be attributed, on the one hand, to the size of the company (small-, medium-, and large-cap) and, on the other hand, to high P/E or P/BV ratios (growth shares) or low P/E or P/BV ratios (value shares).[b] The operational practice following these results consists in segmenting the equity market using a matrix, in which one dimension consists of the company's market capitalisation, while the other focuses on the P/E and/or P/BV ratios.

Even more recently, a new way to achieve strategic asset allocation in equity market is also to adopt a thematic approach, often within core-satellite strategies. Thematic investing is an approach that focuses on expected long-term changes, rather than geographic or industry subdivisions. It allows investors to take advantage of the structural changes taking place, which are capable of modifying the competitive forces of one or more sectors and the drivers of corporate profitability. The thematic approach requires the identification of the so-called mega-trends in progress, such as: technological revolution; rapid urbanisation, climate change, and resource scarcity; increase in the standard of living in some "emerging" areas; demographic development and social changes.

In the case of the bond market, there is a higher level of homogeneity in the definition of the asset classes, both at the theoretical and operational levels. Usually, market segmentation is based on the ascertainment that the return of a bond in a specific currency can be divided into three components:

(continued)

Box 3.1 (continued)

- The risk-free rate on that currency.
- Premium for duration/liquidity risk.
- Premium for credit risk.

Consequently, in the case of bond asset classes, the global and individual geographical market segments (Euro, Europe, USA, Japan, Emerging countries, etc.), or the categories of the issuers (Government, Government-Related, Corporate, or Securitised), are distinguished according to:

- The rating of the securities in the portfolio.
- Duration.

In this case, it is also possible to segment the bond market in a matrix with the following dimensions: creditworthiness (high, medium, or low), identified based on the rating, and exposure to volatility risk (long, intermediate, or short), determined based on the duration.

The table below illustrates the distribution of the Bloomberg Global Aggregate Bonds index, which represents a broad market index of investment-grade bonds.

Bloomberg Global Aggregate: Geographical breakdown				
U.S. and Canadian Aggregate	Pan-European Aggregate		Asian-Pacific Aggregate	
Bloomberg Global Aggregate: Issuer breakdown				
Treasury	Government-related	Corporate	Securitised	
Bloomberg Global Aggregate: Duration breakdown				
Global Aggregate 1–3 years	Global Aggregate 3–5 years	Global Aggregate 5–7 years	Global Aggregate 7–10 years	Global Aggregate 10+ years
Bloomberg Global Aggregate: Credit quality breakdown				
Global Aggregate AAA	Global Aggregate AA	Global Aggregate A	Global Aggregate BBB	
In the case of the money market, asset classes are identified depending on the currency of the financial instruments to which they refer.				

[a]Fama and French (1992).

[b]At times, when shares cannot be assigned univocally to the categories of growth or value securities, there is a third, intermediate security category, known as blend (or core).

At the second level, we need to identify a benchmark index suitable for representing each asset class, and on which to formulate forecasts of the optimisation process inputs.[14] Each benchmark can play a key role in strategic asset allocation via two distinct methods. They can represent:

- A portfolio to replicate, in case of indexed management.
- A reference portfolio to beat, in case of active management.

It should be noted that the benchmark merely reflects the risk-return profile of each asset class in general terms. Instead, it is the type of strategy—active or indexed—that outlines the breadth of the risk profile of the portfolio being managed with respect to the level of market risk, considered "neutral", of the benchmark portfolio. In the case of an active strategy, the benchmark has not a stringent effect on the asset manager's operations: the objective, in this case, is to beat the reference benchmark, achieving a performance, in relative terms, better than the reference parameter. Given this opportunity, however, higher costs for analysis and research of individual securities and for evaluation of the most favourable moment to enter and exit the market arise. Active management, therefore, increases the asset manager's scope of discretion and the breadth of the path of market risk assumed. Conversely, in the case of an indexed strategy, the management company limits itself to replicating the composition of the benchmark which, therefore, itself defines synthetically the level of risk-return of the portfolio. The opportunity to beat the benchmark is absent, but the result is lower costs for research and analysis of individual stocks and for the expected trend of the market. As anticipated in Chap. 2, financial theory does not agree on which investment strategy is to be preferred between indexed management and active management, there being valid arguments in support of both.[15] A somewhat hybrid approach, which is able to grasp, albeit partially, the strengths of both management styles, is the so-called core-satellite strategy. This strategy divides the portfolio into two sub-portfolios: the core portfolio and satellite portfolio. The core portfolio is managed using an indexed (or semi-indexed) strategy, with the objective of minimising both benchmark-related risk and costs. The satellite portfolio is managed with an active management strategy, which aims to reach an over-performance, with respect to the benchmark, and, consequently, with respect to the core portfolio.[16]

At the third level, we need to determine whether the investment strategy is constrained to the long-only investment approach, or whether the long-short approach is possible. In the first case, even in a situation of highly negative forecasts for a given market sector, it would not be possible to open short positions because of

[14] When defining the benchmark index representative of each asset class, we need to define: the inclusion criteria of the securities, depending on the specific asset class; weighting methods, which have a considerable impact on index efficiency; treatment of periodical cash flows, depending on the decision of whether to accumulate or distribute; and index currency.

[15] See Flood and Ramachandran (2000), as well as the bibliography quoted in Chap. 2, regarding this subject.

[16] Gastineau et al. (2007) and Scherer (2015).

the long-only constraint; therefore, it is only possible to zero the weight of that asset class in the portfolio. In the second case, by removing the no-short-selling constraint, it is possible to open short positions for a given sector for which there are highly negative forecasts.[17]

Numerous empirical verifications have demonstrated how strategic asset allocation and the medium-long term weights of the corresponding strategic benchmarks are able to explain most of the level and variability in the returns of an investment portfolio over time, conversely leaving a marginal role to market timing and stock picking in explaining the variance of an investment portfolio's historical returns.[18]

The result of the strategic asset allocation can be summarised in a pie chart, which represents the weights of the various asset classes within the portfolio in the medium-long term (Fig. 3.1). It corresponds, in a logic of mean-variance optimisation, to the composition in terms of asset class of the portfolio selected on the *ex ante* efficient frontier by the individual investor, according to his/her degree of risk aversion.

3.4.2 Tactical Asset Allocation

Tactical asset allocation (or market timing) entails a temporary overweight/underweight of some asset classes, depending on short-term expectations. Therefore, it is composed of the set of actions to manage, in the short-term, the portfolio established during the strategic asset allocation, in order to take advantage of the best market opportunities offered by the evolution of the economic outlook in the near term. In other words, tactical asset allocation refers to the possibility of deviating from the medium-long term portfolio strategy for short or very short periods, with the objective of achieving over-performance with respect to the market.

The possibility of achieving a positive differential return with respect to the benchmark depends on the ability to correctly predict the timing of the upward or downward trends of the market, and the consequent variation of the portfolio's exposure to systematic risk. Operationally, tactical asset allocation involves the fixing of admissible fluctuation bands, with respect to the medium-long term weight defined at the strategic asset allocation stage. In the short-term, therefore, an overweight/underweight is allowed, with respect to the medium-long term strategic weights, in compliance with the range of predefined fluctuations.[19] For example, Fig. 3.2 shows the structure of the previous portfolio shown in Fig. 3.1 in which two short-term changes have been made to take advantage of the expected performance

[17] This second investment approach is generally reserved for special types of institutional investors.

[18] Brinson et al. (1986, 1991), and Ibbotson and Kaplan (2000).

[19] The distinction between the long-only and long-short approaches is important. The latter allows short positions to exploit brief downward expectations.

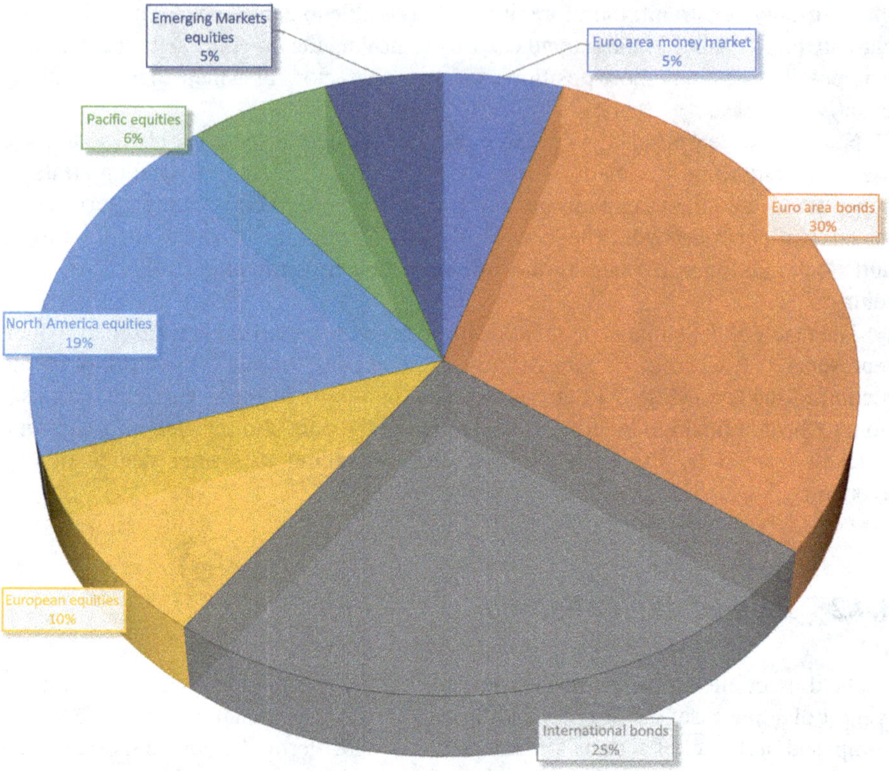

Fig. 3.1 The graphical representation of the strategic asset allocation

of the market: compared to the standard weights defined in the strategic asset allocation, the North American Equities asset class was overweighted by 5%, while the Euro Area Bond asset class was specularly underweighted by the same 5%. This is to take advantage of a positive short-term view on the trend of the North American stock market and a negative expectation on the short-term trend of the euro area bond market. After this phase, the weights will be brought back to the standard ones defined in the strategic asset allocation phase.

This reasoning is valid for the active management approach. However, in the case of indexed management, there is no form of market timing, and deviations from the strategic benchmark are not sought.

The impact of tactical asset allocation on costs and risk requires attention. Of course, management costs increase as the intensity of the use of market timing policies increases. Since it is one of the levers available to the active asset manager to beat the reference benchmark, the use of market timing takes us back to the issue of comparative convenience between costs, market efficiency, and, finally, the investor's degree of risk aversion.

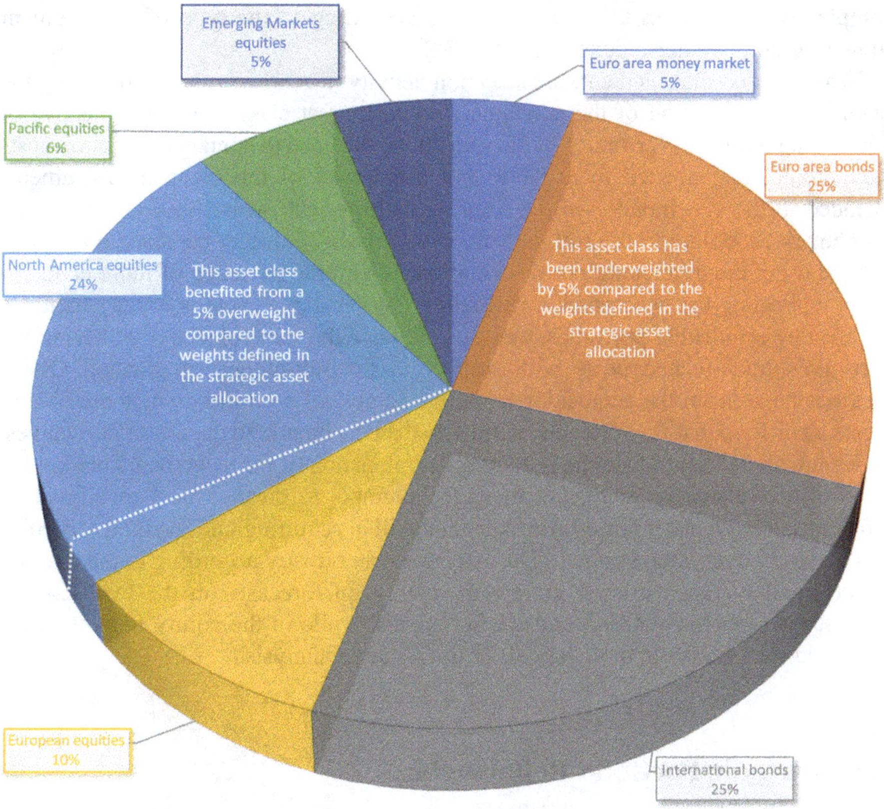

Fig. 3.2 The graphical representation of the tactical asset allocation

In terms of risk, market timing may increase or decrease the total risk, depending on the type of choices made, while it increases, in any case, the relative risk in relation to the benchmark.

A model that is able to simultaneously manage strategic asset allocation—on the basis of a Bayesian approach—and tactical asset allocation—via the use of absolute or relative views, in relation to the evolution expected for the various market sectors in the short-term, and the identification of a degree of trust associated with each estimate—is that developed by Black and Litterman. This model is described in Chap. 4.

3.4.3 *Stock Picking*

Stock picking, also known as security selection, consists of the identification of the individual financial instruments to be included in each asset class. The degree of complexity involved in this activity varies according to the type of management strategy used, whether indexed or active.

With indexed management, the selection activity is conducted by reproducing the risk-return combination of the benchmark of each asset class, with the objective of minimising replication errors.[20] At the same time, the need to contain transaction costs may prompt the indexed manager to not acquire all of the financial instruments included in the benchmark—in the same proportion with which they compose that benchmark according to the full-replication approach—but may persuade him/her to operate using the mimicking portfolio approach, with the objective of more economically replicating the benchmark's risk-return combination. There are many methods for creating mimicking portfolios, and they all have the same objective of saving the portfolio's trade-related costs with respect to the full-replication method. These methods range from the acquisition of only a sample of securities, which are able to create a portfolio that is sufficiently similar to the benchmark, to the use of derivatives, which are able to reproduce the risk-return combination of the same benchmark.

With active management, the manager attempts to create value added via an efficient selection of securities due to his/her skill in acquiring undervalued securities and avoiding overvalued ones.[21] The selection criteria vary according to the market; in the bond markets, criteria are based mainly on forecasts on the evolution of interest rates, their volatility, and credit spreads, while in the equity markets, they are based on fundamental, technical, or quantitative analysis.[22]

3.5 Periodic Portfolio Rebalancing

An investment policy must necessarily define the methods that guide the portfolio's periodic rebalancing, which is aimed at aligning the proportions of the single asset classes to the original weights from the strategic asset allocation. A periodic rebalancing action may also be required solely because of the effect of the portfolio's natural deviation from its strategic combination, due to the higher performance of some asset classes over others. In this situation, the implementation of periodic rebalancing could be interpreted as a so-called contrarian strategy, since it comprises a reduction in the weight of the asset classes subject to relative appreciation and an increase in the weight of those that recorded relative depreciation, with the consequent temptation to opt for a non-rebalancing strategy.

The holding of a non-rebalanced portfolio for a preset time horizon is known as the buy-and-hold strategy, also known as the do-nothing strategy. In this situation, which does not occur very frequently, especially with institutional investors, the

[20] In reality, even full replication does not guarantee a perfect replication of the index, since it is hindered by the various factors described in Chap. 2.

[21] A long-only investment approach is implicit here.

[22] Grinold and Kahn (2000).

investment policy concentrates on the definition of an initial optimum portfolio, identifying the weights of the different asset classes, without any further intervention. Although having the advantage of limiting portfolio trade-related costs, due to the absence of a periodic rebalancing activity, this strategy has three significant consequences on the portfolio:

- The redistribution of the weights of the asset classes with respect to the original asset allocation target.
- Lack of consideration of the risk control function implicit in the rebalancing action.
- Incoherence between the investor's risk aversion and exposure of the portfolio to market volatility, which is increasingly more concentrated in the highest performing asset classes.

In support of periodic rebalancing, it is also worth considering the broad quantitative analysis that documents a mean-reverting behaviour of the asset classes, which is, therefore, in contrast with the idea of the indefinitely high or low returns of a single-asset class.[23]

Thus, what has been said so far supports the expediency of periodic rebalancing through the implementation of a constant-mix strategy (also known as do-something strategy), where interventions to the portfolio are periodically made, in order to restore the weights of each asset class to the original asset allocation target.

From the operational perspective, the adoption of a constant-mix strategy requires the definition of the periodical rebalancing timing and methods. These are important aspects, given that each rebalancing intervention involves transaction costs, which have an inevitable effect on the overall performance. With regard to rebalancing timing, there are three main alternative methods available for implementing a constant-mix strategy:[24]

- Periodic rebalancing (calendar rebalancing) with a predefined time interval (monthly, quarterly, etc.).
- Periodic rebalancing based on exceeding predefined thresholds (per cent range rebalancing), which determines a restoration of the original asset mix when the set fluctuation bands are exceeded.
- Rebalancing of the intervals within an allowed range (rebalancing to the allowed range), which, like the previous case, requires fixing restoration once the set bands have been exceeded; however, in this case, restoration does not re-establish the original asset mix, but restores the upper or lower limit of the threshold.[25]

[23] Fama and French (1988), Lo and MacKinlay (1988), and Poterba and Summers (1988).

[24] Arnott et al. (2007).

[25] If, for example, the fluctuation bands were absolute ±5% and the original weight of the asset class was 30%, exceeding the threshold by 35% or 25% would determine the restoration of the weight of the single asset class to 35% and 25%, respectively, and not to 30%, as in the case of range rebalancing. The underlying logic to this rebalancing method is based on the need to reduce the transaction costs associated with the period recomposition of the portfolio and take into account the "momentum" of each asset class.

An alternative rebalancing method to the above is the so-called constant-proportion strategy, which cannot be considered simply a method for realising periodic rebalancing, but part of a greater framework of strategies aimed at seeking a dynamic form of protection for the investment portfolio value. This strategy is summarised in Box 3.2. It is useful to note that a periodic rebalancing action is not a tactical asset allocation activity, but consists in the elimination of the active positions created by the market or deriving from previous choices to overweight/underweight some of the asset classes, even though its implementation requires the simultaneous consideration of the choices made during tactical asset allocation.

Box 3.2 The Constant-Proportion Portfolio Insurance Strategy
In the constant-proportion strategy, the weight of risky assets in the overall portfolio is a function of the difference between the market value of the portfolio and a predefined minimum portfolio value (floor), below which the portfolio market value must not fall. In practice, once established the floor below which the portfolio value must not fall, the constant-proportion strategy formulates a simple rule for determining the combination of risk-free and risky assets. The former have the objective of achieving the established protection level, while the latter have the task of generating returns. To be more precise, the size of the risky assets is defined according to the following algorithm:

$$\text{Target investment in risky assets} = m \cdot (\text{portfolio value} - \text{floor value}) = m \cdot \text{cushion},$$

where:

m = preset multiple.
portfolio value = portfolio market value.
floor value = minimum value below which the portfolio value must not fall.
cushion = portfolio value − floor value.

The floor value is a function of the investor's risk tolerance and influences, in inverse proportion, the exposure level of the portfolio to risky assets. The cushion value depends on the floor value and dynamically represents the maximum acceptable loss to ensure that the portfolio value does not fall below the minimum safety value. This minimum value can be identified as the present value of the amount the investor has to dispose of at a certain date, calculated by discounting this amount based on the risk-free rate.

The preset multiple m is a function of the investor's risk appetite. This strategy is defined as constant proportion, since risky assets are kept in the portfolio based on a constant multiple of the cushion. If, for example, the

(continued)

Box 3.2 (continued)
cushion were equal to zero, the weight of the risky assets in the portfolio would be zero.

When *m* is greater than one, the constant-proportion strategy is known as the constant-proportion portfolio insurance strategy (CPPI).[a] In the CPPI strategy, risky assets must be sold when their value falls, and bought when their value rises. Therefore, in bullish markets, the increase of the weight of risky assets is greater than that of their value, given the presence of a constant *m* greater than 1, which leads to a progressive reduction in the weight of the risk-free asset, as we move away from the floor value. Conversely, in bearish markets, the decrease of the weight of risky assets is greater than that of their value, producing a rapid increase in the weight of the risk-free asset, as we approach floor value. Thus, the CPPI strategy should produce higher returns when risky assets face a rising market and limit losses when risky assets face a falling market. By contrast, CPPI strategies perform poorly in markets characterised more by reversal than by trends.

In addition, precise rebalancing rules must be set within the CPPI strategies by setting the thresholds over which the portfolio must be rebalanced. As the portfolio's market value changes, it is necessary to measure the size of the cushion and rebalance the portfolio again. If, theoretically, continuous rebalancing should be implemented, in practice, changes to portfolios are periodically made based on established rules that must necessarily consider two key factors: transaction costs and any tax charge consequent to rebalancing.

The CPPI strategy provides results that are very similar to those of a risky assets portfolio, which is protected by the purchase of a put option on the appropriate underlying instruments. However, in a stable or rising market, a put option with a strike price equal to or lower than the current index level reaches maturity with zero value. Conversely, the CPPI strategies follow another path, seeking dynamic protection through the change of the portfolio composition, according to preset rules depending on market trends. In contrast to the options-based insurance strategies, a CPPI strategy does not involve explicit hedging costs, with the exception of the transaction costs associated with the periodic rebalancing of the portfolio. However, it is evident that the CPPI strategy must also be subjected to the fundamental laws that govern the financial markets; high levels of protection are obtained at the price of a lower participation in a rising market, making this lower return potential a sort of implicit hedging cost, to be compared to the explicit costs of the option-based portfolio insurance strategies.

In order to understand the dynamic implementation of the CPPI strategy, we shall consider the following example, based on a portfolio with an initial value of 100 euros. If the floor value below which the investor does not want to

(continued)

Box 3.2 (continued)
fall is, for example, 80 and the multiple m is equal to 2, the initial weight of the risky assets is:

$$\text{Risky assets} = 2 \cdot (100 - 80) = 2 \cdot 20 = 40.$$

Therefore, the initial combination will contain 40 euros in risky assets and 60 in risk-free asset. If the risky assets increased by 10% and the value of the risk-free asset remained at 60, the new value of risky assets would be equal to 44, and that of the portfolio 104. Thus, on the basis of the CPPI algorithm, the new size of the risky assets would be:

$$\text{Risky assets} = 2 \cdot (104 - 80) = 48.$$

This implies that the sale of risk-free asset for an amount of 4 euros, and the further acquisition of a corresponding amount of risky assets, will increase risky assets from 44 to 48. If, soon after rebalancing, the price of the risky assets falls by 20%, their value will fall to 38.4, and the value of the portfolio—assuming a constant value for the risk-free asset—to 94.4 = 38.4 + 56. In this context, the CPPI rule demands a rebalancing of the portfolio, reducing the weight of the risky assets to:

$$\text{Risky assets} = 2 \cdot (94.4 - 80) = 28.8.$$

Thus, we obtain a decrease in the risky assets from 38.4 to 28.8, with a corresponding increase of 9.6 in the risk-free asset. This example highlights the difference between the CPPI and constant-mix strategies, since an increase in the price of the risky assets leads to an increase in their weight in the portfolio, and not to a reduction, as happens with the latter strategy. In contrast, the reduction in price of the risky assets, in the CPPI strategy, leads to a reduction of their weight, and not an increase, as happens in the constant-mix strategy.

The multiplier m is the instrument through which the degree of management aggressiveness is defined; the higher the level of m, the higher the exposure to risky assets and pro-cyclicality of the portfolio.

Moreover, exposure to risky assets is influenced by the size of the cushion and, consequently, by the size of the floor value. In reference to the latter, it is useful to note that the protection of the portfolio is, in reality, uncertain and linked to the downward variation of the market value of risky assets. When risky assets undergo a reduction equal to $1/m$, the quota of risky assets is set to zero, and the value of the portfolio, corresponding to the floor value, consists entirely of risk-free asset. When there is a decline in the market value of risky

(continued)

Box 3.2 (continued)
assets greater than $1/m$, the floor value would no longer be guaranteed, due to the rapid fall of the prices of risky assets, which does not leave time for rebalancing the portfolio.[b]

[a]If we consider a preset multiple equal to 1 and floor value of zero, the constant-proportion strategies will coincide with those of buy-and-hold. In the case of a multiple between zero and 1 ($0 < m < 1$) and floor value of zero, the constant-proportion strategies coincide with those of constant-mix.

[b]Having considered this aspect, the preset multiple m should be defined according to the maximum loss encountered by the risky assets between one rebalancing and another. If, for example, the maximum risk encountered was 20%, m should not be greater than 5, to prevent the sharp decline in the price of the risky assets from causing a reduction in the market value of the portfolio to below the minimum safety margin.

Source: Black and Jones (1987) and Perold and Sharpe (1988)

3.6 Performance Assessment and Risk Control

The final stage of the investment process is performance assessment and risk control. The first step in performance assessment is the calculation of returns, which requires the identification of the most appropriate measure for the objective pursued among the methods of calculation available. As you will see in Chap. 8, the alternatives are:

- Simple, compounded, and continuous returns.
- Arithmetic and geometric returns.
- Time- and money-weighted returns.
- Gross and net returns.

In order to form an opinion on portfolio performance, it is necessary to identify the risk of the investment portfolio in its various forms of absolute, asymmetric, and relative risk and to calculate the resulting risk-adjusted performance measures that allow us to assess the efficiency of the portfolio manager relative to the benchmark, competitors, and *ex ante* risk limits.

When comparing the performance of competitors, it is essential to create homogeneous peer groups consisting of portfolios with the same management approach. To this end, the most common operational solution is to create a peer group based on the "investment style" adopted, using a deductive approach based on the so-called style analysis. According to this logic, the mimicking portfolio is obtained by regressing the portfolio's returns on the returns of a series of benchmark indices representative of the investment universe, with the benchmark weights within the portfolio as unknown quantities and two constraints, so that the sum of the weights is equal to one and each weight has an admissible value between zero and one.[26] This approach is described further in Chap. 9.

In the case of non-indexed strategy, performance assessment requires the evaluation of the ability to realise an effective stock picking activity via the selection of the best securities in a given market. To this end, we have to identify the most suitable measuring model, which, as is discussed in Chap. 8, depends upon the simultaneous presence or absence of a tactical asset allocation activity.

When stock picking and market timing activities are carried out simultaneously, it is essential to measure together both the ability to carry out effective stock selection and the ability to carry out a profitable tactical asset allocation capable of temporarily overweighting asset classes that are expected to outperform the market and, at the same time, underweighting asset classes that are expected to underperform the market.[27]

A separate measurement of the two abilities could be misleading and confuse the results of the first activity with those of the second. An in-depth *ex post* performance assessment requires the identification of the most appropriate performance attribution model, aimed at shedding light on those management choices that generated the gap between the overall result of the portfolio and benchmark, breaking down relative performance into its determinants and attributing it to the various factors that contributed to its generation. This is discussed further in Chap. 10.

The performance assessment activity must therefore be integrated into an appropriate control system designed to verify that the results of the actions taken by the various entities involved in the investment process are in line with the financial objectives set.[28] The control system must refer to management parameters and risk thresholds formalised *ex ante* at portfolio level and, in the case of delegated management, at manager level.

Finally, the consistency between the established investment policy and the investor's financial objectives must be verified, taking into account the constraints and assessing any adjustments that may be required, including those due to changes in external circumstances and financial market trends.

References

Arnott RD, Burns TE, Plaxco L, Moore P (2007) Monitoring and rebalancing. In: Maginn JL, Tuttle DL, Pinto JE, McLeavey DW (eds) Managing investment portfolios. A dynamic process. Wiley, Hoboken, NJ

Black F, Jones RW (1987) Simplifying portfolio insurance. J Portf Manag 14(1):48–51

Bodie Z (1995) On the risk of stocks in the long run. Financ Anal J 51(3):18–22

[26] Sharpe (1992).

[27] Here too, the long-only investment approach is implicit.

[28] Lee et al. (2010).

Brinson GP, Hood LR, Beebower GL (1986) Determinants of portfolio performance. Financ Anal J 42(4):133–138
Brinson GP, Singer BD, Beebower GL (1991) Determinants of portfolio performance II: an update. Financ Anal J 47(3):40–48
Fama EF, French KR (1988) Permanent and temporary components of stock prices. J Polit Econ 96(2):246–273
Fama EF, French KR (1992) The cross-section of expected stock returns. J Financ 47(2):427–465
Farrell JL (1997) Portfolio management. Theory and application, 2nd edn. McGraw Hill, New York
Flood E, Ramachandran N (2000) Integrating active and passive management. J Portf Manag 27(1): 10–19
Gastineau GL, Olma AR, Zielinski RG (2007) Equity portfolio management. In: Maginn L, Tuttle DL, McLeavey DW, Pinto JE (eds) Managing investment portfolios. A dynamic process, 3rd edn. Wiley, Hoboken, NJ
Gibson RC (2013) Asset allocation. Balancing financial risk, 5th edn. McGraw-Hill, New York
Grinold RC, Kahn RN (2000) Active portfolio management, 2nd edn. McGraw Hill, New York
Hull J (2023) Risk management and financial institutions, 6th edn. Wiley, Hoboken, NJ
Ibbotson RG, Kaplan PD (2000) Does asset allocation policy explain 40, 90 or 100 percent of performance? Financ Anal J 56(1):26–33
Kaplan PD (2012) Frontiers of modern asset allocation. Wiley, Hoboken, NJ
Kinlaw W, Kritzman M, Turkington D (2021) Asset allocation. From theory to practice and beyond. Wiley, Hoboken, NJ
Lee CF, Lee AC, Lee J (2010) Handbook of quantitative finance and risk management. Springer, Heidelberg
Lo AW, MacKinlay AC (1988) Stock market prices do not follow random walks: evidence from a simple specification test. Rev Financ Stud 1(1):41–66
Markowitz HM (1952) Portfolio selection. J Financ 7(1):77–91
Perold AF, Sharpe WF (1988) Dynamic strategies for asset allocation. Financ Anal J 44(1):16–27
Poterba J, Summers L (1988) Mean reversion in stock prices: evidence and implications. J Financ Econ 22(1):27–59
Reilly FK, Brown KC (2012) Investment analysis and portfolio management, 10th edn. South-Western Cengage Learning, Mason
Scherer B (2015) Portfolio construction and risk budgeting, 5th edn. Risk Books, London
Sharpe WF (1992) Asset allocation: management style and performance measurement. J Portf Manag 18(2):7–19
Sharpe WF, Chen P, Pinto JE, McLeavey DW (2007) Asset allocation. In: Maginn L, Tuttle DL, McLeavey DW, Pinto JE (eds) Managing investment portfolios. A dynamic process, 3rd edn. Wiley, Hoboken, NJ
Tschampion RC, Siegel LB, Takahashi DJ, Maginn JL (2007) Managing institutional investor portfolios. In: Maginn L, Tuttle DL, McLeavey DW, Pinto JE (eds) Managing investment portfolios. A dynamic process, 3rd edn. Wiley, Hoboken, NJ

Chapter 4
Strategic Asset Allocation with Mean-Variance Optimisation

Maria Debora Braga

4.1 Strategic Asset Allocation with Markowitz's Mean-Variance Optimisation

A key step towards a disciplined and quantitative approach to the "issue" of strategic asset allocation was taken by Harry Markowitz, in his article "Portfolio Selection" published on the *Journal of Finance* in 1952. The ideas expressed in that work led to the definition of a normative theory describing a behavioural standard to be adopted for portfolio construction and constitute the basis of what has come to be known as Modern Portfolio Theory, Mean-Variance Analysis, and Mean-Variance Optimisation.[1]

In addition to the merit of having proposed a framework for portfolio construction (which will be investigated below), Markowitz deserves recognition for two further achievements: the definition in technical and formal terms of the concept of diversification, and the notion of the investor as a decision-maker in a two-dimensional space.

With regard to the former, it should be noted that Markowitz quantifies the concept of diversification with reference to the statistical notion of covariance/correlation between asset classes, thereby defining portfolio variance not only in terms of the stand-alone risk of the asset classes (and their weights), but also in terms

[1] A more extensive and detailed discussion of Markowitz's contribution to the subject of portfolio construction can be found in the book he published some years later. See: Markowitz (1959).

M. D. Braga (✉)
Department of Economics and Political Science, University of "Valle d'Aosta", Aosta, Italy

SDA Bocconi School of Management, Bocconi University, Milan, Italy
e-mail: m.braga@univda.it; mariadebora.braga@sdabocconi.it

I. Basile, P. Ferrari (eds.), *Asset Management and Institutional Investors*, Contributions to Finance and Accounting,
https://doi.org/10.1007/978-3-031-59819-7_4

of how one asset class interacts with another. Therefore, he provides a more convincing and elegant justification for the shift in focus from the selection of a single asset class to the portfolio as a whole, than just the familiar adage "Don't put all your eggs in one basket". Further to this, Markowitz's statement that the concept of diversification cannot be traced back exclusively to numerical terms is particularly interesting: "Not only does the E-V [expected return-variance] hypothesis imply diversification, it implies the 'right kind' of diversification for the 'right reason'. The adequacy of diversification is not thought by investors to depend solely on the number of different securities held [. . .]".[2]

As for the latter, it should be remembered that, unlike most previous authors, Markowitz rejected an investor profile aimed exclusively at maximising expected returns, preferring an investor profile that considers expected return as "something desirable" and variance in returns (standard deviation) as "something undesirable".[3] The decision criterion known as the mean-variance principle derives from this notion of investor preference, according to which, for two portfolios, A and B, with expected return μ_A and μ_B and expected risk σ_A and σ_B, A can be said to dominate (be certainly preferable to) B if:

$$\mu_A > \mu_B \text{ and } \sigma_A \leq \sigma_B$$

with at least one verified strong inequality. Even if it can be argued that quantification of risk as dispersion from the mean and its assessment in terms of uncertainty are not the only possible interpretations, Markowitz's recognition of the trade-off between return and risk as a decision factor certainly remains worthy of consideration.

Given the above, before delving into Markowitz's approach to asset allocation, it is worth providing a mathematical formulation of the descriptive parameters of the characteristics of a portfolio of N risky asset classes (with $N \geq 3$). For the sake of completeness, we shall resort to both classical and matrix algebra.

The composition of a portfolio is given by the vector $\mathbf{w}$ of size $N \times 1$, where w_i is the percentage of asset class i in the portfolio:

---

[2] Markowitz (1952, p. 89).

[3] In his key study of 1952 Markowitz used variance as a measure of risk; in practice and in the subsequent literature, however, standard deviation was generally preferred. Since standard deviation is the square root of variance, it makes no difference which measure is used. Markowitz himself, in a later work, writes: ". . .although the article noted that the same portfolios that minimize standard deviation for given E [expected return] also minimize variance for given E [expected return], it failed to point out that standard deviation (rather than variance) is the intuitively meaningful measure of dispersion", Markowitz (1999, p. 6).

$$\mathbf{w} = \begin{bmatrix} w_1 \\ w_2 \\ \dots \\ \dots \\ w_i \\ \dots \\ \dots \\ w_N \end{bmatrix}$$

In addition:

$\sum_{i=1}^{N} w_i = 1$ or, equivalently, $\mathbf{w}'\mathbf{e} = 1$ with $\mathbf{e}' = [1, 1, \dots, 1]$.

The properties of expected return and risk of the single asset classes can be represented in the vectors $\boldsymbol{\mu}$ and $\boldsymbol{\sigma}$, respectively, both of dimension $N \times 1$, as shown below:

$$\boldsymbol{\mu} = \begin{bmatrix} \mu_1 \\ \mu_2 \\ \dots \\ \dots \\ \mu_i \\ \dots \\ \dots \\ \mu_N \end{bmatrix} \quad \boldsymbol{\sigma} = \begin{bmatrix} \sigma_1 \\ \sigma_2 \\ \dots \\ \dots \\ \sigma_i \\ \dots \\ \dots \\ \sigma_N \end{bmatrix}$$

In contrast, the correlation matrix, **C**, and the covariance matrix, $\boldsymbol{\Sigma}$, both of dimension $N \times N$, show information on the interaction, i.e. the relationship, between possible pairs of asset classes. This can be generically indicated with the symbol ρ_{ij} for correlation and the symbol σ_{ij} for covariance.

$$\mathbf{C} = \begin{bmatrix} \rho_{11} & \rho_{12} & \cdots & \cdots & \rho_{1i} & \cdots & \cdots & \rho_{1N} \\ \rho_{21} & \rho_{22} & \cdots & \cdots & \rho_{2i} & \cdots & \cdots & \rho_{2N} \\ \cdots & \cdots & \cdots & \cdots & \cdots & \cdots & \cdots & \cdots \\ \cdots & \cdots & \cdots & \cdots & \cdots & \cdots & \cdots & \cdots \\ \rho_{i1} & \rho_{i2} & \cdots & \cdots & \rho_{ii} & \cdots & \cdots & \rho_{iN} \\ \cdots & \cdots & \cdots & \cdots & \cdots & \cdots & \cdots & \cdots \\ \cdots & \cdots & \cdots & \cdots & \cdots & \cdots & \cdots & \cdots \\ \rho_{N1} & \rho_{N2} & \cdots & \cdots & \rho_{Ni} & \cdots & \cdots & \rho_{NN} \end{bmatrix}$$

$$\mathbf{\Sigma} = \begin{bmatrix} \sigma_{11} & \sigma_{12} & \cdots & \cdots & \sigma_{1i} & \cdots & \cdots & \sigma_{1N} \\ \sigma_{21} & \sigma_{22} & \cdots & \cdots & \sigma_{2i} & \cdots & \cdots & \sigma_{2N} \\ \cdots & \cdots & \cdots & \cdots & \cdots & \cdots & \cdots & \cdots \\ \cdots & \cdots & \cdots & \cdots & \cdots & \cdots & \cdots & \cdots \\ \sigma_{i1} & \sigma_{i2} & \cdots & \cdots & \sigma_{ii} & \cdots & \cdots & \sigma_{iN} \\ \cdots & \cdots & \cdots & \cdots & \cdots & \cdots & \cdots & \cdots \\ \cdots & \cdots & \cdots & \cdots & \cdots & \cdots & \cdots & \cdots \\ \sigma_{N1} & \sigma_{N2} & \cdots & \cdots & \sigma_{Ni} & \cdots & \cdots & \sigma_{NN} \end{bmatrix}$$

The terms ρ_{ii} and σ_{ii} on the main diagonal of the matrices **C** and **Σ** denote, respectively, correlation and covariance of an asset class with itself and therefore they correspond to the values 1 and σ_i^2. A generic term σ_{ij}, obviously not on the main diagonal of the matrix **Σ**, is instead equivalent to the expression $\rho_{ij}\sigma_i\sigma_j$. As can be observed, both matrices are perfectly symmetrical for the off-diagonal elements. Given this and considering that the elements on the main diagonal refer to a single asset class, the unique, non-redundant terms in the relationship between the two different asset classes in the matrices **C** and **Σ** can be calculated through the expression $[N \times (N - 1)]/2$.

Using the stated parameters and notations, the expected return of portfolio μ_P can be calculated by either (4.1a) or (4.1b):

$$\mu_P = \sum_{i=1}^{N} w_i \mu_i \tag{4.1a}$$

$$\mu_P = \mathbf{w}'\boldsymbol{\mu} \tag{4.1b}$$

Conversely, portfolio risk, here expressed as variance σ_P^2, can be obtained using either (4.2a) or (4.2b):

$$\sigma_P^2 = \sum_{i=1}^{N}\sum_{j=1}^{N} w_i w_j \sigma_{ij} \text{ or } \sum_{i=1}^{N}\sum_{j=1}^{N} w_i w_j \sigma_i \sigma_j \rho_{ij} \text{ or } \sum_{i=1}^{N} w_i^2 \sigma_i^2 + \sum_{i=1}^{N}\sum_{\substack{j=1 \\ j \neq i}}^{N} w_i w_j \sigma_{ij} \tag{4.2a}$$

$$\sigma_P^2 = \mathbf{w}'\boldsymbol{\Sigma}\mathbf{w} \tag{4.2b}$$

We can now move on to look at Markowitz's approach to the "issue" of strategic asset allocation. To this end, let us begin by considering a rational investor who decides at the beginning of a given period which portfolio he will hold until the end of that period (i.e. his/her holding period) with no possibility of changing portfolio composition in the face of gains or losses, and with no concern for what may happen "afterwards". This, then, is an example of myopic behaviour, in which an investor acts in a single period investment horizon with the aim of maximising expected utility.

According to Markowitz, for each level of expected return, such an investor would choose, in the feasible set or, simply, among the feasible portfolios, the portfolio with the minimum variance (therefore, the minimum standard deviation). The portfolios corresponding to this description can be called mean-variance efficient portfolios.

Within the risk-return space, the set of mean-variance efficient portfolios identifies Markowitz's efficient frontier. Essentially, it provides the best trade-off between expected return and risk for each level of expected return or for each level of risk. The portfolios below the efficient frontier can be called either inefficient or dominated portfolios; those above the frontier, instead, are not feasible. Finally, the portfolio on the efficient frontier with the smallest possible variance is often referred to as the global minimum-variance portfolio (GMVP).

To facilitate understanding of the concepts and terms used, Fig. 4.1 shows an example of three asset classes with intermediate positive correlation, specifying positioning in terms of risk and expected return. A significant number (1000) of feasible portfolios have been calculated, which gives an approximate idea of the feasible set region. The curve indicating the efficient frontier highlights those portfolios that expose an investor to the smallest possible standard deviation for each level of expected return or, alternatively, the portfolios that maximise expected return for each level of risk.

The previous pages have clarified the logic that Markowitz's investor would use when making an investment decision. In order to provide an equally clear picture from a practical point of view, we must underline that mean-variance efficient portfolios are not derived by first identifying feasible portfolios (which are indeed innumerable), but rather by the direct application of an algorithm specifically formulated by Markowitz to address the so-called Mean-Variance Optimisation problem.

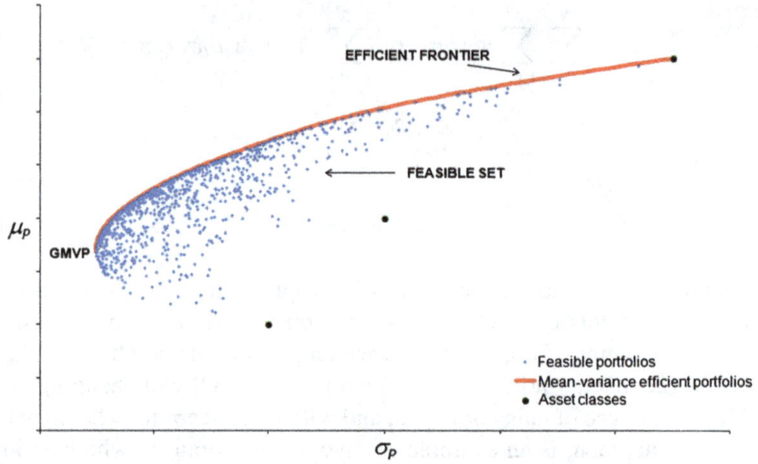

Fig. 4.1 Feasible set and Markowitz's efficient frontier

Generally speaking, an optimisation problem/algorithm has three basic components:

- An objective function, that is, a mathematical expression of what we want to optimise (i.e. minimise or maximise).
- A set of decision variables (by definition, these are initially unknown) on which depends the value of the mathematical expression to be optimised.
- A set of restrictions to be applied to some or all of this set of decision variables, which take the form of equality and/or inequality constraints.

These three components are clearly recognisable in the classic formulation of Mean-Variance Optimisation (MVO) that seeks to establish which weights to assign to asset classes in order to minimise portfolio risk (variance or standard deviation) for a targeted expected portfolio return (μ_P^*),while ensuring that each weight is not negative and that the sum of the weights is equal to 1 (i.e. 100%). Consequently, in algebraic terms, Mean-Variance Optimisation can be expressed alternatively as (4.3a):[4]

[4]For the sake of completeness, we note that in system (4.3a) and (4.3b) the objective function is still written in terms of the portfolio variance to be minimised. It goes without saying that considering the square root, we are thinking in terms of standard deviation. What deserves to be underlined even more is that, whatever the path chosen, the output of the Markowitz optimisation does not undergo alterations or changes. See also note 3 above.

$$
\begin{aligned}
&\min_{w} \sum_{i=1}^{N}\sum_{j=1}^{N} w_i w_j \sigma_{ij} \\
&\text{subject to} \\
&\sum_{i=1}^{N} w_i \mu_i = \mu_P^* \\
&\sum_{i=1}^{N} w_i = 1 \\
&w_i \geq 0
\end{aligned}
\tag{4.3a}
$$

or

$$
\begin{aligned}
&\min_{\mathbf{w}} \mathbf{w}'\boldsymbol{\Sigma}\mathbf{w} \\
&\text{subject to} \\
&\mathbf{w}'\boldsymbol{\mu} = \mu_P^* \\
&\mathbf{w}'\mathbf{e} = 1 \\
&[w] \geq 0
\end{aligned}
\tag{4.3b}
$$

The formulation of Mean-Variance Optimisation entails an equality constraint on portfolio weights, if their total must be 1 (the so-called budget constraint), and inequality constraints, if short selling of any asset class is not permitted (the so-called non-negativity or long-only constraint). The latter constraints preclude the analytical resolution of the Mean-Variance Optimisation problem; to that end, the use of numerical techniques of iterative nature is required.[5]

To classify correctly Mean-Variance Optimisation, it is useful to point out that it poses a Quadratic Programming (QP) problem, as it aims to minimise a quadratic function while respecting linear constraints (of equality and inequality, as stated above). This qualification remains unchanged if, instead of using the original risk minimisation formulation of the optimisation algorithm, two alternative formulations are adopted. One is the expected return maximisation formulation. As shown both in (4.4a) and (4.4b), this formulation considers the problem of optimisation in terms of an initial target level of risk ($\sigma_P^{2,*}$) and of the subsequent identification of weights for the asset classes which, respecting their classic constraints, can maximise expected portfolio return:

[5] In other words, the numerical optimisation techniques work by generating a sequence of approximate solutions which move progressively closer to each other, converging on the true solution. As the latter is not known, and the sequence cannot continue indefinitely, these numerical techniques employ a termination criterion or convergence criterion which, when satisfied, interrupts the iterations.

$$\begin{aligned}
&\max_{w} \sum_{i=1}^{N} w_i \mu_i \\
&\text{subject to} \\
&\sum_{i=1}^{N} \sum_{j=1}^{N} w_i w_j \sigma_{ij} = \sigma_P^{2,*} \\
&\sum_{i=1}^{N} w_i = 1 \\
&w_i \geq 0
\end{aligned} \tag{4.4a}$$

or

$$\begin{aligned}
&\max_{\mathbf{w}} \ \mathbf{w}'\boldsymbol{\mu} \\
&\text{subject to} \\
&\mathbf{w}'\boldsymbol{\Sigma}\mathbf{w} = \sigma_P^{2,*} \\
&\mathbf{w}'\mathbf{e} = 1 \\
&[w] \geq 0
\end{aligned} \tag{4.4b}$$

Another way to pose the problem of optimisation *à la* Markowitz is the risk aversion formulation. Equation (4.5) shows that, in this case, the objective function seeks to maximise utility while continuing to respect the weight constraints:

$$\begin{aligned}
&\max_{\mathbf{w}} \left(\mathbf{w}'\boldsymbol{\mu} - \frac{\lambda}{2} \mathbf{w}'\boldsymbol{\Sigma}\mathbf{w} \right) \\
&\text{subject to} \\
&\mathbf{w}'\mathbf{e} = 1 \\
&[w] \geq 0
\end{aligned} \tag{4.5}$$

Parameter λ represents the risk aversion coefficient (often called the Arrow-Pratt risk aversion index) and determines the magnitude of the penalisation applied to portfolio risk, which is inversely proportional to expected utility. If λ is small, portfolio risk is given a low weight, thereby identifying an aggressive optimised portfolio. If λ is heavy, on the other hand, the high weight given to the variance produces a more cautious optimised portfolio.

The three formulations for the Mean-Variance Optimisation problem are equivalent insofar as they lead to the same efficient frontier, with the common aim of identifying the best trade-off between expected portfolio return and variance. The calculation of mean-variance efficient portfolios in the three cases illustrated above originates from variations in the target level of expected return, in the target level of sustainable risk and in the risk aversion coefficient, respectively.[6]

4.2 The Assumptions of Mean-Variance Optimisation

Resorting to Markowitz's Mean-Variance Optimisation as a criterion and technique to build strategic (policy) portfolios means that those who use it must be aware of (and willing to accept) the underlying simplifications on which it is based. These assumptions concern the behaviour of the series of asset class returns and investors' preferences, or in more general terms, their decision-making process. Specifically, Mean-Variance Optimisation assumes that:

- The returns of the asset classes are normally distributed, i.e. that they conform to the "bell-curve" (Gaussian distribution). To be exact, we assume that they are jointly normal and they follow, therefore, a multivariate normal distribution.
- Investors present a quadratic utility function.
- Investors have a single-period investment horizon.

Each of these hypotheses should be considered in detail, along with an examination of the probable consequences should they not hold, since it cannot be denied that they may differ from reality.

As is known, the assumption of a normal distribution of asset class returns makes possible an exhaustive description of their behaviour by merely estimating the first two statistical moments, i.e. expected return and variance (or standard deviation) or, in the event of joint normality, the covariance matrix as an expression of the relevant co-moments.

---

[6]The calculation of the global minimum variance portfolio is the only one for which returns (μ_i) do not need to be taken into consideration, as in this case, the vector of the weights is totally dependent on the covariance matrix. If there is a budget constraint and a long-only constraint, the allocation is obtained by resolving the following optimisation problem:

$$\begin{aligned}&\min_{\mathbf{w}}\ \mathbf{w}'\boldsymbol{\Sigma}\mathbf{w}\\&\text{subject to}\\&\mathbf{w}'\mathbf{e}=1\\&[w]\geq 0\end{aligned}$$

For the sake of completeness (even if this is not particularly relevant to usual asset allocation activities), it should also be said that if the long-only constraint is removed, the problem could also be resolved directly as:

$$\mathbf{w}_{\mathrm{GMVP}}=\frac{\boldsymbol{\Sigma}^{-1}\mathbf{e}}{\mathbf{e}'\boldsymbol{\Sigma}^{-1}\mathbf{e}}$$

As opposed to non-normal distributions, the assumption of normal distribution allows the easy aggregation of risks across assets and over time.[7] Nevertheless, a number of scholars have provided empirical evidence suggesting that the assumption should be rejected.

A first form of deviation from the normal distribution, i.e. returns asymmetry, is provided in the works of Simkowitz and Beedles, Kon, Singleton and Wingender, Peirò, Prakash et al., and Jondeau and Rockinger.[8] This feature is recorded by measuring skewness, also known as the third central moment.[9] In these cases, the density function of the distribution of returns does not show an equal and mirror-like dispersion of returns to the left and right of the expected value (as in a Gaussian distribution which has no skewness), or the coincidence of mean and median return values; it shows instead a lengthening of the tail on the right (positive skewness or asymmetry) or its extension on the left (negative skewness or asymmetry).

A second form of departure from normal distribution lies in extreme returns which occur much more frequently than expected. An example of the confirmation of this can be found in the works of Fama, Jansen and Devries, and Longin.[10] In practice, the actual distribution of returns is fat-tailed, i.e. has a higher than expected number of values considered to be outliers (which makes the tails of the distribution fatter) and simultaneously reduces the frequency of the values existing between centre and tails. This characteristic of the distribution of asset returns can be observed through the kurtosis parameter, also known as the fourth central moment.[11] Given that in a Gaussian distribution the value of kurtosis is 3, in fat-tailed distributions that value is greater than 3 and distribution is termed leptokurtic.[12,13]

As stated above, the second stringent hypothesis attributable to Markowitz's model regards the "rule" that investors assign a value to their level of satisfaction

---

[7]Note that, in this case, aggregation of risks over time uses the so-called square-root-of-time rule.

[8]Simkowitz and Beedles (1980), Kon (1984), Singleton and Wingender (1986), Peiró (1999), Prakash et al. (2003), and Jondeau and Rockinger (2006).

[9]In its standardised version, the examined parameter is defined as follows: $S = E\left[\left(\frac{x-\mu}{\sigma}\right)^3\right]$. Its sample estimate can instead be calculated by the expression: $\widehat{S} = \frac{1}{T}\sum_{t=1}^{T}\left(\frac{x_t-\widehat{\mu}}{\widehat{\sigma}}\right)^3$.

[10]Fama (1963), Jansen and de Vries (1991), and Longin (1996).

[11]In its standardised version, kurtosis is defined as follows: $K = E\left[\left(\frac{x-\mu}{\sigma}\right)^4\right]$. Its sample estimate, on the other hand, can be calculated by the expression: $\widehat{K} = \frac{1}{T}\sum_{t=1}^{T}\left(\frac{x_t-\widehat{\mu}}{\widehat{\sigma}}\right)^4$.

[12]For the sake of completeness, despite limited empirical evidence, we should also remember the case where the frequency of asset returns considerably above or below the mean is lower than that indicated by the "bell curve" and that distribution is consequently "thin-tailed". In this case, such distribution would be called platykurtic and have a kurtosis value lower than 3.

[13]In empirical studies of the behaviour of asset class returns, as an alternative to kurtosis, the value for excess kurtosis can also be considered, i.e. kurtosis minus 3, which would be the normal value of kurtosis.

depending on the possible choices offered by different portfolios. This rule corresponds to a quadratic utility function in the following form (4.6):

$$U(W) = W - \frac{b}{2}W^2;\ b > 0 \tag{4.6}$$

where W indicates individual investors' final wealth.

The assumption of this particular utility function is intentional, as it allows individual investors' expected utility to depend exclusively on mean and variance. This can be demonstrated by considering an investor seeking to maximise the expected utility of his/her final wealth as a result of a portfolio choice and of the subsequent re-formulation of the utility function by the Taylor series expansion. In formal terms, indicating the vectors of single asset class returns with $\mathbf{R}$, the generic utility function linked to final wealth can be expressed as follows (4.7):[14]

$$U(W) = U(1 + \mathbf{w}'\mathbf{R}) = U(1 + R_P) \tag{4.7}$$

while the expected value of final wealth is given by (4.8):

$$\overline{W} = 1 + \mathbf{w}'\boldsymbol{\mu} = (1 + \mu_P) \tag{4.8}$$

According to the Taylor series expansion, the expected value of the utility function can be re-written using a number, k, approaching infinity, of successive derivatives of the utility function itself, and all the central moments of the distribution of variable W, as described by expression (4.9):

$$\mathrm{E}[\mathrm{U}(W)] = \sum_{k=0}^{\infty} \frac{\mathrm{U}^k(\overline{W})}{k!} \mathrm{E}\left[(W - \overline{W})^k\right] \tag{4.9}$$

And, more extensively, by the expression (4.10):[15]

$$\begin{aligned} E[U(W)] = {} & U(\overline{W}) + U^1(\overline{W})E\left[(W - \overline{W})^1\right] + \frac{1}{2!}U^2(\overline{W})E\left[(W - \overline{W})^2\right] \\ & + \frac{1}{3!}U^3(\overline{W})E\left[(W - \overline{W})^3\right] + \frac{1}{4!}U^4(\overline{W})E\left[(W - \overline{W})^4\right] + \cdots \end{aligned} \tag{4.10}$$

At this point, it should be stressed that if the specific utility function is quadratic, i.e. coincides with (4.6), then only the first and second order derivatives ($1 - bW$ and

[14] Implicitly, an initial wealth equal to 1 was arbitrarily chosen.

[15] It is useful to recall that $E\left[(W - \overline{W})^2\right]$ is the variance and that the expected value of differences from the mean, i.e. $E\left[(W - \overline{W})\right]$, is zero.

$-b$, respectively) are not zero, while all higher order derivatives are equal to zero. This means that expected utility depends exclusively on mean and variance (4.11):

$$E[U(W)] = U(\overline{W}) + \frac{1}{2!}U^2(\overline{W})\sigma_P^2 \tag{4.11}$$

In short, then, restricting the form of utility makes it legitimate to address the problem of asset allocation by looking exclusively at the first two moments of the distribution of portfolio expected returns and/or final wealth.

In the scientific literature, attention has been drawn to problems related to the assumption of a quadratic utility function. Sarnat claimed that such a function did not reflect the "image" of a real investor and criticised it for:

- Violating the principle of non-satiety.
- Implying increasing absolute risk aversion.[16]

Regarding the first point, it is generally accepted that greater utility is always preferred to lower utility, so the first derivative of the utility function must be positive. In the case of a quadratic utility function, since the first derivative is $1 - bW$, this is only true for values $W < 1/b$, while for larger values, exactly the opposite preference would be found. With respect to the second point, considering that absolute risk aversion expresses how the absolute sum invested in risky assets changes as wealth changes, the non-desirable aspect is that increasing absolute risk aversion results in reduced risk-taking and, therefore, fewer risky investments being made as wealth increases—precisely when a typical investor should show decreasing absolute risk aversion, and thus willingness to increase the sums at risk.[17,18]

Apart from the problems relative to the implications of the quadratic utility function, further attacks on the mean-variance framework have been launched by various academic contributions which not only reject the idea that the higher order moments are irrelevant, but also identify a precise direction of preferences for the higher moments. In particular, Horwath and Scott claim that investors typically show a preference for positive movements in skewness and negative movements in kurtosis.[19] Kraus and Litzenberger, and Harvey and Siddique show that investors may be prepared to accept a lower expected return and possibly even greater volatility than that suggested by a mean-variance efficient solution in return for greater skewness and lower kurtosis.[20]

---

[16] Sarnat (1974).

[17] Absolute risk aversion is measured by the ratio $-U^2(W)/-U^1(W)$. With decreasing absolute risk aversion, the first derivative of this ratio is lower than zero, while increasing absolute risk aversion (a condition considered to be contradictory and problematic) is confirmed by a first derivative greater than zero. This latter situation leads, in turn, to increasing relative risk aversion, i.e. a situation in which the proportion of wealth invested in risky assets falls as its absolute value rises.

[18] See Pratt (1964) for an analysis of risk-averse measures.

[19] Scott and Horwath (1980).

[20] Kraus and Litzenberger (1976) and Harvey and Siddique (2000).

All in all, on the basis of the above considerations, it is clear that restrictions on the form of the utility function or of the distribution of asset class returns in order to make Mean-Variance Optimisation the most appropriate approach to portfolio construction may seem (or actually be) scarcely realistic or justifiable. Therefore, we need to ask ourselves whether it is plausible for strategic asset allocation to "clear the table" of these assumptions and adopt a different, equally pragmatic and easy-to-use approach, and the answer is no. Optimal asset allocation under higher moments as an alternative or concomitant approach produces a series of difficulties:

- In addition to the traditional parameters of mean and variance/standard deviation, it requires the estimation of the skewness and kurtosis of the asset classes for the univariate parameters; in the multivariate cases, other forms of interaction have to be estimated besides the linear pair relations in the covariance/correlation matrix to be expressed through co-skewness and co-kurtosis matrices. The need to estimate these cross-moments leads to a dramatic increase in the problem dimensionality.[21]
- It creates the necessity to manage the trade-offs between various conflicting and competing objectives, given that an investor wishes to maximise portfolio expected return and skewness (the so-called odd moments) and minimise portfolio variance and kurtosis (the so-called even moments), and it makes it impossible to preserve the individual objectives.
- It exposes investors to the difficulty of having to specify an exact utility function (of which they may be hardly aware) and of having to decide the order of the Taylor series expansion which allows the approximation of the function for subsequent numerical implementation. Alternatively, it assumes that investors can express the weight/importance of their different objectives.
- It can take advantage of alternative asset class return distribution models, which, although they may provide a better fitting of the real behaviour of the data, at the same time make aggregation of the individual characteristics in portfolio metrics more difficult.

Understandable reluctance to confront these "obstacles" means that Mean-Variance Optimisation continues to be a fundamental point of reference for most of the scientific literature on asset allocation and in the community of practitioners. Moreover, one should not overlook the evidence in support of mean-variance as a satisfactory way of approximating the expected utility derived from direct optimisation (maximisation) of utility functions other than a quadratic expression, even for

[21] In support of this claim, note that if, for example, in an optimisation exercise with ten asset classes, the required number of unique (non-redundant) correlation/covariance parameters is 45, the required number of unique co-skewness and co-kurtosis parameters will be 220 and 715, respectively. These figures are obtained from the calculation of $N(N + 1)(N + 2)/3!$ and of $N(N + 1)(N + 2)(N + 3)/4!$, respectively. For a more detailed examination of the estimation problems in strategic asset allocation when using moments higher than the second, see Jondeau and Rockinger (2006) and Martellini and Ziemann (2010).

asset class returns not distributed as a normal.[22] Obviously, these comforting findings should not detract from the wisdom of determining (prior to the implementation of an optimisation algorithm) the statistical significance of any non-normality in the asset class returns, as well as its persistence.

As stated at the beginning of the present section, the last hypothesis adopted by Mean-Variance Optimisation concerns the myopic, single-period nature of the model. The importance of this element can be illustrated in Mossin's words: "By a single-period model is meant a theory of the following structure: the investor makes his portfolio decision at the beginning of a period and then waits until the end of the period when the rate of return on his portfolio materializes. He cannot make any intermediate changes in the composition of his portfolio. The investor makes his decision with the objective of maximizing expected utility of wealth at the end of the period (final wealth)".[23] In an ideal environment—characterised by stationary asset class return distributions, by the absence of serial auto-correlation and transaction costs, by constant relative risk aversion and no intermediate flows on the part of investors (so that portfolio return can be considered path-independent)—the fact that the Markowitz framework does not recognise the sequential nature of the problem of asset allocation, or the significance for the investor of "over time" dynamics in portfolio return, or the issue of rebalancing would not represent a weakness. This is because in such a context, roll-over in mean-variance efficient single-period portfolios would lead to efficient multiperiod portfolios. However, it is clear that these conditions are not systematically true; consequently, in principle, a multiperiod asset allocation model would be welcome. Once again, Mossin describes its principal characteristics: "By a multiperiod model is meant a theory of the following structure: the investor has determined a certain future point in time (his horizon) at which he plans to consume whatever wealth he has then available. He will still make his investment decisions with the objective of maximizing expected utility of wealth at that time. However, it is now assumed that the time between the present and his horizon can be subdivided into n periods (not necessarily of the same length), at the end of each return on the portfolio held during the period materializes and he can make a new decision on the composition of the portfolio to be held during the next period".[24] Regarding these decisions, Mossin adds: ". . ., any sequence of portfolio decisions must be contingent upon the outcomes of previous periods and at the same time take into account information on future probability distributions".[25]

Finally, to clarify why, despite having justified this approach, no further space will be devoted to the multiperiod framework in the present chapter, one should consider that the various attempts to develop such an approach have faced bigger problems of practical implementation than those arising in a single-period model, especially for portfolios that can include a realistic (not too limited) number of asset

---

[22] Kroll et al. (1984), Levy and Markowitz (1979), Chamberlain (1993), and Simaan (1993).

[23] Mossin (1968, p. 216).

[24] Mossin (1968, p. 220).

[25] Mossin (1968, p. 221).

classes, and have used even more stringent, albeit different, hypotheses than those applied in a single-period framework.[26]

4.3 Estimation Risk and Practical Problems with Mean-Variance Optimisation

Section 4.2 has shown that concerns regarding the assumptions underlying the Markowitz framework are not sufficient to discourage its use in strategic asset allocation. However, while theory may not represent an impediment to its application, problems may arise with the operational practice: as Michaud has pointed out, "Markowitz MV-optimisation technology is not easy to use properly".[27] We should, therefore, examine a classic implementation of MVO and the possible dangers involved.

Portfolio construction following Markowitz occurs in two phases:

- Estimation of the inputs.
- "Activation" of the optimisation.

In the original application of Mean-Variance Analysis, the first step involves the calculation of the necessary parameters from time series of T returns of N asset classes over a given past period. Thus, following one of the hypotheses outlined in the previous section, according to which asset class returns are random variables R_i (with $i = 1, \ldots, N$) with a multivariate normal distribution, as shown in (4.12):[28]

$$f(R) = f(R_1, R_2, \ldots, R_N) = \frac{1}{\sqrt{|\boldsymbol{\Sigma}|(2\pi)^N}} \exp\left(-\frac{1}{2}(\mathbf{R}-\boldsymbol{\mu})'\boldsymbol{\Sigma}^{-1}(\mathbf{R}-\boldsymbol{\mu})\right) \tag{4.12}$$

The statistics required from the historical returns for a generic asset class i or a pair of asset classes ij are given by (4.13) and (4.14):

$$\widehat{\mu}_i = \frac{1}{T}\sum_{t=1}^{T} R_t \tag{4.13}$$

[26]Even though, as we have said, the present study does not aim to offer a review of the attempts made to develop multiperiod asset allocation models, it is useful to point out that the latter have exploited stochastic programming, dynamic programming, and robust optimisation techniques. A brief discussion of the methodology can be found in Chap. 10 of Fabozzi et al. (2007).

[27]Michaud (1989, p. 40).

[28]As an argument in the square root in (4.12), $|\boldsymbol{\Sigma}|$ indicates the determinant of the covariance matrix.

Table 4.1 The multivariate probability distribution

A multivariate probability distribution, **X**, of dimension N can be seen as a representation of the probability of events measured jointly with reference to N random variables, X_i (with $i = 1, \ldots, N$). In a Gaussian framework, generalisation from a univariate to a multivariate case results in a multivariate normal distribution which depends on two parameters: a vector of the means and a covariance matrix. This explains the commonly used notation to indicate such a distribution, i.e. $\mathbf{X} \sim N(\boldsymbol{\mu}; \boldsymbol{\Sigma})$. To obtain a three-dimensional representation of the probability density function (pdf) of a multivariate normal distribution, we take as an example the case of a bivariate (standard) normal distribution for which $N = 2$. This gives $\mathbf{X} \sim \begin{pmatrix} 0 & 1 & 0 \\ 0 & 0 & 1 \end{pmatrix}$; the vector of the means contains zeros, and the covariance matrix coincides with an identity matrix. The pdf is shown below:

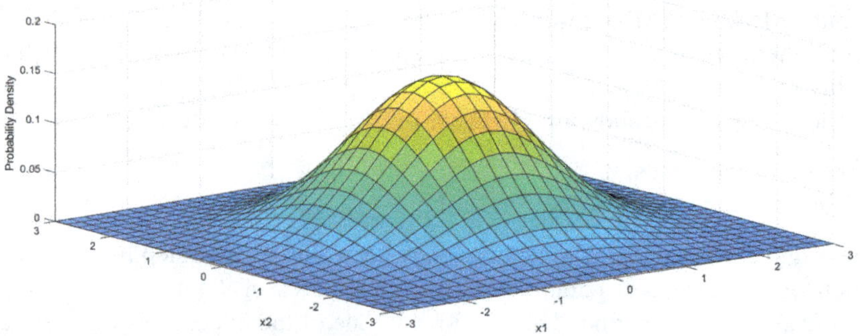

$$\widehat{\Sigma}_{ij} = \frac{1}{T-1} \sum_{t=1}^{T} (R_{t,i} - \widehat{\mu}_i) \cdot (R_{t,j} - \widehat{\mu}_j) \tag{4.14}$$

which describe, respectively, the elements of the vector of the (historical) sample means, and the (historical) sample variance-covariance matrix. For an immediate visual representation of the multivariate (normal) nature of a phenomenon, see Table 4.1. The second step involves resolving the quadratic programming problem described, alternatively, by (4.3a), (4.3b), (4.4a), (4.4b), or (4.5), in order to obtain the optimal portfolio weights. These estimated parameters shall be inputs in the quadratic programming problem.

As manifest, this approach ultimately equates the sample estimates with the actual parameters of the optimisation model, and thus assumes the availability of inputs able to forecast exactly and reliably the returns, risks, and correlations/covariances of a future investment period. This inevitably means that traditional Mean-Variance Optimisation is seen as a deterministic approach to strategic asset allocation which obliges potential users to follow the utopian and imprudent behaviour of those who do not even consider (or are completely unaware of) the uncertainty in the inputs. This is obviously not the real situation. The parameters derived from the data simply describe probability distributions of R_i that are not perfectly known. Consequently, these parameters are subject to uncertainty.

Given these considerations, we can state that the standard implementation of Mean-Variance Optimisation overlooks an additional form of risk determined by the need to estimate parameters. This is known as estimation risk.[29] In order to determine whether this omission constitutes a serious problem, and a contraindication to the use of the Markowitz approach, we must first define the risk itself and then explore the impact that any failure to recognise (and above all, manage) this risk may have on mean-variance efficient portfolios.

First, estimation risk can be defined as the probability of committing an estimation error ($\varepsilon_{\text{error}}$). It is the difference between the sample estimate for a generic parameter ($\widehat{\theta}$) and the true population value of that parameter (θ):

$$\varepsilon_{\text{error}} = \widehat{\theta} - \theta \tag{4.15}$$

In simple terms, estimation risk defines the far from remote possibility that the input estimates might be inaccurate. Resolving the problem of asset allocation with inputs that are rashly or erroneously considered to be accurate can have various, inter-related effects. To start with, as Michaud has observed, Markowitz portfolios are not intuitive: "MV optimizers have serious financial deficiencies, which will often lead to financially meaningless 'optimal portfolios'."[30]

Clues to, or rather, actual proof of the modest investment value and unintuitive character of mean-variance efficient portfolios can be seen in the limited diversification of their compositions, which include only a few of the available asset classes, and in the sharp and extreme shifts of the allocations along the efficient frontier. Indeed, it is precisely in the extreme, concentrated nature of the optimal portfolios that Michaud sees the explanation for what he calls the "Markowitz Optimisation Enigma", i.e. the initial and apparently incomprehensible situation that has long seen practitioners of the investment community avoid the Markowitz approach, despite the fact that it provides an easy-to-use algorithm to calculate efficient portfolios and a conceptually accessible framework.[31] Considering the inputs as exact parameters,

---

[29] Various authors have "exposed" the existence and criticality of estimation risk. These include: Garlappi et al. (2007), Frankfurter et al. (1971), Frost and Savarino (1986a, b), and Herold and Maurer (2003, 2006).

[30] Michaud (1989, p. 33).

[31] In his well-known study "The Markowitz Optimisation Enigma: is "Optimized" Optimal?", Michaud (1989) wrote (p. 31): "Given the success of the efficient frontier as a conceptual framework, and the availability for nearly 30 years of a procedure for computing efficient portfolios, it remains one of the outstanding puzzles of modern finance that MV optimisation has yet to meet with widespread acceptance by the investment community, particularly as a practical tool for active equity investment management. Does this Markowitz optimisation enigma reflect a considerable

that is, rejecting estimation risk, lies at the origin of the counter-intuitive composition of efficient portfolios, because it means that the optimiser tends to become very discriminatory, preferring attractive asset classes (with high returns, and low risk and correlation) and rejecting the less appealing alternatives, rather than seeking to exploit them as complementary.

However, it is precisely those asset classes which present particularly attractive features in the light of the sample estimates which are most liable to significant estimation errors. As a result, the fact that these asset classes are overweighted in an *MVO* maximises the impact of estimation errors on optimal portfolio weights. Let us again quote Michaud, who succinctly summarises these considerations as follows: "The unintuitive character of many optimized portfolios can be traced to the fact that MV optimizers are, in a fundamental sense, estimation error maximizers".[32]

To support this point, we may add a possibly banal, yet extremely relevant observation for those involved in strategic asset allocation operationally: portfolios which fail to take account of the problem of estimation risk will turn out to be the most vulnerable to estimation errors (which will emerge), since their counter-intuitive and extremely concentrated composition leaves them without any possible form of defence![33]

When considering the consequences of neglecting estimation risk, we must also look at the problem of the instability of optimal portfolios. This expression refers to the extreme sensitivity of portfolio allocations to even small variations in inputs (particularly in the estimates of expected returns). In other words, it defines the distortion of portfolio weights due to the fact that Mean-Variance Optimisation responds excessively to changes in the set of parameters that serve as inputs. Although this instability is a weakness attributable, at least in logical terms, to all Markowitz portfolios, it becomes more tangibly perceptible if specific conditions subsist, more specifically, if the investment universe includes at least pairs of asset classes with very similar risk-return combinations $(\widehat{\sigma}, \widehat{\mu})$. It is easy to explain why this is the case. An algorithm which uses point estimates as inputs and treats these as known values reacts disproportionately to small shifts in these estimates if these are sufficient to modify the dominant and dominated asset classes, because such a variation is seen as substantial and indubitable. The reaction of the optimiser is so cold, detached, and objective that, unlike any other human being, it does not realise that the discrepancy may simply be due to a measurement error (caused, for example, by the limited number of additional observations in the historical sample of returns or the roll-over of the historical sample). Obviously, optimal weights instability worries asset managers and could cause them to abandon a quantitative approach to asset allocation. In effect, they could understandably be uncertain about proposing

judgment (by the investment community) that such methods are not worthwhile, or is it merely another case of deep-seated resistance to change?".

[32] Michaud (1989, p. 33).

[33] Implicitly, the claim implies the belief that asset diversification should be considered positively, not only as a way to manage financial risk, but also as a form of estimation risk management.

optimised strategic portfolios if they feel that the investment proposals may be "overturned" by changes considered to be insignificant in some parameters. As a result, managers may end up rejecting the entire process. For an example of instability in Markowitz portfolios, see Table 4.2.

A third implication connected to the failure to recognise possible estimation errors in the inputs is the popularity of the misleading idea of optimal portfolio uniqueness. Here, again, the following quotation from Michaud offers valuable insight: "Optimizers, in general, produce a unique 'optimal' portfolio for a given level of risk. This appearance of exactness is highly misleading, however. The uniqueness of the solution depends on the erroneous assumption that the inputs are without statistical estimation error".[34]

In fact, the notion of optimal portfolio is an ambiguous, poorly defined concept if we admit the existence of estimation risk. This new concept can be illustrated visually by showing a region of points around any point on the classic efficient frontier to indicate portfolios which are statistically equivalent to mean-variance efficient portfolios.[35] A diagram of this type effectively reminds us that the sample estimates that feed the optimiser are just one specific realisation of random variables among the many possible alternatives.

The last "charge" brought against Markowitz portfolios is the poor out-of-sample performance that has been noted by various authors.[36] This characteristic expresses the drastic and worrying worsening in performance (even in risk-adjusted terms) of mean-variance efficient portfolios when measured against a holding period outside (or more precisely, subsequent to) the time period used to estimate mean returns, risks, and correlations/covariances. The level of approximation with respect to the "true inputs" shown in the sample parameters does not allow any *ex post* confirmation of the efficiency of the asset mix that was considered optimal a priori. The occurrence of estimation errors therefore makes Markowitz portfolios false optimal portfolios that even a naïve diversification rule (e.g. the "1 over N rule") could beat.

It should be noted that there is generally an inverse relation between the degree of portfolio diversification and the level of exposure to the danger of poor out-of-sample performance. In other words, the possibly ill-fated consequences of inaccurate input estimates are probably worse when extremely large portions of the asset allocation are absorbed by just one, or a very small number of, asset class (es). To better understand the fact that Markowitz optimal portfolios are not necessarily good portfolios in practice, consider the contribution of Ceria and Stubbs which, on a risk-return space, shows the actual frontier (i.e. the true risk-return combinations of portfolios on the estimated efficient frontier whose asset allocation is derived from prior optimisation of estimated expected returns) well below the true

---

[34] Michaud (1989, p. 35).

[35] For a full discussion of statistically indifferent or equivalent portfolios, see Sect. 4.7.1.

[36] Michaud (1989), Jorion (1985, 1986, 1991, 1992), DeMiguel et al. (2009), and Kan and Zhou (2007).

Table 4.2 Instability of Markowitz portfolios

The example represented below is a didactic illustration of the problem of instability in portfolios obtained by Mean-Variance Optimisation. As is well-known, instability is induced by marked optimisation sensitivity to marginal changes in inputs. In the example, the investment universe comprises four asset classes with the following original estimated values for the significant parameters (mean returns, standard deviations, correlations):

	Mean return (%)	**Standard deviation (%)**		**Asset class 1**	**Asset class 2**	**Asset class 3**	**Asset class 4**
Asset class 1	3.20	5.00	Asset class 1	1	0.85	0.50	0.45
Asset class 2	3.12	5.00	Asset class 2	0.85	1	0.50	0.45
Asset class 3	8.00	22.00	Asset class 3	0.50	0.50	1	0.90
Asset class 4	8.20	22.00	Asset class 4	0.45	0.45	0.90	1

For the same investment universe, the table below shows two variations in mean returns that an asset manager would probably consider either not very significant or crucial.

	Mean return (%)	**Standard deviation (%)**		**Asset class 1**	**Asset class 2**	**Asset class 3**	**Asset class 4**
Asset class 1	3.20	5.00	Asset class 1	1	0.85	0.50	0.45
Asset class 2	**3.22**	5.00	Asset class 2	0.85	1	0.50	0.45
Asset class 3	**8.40**	22.00	Asset class 3	0.50	0.50	1	0.90
Asset class 4	8.20	22.00	Asset class 4	0.45	0.45	0.90	1

The parameters from the tables are then used to "launch" two optimisations following the Markowitz model. The charts below show the composition of efficient portfolios obtained with the first and second set of inputs, respectively.

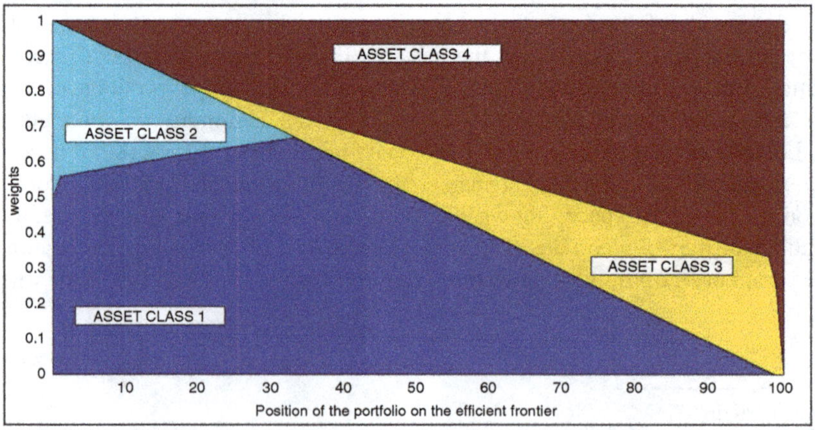

(continued)

Table 4.2 (continued)

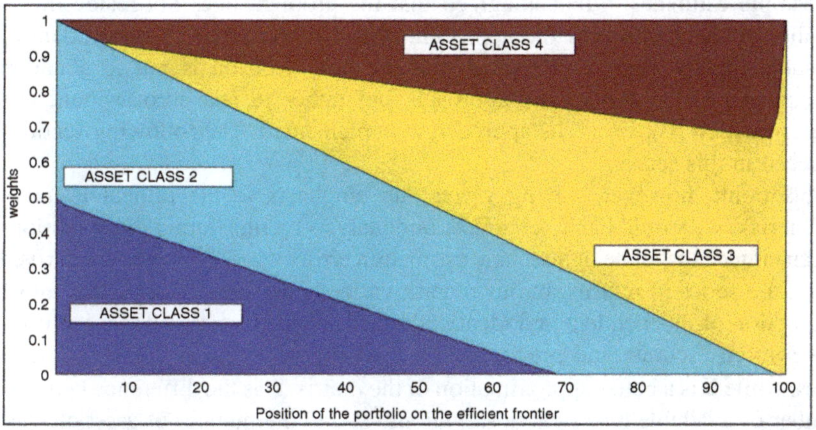

As can be easily seen, portfolio weights are substantially modified by variations in the inputs, even if considered almost irrelevant by a decision-maker. In the chart on the left, the allocations of optimal portfolios strongly favour asset classes 1 and 4; the weight of asset class 2 gradually declines in the optimal portfolios and the class is not present beyond the 35th position; asset class 3, especially in the central portfolios, is very limited. In contrast, the chart on the right shows a clearly predominant preference for asset classes 2 and 3, while asset classes 1 and 4 play only a secondary role. The reason for such an evidently different composition of the efficient frontier in the two cases is easily identifiable: for the Markowitz optimisation algorithm, the very slight changes in the inputs are perceived as substantial and, by creating an alternation between dominant and dominated asset classes, they end up distorting the weight vectors that describe the portfolio structures.

frontier (i.e. the risk-return combinations of portfolios calculated with true expected returns which were obviously unobservable initially).[37]

There is no doubt that the non-intuitiveness, instability, ambiguity, and poor out-of-sample performance of portfolios generated by the traditional implementation of Markowitz's Mean-Variance Optimisation can quite understandably discourage practitioners/experts from the practical implementation of the model. To avoid misunderstandings or rash decisions, the origin of these shortcomings must be explained. In effect, it is not a problem of any intrinsic, structural, or scientific limitation in the Markowitz approach, but rather a matter regarding the way the approach uses the information provided. As mentioned above, the model completely ignores the risk associated with optimisation inputs, as well as the possible effects of estimation errors. We might even go as far as to say that Markowitz's portfolio optimisation algorithm is far too powerful, given the degree of validity and reliability

[37] Ceria and Stubbs (2006).

of information on the prospective financial properties of the asset classes expressed by the sample estimates.[38] If it is agreed that the problem does not reside in the irrationality or incoherence of the approach, but in its potential to overfit data of questionable quality, then it is logical that the best decision is not to abandon operational use of the Markowitz approach, but rather to take into account and manage estimation risk when the approach is implemented. The following sections will proceed in this sense.

At this point, however, having given due emphasis to the critical issue of estimation risk, we would like to complete our analysis by illustrating the conditions which can either exasperate or attenuate estimation errors when estimates are derived from the time series of returns. In this regard, various contributors agree that under the assumption of independent and identically distributed (i.i.d.) asset class returns, $\widehat{\boldsymbol{\mu}}$ gives a more accurate and precise estimate of the vector $\boldsymbol{\mu}$ as the sample size increases, while $\widehat{\boldsymbol{\Sigma}}$ is a better approximation of the matrix $\boldsymbol{\Sigma}$ as the difference between the number of available historical observations, T, and the number of asset classes, N, becomes greater. Briefly and simply, the impact of estimation risk (and the number of possible estimation errors) tends to grow as the number of asset classes involved in the process of strategic asset allocation increases, but then it declines as the extension of the time series from which the sample estimates are taken becomes larger. This claim obviously identifies sampling error as the cause of estimation risk. Consequently, the problem may appear to be easily resolvable by increasing the number of observations, T, e.g. by using longer time series.[39] If, on the other hand, we question the i.i.d. assumption, it is clear that non-stationarity of time series of returns is another source of estimation risk that could be intensified by lengthening the time series used.

Previously, we correctly referred to estimation risk as a problem concerning the entire set of inputs needed for Mean-Variance Optimisation. Nevertheless, it is completely reasonable to consider the possibility of attributing a different relative impact to the estimation errors regarding return, volatility, and correlation/covariance parameters. As regards this issue, the literature maintains that the sacrifice, in terms of unattainable expected out-of-sample performance, resulting from estimation errors in expected returns is much worse than that caused by errors in the covariance matrix. Chopra and Ziemba, for example, estimate that the impact of errors in means is ten times greater than that of errors in variances (standard

[38] To use a common expression in the financial community, it is a matter of "garbage in, garbage out".

[39] On this point, Jobson and Korkie (1981, p. 72) write: "The number of historical observations of monthly returns required to give reasonably unbiased estimates of the optimal risk and return is at least two hundred" and (p. 72) "The traditional Markowitz procedure for predicting the optimal risk-return is extremely poor with conventional sample sizes of four to seven years". Note that the frequent reference in the literature on asset allocation to monthly rather than daily or weekly sample observations is not accidental. The reason is that in the first case there is greater probability of finding a stationary (as well as normal) series than in the other two cases. On the topic, see also Jobson and Korkie (1980).

deviations), while the latter is twice as relevant as the impact of errors in correlations.[40] In the light of these issues, it is no surprise that, specifically concerning estimates of expected returns, considerable efforts have been made by the scientific community to ensure that the optimisation algorithm is not "fed" exclusively with a mean extrapolated from a sequence of historical returns.

4.4 The Application of the Markowitz Approach: An Example

The previous pages have provided a detailed description of the structure, assumptions, and critical elements of the Markowitz approach. To give the reader some practical "hands-on experience", the present section presents an example of the application of this well-known model.

In effect, the process of portfolio construction starts with the asset manager identifying asset classes, i.e. the markets that he/she feels can be included in final strategic (policy) portfolios. This first step defines the borders of the investment universe. In doing so, the asset manager should take account of investor/end user preferences and/or encourage their involvement in decisions. In reply to any surprise that this claim might generate, we need to only consider that the pension fund managers (e.g. the management board) should share the selection of investment categories used to run the assets under management of each division of the fund, or that financial advisors (specifically of high-standing clients) should prevent being excluded from defining the markets that could be included in their clients' financial portfolios.

It is, therefore, easy to imagine that this first step does not follow any strict rule that could be listed in a kind of *vademecum*, and that it includes elements of subjectivity. Nevertheless, some basic indications can be suggested which, it is hoped, would not be completely ignored. Specifically, it is important to:

- Highlight the supposed main driver of diversification (country effect versus industry effect), with the identification of the strategic asset mixes in geographic or sector-based terms.[41]
- Avoid or limit any partial overlap between the selected asset classes.

---

[40] Chopra and Ziemba (1993).

[41] Choosing a geographical criterion as a diversification driver implicitly sees factors such as monetary and fiscal policy, exchange-rate dynamics, the state of public finances, the evolution of GDP, and the institutional and political framework as features able to generate at one and the same time similarities between the securities belonging to a given geographic area/country and differences compared to those in other areas/countries. In contrast, distinction by sector sees a different set of factors, such as availability of certain raw materials, presence of technological innovation, demand for certain goods/products, as features able to create a certain level of similarity among the securities used in the same sector and, at the same time, heterogeneity with respect to those found in other sectors.

- Choose a limited number of asset classes, to avoid magnifying the "size" of the problem of calculating the optimal portfolios, and, in some situations, to avoid recommending investments in marginal markets that would be difficult to implement.
- Preserve investors' access to international diversification, avoiding exclusive concentration on domestic asset classes which would give any strategic portfolio a heavy, pre-determined home bias.
- Make a conscious decision, once international diversification has been agreed, on how to treat the currency exposure implicit in foreign investments. Put simply, the alternative is hedging versus non-hedging, i.e. either considering foreign investments protected from exchange rate risk or exposed to exchange rate risk.[42]

Assuming that the asset manager has followed these general principles in identifying the asset classes, the next task is to associate a benchmark or reference index with each class. Asset class/benchmark association is an essential requirement for a number of reasons:

- To eliminate any ambiguity in identifying and defining an asset class.
- To provide elements which are shared, objective, and accessible (the time series of benchmark returns) on which to base the estimates of the parameters required for the optimisation procedure.
- To support the implementation phase of strategic (policy) portfolios, i.e. the phase in which, among other things, asset managers and financial advisors select investment vehicles and mandates to realise the suggested investments in the asset classes, since this indicates from the outset the objective that portfolio management will pursue or against which it will be measured, depending on whether a passive or an active strategy is preferred.

In the practical application proposed here, it is assumed that the strategic asset allocation process is undertaken for an investor residing in the Eurozone. To this end, the set of financial markets that the asset manager is willing to combine includes nine asset classes covering cash, bond, and equity investments. As explained above, a benchmark is associated with each of these asset classes. In all cases, preference has been given to gross total return and weighted value/capitalisation indices. Table 4.3 shows the association between asset class and benchmark.

The strategic asset allocation procedure continues with the estimates of Mean-Variance Optimisation inputs (expected returns, expected risks, expected covariance/correlations). In accordance with the classic procedure discussed in the previous section, these parameters are "replaced" by values extrapolated from a sample of historical data. Specifically, the monthly returns of each asset class in the 20-year

[42] From a logical and practical point of view, the alternative between hedging and non-hedging means, in simple terms, accepting an a priori forward premium or discount, on the one hand, and an unknown profit or loss as a result of the exchange rate risk, on the other. On this topic, see Chap. 18 and Schmittmann (2010).

Table 4.3 Asset classes and selected benchmarks

Asset classes	Selected benchmarks
Euro Cash	BARCLAYS 3 MONTH EURIBOR CASH INDEX
Eurozone Government Bonds	BLOOMBERG EURO AGGREGATE TREASURY BOND
US Government Bonds	BLOOMBERG US TREASURY BOND
Emerging Markets Government Bonds	J.P. MORGAN EMBI GLOBAL DIVERSIFIED COMPOSITE
Global Corporate Bonds	BLOOMBERG GLOBAL AGGREGATE CORPORATE
European Equity (excluding Eurozone)	MSCI EUROPE
North America Equity	MSCI NORTH AMERICA
Pacific Equity	MSCI PACIFIC FREE
Emerging Markets Equity	MSCI EMERGING MARKETS

Table 4.4 Sample (historical) estimates of returns and risk ($\widehat{\boldsymbol{\mu}}$ and $\widehat{\boldsymbol{\sigma}}$)

	(Annualised) Mean return (%)	(Annualised) Standard deviation (%)
BARCLAYS 3 MONTH EURIBOR CASH INDEX	1.07	0.46
BLOOMBERG EURO AGGREGATE TREASURY BOND	2.80	4.65
BLOOMBERG US TREASURY BOND	3.00	9.92
J.P. MORGAN EMBI GLOBAL DIVERSIFIED COMPOSITE	6.38	9.65
BLOOMBERG GLOBAL AGGREGATE CORPORATE	4.10	9.31
MSCI EUROPE	7.24	14.44
MSCI NORTH AMERICA	8.22	14.01
MSCI PACIFIC FREE	7.12	13.95
MSCI EMERGING MARKETS	9.85	17.58

period from January 2003 to December 2022 were used. All returns are expressed in Euros, which means that even for those asset classes not denominated in Euros, performance is expressed from the point of view of an investor residing in the Eurozone. Essentially, regarding the issue of currency exposure, an unhedged approach against exchange rate risk was taken. This explains the high estimated volatility for the Bloomberg US Treasury Bond index, which, instinctively, does not appear appropriate for an asset class representing government bonds. The annualised sample estimates are given in Tables 4.4, 4.5, and 4.6.

In the light of our discussions in Sect. 4.2, we know that exclusive reliance on these estimated parameters when building a portfolio takes advantage of a simplification of investor preference, or of the acceptance of behaviour of asset class returns that is functional for the portfolio but is certainly not systematically satisfied by real

Table 4.5 Sample (historical) estimates of variances and covariances $\left(\widehat{\boldsymbol{\Sigma}}\right)$

VARIANCE-COVARIANCE MATRIX (annualised values)	BARCLAYS 3 MONTH EURIBOR CASH INDEX	BLOOMBERG EURO AGGREGATE TREASURY BOND	BLOOMBERG US TREASURY BOND	J.P. MORGAN EMBI GLOBAL DIVERSIFIED COMPOSITE	BLOOMBERG GLOBAL AGGREGATE CORPORATE	MSCI EUROPE	MSCI NORTH AMERICA	MSCI PACIFIC FREE	MSCI EMERGING MARKETS
BARCLAYS 3 MONTH EURIBOR CASH INDEX	**0.000021**	0.000017	0.000008	−0.000021	−0.000034	−0.000090	−0.000129	−0.000086	−0.000047
BLOOMBERG EURO AGGRE-GATE TREA-SURY BOND	0.000017	**0.002164**	0.001457	0.001764	0.001582	0.000414	0.000634	0.000627	0.000357
BLOOMBERG US TREASURY BOND	0.000008	0.001457	**0.009845**	0.005592	0.008148	−0.003581	0.001196	0.000602	−0.003607
J.P. MORGAN EMBI GLOBAL DIVERSIFIED COMPOSITE	−0.000021	0.001764	0.005592	**0.009304**	0.007261	0.005004	0.007286	0.007082	0.007124
BLOOMBERG GLOBAL AGGREGATE CORPORATE	−0.000034	0.001582	0.008148	0.007261	**0.008667**	0.001148	0.005065	0.004390	0.001663
MSCI EUROPE	−0.000090	0.000414	−0.003581	0.005004	0.001148	**0.020849**	0.016304	0.014789	0.018849
MSCI NORTH AMERICA	−0.000129	0.000634	0.001196	0.007286	0.005065	0.016304	**0.019620**	0.014620	0.016007
MSCI PACIFIC FREE	−0.000086	0.000627	0.000602	0.007082	0.004390	0.014789	0.014620	**0.019450**	0.018340
MSCI EMERG-ING MARKETS	−0.000047	0.000357	−0.003607	0.007124	0.001663	0.018849	0.016007	0.018340	**0.030919**

Table 4.6 Sample (historical) estimates of correlations

CORRELATION MATRIX	BARCLAYS 3 MONTH EURIBOR CASH INDEX	BLOOMBERG EURO AGGREGATE TREASURY BOND	BLOOMBERG US TREASURY BOND	J.P. MORGAN EMBI GLOBAL DIVERSIFIED COMPOSITE	BLOOMBERG GLOBAL AGGREGATE CORPORATE	MSCI EUROPE	MSCI NORTH AMERICA	MSCI PACIFIC FREE	MSCI EMERGING MARKETS
BARCLAYS 3 MONTH EURIBOR CASH INDEX	**1.00**	0.08	0.02	−0.05	−0.08	−0.14	−0.20	−0.13	−0.06
BLOOMBERG EURO AGGRE-GATE TREA-SURY BOND	0.08	**1.00**	0.32	0.39	0.37	0.06	0.10	0.10	0.04
BLOOMBERG US TREASURY BOND	0.02	0.32	**1.00**	0.58	0.88	−0.25	0.09	0.04	−0.21
J.P. MORGAN EMBI GLOBAL DIVERSIFIED COMPOSITE	−0.05	0.39	0.58	**1.00**	0.81	0.36	0.54	0.53	0.42
BLOOMBERG GLOBAL AGGREGATE CORPORATE	−0.08	0.37	0.88	0.81	**1.00**	0.09	0.39	0.34	0.10
MSCI EUROPE	−0.14	0.06	−0.25	0.36	0.09	**1.00**	0.81	0.73	0.74
MSCI NORTH AMERICA	−0.20	0.10	0.09	0.54	0.39	0.81	**1.00**	0.75	0.65
MSCI PACIFIC FREE	−0.13	0.10	0.04	0.53	0.34	0.73	0.75	**1.00**	0.75
MSCI EMERG-ING MARKETS	−0.06	0.04	−0.21	0.42	0.10	0.74	0.65	0.75	**1.00**

Table 4.7 Estimates based on monthly data of the first four statistical moments

	Mean (%)	Standard deviation (%)	Skewness	Kurtosis
BARCLAYS 3 MONTH EURIBOR CASH INDEX	0.089	0.133	0.993	3.003
BLOOMBERG EURO AGGREGATE TREASURY BOND	0.234	1.343	−0.516	4.619
BLOOMBERG US TREASURY BOND	0.250	2.864	0.665	4.305
J.P. MORGAN EMBI GLOBAL DIVERSIFIED COMPOSITE	0.531	2.784	−0.276	5.700
BLOOMBERG GLOBAL AGGREGATE CORPORATE	0.342	2.688	0.464	3.719
MSCI EUROPE	0.603	4.168	−0.435	4.472
MSCI NORTH AMERICA	0.685	4.043	−0.307	3.785
MSCI PACIFIC FREE	0.593	4.026	−0.092	3.835
MSCI EMERGING MARKETS	0.821	5.076	−0.475	4.428

performance. One way to demonstrate the simplified, agile framework of the Markowitz model would be to take a closer look at the statistical properties of the returns of the selected benchmarks. To begin with, the first four statistical moments are estimated from the monthly time series (Table 4.7).

Second, we verify the null hypothesis that these come from a normal distribution, using the Jarque-Bera and Lilliefors tests. The former is a test based on moments. It relies on the comparison between the sample values for skewness and kurtosis and the expected values for these parameters in a Gaussian distribution.[43] Although widely used, this test is more suitable for use with a very large data sample. The Lilliefors test, on the other hand, uses the comparison between the empirical cdf (the so-called cumulative distribution function of returns) and the normal cumulative distribution with mean and standard deviation comparable to the sample values.[44] The results of these tests, using a statistical significance level of 1%, are given in Tables 4.8 and 4.9.

---

[43]The Jarque-Bera test is based on the (JB) statistic, defined as follows: $\text{JB} = T\left[\frac{\widehat{S}}{6} + \frac{\left(\widehat{K} - 3\right)^2}{24}\right]$, which has an asymptotic chi-squared distribution with 2 degrees of freedom. As is well-known, the terms $\widehat{S}$ and $\widehat{K}$ are the sample values of skewness and kurtosis, which, in a normal distribution, are 0 and 3, respectively.

[44]Essentially, the Lilliefors test involves:

- Estimating the mean and variance (or standard deviation) based on the data available.
- Searching for the maximum discrepancy between the empirical distribution function and the cumulative distribution function (cdf) of the normal distribution, the mean and variance (or standard deviation) having been previously identified from the data.
- Verifying whether or not this maximum discrepancy is big enough to be statistically significant and so reject the null hypothesis of a normal distribution.

Table 4.8 The Jarque-Bera test on selected benchmarks

	JB statistic	Critical value	*p*-value (%)	Normality assumption
BARCLAYS 3 MONTH EURIBOR CASH INDEX	39.405	11.642	0.100	Rejected
BLOOMBERG EURO AGGREGATE TREASURY BOND	36.855	11.642	0.100	Rejected
BLOOMBERG US TREASURY BOND	34.695	11.642	0.100	Rejected
J.P. MORGAN EMBI GLOBAL DIVERSIFIED COMPOSITE	75.938	11.642	0.100	Rejected
BLOOMBERG GLOBAL AGGREGATE CORPORATE	13.802	11.642	0.648	Rejected
MSCI EUROPE	29.222	11.642	0.100	Rejected
MSCI NORTH AMERICA	9.923	11.642	1.486	Accepted
MSCI PACIFIC FREE	7.314	11.642	2.959	Accepted
MSCI EMERGING MARKETS	29.409	11.642	0.100	Rejected

Table 4.9 The Lilliefors test on selected benchmarks

	Lilliefors statistic	Critical value	*p*-value (%)	Normality assumption
BARCLAYS 3 MONTH EURIBOR CASH INDEX	0.175	0.068	0.100	Rejected
BLOOMBERG EURO AGGREGATE TREASURY BOND	0.070	0.068	0.584	Rejected
BLOOMBERG US TREASURY BOND	0.067	0.068	1.153	Accepted
J.P. MORGAN EMBI GLOBAL DIVERSIFIED COMPOSITE	0.045	0.068	28.910	Accepted
BLOOMBERG GLOBAL AGGREGATE CORPORATE	0.054	0.068	9.578	Accepted
MSCI EUROPE	0.076	0.068	0.187	Rejected
MSCI NORTH AMERICA	0.062	0.068	2.530	Accepted
MSCI PACIFIC FREE	0.077	0.068	0.162	Rejected
MSCI EMERGING MARKETS	0.052	0.068	11.084	Accepted

Actually, the test results show some cases in which the Gaussian distribution is rejected. However, we must not forget that normality of the individual series does not ensure multivariate normality.[45] In concrete terms, this reveals, then, that there may be other aspects/market properties that are not completely irrelevant.

Returning to the practical implementation of Markowitz asset allocation, the next crucial step is to "launch" the optimisation. In the present case, the procedure leads to the identification of the expected risk–expected return combinations of the optimal

[45] To test the hypothesis of joint normality, we suggest the Mardia test described in Mardia (1970) and in Doornik and Hansen (2008), based on multivariate measurements of skewness and kurtosis.

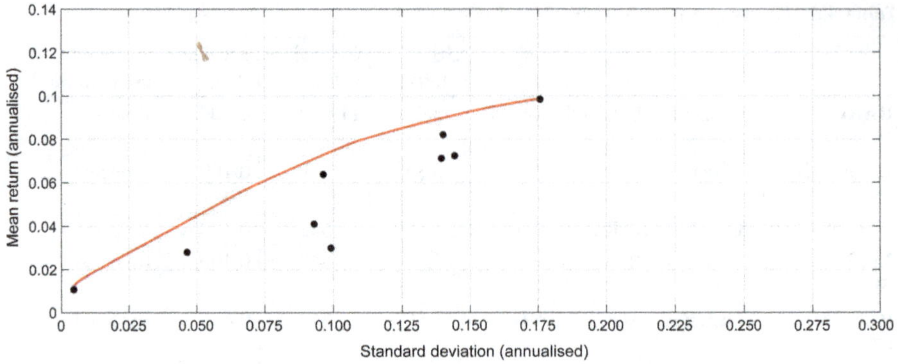

Fig. 4.2 Markowitz's efficient frontier and positioning of the asset classes

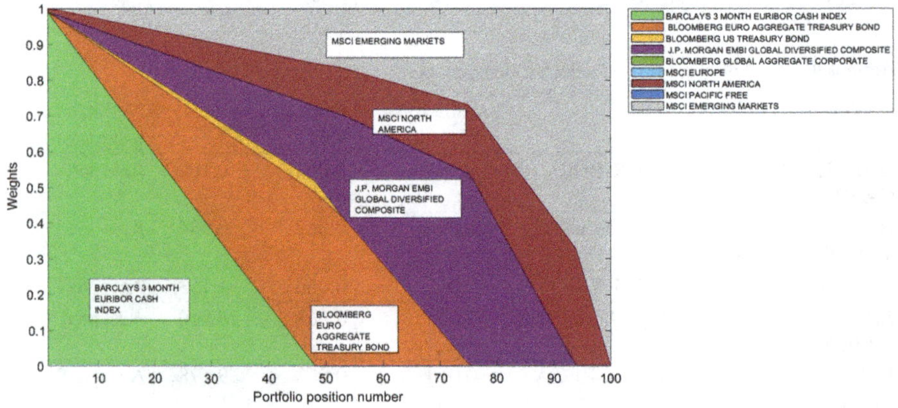

Fig. 4.3 Composition of Markowitz's efficient frontier portfolios

portfolios that describe the mean-variance efficient frontier illustrated in Fig. 4.2, which also shows the position of the individual asset classes.

On the basis of Fig. 4.2, the asset manager would propose strategic (policy) portfolios offering an expected annual return in a range from 1.13% to 9.85% and an expected risk in an annualised standard deviation range from 0.45 to 17.58%. Figure 4.3 highlights the composition of the portfolios (a total of 100) constituting the efficient frontier.

The area graph immediately brings to mind the question posed by Michaud (1989) in his well-known study, "Is 'optimized' optimal?".[46] The only case (and an altogether imaginary one) in which an asset manager would be prepared to answer "yes" to this question would be if she/he was certain that the parameters in Tables 4.4

[46] Michaud (1989).

and 4.5 are an exact representation of the returns, the volatilities and the covariances which will occur annually over the investment period! In all other (realistic) cases, managers would quite understandably find it difficult to accept that the portfolios resulting from MVO are dominant and better than any other asset mix, given that Fig. 4.3 offers evident proof of some of the undesirable properties of these portfolios reported in Sect. 4.3 above. As manifest, Fig. 4.3 reveals:

- The limited diversification of the optimised portfolios; this is particularly true for those beyond the 52nd position, which often include no more than 3/4 asset class and, starting from the 94th position only two asset classes (North America Equity and Emerging Markets Equity) within and investment universe that includes 9 asset classed. It is worth adding that no mean-variance efficient portfolio includes more than 6 asset classes.
- The mono-asset class nature of the maximum return (variance) portfolio, totally concentrated on the MSCI Emerging Markets index, which is reason enough to make the portfolio unfeasible in practical terms.
- The systematic absence or almost total absence of some indices in the mean-variance efficient compositions, e.g. MSCI Europe, MSCI Pacific Free and Bloomberg Global Aggregate Corporate.[47] The aggregation of MSCI North America and MSCI Emerging Markets is preferred by the MVO compared to a greater variety of equity asset classes when higher risk tolerance levels are considered because it benefits from a greater return estimate (in the case of MSCI North America) associated with a volatility lower than that of an equity component excluded from the portfolios (MSCI Europe) and also to a correlation with the MSCI Emerging Markets lower than the one existing between other potential equity components and the Emerging Countries. However, where there is a more pronounced risk aversion, the risk-return combination of the Bloomberg Euro Aggregate Treasury Bond (4.65%; 2.80%) makes it preferable to the Bloomberg Global Aggregate Corporate (9.31%; 4.10%) as an element to complement MSCI North America, MSCI Emerging Markets, and also J.P. Morgan EMBI Global Diversified Composite. Furthermore, the Bloomberg Euro Aggregate Treasury Bond has lower correlations, although positive, with both the selected equity asset classes and with the J.P. Morgan EMBI Global Diversified Composite compared to Bloomberg Global Aggregate Corporate.
- The particularly violent and drastic shifts in the compositions along the efficient frontier.
- The counter-intuitive and illogical nature of the mean-variance efficient portfolios for the decision-maker and/or the final investor. As proof of this claim, let us ask the simple question: what practical value would a European investor attach to a proposed strategic asset allocation that completely omits the European Equity

[47] Actually, the Bloomberg Global Aggregate Corporate is included in the minimum variance portfolio and in the one immediately following it with a very marginal, almost imperceptible exposure, equal to 0.216% and 0.179%, respectively.

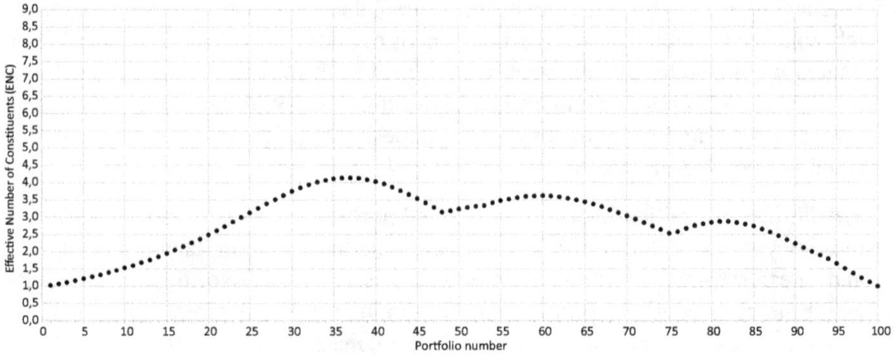

Fig. 4.4 The ENC metric behaviour for Markowitz's efficient frontier portfolios

asset class from the equity component of the portfolio? The probable answer is: none.

In light of the comments above, regarding the behaviour of the portfolio weights suggested by the MVO, it may be interesting to confirm the problem of their poor degree of diversification through a measure called "Effective Number of Constituents" (in short, ENC) by Deguest et al. (2013). This metric is of rather easy interpretation, given the fact that it takes on the minimum value of 1 in the case of a highly concentrated portfolio (as is the case of a single-asset portfolio) and reaches the maximum value of N when the portfolio has a uniform distribution of weights (as is the case of the equally weighted portfolio).[48] Figure 4.4 shows that, for the 100 portfolios of the Markowitz efficient frontier, the ENC evidently remains far from the maximum possible value even (and this is a cause for concern) for the portfolios that make up the second part of the efficient frontier, i.e. for portfolios from the 50th position, never going beyond the average value, thus demonstrating that classic portfolios are far from an idea of non-concentration of portfolio exposures.

The above-mentioned points can be taken as a good summary of the implications of an exercise in strategic asset allocation carried out in a deterministic context, in which faith is blindly and incautiously placed in the past performance of the asset classes. It is well-known that past performance has little predictive power, especially with regard to expected returns. Estimates, corresponding to the sample average, are significantly influenced by outliers. These observations can be supported by considering powerful, empirical evidence showing the extremely erratic portfolio risk-return combinations generated using rolling windows of fixed length to calculate the efficient frontier. To this end, Figs. 4.5 and 4.6 show the estimated efficient frontiers

[48] The measure called ENC is defined as follows: $\frac{1}{\sum_{i=1}^{N} w_i^2}$. The reciprocal of the so-called Herfindahl Hirschman index is recognizable in it.

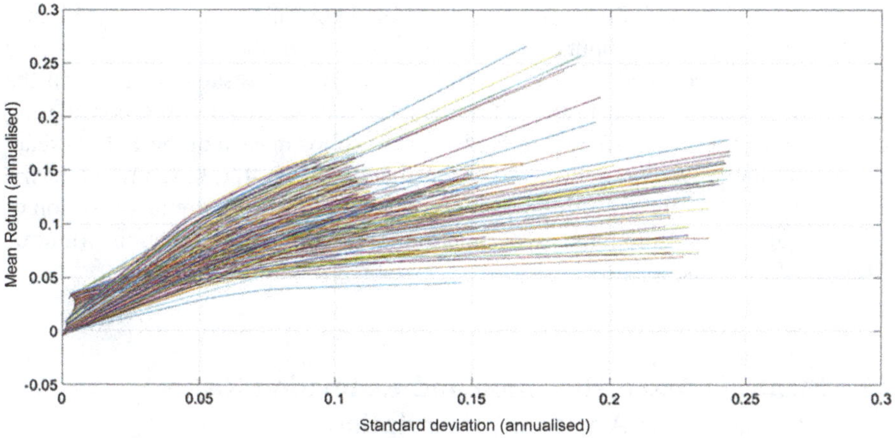

Fig. 4.5 Instability of the historical efficient frontier based on 5-year historical series

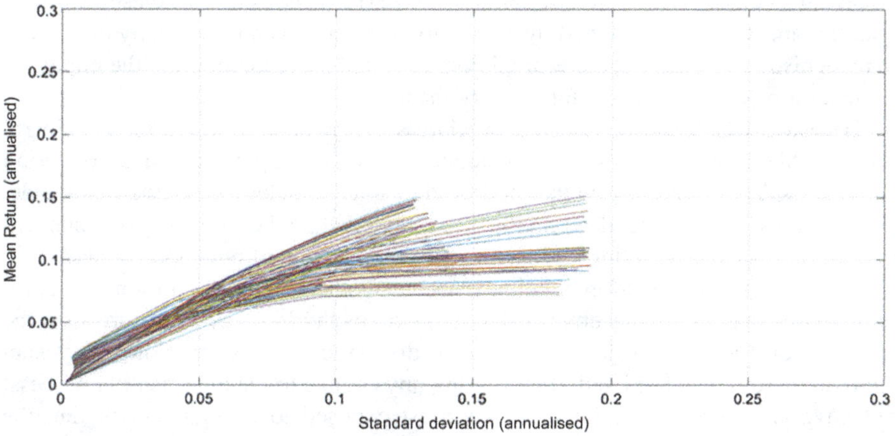

Fig. 4.6 Instability of the historical efficient frontier based on 12-year historical series

for the asset classes above, using historical samples of monthly data for rolling periods of 5 and 12 years, respectively. The availability of monthly data for the period January 2003 to December 2022 allows us to draw 181 efficient frontiers in the first case and 97 in the second. It is wholly legitimate to interpret the considerable variability of the historical efficient frontiers, in particular with regard to the returns of the proposed portfolios and as sample size diminishes, as a sign of poor reliability from an out-of-sample point of view. More explicitly, Figs. 4.5 and 4.6 are evidence of the non-robustness/non-resilience of classic efficient portfolios to scenarios different from that single historical scenario that determined them, especially if the latter turns out to be characterised by a short time length.

Further confirmation of the ambiguity of optimal portfolios due to the uncertainty affecting the estimated inputs can be obtained from the observation of the asset allocation associated with the portfolios occupying the same position on the previously estimated efficient sample frontiers. For illustrative purposes, considering the asset allocation of the 25th, 50th, and 75th portfolios in each of the 181 "5-year" efficient frontiers, we can notice, as demonstrated by Fig. 4.7, how important the differences are in terms of portfolio weights and, therefore, the extreme distortion of the asset allocation solutions as a function of a different estimation window, although of the same length.

4.5 Strategic Asset Allocation and Estimation Risk Management: An Initial Description

Investors and asset managers use strategic asset allocation to identify stable, diversified, and financially efficient portfolios. As seen in Sect. 4.3 and given the empirical evidence provided in Sect. 4.4, however, it is clear that these desirable qualities are severely threatened by estimation risk and, even more so, by the failure to recognise such risk. It is, therefore, imperative that the recognition of the impact of estimation risk leads to its being actively managed.

The ways which can be adopted to address this issue correspond to two clearly recognisable approaches in the literature: heuristic approaches and Bayesian approaches. The former act on the model-side while the latter on the estimation-side. Here we present the basic idea underling each case, while the following sections will provide a detailed description and an empirical application of both approaches.

Heuristic approaches to estimation risk management exploit intuition and common sense to identify the empirical criteria or methods so as to ensure that the optimisation process produces as much diversification as possible. Bayesian approaches, on the other hand, do not bring any change to the optimisation process, but have an effect on the input that is most exposed to estimation risk, i.e. the expected returns vector. They exploit estimation techniques other than the sample means obtained from historical data sets, so as to calculate "better" estimators. Put simply, the Bayesian approaches prefer to use and "combine" information from a number of sources into a single estimate, rather than relying on estimates generated by a single source.

If we consider the ways of incorporating estimation risk into the process of strategic asset allocation associated with the two approaches, it is easy to see how the heuristic and Bayesian approaches are based on two different statistical "traditions": frequentist in the former case and Bayesian (obviously) in the latter. An important difference between the two lies with the interpretation of the concept of uncertainty/probability. In the heuristic approaches, uncertainty derives from the random realisations of a random variable whose probability distribution is known. The probability of an event is understood as the limit to which its relative frequency tends, i.e. the frequency revealed as the mass of data, with which to calculate

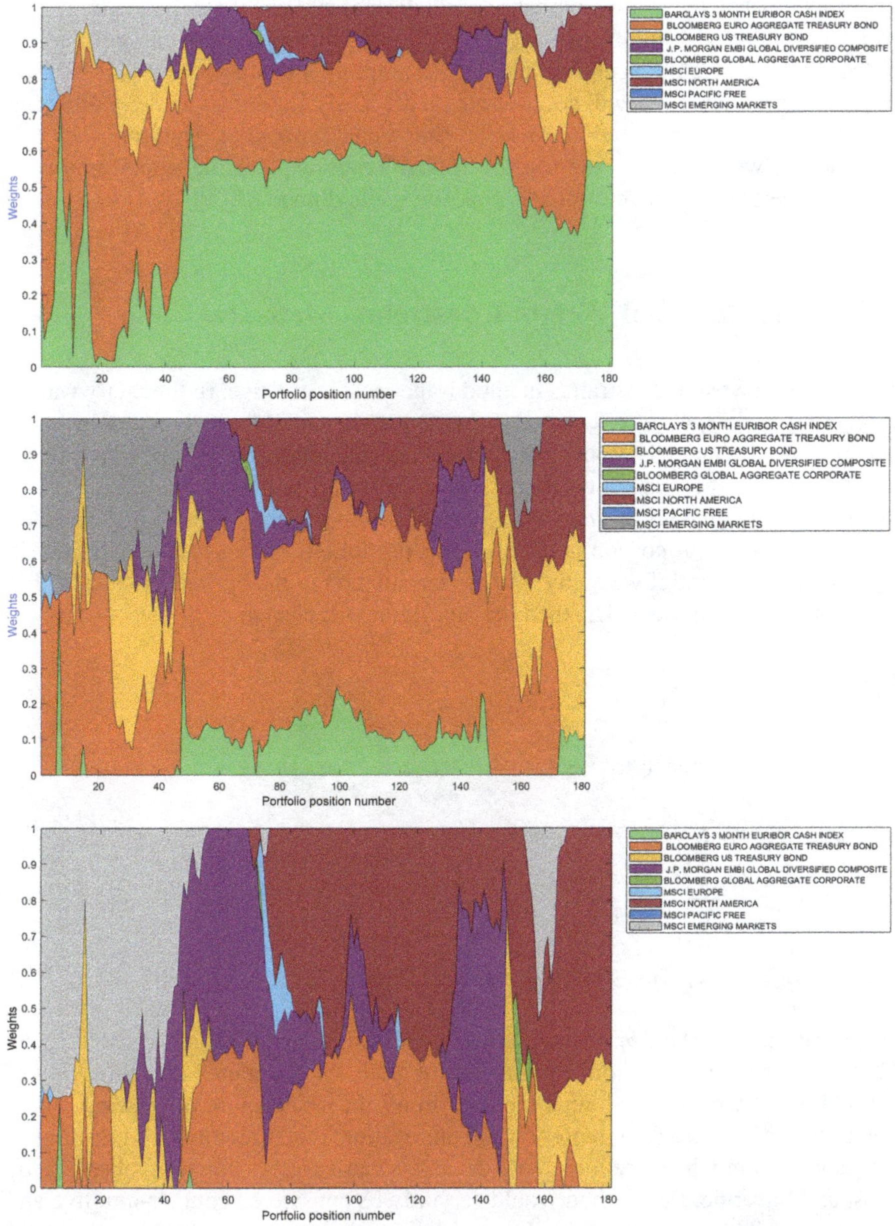

Fig. 4.7 Composition of the 25th, 50th, and 75th portfolio in the 181 "5-year" efficient frontiers

estimates, increases. In the Bayesian approaches, on the other hand, the exact probability distributions are unknown and uncertainty refers to a subjective estimate of the probability of events; such estimate might be updated as more information becomes available.

As anticipated, the following sections will discuss in detail the different "ways" in which estimation risk can be managed. For the heuristic approaches, we will look at the Additional weight constraints and the Resampling methodologies, while for the Bayesian approaches, we will concentrate on the Black-Litterman model. Following this introductory presentation, despite the clear separation between the two approaches, we shall also suggest that practitioners explore the opportunities to integrate the two approaches into "composite methodologies".

4.6 The Additional Weight Constraints Method

The additional weight constraints method is the most instinctive, rudimentary way of countering the negative effects of estimation risk, and is widely used by asset managers.[49] Technically, the method involves integrating a set of further linear constraints into the quadratic programming algorithm proposed by Markowitz. If Lb_i is the minimum allocation of asset i in a portfolio (i.e. the lower bound for i) and Ub_i the maximum proportion of the asset in a portfolio (i.e. the upper bound for i) in a given portfolio, and if we apply it to the formulation of the optimisation algorithm as given in (4.3a) in Sect. 4.1, then the algorithm will change as follows (4.16):

$$\begin{aligned}
&\min_{w} \sum_{i=1}^{N}\sum_{j=1}^{N} w_i w_j \sigma_{ij} \\
&\text{subject to} \\
&\sum_{i=1}^{N} w_i \mu_i = \mu_P^* \\
&\sum_{i=1}^{N} w_i = 1 \\
&Lb_i \leq w_i \leq Ub_i \ \text{ with } Lb_i \leq Ub_i \ \text{ and } Lb_i \geq 0, Ub_i \leq 1
\end{aligned} \tag{4.16}$$

Clearly, if we take $Lb_i = 0.00$ and $Ub_i = 1.00$, then we return to the original Mean-Variance Optimisation, yet this cannot really be considered an application of the additional weight constraints method. In all the other, more interesting, cases, however, the method does represent an "innovation", as it requires a tightening of the range within which the weight of the assets can vary. As correctly observed by Frost and Savarino, the method would certainly be considered counter-intuitive and meaningless if asset managers implemented the optimisation process under certainty conditions, because it would curb the potentialities of the algorithm. On the other hand, it would be logical and extremely appealing if we were to admit—as in this and many other contributions—that the estimated parameters used in optimisation are

[49] Amenc et al. (2011).

subject to uncertainty and estimation error.[50] Under these circumstances, the additional weight constraints method identifies its target in the unreasonable, polarised, and extreme nature of the classical compositions of portfolios. Consequently, it is clear that the method is favourably received by asset managers because it can ensure that an algorithm fed with inputs of "dubious quality" does not liberally unleash its power and give rise to significant estimation bias. Essentially, managers see in the additional weight constraints method a welcome form of coercion of the Markowitz algorithm which aims to ensure that the latter provides a more diversified composition of optimal portfolios.

In support of this claim, we can return to the strategic asset allocation exercise presented in Sect. 4.4 and, by making use of the input parameters estimated there, extend that example by applying the above-described method. In a first case, we re-write the last constraint in the algorithm in (4.17) as follows:

$$0 \leq w_i \leq 0.30 \tag{4.17}$$

and, in the second case, as (4.18):

$$0 \leq w_i \leq 0.20 \tag{4.18}$$

The inclusion in the two cases of a maximum holding constraint for each of the nine asset classes—of 30% and 20% of the portfolio allocations, respectively—implicitly entails a minimum number of four or five asset classes, respectively, in each portfolio. The consequences of these individual constraints on the composition of the optimal portfolios can be seen in Figs. 4.8 and 4.9.

The $[\sigma_P;\ \mu_P]$ plots on the graph in Fig. 4.10 show the constrained efficient frontiers derived from the risk-return combinations of the portfolios in the two cases compared with the classic Markowitz efficient frontier.

Taken together, Figs. 4.8, 4.9, and 4.10 show the various effects of the additional weight constraints method on the strategic (policy) portfolios.

In line with what has been said so far, the allocations in the constrained efficient portfolios reveal an increase in the diversification effect. This change is extremely evident for the portfolios in the higher positions, from approximately the 50th position on and is even more evident for portfolios after the 75th position. This is because, with the constraints introduced, the aggregation of J.P. Morgan EMBI Global Diversified Composite, MSCI North America, and MSCI Emerging Markets is no longer feasible and sufficient. Then, if we consider the portfolios with a very high position number, approximately from the 94th position, the transformation is extraordinary. The constraints introduced in the optimisation algorithm no longer allow the exclusive combination of MSCI North America and MSCI Emerging Markets as happened with the traditional MVO and impose an inclusion of further equity components; in fact, the inclusion of MSCI Europe and MSCI Pacific Free

[50] Frost and Savarino (1988).

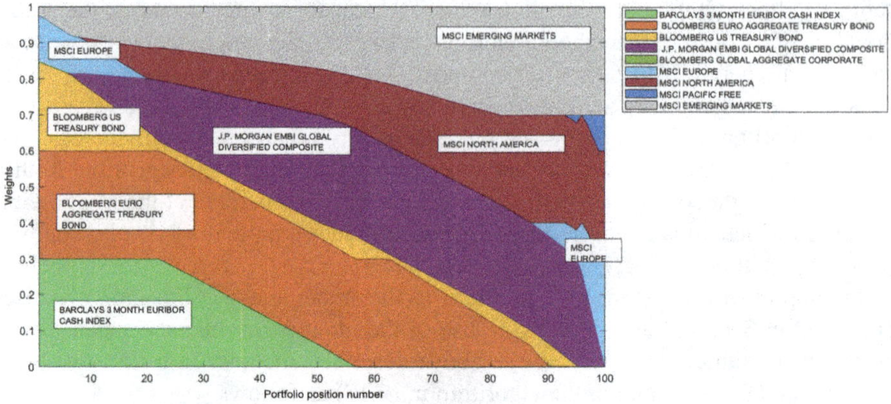

Fig. 4.8 Composition of constrained portfolios (individual maximum holding = 30%)

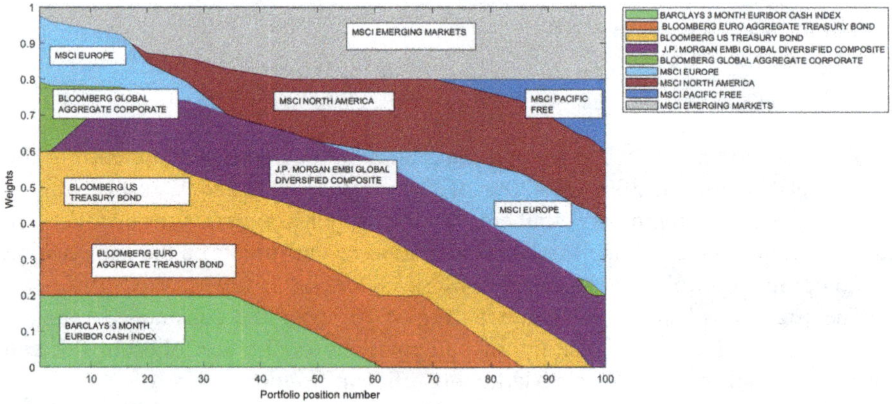

Fig. 4.9 Composition of constrained portfolios (individual maximum holding = 20%)

which were previously absent can be observed. It should be noted that this methodology also opposes "exceptional" exposure to asset classes which are probably less subject to estimation risk. In the example, this is the case with Barclays 3-month Euribor Cash Index. With the additional weight constraints method, this class no longer predominates in those initial classic portfolios (approximately the first fifteen/twenty) that constitute allocation proposals for investors with a very low risk tolerance. At present, then, the impression is that the methodology employed is not able to discriminate between different levels of intensity regarding the need for diversification. Furthermore, a second characteristic of constrained asset mixes compared to traditional unconstrained mixes is reduced irregularity in the behaviour of portfolio weights.

If we look at the position $[\sigma_P; \mu_P]$ of the combinations of the constrained efficient portfolios, the method described produces two fundamental results. The first is a

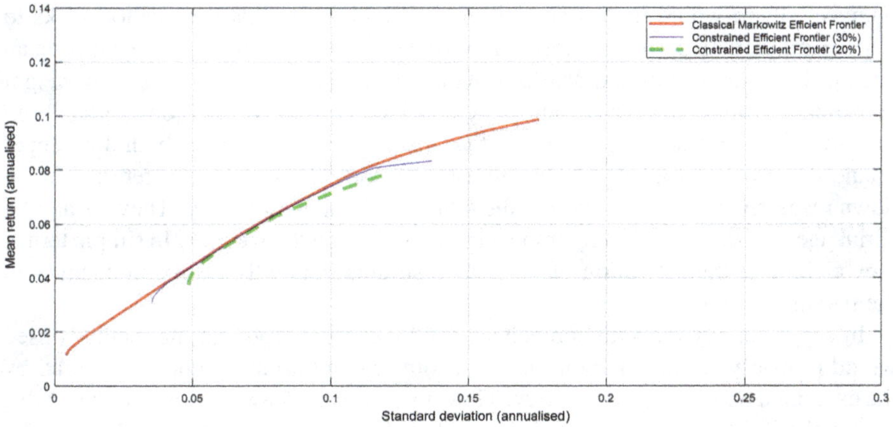

Fig. 4.10 Constrained efficient frontiers and classic Markowitz efficient frontier

reduction in the extension of the opportunity set. The uniform maximum holding constraints introduced for the asset classes have caused a shift to the right of the constrained efficient frontier, as the original portfolios which included mainly the Eurozone Cash asset class can no longer be proposed. At the same time, the constrained efficient frontier is limited by a maximum level of risk below the one found in the classic efficient frontier. Again, the explanation is simple: concentration on a single asset class at the right extreme of the Markowitz efficient frontier is no longer possible. Mixing the originally preferred asset class with others (e.g. MSCI North America, MSCI Pacific Free, and MSCI Europe) which offers lower risks and lower returns obviously cannot "outclass" the initial, single asset class portfolio. The second important fact is that the constrained efficient frontier is positioned below the classic efficient frontier. This means that, compared to a portfolio belonging to the classic mean-variance efficient frontier with an equal level of risk, any portfolio belonging to the constrained efficient frontier offers a lower return. This "distance" is generally indicated as the cost (in terms of reduced performance) of constrained optimisation. However, to avoid any misleading or confusing interpretations, it should be emphasised that this is clearly an in-sample sacrifice in performance, while what matters is the out-of-sample performance of the portfolios. From this point of view, it is reasonable to assume that, compared to classic portfolios, constrained portfolios have adopted a better strategy to hedge estimation risk, namely greater diversification.

So far, our examination of the additional weight constraints method has deliberately ignored a question that must not be underestimated, which is the criteria used to determine the constraints, for the obvious reason that it is simply not possible to provide clear-cut responses and indications, since the method is highly subjective. We need only to look back to the example given above. Why were 30% and 20% used as the upper bounds, and not 40%, 25%, or 15%? There is no straightforward answer to this question. Moreover, had the exercise even used these other upper bounds,[51] there would understandably have been a sense that the method seeks to impose, rather than derive, solutions. It must, however, be said that there has been an attempt by Jagannathan and Ma to provide a scientific basis for the use of weight constraints—although no rigorous and exclusive criterion for their design has been identified.[52] These authors recognise that mathematically a lower bound or upper bound constraint on asset weights adjusts the corresponding mean returns, up or down respectively, in proportion to the related Lagrange multiplier. They speak of a shrinkage-like effect caused by imposing constraints on the weights. In simple terms, they claim that the activation of weight constraints implicitly results in changes to the parameter estimates.

In any case, anyone who appreciates solid theoretical roots in the methods used would probably complain about these customised and arbitrary choices offered by the examined heuristic methodology. They do, however, draw us closer to the modus operandi of asset managers and mitigate their fear that quantitative tools might be "taking over" their role. As noted by Eichhorn, Gupta, and Stubbs, asset managers see the setting of constraints as a way to:[53]

- Allow portfolio construction to be influenced by the different levels of trust they have in the estimated parameters. In this sense, more severe weight constraints would be applied to asset classes for which the estimated statistical moments are considered less reliable (above all, expected returns).
- Introduce preferences for certain asset classes rather than others into the optimisation tool. Note that weight constraints could prevent the elimination from the asset mix of those asset classes which managers and/or investors are more familiar with, even where they would be eliminated by traditional Mean-Variance Optimisation.[54]
- Ensure respect of external constraints imposed by law or regulation able to (for example) prohibit exposure to certain markets beyond given limits.

These points reveal that the flexibility of the additional weight constraints method is not limited to the setting of upper and lower bounds, but also involves how the constraints are applied. For the sake of simplicity, the example above used individual constraints that defined one (in that case, identical) maximum and/or minimum for each asset class. However, for a more complete representation of the method under examination, it should be pointed out that one could also (or alternatively) apply group or infra-group constraints: the former are the minimum or maximum values that a sub-set of the asset classes must respect in portfolios; the latter define the minimum or maximum of an asset class compared to the overall weight of its group in the strategic portfolios as defined by the asset manager. Note that these last two types of constraints can prove particularly useful in responding to the need to gradually increase the diversification effect along the (constrained) efficient frontier, a need which individual constraints have difficult responding to, especially if they are uniform.

---

[51] For example, it might have tested various solutions aimed at balancing the wish for portfolio diversification with limited disadvantage and reduction in the proposed risk-return combinations.

[52] Jagannathan and Ma (2003).

[53] Eichhorn et al. (1998).

[54] In the example, this would be the case for MSCI Europe.

With these three variations in the application of the constraints, we can present a third example of the application of the additional weight constraints method using the same inputs as before. Specifically, this new example entails:

– An individual constraint requiring that each optimal portfolio has an exposure of at least 5% but no more than 90% to the Eurozone Cash asset class. The constraint assumes the form given in (4.19):

$$0.05 \leq w_{\text{BARCLAYS 3 MONTH EURIBOR CASH INDEX}} \leq 1 \tag{4.19}$$

– A group constraint setting a maximum limit of 75% for equity asset classes in each constrained efficient portfolio. This constraint is expressed formally as (4.20):

$$\begin{aligned} 0.00 \leq\ & w_{\text{MSCI EUROPE}} + w_{\text{MSCI NORTH AMERICA}} + w_{\text{MSCI PACIFIC FREE}} \\ & + w_{\text{MSCI EMERGING MARKETS}} \leq 0.75 \end{aligned} \tag{4.20}$$

– Four infra-group constraints. The first one sets the proportion of Global Corporate Bonds within the total bonds component to be no less than 10%. The second constraint states that Emerging Countries Government Bonds cannot take on a weight greater than 10% of the bond component of the portfolios. The third constraint states that Emerging Markets Equity cannot constitute more than 15% of the equity component of the portfolios. The last one set a minimum of 25% for Eurozone Equity in the overall equity component. These infra-group constraints are expressed formally with the following inequalities (4.21), (4.22), (4.23), and (4.24).

$$\begin{aligned} 0.10 \leq & \frac{w_{\text{BLOOMBERG GLOBAL AGGREGATE CORPORATE}}}{\begin{array}{c} w_{\text{BLOOMBERG GLOBAL AGGREGATE CORPORATE}} + w_{\text{J.P.MORGAN EMBI GLOBAL DIVERSIFIED C.}} + \\ w_{\text{BLOOMBERG US TREASURY B.}} + w_{\text{BLOOMBERG EURO AGGREGATE TREASURY B.}} \end{array}} \\ & \leq 1 \end{aligned} \tag{4.21}$$

$$\begin{aligned} 0.00 \leq & \frac{w_{\text{J.P.MORGAN EMBI GLOBAL DIVERSIFIED C.}}}{\begin{array}{c} w_{\text{BLOOMBERG GLOBAL AGGREGATE CORPORATE}} + w_{\text{J.P.MORGAN EMBI GLOBAL DIVERSIFIED C.}} + \\ w_{\text{BLOOMBERG US TREASURY B.}} + w_{\text{BLOOMBERG EURO AGGREGATE TREASURY B.}} \end{array}} \\ & \leq 0.10 \end{aligned} \tag{4.22}$$

$$\begin{aligned} 0.00 \leq & \frac{w_{\text{MSCI EMERGING MARKETS}}}{w_{\text{MSCI EUROPE}} + w_{\text{MSCI NORTH AMERICA}} + w_{\text{MSCI PACIFIC FREE}} + w_{\text{MSCI EMERGING MARKETS}}} \\ & \leq 0.15 \end{aligned} \tag{4.23}$$

$$0.25 \leq \frac{w_{\text{MSCI EUROPE}}}{w_{\text{MSCI EUROPE}} + w_{\text{MSCI NORTH AMERICA}} + w_{\text{MSCI PACIFIC FREE}} + w_{\text{MSCI EMERGING MARKETS}}} \leq 1 \quad (4.24)$$

To summarise, the third example of the additional weight constraints method is an exercise in strategic asset allocation in which, by using individual, group, and infra-group constraints, the aim is to ensure that each strategic portfolio has a minimum component of precautionary and extremely liquid savings, while avoiding being over-aggressive so as to prevent all-equity solutions. The strategy also takes into account investors' habits and a familiarity with certain asset classes that may lead these to be seen as "anchors" for their financial portfolio, while they may wish to limit (with respect to their capitalisation of other asset classes) exposure to less common categories which they are not so accustomed to, and where research and analysis might prove more difficult. This last point hints at the criterion underlying the constraint (4.23), which we need to consider in further detail.

In this case, the upper bound (15%) of the infra-group constraint is fixed by subtracting approximately 2.5% to the market capitalisation value of MSCI Emerging Markets compared to the market capitalisation value of all the other equity asset classes at the end of December 2022. This approach is based on Grauer and Shen who suggest defining the field of variation of an asset class within a risky group by adding or subtracting an (arbitrary) quantity, c, to the target-weight constraint ($w_{\text{target\_}i}$), which the authors define as the proportion of market capitalisation associated with that asset class compared to the total figure for the risky group.[55] Using this method, a lower bound and an upper bound for an asset class, i, in an infra-group constraint are defined, respectively, as follows (4.25), (4.26):

$$\text{Lb}_i = \max\left(0, \; -c + w_{\text{target\_}i}\right) \quad (4.25)$$

$$\text{Ub}_i = \min\left(1, c + w_{\text{target\_}i}\right) \quad (4.26)$$

Analytically and in general terms, the infra-group constraint is given as (4.27):

$$\text{Lb}_i \leq \frac{w_i}{w_{\text{Risky group}}} \leq \text{Ub}_i \quad (4.27)$$

Grauer and Shen's proposal is interesting because of its strong consistency with one of the central results of the CAPM (Capital Asset Pricing Model) equilibrium

[55] Grauer and Shen (2000).

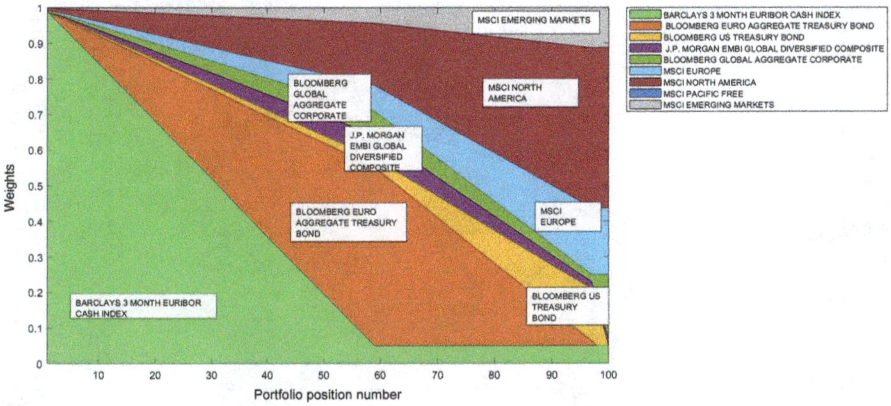

Fig. 4.11 Composition of constrained portfolios with individual, group, and infra-group constraints

model, according to which a rational investor should hold all risky assets in proportion to their market capitalisation. This is useful for explaining how, in the formulation of upper and lower bounds presented above, the quantity c will reasonably tend to zero when the level of reliability assigned by the asset manager to the inputs is lower.

Having completed the discussion of the additional weight constraints method in the third example, we need to consider the composition and risk-return combination of the resulting efficient portfolios. These are illustrated in Figs. 4.11 and 4.12.

Compared to the first two examples, the original and relevant aspects that emerge clearly from this further application can be summarised as follows:

- Moving along the efficient frontier from left to right, there is a gradual progression towards portfolios that exclude clearly predominant exposure to certain asset classes. This no longer occurs uniformly along the plot, as observed when only homogeneous, individual constraints are used.
- An absence of a marked shift to the right of the starting point of the constrained efficient frontier compared to the classical Markowitz efficient frontier. This maintains the possibility of satisfying the preferences of extremely risk-averse investors.

It goes without saying that in both respects the merit resides in the successful determination of a more disciplined and conscious search for diversification through the combined use of different criteria to apply weight constraints.

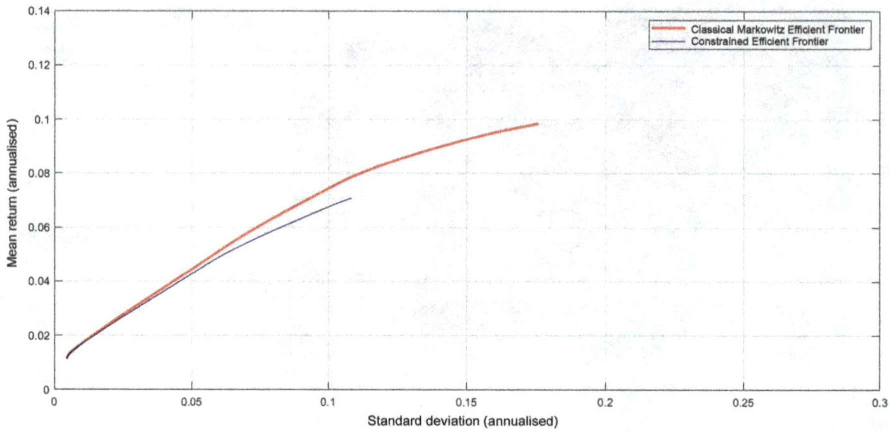

Fig. 4.12 Constrained efficient frontier (with individual, group, and infra-group constraints) and classical Markowitz efficient frontier

4.7 The Resampling Method

The Resampling method invented by Richard Michaud and Robert Michaud[56] seeks to improve Markowitz optimisation by introducing Monte Carlo simulations into the portfolio construction process, in order to reflect in visible and tangible way the uncertainty associated with the estimated parameters and avoid their "literal" use.[57] In this perspective, the power of Monte Carlo simulations consists of generating new random series of asset class returns in addition to the single series seen in the historical data on which the sample estimates of the inputs are based. This means that the Resampling method demonstrates the need and usefulness for a stochastic (or statistical) rather than a deterministic "concept" of portfolio efficiency.

Moving on to a detailed description of Resampling, the methodology involves the following sequence of steps:

1. Estimating $\widehat{\boldsymbol{\mu}}$ and $\widehat{\boldsymbol{\Sigma}}$ by using T observations of the historical return for each asset class, which means setting up the inputs for classical portfolio optimisation.
2. Calculating the traditional Markowitz Efficient Frontier of M portfolios starting from the minimum return (variance) portfolio and continuing to the maximum return (variance) portfolio. Each portfolio is given a position index or rank, k ($k = 1, \ldots, M$).

[56]The procedure described below forms the object of a patent from December 1999 (US patent 6,003,018) entitled "Portfolio Optimisation by Means of Resampled Efficient Frontier" granted to Richard and Robert Michaud. The licensing rights belong exclusively to New Frontier Advisors LLC, Boston, USA.

[57]Michaud and Michaud (2008).

3. Using Resampling to create a new data set. Specifically, we need to simultaneously simulate a random series of returns of length L for each asset class through draws from a multivariate normal distribution with mean vector $\widehat{\boldsymbol{\mu}}$ and covariance matrix $\widehat{\boldsymbol{\Sigma}}$ estimated in the first step. Generally, L coincides with T. It is worth stressing that what has just been stated is the step of the procedure in which the use made of what has been learned from history is best demonstrated.
4. Using the simulated data to calculate a new set of inputs for the optimisation $(\widehat{\boldsymbol{\mu}}_{\mathrm{SIM}\ 1}; \widehat{\boldsymbol{\Sigma}}_{\mathrm{SIM}\ 1})$.
5. Using $(\widehat{\boldsymbol{\mu}}_{\mathrm{SIM}\ 1}; \widehat{\boldsymbol{\Sigma}}_{\mathrm{SIM}\ 1})$ to determine a new efficient frontier and, therefore, the sequence of optimal portfolios resulting from the new set of estimated inputs. This frontier is termed either the simulated efficient frontier or the statistically equivalent efficient frontier and can be represented symbolically as $FE_{\mathrm{SIM}\ 1}$. The latter, therefore, represents the set of efficient portfolios for the specific scenario simulated for the asset classes.
6. Repeating steps (3) to (5) a large number, H, of times. Reiteration results, first, in H new sets of optimisation inputs (from $\widehat{\boldsymbol{\mu}}_{\mathrm{SIM}\ 1}; \widehat{\boldsymbol{\Sigma}}_{\mathrm{SIM}\ 1}$ to $\widehat{\boldsymbol{\mu}}_{\mathrm{SIM}\ H}; \widehat{\boldsymbol{\Sigma}}_{\mathrm{SIM}\ H}$) and, subsequently, in a similar number of simulated efficient frontiers or statistically equivalent efficient frontiers (from $FE_{\mathrm{SIM}\ 1}$ to $FE_{\mathrm{SIM}\ H}$), each including M portfolios.
7. Associating a portfolio from the classical efficient frontier with each of the M portfolios on each of the H simulated efficient frontiers. These are called rank associated portfolios, as the association is based on the similar position of the simulated and the classic portfolios, given that both sets are distributed along their respective efficient frontiers so that they are equidistant by return.
8. Calculating the mean weight assigned to each of the N asset classes for each of the M portfolios with the same rank on the H simulated efficient frontiers. This gives the composition of the M resampled portfolios.
9. Using the classical inputs $(\widehat{\boldsymbol{\mu}}; \widehat{\boldsymbol{\Sigma}})$, calculate the risk-return to associate with each of the M resampled efficient portfolios.
10. Indicating in the risk-return space the Resampled Efficient Frontier (REF) as the set or sequence of the M resampled portfolios.

As a whole, a close reading of the above steps points to the following description, in essential terms, of the basic idea of Resampling: portfolio construction should not only make the level of randomness of the inputs visible (by virtue of the simulated efficient frontiers), but also seek to exploit this uncertainty taking into account the resulting greater dispersion and variation of the allocation vectors by averaging the weights of the rank associated portfolios so as to arrive at less extreme final compositions.

4.7.1 The Application of Resampling: An Example

This section provides an example of the application of Resampling. To this end, note that the classic inputs will once again be the sample estimates from Sect. 4.4, as summarised in Tables 4.4, 4.5, and 4.6.

Given that the initial steps of the Resampling procedure clearly seek to highlight the influence of estimation error in rendering optimal portfolios ambiguous, the present example must start by generating statistically equivalent efficient frontiers. We therefore repeat 500 times the simulation of random paths consisting of 240 monthly observations for the nine asset classes in the investment universe, drawing these from a multivariate normal distribution $N\ (\widehat{\boldsymbol{\mu}};\ \widehat{\boldsymbol{\Sigma}})$. This results in 500 new sets of inputs for optimisation, with which we can compute the 500 simulated efficient frontiers (each including 100 portfolios) shown in Fig. 4.13. In effect, the plot in the risk-return space of the alternative portfolios appears to be a convincing proof of the ambiguity of the Mean-Variance optimality concept: some simulated efficient frontiers do not cover the range of risk of the original frontier, while others go visibly beyond it; the observed range for returns is even larger, stretching from a value about 1% to values close or slightly higher than 22%/23% for the simulated efficient frontiers, compared to a range of 1.13–9.85% for the original frontier.

The impact of estimation error can be shown even more clearly by comparing the representation of the simulated efficient frontiers with that of the statistical equivalence region. This can be done by evaluating the mean return and standard deviation of each of the simulated efficient frontier portfolios on each simulated efficient frontier relative to the original inputs, thus interpreting their allocations vector as a set of statistically equivalent weights.

As a result of this procedure, the statistically equivalent region shown in Fig. 4.14 is populated by 50,000 portfolios (100 × 500) lying below the classical efficient frontier. The dispersion of these portfolios compared to the classic efficient frontier

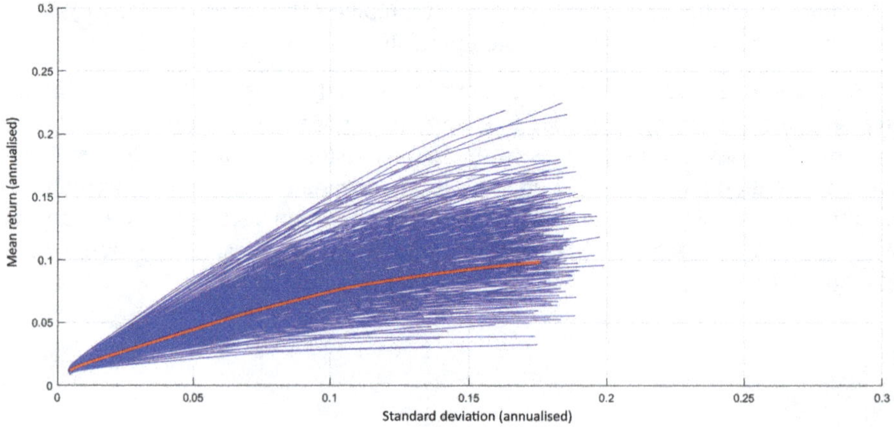

Fig. 4.13 Classical efficient frontier and 500 simulated or statistically equivalent efficient frontiers

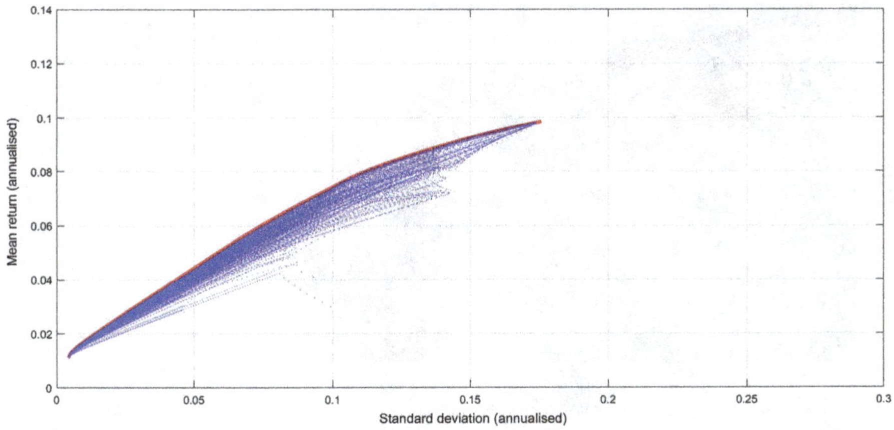

Fig. 4.14 Classic efficient frontier and statistically equivalent region

increases as we move from bottom to top of the frontier, i.e. proceeding from left to right. This is in line with the idea that estimation risk affects intermediate and extreme portfolios more than low risk portfolios.

At this point in the application of the Resampling procedure, it is natural to feel the need to exploit the uncertainty associated with the optimal portfolios among the mean-variance efficient examples, highlighted in both Figs. 4.13 and 4.14, in order to identify reasonable and feasible portfolios. This is the same as asking how to convert the simulated efficient frontiers into a single efficient frontier, to arrive at a single opportunity set that does not overlook information uncertainty. Considering that the simulated efficient frontiers are equally likely and that each simulated efficient frontier expresses the correct asset mixes for a given set of simulated inputs, it is no surprise that a new and unique Resampled Efficient Frontier is identified, as suggested in step 8 of the Resampling sequence, by the mean of the weights vectors of the rank associated portfolios. The next step in the present example is, therefore, the calculation (4.28):

$$\mathbf{w}_{h,k}^{\text{resampled}} = \frac{1}{500}\sum_{h=1}^{500}\mathbf{w}_{h,k} \tag{4.28}$$

which determines the allocations vector of the kth resampled portfolio ($\mathbf{w}_k{}^{resampled}$) as the average ($\mathbf{w}_k$) of portfolio weights of all properly associated optimal portfolios (portfolios in position k) on the H simulated efficient frontiers. The compositions of the resampled portfolios thus obtained are shown in Fig. 4.15. The figure illustrates how the use of Monte Carlo simulations in Resampling succeeds in curbing the excessive strength (compared to the quality of the inputs) of classic Markowitz optimisation, generating portfolios with a greater degree of diversification and with fewer unexpected shifts in the allocations as expected return varies, compared to those found in classic mean-variance efficient portfolios. Both these aspects are

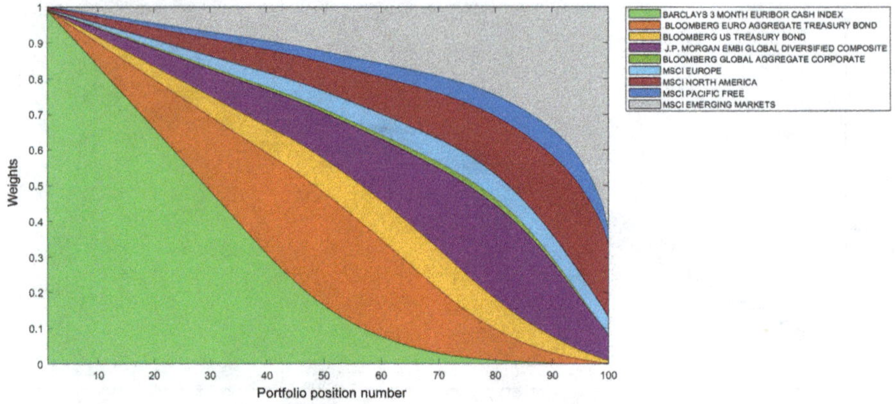

Fig. 4.15 Composition of portfolios on the Resampled Efficient Frontier

taken up again in Sect. 4.7.2. Here, we need only to note that asset classes totally excluded from the traditional Markowitz model (MSCI Europe, MSCI Pacific Free and Bloomberg Global Aggregate Corporate) are present in the 100 resampled portfolios, respectively, with a mean allocation weight of 4.43%, 3.45%, and 0.86%. In the same portfolios, the range for the bond component is from 99.29 to 8.20% and for the equity component from 0.71 to 91.80%. However, unlike in the classic approach, no asset allocation solution uses exclusively MSCI Emerging Markets.

The expected risk-expected return combinations of the resampled portfolios are shown in Fig. 4.14 together with those of traditional portfolios. The Resampled Efficient Frontier (REF) clearly lies below the classic efficient frontier. In this respect, it should be remembered that the lower position of the REF is due to the fact that these portfolios are evaluated using sample inputs; this is therefore an in-sample result. If we look at the out-of-sample realisations, the portfolios on the Resampled Efficient Frontier offer more appropriate investments compared to Markowitz's efficient frontier, for those that do not assign a probability of 100% to their estimates. Figure 4.16 also shows a common occurrence with Resampling, i.e. a significant tightening of the range of risk associated with resampled portfolios. The maximum standard deviation of the Resampled Efficient Frontier is indeed significantly lower than the one reached by the classic efficient frontier (14.35% versus 17.58%). This is caused by the greater diversification of the resampled portfolios (obtained by portfolio optimisation in a situation of uncertainty) which, evaluated on the basis of the original inputs, makes it impossible, with increases in return, to achieve the level of risk corresponding to a 100% investment in the asset class with the highest return (the MSCI Emerging Markets) as it happened with the traditional Mean-Variance Optimisation.

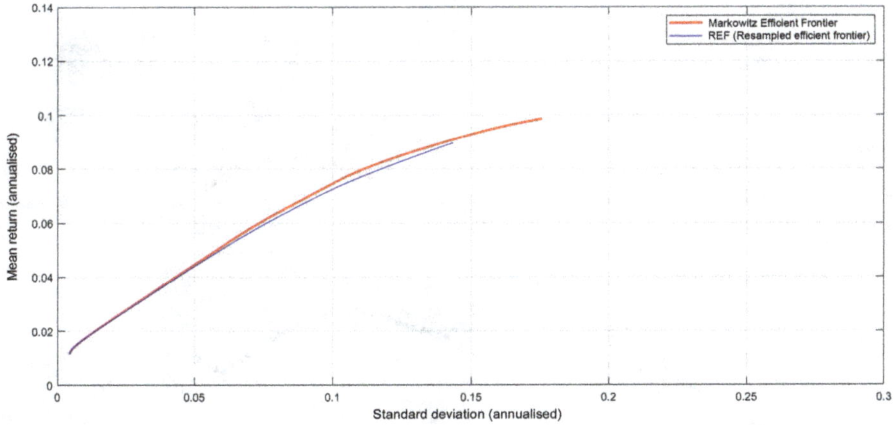

Fig. 4.16 Resampled Efficient Frontier and classical efficient frontier

4.7.2 The Properties of Resampled Portfolios

Continuing and expanding on the discussion about the application of Resampling, this section provides a detailed analysis of the specific characteristics of resampled portfolios. Having introduced the methodology as a way of managing estimation risk, first we must outline the desirable qualities of resampled portfolios. It should be remembered that Resampling generates portfolios:

- With a higher level of diversification.
- With a smooth transition in the vector of weights as risk increases.

Regarding the first point, the fact that "... the REF is based on averages of all properly associated optimal portfolios on the simulated MV efficient frontiers"[58] means that it is natural to include a larger number of asset classes in the vectors of resampled portfolio weights. The attitude of Resampling for greater inclusiveness towards the components of the investable universe, as well as for "moderating" very concentrated positions where they would appear more daring and "braze", i.e. in correspondence with the most rightward section of the efficient frontier, can also be deduced from the comparison between the ENC metric introduced in Sect. 4.4 calculated for the portfolios of the classic efficient frontier and for the resampled portfolios. In fact, from Fig. 4.17 the higher values of the ENC for the resampled portfolios located in that area of the opportunity set are evident.

It is also worth noting that the dissimilarity between the composition of resampled and the corresponding classic portfolios increases as we move to the right along the respective efficient frontiers. By way of example, Table 4.10 shows the composition of the first, the fiftieth, and the last portfolios on both the resampled and the classic

[58]Michaud and Michaud (2008, p. 44).

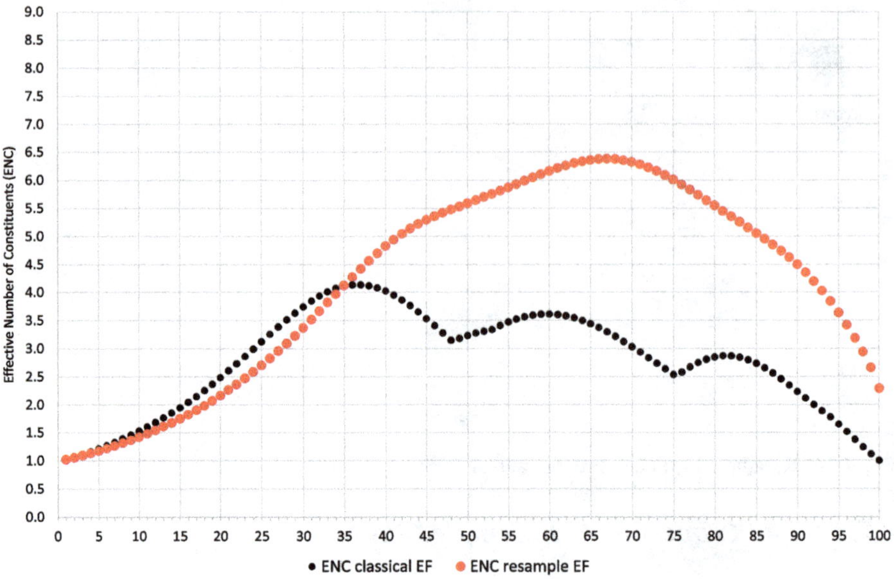

Fig. 4.17 Comparison of the ENC metric for the classical efficient frontier and the resampled efficient frontier

Table 4.10 Comparison of the asset allocations on the classic and on the Resampled Efficient Frontiers

	Portfolio number 1		Portfolio number 50		Portfolio number 100	
	REF (%)	Classic EF (%)	REF (%)	Classic EF (%)	REF (%)	Classic EF (%)
BARCLAYS 3 MONTH EURIBOR CASH INDEX	98.868	99.088	16.361	0.000	0.000	0.000
BLOOMBERG EURO AGGREGATE TREASURY BOND	0.153	0.000	31.764	46.009	0.200	0.000
BLOOMBERG US TREASURY BOND	0.017	0.000	9.387	1.900	0.800	0.000
J.P. MORGAN EMBI GLOBAL DIVERSIFIED COMPOSITE	0.004	0.000	13.097	23.893	7.000	0.000
BLOOMBERG GLOBAL AGGREGATE CORPORATE	0.253	0.216	0.911	0.000	0.200	0.000
MSCI EUROPE	0.092	0.000	5.013	0.000	4.600	0.000
MSCI NORTH AMERICA	0.537	0.697	8.046	12.485	21.000	0.000
MSCI PACIFIC FREE	0.074	0.000	3.065	0.000	4.200	0.000
MSCI EMERGING MARKETS	0.003	0.000	12.355	15.713	62.000	100.000

efficient frontiers: the differences in the optimal compositions are indeed more marked for the 50th and the 100th portfolios. In both cases, the REF recommends more moderate bets on individual asset classes and, generally, less extreme weights.

As far as the gradual transition of weights from one asset class to another is concerned, it should be kept in mind that this is important since it makes optimal portfolios more reasonable and thus easier to propose and of greater practical significance. The absence of sharp changes is explained by the portfolios' more limited dependence on a specific set of inputs. This, in turn, results from the process of averaging the weights of portfolios with the same rank on the simulated efficient frontiers, which, as well as ensuring that resampled portfolios weights sum to 1, makes the latter robust and stable and no longer hypersensitive to slight changes in sample estimates.

Alongside these strengths, Scherer attributes some pitfalls to resampled portfolios.[59] The first of these is the possibility that Resampling may lack a fundamental requirement of a set of optimal portfolios, namely monotonous increase in expected returns as risk increases. Indeed, it cannot be denied that the Resampled Efficient Frontier may present convex parts. Although this situation may be difficult to accept (because if it occurs, it means that one could construct portfolios superior to the resampled portfolios in terms of risk-return trade-offs by a linear combination of two resampled portfolios), it is easily explained. Specifically, we must remember that the optimal weights derived from multiple sets of simulated inputs are not optimal for the original inputs: this means that the "responsibility" lies with the mechanism used to evaluate the parameters of the resampled portfolios.

The second weakness that Scherer attributes to resampled portfolios is the possibility that their composition may be significantly influenced by a few "lucky draws". In other words, the average allocation (and, consequently, the "resampled weight") of an asset class in a given resampled portfolio may be markedly conditioned by a small number of cases (compared to the total number of simulations) characterised by particularly high allocations. In this respect, however, we should note that the level of exposure of the average of portfolio weights to outlier extractions can be verified by the user of the Resampling procedure. In practice, checks for lucky draws can be made by examining the distribution of weights by single asset class and by portfolio with a given rank across the simulations performed. This implies that various statistics can be calculated on an asset-by-asset basis besides the average weight, for example the median, percentiles, and inter-percentile ranges. By way of example, a test of this type on the asset class MSCI Europe in the portfolio ranked 95th on the REF shown in Fig. 4.15, which has a resampled weight of 6.28%, reveals 89 cases (out of a total of 500 simulations) in which the weight of European Equity is greater than zero. Taking only those cases with positive weights, the 5th and 95th percentiles are represented by the weights 1.24% and 89.31%, respectively, with an average value of 35.30%. With reference to the same portfolio, also for the MSCI Pacific equity, which has a resampled weight of 7.37%, a more in-depth investigation would lead to ascertaining that only in 96 of the 500 simulations the associated weight is greater than zero. Limited to the

[59] Scherer (2002, 2015).

aforementioned 96 simulations, the average weight associated with the MSCI Pacific would be equal to 38.38%.

Scherer calls the third shortcoming or unwanted feature of Resampling the "volatility option".[60] This describes the increase in the average allocation of an asset class by the Resampling procedure as a result of increasing its volatility while leaving the other parameters unchanged. Paradoxically, then, a worsening in the quality of an asset class leads to an increased weight. However, this apparently not plausible situation can be easily explained: it is the result of the average rule of the Resampling procedure in combination with the long-only constraint typical of Markowitz optimisation. Indeed, the simulations that benefit from a hypothesis of greater volatility, suggesting scenarios of high expected returns for the given asset class, propose strongly positive allocations. In contrast, because of the long-only constraint, those simulations that suffer as a result of supposed greater volatility and present scenarios with strongly negative expected returns cannot propose short positions, but only set the weight to zero. Obviously, the subsequent calculation of the average weight cannot benefit from the compensation between positive and negative exposures that would occur with unconstrained optimisation.

As an example of the "volatility option", we will vary the volatility of one poor performing equity asset class using the classic Markowitz inputs given in Table 4.4. Specifically, this is the Pacific Equity asset class, for which, in addition to the sample value (annualised) of $\widehat{\sigma}$ that equals 13.95%, we also set the values15%, 20%, and 25%. Next, using the assumed new values of $\widehat{\sigma}$ and implementing the Resampling procedure with the same operational choices as in Sect. 4.7.1, we obtain the new compositions of the maximum-variance portfolio on the Resampled Efficient Frontier.

Table 4.11 shows that, although Pacific Equity is an equity asset class with low average return, its percentage weight in the maximum-variance portfolio increases as the assumed level of dispersion of its returns grows. Consequently, the "volatility option" is really a possible weakness of Resampling. Nevertheless, imagining changes in only one single input (as volatility), while all else remains unaffected is stretching the point a bit.

The above-mentioned author also indicates one final problem with Resampling: the assumption that the multivariate distribution generating the possible paths of asset returns for N random variables is perfectly well-known, and even more so, that it can be described by $\left(\widehat{\boldsymbol{\mu}};\widehat{\boldsymbol{\Sigma}}\right)$ derived exclusively from historical returns. As a starting point, exclusive trust in sample data is considered a limitation since it means that the inputs for the resampling are themselves also very uncertain. To be honest, this claim should not be considered an attack on the effectiveness and the acumen found in the heuristic approach adopted by Resampling so as to ensure that the problem of estimation risk is not neglected, but rather as a stimulus to explore possible integration between heuristic and Bayesian approaches, as suggested in Sect. 4.5.

[60] Scherer (2002, 2015).

Table 4.11 Resampling and "volatility option"

	Portfolio number 100	Portfolio number 100	Portfolio number 100	Portfolio number 100
	ST. DEV = 13.95%	ST. DEV = 15.00%	ST. DEV = 20.00%	ST. DEV = 25.00%
BARCLAYS 3 MONTH EURIBOR CASH INDEX	0.000%	0.000%	0.000%	0.000%
BLOOMBERG EURO AGGREGATE TREASURY BOND	0.200%	0.200%	0.200%	0.000%
BLOOMBERG US TREASURY BOND	0.800%	0.800%	1.200%	1.800%
J.P. MORGAN EMBI GLOBAL DIVERSIFIED COMPOSITE	7.000%	6.800%	6.400%	7.000%
BLOOMBERG GLOBAL AGGREGATE CORPORATE	0.200%	0.200%	0.400%	0.000%
MSCI EUROPE	4.600%	4.600%	3.600%	4.400%
MSCI NORTH AMERICA	21.000%	19.800%	19.200%	12.600%
MSCI PACIFIC FREE	**4.200%**	**8.400%**	**14.800%**	**25.600%**
MSCI EMERGING MARKETS	62.000%	59.200%	54.200%	48.600%
BARCLAYS 3 MONTH EURIBOR CASH INDEX	0.000%	0.000%	0.000%	0.000%

4.7.3 *Discretionary Choices in Resampling*

The description of the steps of the Resampling process, together with the practical example given in Sects. 4.7 and 4.7.1, deliberately suggests that the method is completely standardised and leaves no room for adaptation and customisation. At this point, however, we must admit that this is not wholly correct, as there is certainly one aspect of Resampling which does require the user to make a choice: the length (L) of simulated paths of returns for N asset classes that are necessary to compute the simulated efficient frontiers. Previously, L was set at 240 simply because the original inputs describing the multivariate normal distribution from which the simulated returns were generated are sample estimates based on time series of monthly returns over a period of

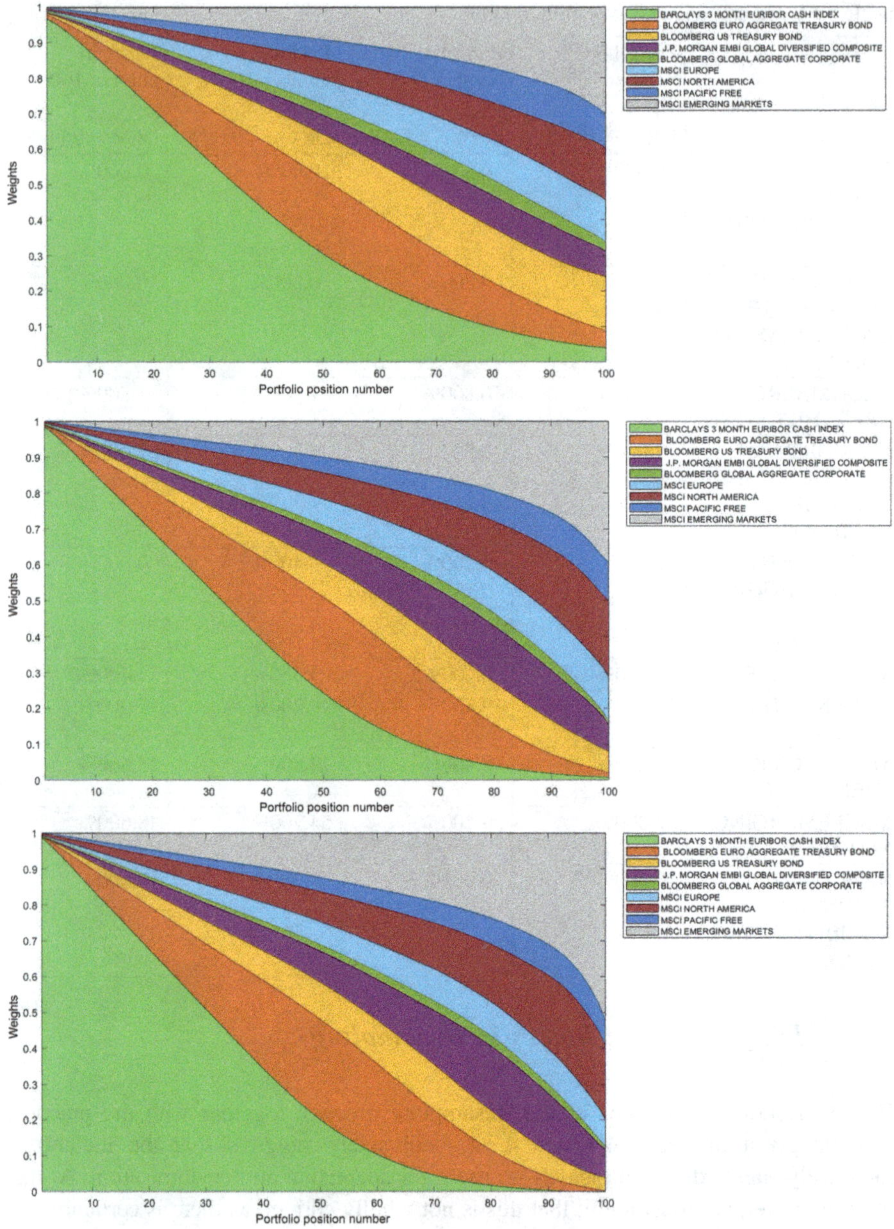

Fig. 4.18 Composition of the Resampled Efficient Frontier (REF) with simulated paths of 12, 60 and 120 observations

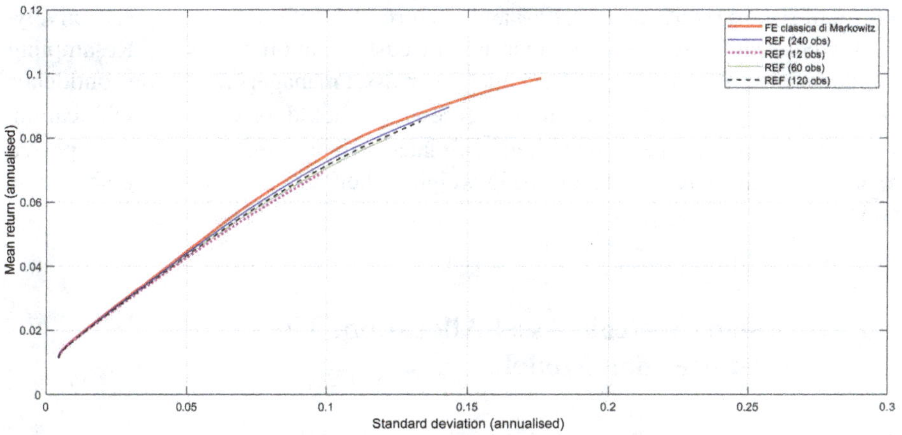

Fig. 4.19 Resampled Efficient Frontier (REF) with simulated paths of 12, 60 and 120 observations

20 years. In this way, we associated the level of imprecision of the sample estimates with the level of information uncertainty that Resampling is called upon to reveal.

The flexible nature of parameter L is clear from the words of Michaud and Michaud: "The number of simulated observations used to compute the simulated MV efficient frontiers in each resampling of risk-return estimates is a free parameter of the RE optimisation process".[61]

That said, the key question is how this flexibility should be used. The recommendation might be that users of the Resampling procedure see L as the way to consider the level of trust and reliability afforded to their estimates.

To facilitate the understanding of this idea, it is useful to provide concrete examples of the impact that a different value of L could have on the composition of the resampled portfolios and on their risk-return combinations. To this end, the Resampling procedure described in Sect. 4.7 was, therefore, repeated with the same original inputs $(\widehat{\boldsymbol{\mu}}; \widehat{\boldsymbol{\Sigma}})$ but with L equal to 12, 60, and 120. The composition of the allocation vectors for the 100 portfolios on the Resampled Efficient Frontier in each of the three cases is shown in Fig. 4.18. Careful observation reveals that a very low value for L results in more diversified compositions, above all for the more extreme portfolios expressing solutions that tend towards the equally weighted allocation and, consequently, towards forms of naïve diversification. On the other hand, high values for L lead to more defined compositions. The impacts of different values of L on the descriptive parameters of resampled portfolios are illustrated in Fig. 4.19, which shows that the smaller the value of L, the lower and shorter the Resampled Efficient Frontier. With, for example, $L = 12$, the minimum and maximum risk-return portfolios have standard deviations of 0.47% and 9.87% and expected returns of 1.19% and 6.91%, respectively, while with $L = 60$, the equivalent portfolios have

[61] Michaud and Michaud (2008, p. 16).

a level of risk of 0.45% and 12.29% and a return of 1.14% and 8.02%, respectively. At this point, it is clear how the margins for customisation offered by Resampling should be used. A low value of L is better when asset managers have little confidence in their $(\widehat{\boldsymbol{\mu}}; \widehat{\boldsymbol{\Sigma}})$ estimates, as in this case they should privilege diversification. Whereas when managers trust their estimates and are seeking higher expected returns through more defined portfolio weights, then a higher value of L should be preferred.

4.8 Bayesian Strategic Asset Allocation: The Black-Litterman Model

As stated in Sect. 4.5, the use of Bayesian approaches for portfolio construction implies that estimation risk will be managed by adjusting the inputs rather than doing something on the optimisation process. It should also be noted that Bayesian methods focus mainly on those parameters where estimation errors have more serious consequences for portfolios, i.e. those parameters representing expected returns.

A necessary prerequisite for an understanding of how Bayesian methods work in strategic asset allocation problems is awareness of the notion of parameter accepted by these methods. It is correct to claim that in a Bayesian framework, a parameter θ is an unknown random variable, knowledge of which is to be improved by virtue of different information sets/sources. This is a fundamental distinctive element compared to the classic framework used in Sect. 4.4 and the following sections, since the classic framework views a parameter as a single (i.e. fixed), yet unknown value, and seeks to ascertain the true value of the parameter by relying exclusively on one source of information. These essential elements obviously lead to a notable difference between the Bayesian and the classic estimation processes in terms of selected and treated inputs and resulting outputs. As is well-known, the classic statistical framework uses only sample data and the unknown parameter is estimated directly by applying an appropriate function (mean, variance, etc.) to those data. In contrast, a Bayesian framework seeks to include in the estimation process both past knowledge and information, generally identified as prior information, and relevant new and original information, for example, the subjective opinions of asset managers resulting from their own analyses. As regards outputs, the estimation process does not return a single number, but a probability distribution of θ, namely a probability density function (pdf).

The formal, theoretical mechanism that allows existing knowledge to be combined with and updated by new information is given by the Bayes' theorem (or Bayes' rule), according to which the "goal", here indicated as $p(\theta|\mathbf{X})$, is defined as θ's posterior distribution and describes the level of uncertainty of possible values of θ after new information and/or new data, **X**, have been considered. The logical procedure to get the posterior distribution can be written as:

$$p(\theta|\mathbf{X}) = p(\theta)\frac{p(\mathbf{X}|\theta)}{p(\mathbf{X})} \tag{4.29}$$

where the prior (unconditional) distribution of θ, $p(\theta)$, is multiplied by an adjustment factor corresponding to the ratio between the likelihood function of **X**, $p(\mathbf{X}|\theta)$, expressing the probability of observing information **X**, assuming a certain value of θ to be true, and the unconditional probability of **X**. Ultimately, we can state that the Bayes' rule provides a theoretically "healthy" way either to blend two probability distributions and obtain a third (the posterior distribution) or to modify the level of uncertainty or belief in possible values of parameter θ by updating and then transforming an unconditional probability distribution of a random variable into a conditional probability distribution.

Having shown the mechanism that lies at the heart of the Bayesian framework, it must be emphasised what a precious tool it can be for asset managers. Managing to combine a quantitative approach with personal judgements and assessments is one of the most difficult problems they face, and this can lead to resistance or even rejection of a quantitative tool seen as unable to make effective use of the added value that they, the managers, can provide. Bayesian statistics can indeed overcome this conflict by providing asset managers with a logical method, as shown in (4.29), allowing them to incorporate their own judgements into a quantitative framework and so update the optimisation inputs. Blending a number of information sets supports the modus operandi of decision-makers, but also guarantees a rationalisation of their subjectivity. Some sort of favourable opinion of this approach can also be found in the authoritative words of Markowitz himself: "the rational investor is a Bayesian".[62]

The Bayesian method described here finds its most prominent application in the Black-Litterman model, developed by its two creators in the Quantitative Resources Group at Goldman Sachs in the early 1990s.[63] While the sections below will provide a reasoned and exhaustive presentation of the statistical and/or analytical aspects of the model, we would like to indicate here, albeit briefly, the main and decisive output that the method provides to asset managers (in an investment committee or an

[62] Markowitz (1987, p. 57). A close reading of Markowitz's (1952) contribution reveals that, while he used sample estimates as inputs, he also suggested that different methods should be used to form the inputs: "My feeling is that the statistical computations should be used to arrive at a tentative set of μ_i and σ_{ij}. Judgment should then be used in increasing or decreasing some of these μ_i and σ_{ij} on the basis of factors or nuances not taken into account by the formal computations". He added: "One suggestion as to tentative μ_i, σ_{ij} is to use observed μ_i, σ_{ij} for some period of the past. I believe that better methods, which take into account more information, can be found. I believe that what is needed is essentially a "probabilistic" reformulation of security analysis. I will not pursue this subject here, for this is another story". Markowitz (1952, p. 91).

[63] Black and Litterman (1991, 1992).

asset allocation team, for example). This is a vector of expected returns to be used as an input in the optimisation process that represents the mean of posterior distribution, obtained by combining the two elements that are considered the fundamental ingredients of the Black-Litterman model:

- The prior information legitimated by an asset pricing and equilibrium model such as CAPM that can serve, in the words of the authors, as a "centre of gravity" for the final expected returns and, consequently, represent an initial and unconditional forecast of returns; and
- The private information expressing exclusively the more or less robust beliefs and opinions of the asset managers which may cause an adjustment in final expected returns compared to initial forecasts. The private information is obviously exogenous to the asset pricing model.

4.8.1 The First Information Set of the Black-Litterman Model: Equilibrium or Implicit Returns

In their well-known contribution, "Global Portfolio Optimisation", published in the *Financial Analysts Journal* in 1992, Fisher Black and Robert Litterman clearly defined the prior information considered in their model: "In our model equilibrium risk premiums provide a neutral reference point for expected returns" and, as already mentioned above: "Equilibrium risk premiums provide a center of gravity for expected returns". However, this default response to the question "What are the future expected returns?" is not something that can be seen directly, and therefore must be sought out or deduced.

The technique that can be used to this end which ensures that equilibrium risk premiums (and returns) can also be called implied risk premiums (and returns) is that of reverse optimisation introduced by Sharpe.[64] Its name derives from the fact that, compared to a typical portfolio optimisation procedure, the technique envisages the reversal of input and output: the departure point is a portfolio and the arrival point is a vector of returns. Starting from the premise that in the absence of particular private information and subjective views, it makes sense for asset managers to recommend holding a portfolio whose composition is in line with the weights that the asset classes, given their capitalisation, have in the market, the authors propose a market portfolio as an appropriate reference portfolio. Considering that the market portfolio consists by definition of a set of positive weights adding up to 1, it is possible to undertake reverse optimisation, and so obtain the implicit excess return (or risk premiums), starting from a quadratic utility function with a given covariance matrix of asset class returns, market capitalisation weights, and the risk aversion parameter.

[64] Regarding reverse optimisation, see: Sharpe (1974), Best and Grauer (1991), and Sharpe (2010).

Next, the utility function can be maximised in closed form, given the absence of constraints. Formally, then, reverse optimisation starts with (4.30):

$$\max_{\mathbf{w}} \ (U) = \max_{\mathbf{w}}\left(\mathbf{w}'_{Eq.}\mathbf{\Pi}_{Ex\_eq.} - \frac{\delta}{2}\mathbf{w}'_{Eq.}\mathbf{\Sigma}\ \mathbf{w}_{Eq.}\right) \tag{4.30}$$

where:

$\mathbf{w}_{Eq.} = N \times 1$ vector of the market capitalisation weights for an investment universe of N asset classes.

$\mathbf{\Pi}_{Ex\_eq.} = N \times 1$ vector of the equilibrium risk premiums or the equilibrium excess returns.

$\mathbf{\Sigma} = N \times N$ covariance matrix of returns of the asset classes.

$\delta =$ a scalar obtained by the expression $\frac{\overline{R}_{P\_mkt} - R_f}{\sigma^2_{P\_mkt}}$, where R_f is the risk-free return, $\overline{R}_{P\_mkt}$ is the mean return of the market portfolio calculated from the series of returns that it would have generated over time and calculated as $\mathbf{R} \times \mathbf{w}_{Eq}$, where $\mathbf{R}$ is the $T \times N$ matrix of historical returns of the asset classes in the market portfolio, and $\sigma^2_{P\_mkt}$ is the variance of the portfolio given by $\mathbf{w}'_{Eq.}\mathbf{\Sigma}\ \mathbf{w}_{Eq.}$. In qualitative terms, δ can be defined as the average risk aversion coefficient of financial market participants and can be interpreted as the market price of risk, i.e. the excess return demanded by investors per unit of variance. This does not rule out the possibility, when applying the model, of using an expected market risk premium to identify the numerator of δ, rather than the risk premium that would have characterised the equilibrium portfolio in the past.

The unconstrained maximisation problem is solved calculating the derivative of (4.30) with respect to $\mathbf{w}$ and requiring that it be equal to zero. In this way, we obtain (4.31):

$$\mathbf{\Pi}_{Ex\_eq.} - \delta\mathbf{\Sigma}\mathbf{w}_{Eq.} = 0 \tag{4.31}$$

from which, resolving for the unknown variable $\mathbf{\Pi}_{Ex\_eq.}$, we have (4.32):

$$\mathbf{\Pi}_{Ex\_eq.} = \delta\mathbf{\Sigma}\mathbf{w}_{Eq.} \tag{4.32}$$

The equilibrium or implicit returns are obtained by adding the estimated risk-free return to the value in (4.32):

$$\mathbf{\Pi}_{Eq.} = \mathbf{\Pi}_{Ex\_eq.} + R_f\mathbf{e} \tag{4.33}$$

where $\mathbf{e}$ is an $N \times 1$ vector of ones. The formula (4.33) is fully equivalent to (4.34):

$$\mathbf{\Pi}_{Eq.} = R_f\mathbf{e} + \frac{\overline{R}_{P\_mkt} - R_f}{\sigma^2_{P\_mkt}}\mathbf{\Sigma}\ \mathbf{w}_{Eq.} \tag{4.34}$$

Given the procedure followed to obtain it, the prior information indicated by Black and Litterman can legitimately be considered a plausible, and perhaps the best possible forecast of expected returns if public information alone is available, since by virtue of the inverse optimisation starting from a market portfolio, it is as if the views of all market participants were considered.[65]

Obviously, if this prior information is used in a Mean-Variance Optimisation, it will give the market portfolio weights which were used to obtain the equilibrium returns. This proves that market portfolio provides a rational, convincing indication of the best diversification strategy in the absence of actual or supposed information advantages from the asset manager. Consistency of prior information with the CAPM equilibrium model is confirmed not only by the starting point of the reverse optimisation procedure, which allows to extrapolate returns that, in the words of the authors "would clear the market",[66] but also by the correspondence of (4.34) with the theory of "correct expected return" for an asset according to the CAPM.[67]

Having clarified the first information set that helps to obtain expected return inputs according to the Black-Litterman model, the Bayesian framework requires that we formulate their prior distribution. Obviously, the output of reverse optimisation, i.e. the equilibrium or implied risk premiums (and returns) contributes to the formula, which is given by (4.35):

[65] Black and Litterman (1991, 1992) and Cheung (2010).

[66] Indeed, implied equilibrium risk premium (and return) are the returns that, in a given instance, provide a balance between the supply and demand of the asset.

[67] In algebraic terms, according to this model, for a generic asset class i among the N considered can be written as:

$$R_i = R_f + \beta_i\left(\overline{R}_{P\_\mathrm{mkt}} - R_f\right)$$

which, considering the definition of β_i, can be re-written as:

$$R_i = R_f + \frac{\mathrm{cov}(R_i, R_{P\_\mathrm{mkt}})}{\sigma^2_{P\_\mathrm{mkt}}}\left(\overline{R}_{P\_\mathrm{mkt}} - R_f\right)$$

Algebraically, the latter is identical to:

$$R_i = R_f + \frac{\left(\overline{R}_{P\_\mathrm{mkt}} - R_f\right)}{\sigma^2_{P\_\mathrm{mkt}}}\mathrm{cov}(R_i, R_{P\_\mathrm{mkt}})$$

The re-writing extended to N asset classes in matrix form coincides with (4.34):

$$\mathbf{\Pi}_{\mathrm{Eq.}} = R_f\mathbf{e} + \frac{\left(\overline{R}_{P\_\mathrm{mkt}} - R_f\right)}{\sigma^2_{P\_\mathrm{mkt}}}\mathbf{\Sigma}\mathbf{w}_{\mathrm{Eq.}}$$

$$\mu \sim N\left(\mathbf{\Pi}_{Eq.}, \tau\mathbf{\Sigma}\right) \tag{4.35}$$

Thus, in the Black-Litterman model, expected returns are initially described by a multivariate normal distribution with a mean vector given by the implicit returns and a covariance matrix (estimated by sampling) proportional to $\mathbf{\Sigma}$ with a proportionality constant given by the scale parameter τ. This last parameter reflects the level of confidence or uncertainty in the estimated equilibrium returns, in the sense that the closer the parameter is to zero, the greater the confidence in the equilibrium estimates. Despite this common interpretation, differing and conflicting choices can be found in the literature. Satchell and Scowcroft and Meucci opt for a value equal to 1, while Black and Litterman, He and Litterman and Lee recommend values very close to zero, since the uncertainty of the mean is lower than the uncertainty of the returns themselves.[68] Although the choices expressed by the various authors tend to support a preference for τ in the range (0,1], a calibration method is needed to calculate the parameter. Considering that the covariance matrix is estimated on the basis of historical data, and that the estimation error tends to decline as the sample size increases and to increase as the number of asset classes considered rises, acceptable calibration criteria are the following (4.36) and (4.37):

$$\tau = \frac{1}{T} \tag{4.36}$$

$$\tau = \frac{1}{T-N} \tag{4.37}$$

The illustration given in this section of prior information and prior distribution of expected returns fully conforms to Black and Litterman's original contributions.[69] This does not, however, categorically exclude the adoption of possible variations that sometimes have been used by practitioners. To be more precise, one case is the possibility that the starting point for the reverse optimisation procedure may not be the market portfolio, but a portfolio that an investor in a certain geographical area considers neutral/normal;[70] or another may be the possibility of making the calculation of implicit returns correspond to a target Sharpe Ratio, i.e. the ratio between portfolio differential return with respect to the risk-free rate and its volatility, which in part differs from the preceding definition of δ. This last option can be found in Bevan and Winkelmann.[71]

---

[68] Black and Litterman (1991, 1992), He and Litterman (1999), Lee (2000), Satchell and Scowcroft (2000), and Meucci (2010).

[69] Black and Litterman (1991, 1992).

[70] Adoption of a departure point for reverse optimisation other than the market portfolio is usual in cases where the Black-Litterman model is used to support tactical rather than strategic asset allocation. On tactical asset allocation, which is not the object of this work, see: Lee (2000), and Braga and Natale (2012).

[71] Bevan and Winkelmann (1998) and Allaj (2013).

4.8.2 The Second Information Set of the Black-Litterman Model: The Views

The second important source of information envisaged in the Black-Litterman model (which later will interact with implied returns)[72] are the asset manager's views. Obviously, this is extra information of a private and personal nature with which a manager can express expectations deviating from prevailing information/belief by market participants or, more generally, from implied returns.

The presence in the model of this second information set implies that attention must be paid to a number of issues. First, we must consider the fundamental aspects concerning the views, whereas the problem of their formal representation is dealt with below. This is because, from an asset manager's point of view, it is crucial first of all to understand what can be expressed by means of the Black-Litterman model. In this respect, it should be made clear that managers can use the model to formulate:

- Absolute views, i.e. subjective performance expectations for an asset class (e.g. "Asset class 1 is believed to be quite likely to produce an annual mean return of y %").
- Relative views, i.e. subjective differential performance expectations of one or more asset classes compared to one or more others (e.g. "Asset class 1 is strongly believed to be able to beat asset class 2 by y % over a year period").
- A number K of views with the only constraint that $0 \leq K \leq N$. So the model does not require the asset manager to express a complete set of views for all asset classes in the investment universe.

With reference to the formal representation of the views, we must remember that the second information set is uncertain too, which means that the expected returns suggested by the views need to be expressed as probability distributions. To do so, a number of technical requirements must be fulfilled:

- Identification of the asset classes for which the asset manager has a view.
- Statement of the value of the views.
- Inclusion of uncertainty or confidence level associated with the views, since (as stated above) they are probability statements and not deterministic claims.

The first need is satisfied by the $K \times N$ matrix $\mathbf{P}$, commonly called view portfolio (and sometimes also projection matrix). Each row $\mathbf{p}$ of this matrix identifies the asset class involved in the view. If it is an absolute view, the row shows the value +1 in the position corresponding to the asset class involved, while the remaining elements are set to zero. If it is a relative view involving just one over-performing asset class and one under-performing asset class, +1 and −1 need to be inserted in the corresponding positions. Finally, when relative views with more than one over-performing and under-performing asset classes are expressed, a weighting scheme

[72] See Sect. 4.8.3 below.

has to be used in order to incorporate this subjective opinion in a row of matrix **P**. The literature provides the equal weighting scheme proposed by Satchell and Scowcroft, and He and Litterman's market capitalisation weighting scheme.[73] In the first case, in the position of an over-performing asset class a value of +1 divided by the number of over-performing markets has to be written, while the value for the position of an under-performing asset class is -1 divided by the number of under-performing asset classes. In the second scheme, the position of an over-performing asset class must show a positive value indicating its relative capitalisation weight with respect to the total capitalisation of the over-performing markets, while the position of an under-performing market must show a negative value indicating the incidence of its capitalisation with respect to that of all under-performing assets. Obviously, compiling a row of matrix **P** is a little more complicated with the latter type of view. It must also be said, however, that asset managers usually do not express their opinions in this way, but rather with views of the first or, even more often, of the second type. Given the way that matrix **P** is compiled, it is clear that the elements of a row in the matrix sum up to zero when a relative view is expressed, while they sum up to +1 if an absolute view is recorded.

The necessary specification of the size of the views is performed simply by inserting their value in a $K \times 1$ vector, indicated as **Q**. What remains is the need to include the uncertainty associated with the view. For this, a second vector, $\boldsymbol{\varepsilon}$, of similar size and containing the error associated with each view must come with vector **Q**. The Black-Litterman model assumes that $\boldsymbol{\varepsilon}$ is a normally distributed random variable with mean zero and $K \times K$ covariance matrix $\boldsymbol{\Omega}$. The diagonal elements of $\boldsymbol{\Omega}$ reflect the variance of error terms while the off-diagonal elements are set equal to zero. This means that views are assumed to be uncorrelated, independent of one another. Regarding the values on the main diagonal of $\boldsymbol{\Omega}$, it is evident that the higher they are, the greater the uncertainty associated with the views. Therefore, it can be claimed that these values establish an inverse relation with the confidence level that asset managers include in their market outlook.

Recalling the points illustrated in Sect. 4.8, it is clear that in the Bayesian model, this second information set will allow an "updating" of expected returns with respect to their corresponding prior estimate. For this reason, it is important to know its representation in the form of a random variable, as shown in (4.38):

$$\mathbf{P}\boldsymbol{\mu} \sim N(\mathbf{Q}; \boldsymbol{\Omega}) \tag{4.38}$$

It is equally important to express its content and characteristics, so that—as required by the model—it can influence the final estimate of expected returns (posterior estimate). The authors suggest to write the second information set by the matrix form (4.39):

[73] Satchell and Scowcroft (2000) and He and Litterman (1999).

$$\mathbf{P}\boldsymbol{\mu} = \mathbf{Q} + \boldsymbol{\varepsilon} \tag{4.39}$$

which describes a system of linear equations. In a more extended form, (4.39) can be expressed as (4.40):

$$\begin{bmatrix} p_{1,1} & \cdots & \cdots & \cdots & p_{1,N} \\ \cdots & \cdots & \cdots & \cdots & \cdots \\ \cdots & \cdots & \cdots & \cdots & \cdots \\ p_{K,1} & \cdots & \cdots & \cdots & p_{K,N} \end{bmatrix} \times \begin{bmatrix} \mu_1 \\ \cdots \\ \cdots \\ \cdots \\ \mu_N \end{bmatrix} = \begin{bmatrix} Q_1 \\ \cdots \\ \cdots \\ Q_K \end{bmatrix} + \begin{bmatrix} \varepsilon_1 \\ \cdots \\ \cdots \\ \varepsilon_K \end{bmatrix} \tag{4.40}$$

The random nature of the error terms is confirmed by (4.41):

$$\boldsymbol{\varepsilon} \sim N\left(\begin{bmatrix} 0 \\ 0 \\ 0 \\ 0 \end{bmatrix}, \begin{bmatrix} \omega_{1,1} & \cdots & \cdots & 0 \\ 0 & \cdots & 0 & 0 \\ 0 & 0 & \cdots & 0 \\ 0 & 0 & 0 & \omega_{K,K} \end{bmatrix} \right) \tag{4.41}$$

where ω_{ij} (with i and $j = 1, \ldots, K$ in this case) shows the variances of the individual error terms that form the diagonal elements of matrix $\boldsymbol{\Omega}$. Computation of these elements is one of the most complicated and controversial aspects of the Black-Litterman model, also because the authors did not provide any clear methodological indications on how to set up $\boldsymbol{\Omega}$ in their original contribution. For the sake of completeness, here is an overview of the different methods found in the literature:[74]

- He and Litterman's "endogenous" approach.
- Drobetz's approach based on the use of a confidence interval.
- Idzorek's approach.
- Meucci's approach.

The proponents of the first approach, He and Litterman (1999), have stated that: "... the confidence level on a view is calibrated so that the ratio between the parameters ω and τ is equal to the variance of the portfolio in the view".[75] In the light of these words, every term ω_{ij} (when $i = j$) is determined by the following formula (4.42):

$$\omega_{ij} = \mathbf{p}(\tau\boldsymbol{\Sigma})\mathbf{p}' \qquad \forall i = j \tag{4.42}$$

where $\mathbf{p}$, as known, is one of the K rows in matrix $\mathbf{P}$ (view portfolio).

[74] He and Litterman (1999), Drobetz (2001), Idzorek (2002), and Meucci (2005).

[75] He and Litterman (1999, p. 6).

This method is called "endogenous" because users of the Black-Litterman model are not required to express their own confidence level in the expressed views, since the variance of the error terms associated with the views is hypothesised as being the same as the one estimated for prior returns.

The remaining three approaches have a feature in common: they all need an "exogenous" value which expresses the asset manager's confidence level in the stated views. The method based on the use of a confidence interval suggests declaring the value q of the view as the central moment of a normal distribution and accompanying this with an indication of a symmetrical range, M_{Range}, reflecting the level of probability, α, which the asset manager feels appropriate for the stated view. Starting from this information, it is possible to calculate the standardised distance from the mean that in a Gaussian distribution isolates the complement to 1 of the confidence level, dividing it equally between the tails to the right and the left. At this point, the only remaining unknown value is the volatility to which the above standardised distance has to be applied as a multiplier. Once identified, it is to be squared to arrive at ω_{ij} (with $i = j$). Formally, the procedure described involves writing the equality as in (4.43), with the unknown value ω_{ij} (with $i = j$):

$$q \pm N^{-1}\left((1-\alpha)/2\right) \times \sqrt{\omega_{ij}} = q \pm M_{\text{Range}} \tag{4.43}$$

which, when resolved, gives (4.44):

$$\omega_{ij} = \left(\frac{\pm M_{\text{Range}}}{\pm N^{-1}\left((1-\alpha)/2\right)}\right)^2 \tag{4.44}$$

where N^{-1} indicates the inverse function of a standard normal distribution.

To facilitate understanding, we can hypothesise an example view as: "asset 1 will have an expected return of 6 % with a probability of 90 % of being between 5 % and 7 %". This case will lead to the following Eq. (4.45):

$$6\% \pm N^{-1}\left((1-90\%)/2\right) \times \sqrt{\omega_{ij}} = 6\% \pm 1\% \tag{4.45}$$

which when resolved for ω_{ij} gives 0.00003696.

Idzorek's approach requires that the diagonal elements of matrix $\mathbf{\Omega}$ are calculated as reciprocals of the confidence level multiplied by a quantity called calibration factor (CF).[76] In this case, the variance matrix of the views assumes the form shown in (4.46):

[76] Idzorek (2002).

$$\mathbf{\Omega} = \begin{bmatrix} \frac{1}{\alpha_1} \times \mathrm{CF} & 0 & 0 & 0 \\ 0 & \frac{1}{\alpha_2} \times \mathrm{CF} & 0 & 0 \\ 0 & 0 & \cdots & 0 \\ 0 & 0 & 0 & \frac{1}{\alpha_K} \times \mathrm{CF} \end{bmatrix} \tag{4.46}$$

The specification of the calibration factor is based on two quantities. The first is given by the algebraic sum of the terms in the matrix identifying, in the author's words, the variance of the view portfolio. The second quantity is the expected average level of confidence in the views ($\overline{\mathrm{LC}}$) assumed equal to 50%. This is correctly expressed in formal terms as (4.47):

$$\mathrm{CF} = (\mathbf{l}'\mathbf{P}\mathbf{\Sigma}\mathbf{P}'\mathbf{l})\overline{\mathrm{LC}} \tag{4.47}$$

where $\mathbf{l}$ is a $K \times 1$ column vector containing ones, so that the expression in round brackets in (4.47) results in a scalar.[77]

Finally, the approach used by Meucci to define matrix $\mathbf{\Omega}$ is shown in (4.48):

$$\mathbf{\Omega} = \begin{bmatrix} \left(\frac{1}{\alpha_1} - 1\right) \times \mathbf{p}_1\tau\mathbf{\Sigma}\mathbf{p}' & 0 & 0 & 0 \\ 0 & \left(\frac{1}{\alpha_2} - 1\right) \times \mathbf{p}_2\tau\mathbf{\Sigma}\mathbf{p}' & 0 & 0 \\ 0 & 0 & \cdots & 0 \\ 0 & 0 & 0 & \left(\frac{1}{\alpha_K} - 1\right) \times \mathbf{p}_K\tau\mathbf{\Sigma}\mathbf{p}'_K \end{bmatrix} \tag{4.48}$$

As is clearly evident, this solution re-uses the connection between the variance of the error terms in the views and the variance in the prior estimates already seen in He and Litterman, but also adds the subjective aspect of the asset manager's different level of conviction and trust in the views through the quantity given in the round brackets.[78] The latter returns higher values as the level of confidence in the views falls, thus generating greater dispersion for the random variable $\boldsymbol{\varepsilon}$ described in (4.41) and, obviously, lower values with higher terms for α.

---

[77] For the sake of completeness only, we note that in the definition of matrix $\mathbf{\Omega}$, the author of this approach follows the logic in (4.4.7) and suggests a way to identify τ. This should result from the ratio between the variance of the view portfolio and the quantity $\frac{\sum_{k=1}^{K}\left(\frac{1}{\alpha_k}\mathrm{CF}\right)}{K}$ which is equal to the average value of the diagonal elements in $\mathbf{\Omega}$.

[78] Meucci (2005), Fuhrer and Hock (2023).

4.8.3 Blending Information Sets and Expected Returns in the Black-Litterman Model

Sections 4.8.1 and 4.8.2 provided an in-depth illustration of the two distinct sources of information that are considered useful in the Black-Litterman model in order to obtain the expected returns inputs $\boldsymbol{\mu}$. Both have been described by probability distributions, as both admit their unknown and random nature. At this point, the next step is to identify the expected returns that are as consistent as possible with both sources of information examined. In practice, we need an answer to the fundamental question: how can the expected returns be derived, taking into account both information sets at the same time?

To begin with, since we are in a Bayesian context, it should be clear that the answer will also take the form of a probability distribution. To be precise, we will obtain a posterior distribution (a multivariate normal) which can be considered as a conditional probability distribution, derived from the prior distribution and its updates by means of the probability distribution of the views. Although we do not intend to repeat all the analytical steps leading to the estimate of the final expected returns according to the Black-Litterman model, it must be emphasised that the mean vector of the posterior distribution derives from the solution of the following problem of constrained optimisation (4.49):

$$\begin{aligned} &\min_{\boldsymbol{\mu}} \ (\boldsymbol{\mu} - \boldsymbol{\Pi}_{Eq.})' \ \tau\boldsymbol{\Sigma}^{-1}(\boldsymbol{\mu} - \boldsymbol{\Pi}_{Eq.}) \\ &\text{subject to} \\ &\mathbf{P}\boldsymbol{\mu} = \mathbf{Q} + \boldsymbol{\varepsilon} \end{aligned} \tag{4.49}$$

The formulation of the problem demonstrates that the authors' goal of combining prior information and views is to obtain expected returns that reflect asset managers' forecasts, while implying the least possible deviation from implicit/equilibrium returns.[79] This is how the best possible integration between the two information sets is to be understood. The mean of the posterior distribution that we obtain is shown in (4.50). This is often called Black-Litterman's "master formula":

$$\boldsymbol{\mu}_{\mathrm{BL}} = \left[(\tau\boldsymbol{\Sigma})^{-1} + \mathbf{P}'\boldsymbol{\Omega}^{-1}\mathbf{P}\right]^{-1} \times \left[(\tau\boldsymbol{\Sigma})^{-1}\boldsymbol{\Pi}_{Eq.} + \mathbf{P}'\boldsymbol{\Omega}^{-1}\mathbf{Q}\right] \tag{4.50}$$

In (4.51), we show the variance of the posterior distribution:

$$\boldsymbol{\Sigma}_{\mathrm{BL}} = \left[(\tau\boldsymbol{\Sigma})^{-1} + \mathbf{P}'\boldsymbol{\Omega}^{-1}\mathbf{P}\right]^{-1} \tag{4.51}$$

[79] In support of this, note that the objective function in (4.49) envisages the minimisation of the distance (the so-called Mahalanobis distance) between two information sets.

On the basis of these location and dispersion parameters, posterior distribution can be written as (4.52):

$$\boldsymbol{\mu} \sim N(\boldsymbol{\mu}_{\mathrm{BL}}; \boldsymbol{\Sigma}_{\mathrm{BL}}) \tag{4.52}$$

Special attention must undoubtedly be paid to the "master formula" (4.50), as it provides the inputs of the returns for a conventional, Markowitz-like optimisation algorithm. The essential considerations and/or reflections regarding this formula or, similarly, the posterior mean can be presented schematically as follows:

- In the Black-Litterman model, expected returns are a combination or, more precisely, a weighted-average of implied/equilibrium returns and views. Therefore, the model, both in its formal design and in its logical structure, categorically excludes that the views can be used in a banal and direct way to substitute prior returns.
- The weighting factors can be easily identified by noting that the expression in the first round bracket in (4.50) is a common multiplicative factor for the terms in the second round bracket. This means that we can write the "weight" $\mathbf{w}_{\Pi_{Eq.}}$ associated to the implied returns and the weight $\mathbf{w}_Q$ associated to the views as $[(\tau\boldsymbol{\Sigma})^{-1} + \mathbf{P}'\boldsymbol{\Omega}^{-1}\mathbf{P}]^{-1}(\tau\boldsymbol{\Sigma})^{-1}$ and $[(\tau\boldsymbol{\Sigma})^{-1} + \mathbf{P}'\boldsymbol{\Omega}^{-1}\mathbf{P}]^{-1}\mathbf{P}'\boldsymbol{\Omega}^{-1}\mathbf{P}$, respectively. The weighting factors sum up to 1 ($\mathbf{w}_{\Pi_{Eq.}} + \mathbf{w}_Q = 1$).
- On the basis of the weighting factors considered, the importance of the two information sets is directly linked to their level of precision, which presents an inverse relation with their corresponding variance. In practice, all other circumstances being the same, the importance of the equilibrium returns will be greater the smaller the variance $\tau\boldsymbol{\Sigma}$, because that will make $(\tau\boldsymbol{\Sigma})^{-1}$ bigger and thus the information set more precise. Similarly, all other circumstances being the same, the importance of the views will be greater the smaller the value of $\boldsymbol{\Omega}$, because it will make $\mathbf{P}'\boldsymbol{\Omega}^{-1}\mathbf{P}$ bigger.
- In the absence of subjective opinions from asset managers, the "master formula" provides the equilibrium returns as input for expected returns. Consequently, if there is no private information, this formula indirectly confirms holding the market portfolio as the best investment solution.

In essence, what emerges from the previous list is that, by means of the equilibrium returns, the model aims and is able to forecast a common departure point for the construction of a strategic portfolio (the market portfolio), from which deviations are possible if the asset manager has a more bullish or bearish view than market participants in some investment categories or has noted the presence of mispricing in the market.

The final strategic portfolio will differ more or less evidently from the market portfolio depending on the extent to which asset managers trust their own private information and on the level of uncertainty associated to the prior estimates. A distinctive effect of the Black-Litterman "master formula" is that it can produce changes in the expected returns of all asset classes with respect to the equilibrium returns, even if the views refer to some classes only. This is a strength, as it indicates that, by virtue of the structure of correlations and covariance in the investment universe, the model appreciates how an explicit view regarding one or more asset classes can entail an implicit view on other asset classes.[80] In the reality of facts, this is a mechanism that rationalises the implementation of the views especially when there is greater correlation between asset classes.

In mathematical terms, it is the result of the matrix product of $\mathbf{P}$, $(\tau\mathbf{\Sigma})^{-1}$, $\mathbf{\Omega}^{-1}$, and $\mathbf{Q}$, which allows indeed the views (inserted in a $K \times 1$ vector) to propagate to N asset classes with the additional advantage of making the expected returns vector less sensitive to the estimation error of the individual view. On this feature of the model, Lee's favourable comment is of interest: "It is this characteristic that makes the Black-Litterman model a much more practical model to use than mean-variance optimisers. While the latter simply computes the final optimised portfolio weights based on inputs with errors, the Black-Litterman model 'spreads out' the errors of the inputs to all others through the covariance matrices of returns and views so that the error-maximization problem of an optimiser is largely mitigated. This redistribution of errors is made to happen in a controlled fashion, consistent with the way these assets move with one another as well as the investor's relative confidence about its views".[81]

The emphasis granted so far to the mean of posterior distribution can be explained with the simple observation that it represents the most valid and best-known support for those wishing to address estimation risk by improving the way in which the most influential inputs—expected returns—are estimated. The other descriptive element of posterior distribution, namely posterior variance as indicated in (4.51), is instead almost completely disregarded, as portfolio optimisation is generally completed with the original sample covariance matrix. To provide an exhaustive representation of the model, however, it must be pointed out that posterior variance can also contribute to the optimisation process. To this end, it must not be considered in isolation, as an expression of the uncertainty associated to the posterior mean and not to the variance of the returns, but in conjunction with the original variance. In practice, the expected covariance of returns of the asset class ($\mathbf{\Sigma}_{\mathrm{EX}}$) can be identified with the following expression (4.53):

$$\mathbf{\Sigma}_{\mathrm{EX}} = \mathbf{\Sigma} + \left[(\tau\mathbf{\Sigma})^{-1} + \mathbf{P}'\mathbf{\Omega}^{-1}\mathbf{P} \right]^{-1} \tag{4.53}$$

which is equivalent to (4.54):

[80] Idzorek (2002, p. 10) underlines this aspect of the Black-Litterman model, observing that: "...it is not uncommon for a single view to cause the return of every asset in the portfolio to change from its Implied Equilibrium return, since each individual return is linked to the other returns via the covariance matrix of returns ($\mathbf{\Sigma}$)".

[81] Lee (2000, p. 130).

Table 4.12 A synthetic overview of the Black-Litterman model

This table provides a schematic representation of the main steps in the Black-Litterman model:

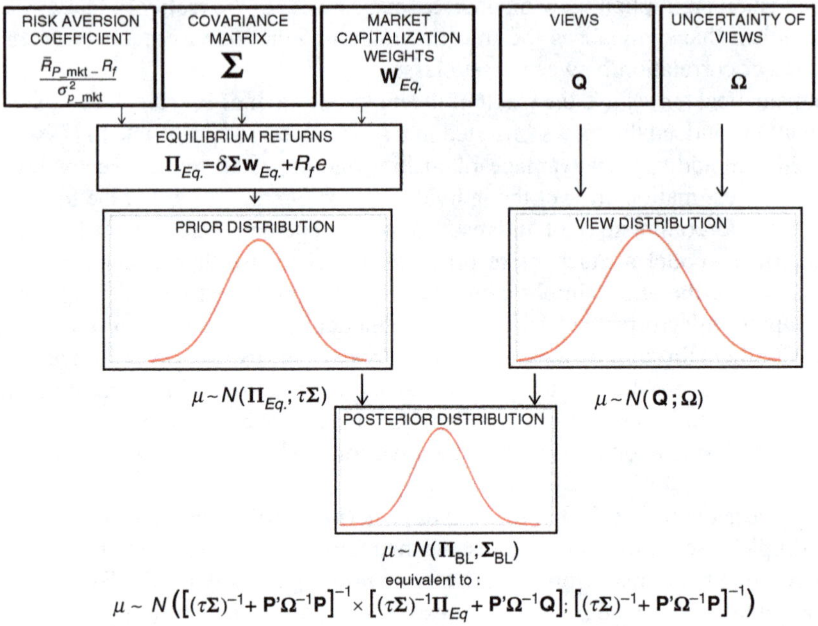

$$\boldsymbol{\Sigma}_{\mathrm{EX}} = \boldsymbol{\Sigma} + \boldsymbol{\Sigma}_{\mathrm{BL}} \tag{4.54}$$

Among the authors who have worked on the Black-Litterman model, only He and Litterman address this issue.[82]

We have now completed both the exploration of the two information sets involved in the Black-Litterman model and the examination of the formal and logical process that makes possible their combination. As a simple, schematic representation of the model, Table 4.12 summarises the main steps in its practical application adapting a layout proposed by Idzorek.[83]

[82] He and Litterman (1999).

[83] Idzorek (2002).

4.8.4 *The Application of the Black-Litterman Model: An Example*

This section gives a practical illustration of the way in which the Black-Litterman model achieves the goal of providing an expected returns vector able to integrate "common knowledge" with an asset manager's "private knowledge".

Based on what has been outlined in the previous sections, an asset manager's first task is to illustrate the composition of the portfolio that he/she would recommend if he/she had no particular private opinion and, therefore, no reason to "bet". We know that the composition should consist of a market capitalisation-weighted portfolio including the asset classes in the investment universe. Considering the initial set of markets selected in Table 4.3 and the corresponding capitalisations at the end of December 2022, this criterion leads to the following composition of the market portfolio (4.55):

$$
w_{\text{Eq.}} = \begin{bmatrix} \text{BARCLAYS 3 MONTH EURIBOR CASH INDEX} \\ \text{BLOOMBERG EURO AGGREGATE TREASURY BOND} \\ \text{BLOOMBERG US TREASURY BOND} \\ \text{J.P.MORGAN EMBI GLOBAL DIVERSIFIED COMPOSITE} \\ \text{BLOOMBERG GLOBAL AGGREGATE CORPORATE} \\ \text{MSCI EUROPE} \\ \text{MSCI NORTH AMERICA} \\ \text{MSCI PACIFIC FREE} \\ \text{MSCI EMERGING MARKETS} \end{bmatrix} = \begin{bmatrix} 3.00\% \\ 13.71\% \\ 19.00\% \\ 4.75\% \\ 14.54\% \\ 7.37\% \\ 28.61\% \\ 3.95\% \\ 5.07\% \end{bmatrix} \tag{4.55}
$$

Next, we estimate the average risk-aversion coefficient of the market participants, δ, assuming an annual expected risk premium of 2.50% and a risk-free return of 1.50%, and using the historical variance of the market portfolio described above which is equal to 0.00588 (approximately corresponding to a standard deviation of 7.67%). The final value obtained for δ is 4.25. At this point, knowing the composition of the market portfolio, the parameter δ, and the historical covariance matrix, we can identify the equilibrium or implied returns using the formula (4.32) indicating the reverse optimisation procedure and adding the risk-free returns to the results obtained. The implementation of these mathematical steps is given in (4.56), which also shows the estimated prior returns for the specific case:

$$
4.25\times\begin{bmatrix}
0.000021 & 0.000017 & 00.00008 & -0.000021 & -0.000034 & -0.000090 & -00.00129 & -0.000086 & -0.000047\\
0.000017 & 0.002164 & 0.001457 & 0.001764 & 0.001582 & 0.000414 & 0.000634 & 0.000627 & 0.000357\\
0.00008 & 0.001457 & 0.009845 & 0.005592 & 0.008148 & -0.003581 & 0.001196 & 0.000602 & -0.003607\\
-0.000021 & 0.001764 & 0.005592 & 0.009304 & 0.007261 & 0.005004 & 0.007286 & 0.007082 & 0.007124\\
-0.000034 & 0.001582 & 0.008148 & 0.007261 & 0.008667 & 0.001148 & 0.005065 & 0.004390 & 0.001663\\
-0.000090 & 0.000414 & -0.003581 & 0.005004 & 0.001148 & 0.020849 & 0.016304 & 0.014789 & 0.01849\\
-0.000129 & 0.000634 & 0.003581 & 0.007287 & 0.005065 & 0.016304 & 0.019620 & 0.014620 & 0.016007\\
-0.000086 & 0.000627 & 0.000602 & 0.007082 & 0.004390 & 0.014789 & 0.014620 & 0.019450 & 0.018340\\
-0.000047 & 0.000357 & -0.003607 & 0.007124 & 0.001663 & 0.018849 & 0.016007 & 0.018340 & 0.030919
\end{bmatrix}
$$

$$
\times\begin{bmatrix} 3.00\%\\ 13.71\%\\ 19.00\%\\ 4.75\%\\ 14.54\%\\ 7.37\%\\ 28.61\%\\ 3.95\%\\ 5.07\% \end{bmatrix} + 1.50\%\times\begin{bmatrix} 1\\1\\1\\1\\1\\1\\1\\1\\1 \end{bmatrix} = \begin{bmatrix} 1.48\%\\ 1.99\%\\ 2.96\%\\ 4.01\%\\ 3.70\%\\ 4.70\%\\ 5.58\%\\ 4.96\%\\ 4.99\% \end{bmatrix} = \mathbf{\Pi}_{Eq} \tag{4.56}
$$

Given the space dedicated to the theoretical analysis of the Black-Litterman model, we do not need to reiterate that, in the model, prior returns do not enjoy an exclusive prerogative in conditioning expected returns; indeed, the views play a role too. In the present example, we assume the asset manager expresses two views, one relative and one absolute, as follows:

- Emerging Markets Equity will over-perform North America Equity by 0.25% annually.
- Emerging Government Bonds will show an annual return of 3%.
- The asset manager associates a confidence level of 80% and 60%, respectively, to these views.

The views certainly deviate from the equilibrium returns, for which MSCI Emerging Markets is 0.59% behind the MSCI North America and the J.P. Morgan Embi Global Diversified has an expected return of 4.10%.With respect to the views, we first need to formulate the system of constraints they represent in the determination of final expected returns[84] and then compile the matrix $\mathbf{\Omega}$ (4.57).

[84] See formula (4.40) in Sect. 4.8.2.

$$\begin{bmatrix} 0\,0\,0\,0\,0\,0\,-1\,0\,1 \\ 0\,0\,0\,1\,0\,0\,\;0\;\,0\,0 \end{bmatrix} \times \begin{bmatrix} \mu_{\text{BARCLAYS 3 MONTH EURIBOR CASH INDEX}} \\ \mu_{\text{BLOOMBERG EURO AGGREGATE TREASURY BOND}} \\ \mu_{\text{BLOOMBERG US TREASURY BOND}} \\ \mu_{\text{J.P.MORGAN EMBI GLOBAL DIVERSIFIED COMPOSITE}} \\ \mu_{\text{BLOOMBERG GLOBAL AGGREGATE CORPORATE}} \\ \mu_{\text{MSCI EUROPE}} \\ \mu_{\text{MSCI NORTH AMERICA}} \\ \mu_{\text{MSCI PACIFIC FREE}} \\ \mu_{\text{MSCI EMERGING MARKETS}} \end{bmatrix}$$

$$= \begin{bmatrix} 0.25\% \\ 3.00\% \end{bmatrix} + \begin{bmatrix} \varepsilon_1 \\ \varepsilon_2 \end{bmatrix} \tag{4.57}$$

In our example, $\boldsymbol{\Omega}$ is a 2×2 matrix, as two views are expressed. We use Meucci's (2005) method described in (4.48) above to construct the matrix.[85] This involves the use of the parameter τ, defined using the calibration criteria suggested in (4.37). Consequently, as the estimate of the sample covariance matrix is based on time series of 240 monthly data, and given that the investment universe comprises nine asset classes, τ is given by (4.58):

$$\tau = \frac{1}{240 - 9} = 0.00433 \tag{4.58}$$

Now that both the information sets used in the Black-Litterman model have been described and explained, we only need to apply the "master formula". Table 4.13 shows the resulting blended returns compared to the equilibrium returns.

The comparison between Black-Litterman expected returns, which represent the posterior mean, and prior (or equilibrium/implied) returns allows us to suggest a number of important considerations for users of the model:

- The returns of the asset classes involved in the views move in the direction suggested by the views, but with an intensity such as to not imply that the final returns will perfectly reflect these forecasts. The reason for this is the effect of uncertainty associated to the views (with confidence levels below 100%). In confirmation of this behaviour, the "master formula" determines a decrease in the expected return for the asset class Emerging Government Bonds from 4.01% to 3.40%, compared to a view suggesting 1% fall, while the expected over-performance of MSCI Emerging Markets against MSCI North America is

[85] Meucci (2005).

Table 4.13 Comparison between Black-Litterman expected returns and prior returns

	Black-Litterman expected returns (%)	Prior returns (%)	Δ (%)
BARCLAYS 3 MONTH EURIBOR CASH INDEX	1.48	1.48	0.00
BLOOMBERG EURO AGGREGATE TREASURY BOND	1.86	1.99	−0.12
BLOOMBERG US TREASURY BOND	2.43	2.96	−0.53
J.P. MORGAN EMBI GLOBAL DIVERSIFIED COMPOSITE	3.40	4.01	−0.61
BLOOMBERG GLOBAL AGGREGATE CORPORATE	3.10	3.70	−0.59
MSCI EUROPE	4.47	4.70	−0.23
MSCI NORTH AMERICA	4.98	5.58	−0.60
MSCI PACIFIC FREE	4.64	4.96	−0.32
MSCI EMERGING MARKETS	5.07	4.99	0.08

estimated at only 0.09%, when the stated view outlined an advantage of 0.25% starting from an equilibrium according to which it was the MSCI North America and not the MSCI Emerging Markets to boast a performance advantage of approximately 60 basis points.

– The Black-Litterman expected returns show a deviation from the equilibrium returns for all asset classes in the investment universe, and not only those explicitly involved in the views, demonstrating that the model records the interactions between asset classes. For example, the weakening, according to the asset manager's opinions, of the relative position of North American equities is reflected not only with a slight increase in the performance of Emerging Markets equity and a more perceptible lowering of the expected return on North America equities, but also with that of the asset class with which the latter has a robust positive correlation, namely European Equity. Likewise, the absolute view regarding Emerging Markets Government Bonds has impacts, albeit in less significant absolute terms, on the expected returns of the other bond components. For all of them, a decrease in return is expected compared to the prior returns.

In Sect. 4.5 we underlined the tendency of the Bayesian approaches to operate on the estimation-side rather than on the model side. In the case of the Black-Litterman model, the output of such estimation effort produces the new set of expected returns discussed above. Although this output was identified by means of a process starting with reverse optimisation (implying the use of a given coefficient, δ), its practical value is certainly not limited to performing a Mean-Variance Optimisation to get the unique portfolio consistent with that coefficient δ and to highlight its deviations from the original market portfolio. It is indeed clear that the expected returns obtained by the Black-Litterman model are a new input for an optimisation process aimed to construct an entire set of efficient portfolios. For this reason, Figs. 4.20 and 4.21 show the composition of the portfolios that a classical Mean-Variance Optimisation

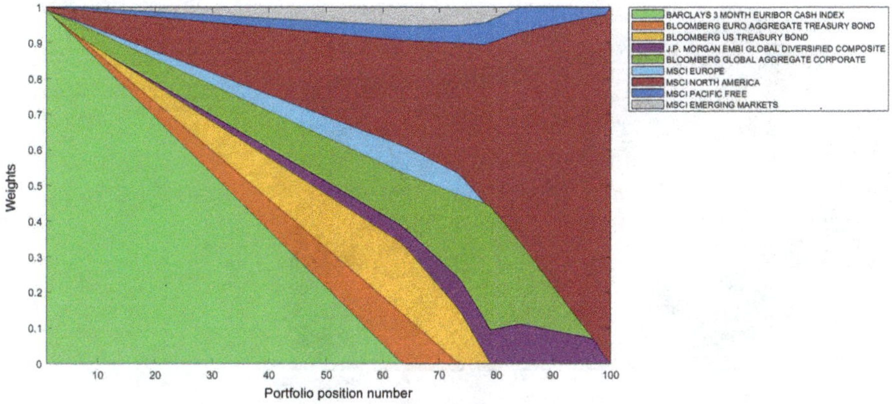

Fig. 4.20 Composition of the efficient frontier portfolios based on the use of equilibrium or implied returns

would have produced if it had been based, in the first case on equilibrium or implied returns, and, in the second, on Black-Litterman expected returns.

On the basis of the example of the application of the Black-Litterman model and the area graphs presented here, two observations can be made. First, it is confirmed that in an optimisation process the Black-Litterman posterior mean can move the efficient compositions in the direction indicated by the views, taking account of the respective level of uncertainty. Indeed, if we look carefully at Figs. 4.20 and 4.21, we can see a notable strengthening of the role played by MSCI Emerging Markets within the most extreme portfolios as it has become, due to the relative view, the asset with the best expected return. In addition to this, it can be observed the removal of any exposure to the J.P. Morgan Embi Global Diversified Composite following the penalisation of its expected return with the consequent expansion of the allocations assigned to the Bloomberg US Treasury Bond and the Bloomberg Global Aggregate Corporate. It is, therefore, true that the Black-Litterman model allows asset managers "to make their opinions count" (more or less incisively). The second and more crucial consideration is that the proposed graphs, in particular, provide the proof or confirmation of the need to integrate heuristic and Bayesian approaches in the strategic asset allocation process.[86] Even if the first obvious sensation for an asset manager looking at Fig. 4.21 is the feeling of greater involvement in the process, it is undeniable that the same graph highlights once again some of the undesirable properties of the portfolios examined in Sect. 4.3.

In order to persuade the adoption of a hybrid approach, it is interesting to show in Fig. 4.22 how the optimal portfolio compositions would have changed if they had been constructed through the combined use of equilibrium returns and Resampling and if, then, the asset manager had expressed the same views previously assumed in

[86]On the issue of the combination between the heuristic and the Bayesian methods, see Barros Fernandes et al. (2012).

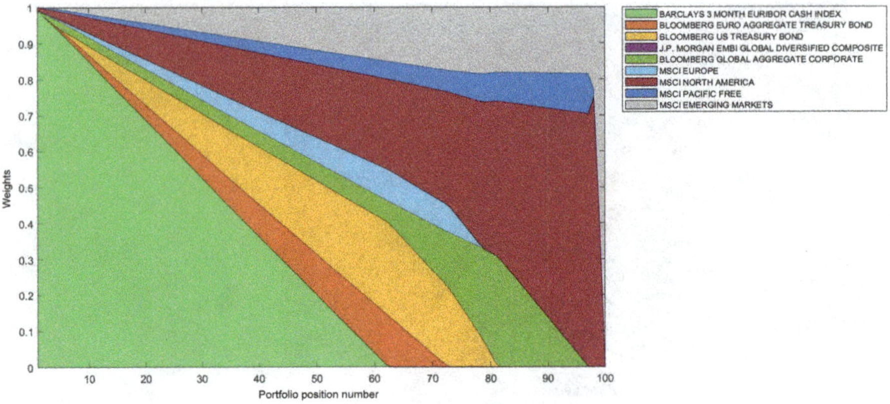

Fig. 4.21 Composition of the efficient frontier portfolios based on the use of Black-Litterman expected returns

this paragraph and, therefore, if the same final expected returns had been identified according to the Black-Litterman model, but the latter had been used as a vector of the means of the multivariate normal used in the Resampling procedure, rather than being used to implement a classic Mean-Variance Optimisation. It can be observed that the portfolios obtained by integrating the final Black-Litterman expected returns and Resampling overcome the undesirable characteristics of the portfolios previously represented in Fig. 4.21 following, at the same time, the asset manager's views in a more disciplined way. As proof of this, it is worth pointing out that, in the proposed application, the average exposure of the portfolios towards the Emerging Markets Government Bonds drops from 6.40% to 4.81% from the case in which they are deducted from "Equilibrium returns + Resampling" to the case in which the portfolio construction exercise is based on "Final expected Black-Litterman returns + Resampling". At the same time, for the Emerging Markets Equity, involved as an asset class outperforming North American Equity in the relative view, there is an increase to 10.33% from 6.57% of the average exposure within the resampled portfolios.

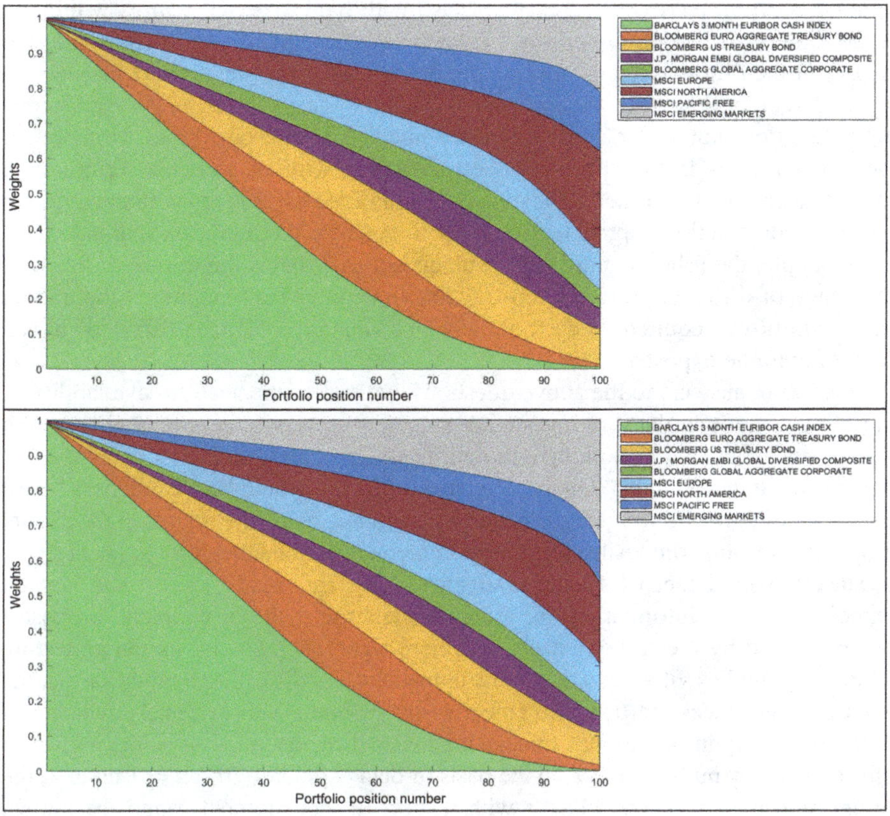

Fig. 4.22 Composition of portfolios based on "Equilibrium returns + Resampling" versus portfolios based on "Black-Litterman returns + Resampling"

4.9 Out-of-Sample Analysis to Evaluate Investment Strategies

In the previous pages of this chapter, a rather extensive and detailed examination of the portfolio construction methodologies having their roots in the Mean-Variance Optimisation has been completed also ending up, it is worth repeating, to make refinements and/or additions to it, with the final effect of arriving at more reasonable, more diversified, and more intuitive strategic asset allocation proposals. It is immediate to note that, in doing this, a dataset of T monthly returns was entirely used, with $t = 1, 2, \ldots, T$, where $T = 240$, ordered from the least recent (January 2003) to the most recent (December 2022) available for each asset class i with $i = 1, \ldots, N$ where $N = 9$, which allowed us to formulate the sample estimates to "launch" the classical optimisation, to characterise the multivariate normal distribution essential for Resampling, to represent the risk structure (variance—covariance matrix) associated with equilibrium returns but, generally, also with final expected returns within the Black-Litterman model. Given the fundamental goal, up to now, of showing the impact of the different approaches on the behaviour of the weight vectors $[w_1, \ldots, w_i, \ldots, w_N]$ representing the proposed strategic asset allocation solutions, this procedure does not give rise to particular criticisms and/or objections. However, if we put ourselves in the shoes of asset managers, without particular efforts and, indeed, instinctively, further interesting questions arise when faced with the different portfolio construction approaches, in order to evaluate them in a comparative way. For example, the following are legitimate questions: what is the degree of financial efficiency of different portfolio construction strategies? How intense is the trading activity/turnover required by each of them? To what type of crash (drawdown) can each of them be exposed?[87] Etc.

In seeking answers to the above questions, we realise the need for availability of the periodic returns that the investment strategies achieve, the cumulative wealth they achieve, but having exploited the entire dataset available for input estimates it is not feasible to have all this. On the one hand, in fact, it would be meaningless to test the vector of portfolio weights suggested by a strategy using the returns in the dataset as if that portfolio composition had been implemented at the beginning, i.e. at $t = 1$, because it would mean evaluate a strategy using an already known and already exploited set of information in order to determine the investment proposals recommended by the strategy itself. Put more rigorously, setting up the procedure in this way implies vitiating it due to the unrealistic and distorting hypothesis that the asset manager faces the portfolio construction problem with perfect knowledge of forward looking information. Conversely, considering, as in fact it is, the portfolios of each strategy built at time T on the basis of data extending from $t = 1$ up to T, the future returns of the asset classes with respect to that moment cannot be known (i.e. the returns in $T + 1$, $T + 2$,...) and therefore the evaluation of investment strategies is once again unfeasible, unless we strangely and inexplicably accept a moment T that dates back a long time compared to the current moment in which we find ourselves having to identify some strategic asset allocation solutions.

The critical issues just highlighted help in bringing out the desirable and appropriate characteristics for a valid evaluation procedure of different approaches for strategic asset allocation. Schematically, they can be described as follows:

- Consider a plurality of asset allocation experiments coming from each asset allocation approach.
- Avoid assuming that the asset manager has perfect information on the future behaviour of the asset classes which, on a practical level, means avoiding evaluating the behaviour of the recommended portfolios on the basis of the same information that contributed to their construction.

[87] For measures or indicators of financial efficiency, please refer to Chap. 8, while for the concept and measurement of drawdown, please refer to Chap. 5.

– Simulate the behaviour of the portfolios recommended by a specific portfolio construction methodology on the basis of unknown information, that is, in practice deliberately ignored during the identification phase of their structure.

It is a shared idea in financial literature that the procedure that can comply with these requirements must be identified with that proposed by DeMiguel et al.[88] The latter can be correctly described as an out-of-sample analysis procedure based, as regards the estimates required, on the use of a moving window mechanism or technique, usually known as rolling window technique.

For a better understanding, it is good to propose an exposition in the following terms: given a dataset composed, for the N asset classes of the investment universe, of T periodic returns (for example, in this contribution, as already known, $T = 240$ monthly returns), a time interval of extension equal to M periods is used where, peremptorily, $M < T$ and, for example, where $M = 60$ months, 120 months... and, in general, where M identifies a time interval that is not excessively narrow (normally not less than 5/7 years) to obtain the sample estimates (therefore in sample data) of the inputs required for the application of a specific portfolio construction approach and, consequently, to determine the vectors of weights for the optimal portfolios. The asset class exposures of the portfolios thus obtained are used to calculate the returns that those portfolios would have achieved in the subsequent F periods (therefore out of sample), where F is significantly lower than M, for example, $F = 3$, 6, 12 months. Once the interval of length F has elapsed, the procedure is repeated in the sense that a new window for estimating the inputs, again of extension M and with the starting point moved forward compared to the previous one precisely by F periods, allows us to arrive at new estimated inputs and, therefore, to the identification of new optimal portfolio weights for a specific portfolio construction approach. For them too, it will be verified which portfolio returns would have been achieved in the subsequent F periods. The mechanism based on rolling windows is reiterated and continues until the rolling window used for the estimates also exploits the most recent periodic return for the asset classes belonging to the investment universe; it goes without saying that with regard to the dataset available for this contribution, this condition occurs when we reach the estimation window that includes the 240th return (that of December 2022).

What most deserves to be emphasised about DeMiguel's described procedure, constituting a real strong point, is the ability to produce, for portfolios derived from a specific portfolio construction methodology, out-of-sample returns, that is, returns referring to holding periods subsequent to the time horizon on the basis of which the same portfolios were estimated; therefore, returns truly characterised by a prospective, forward looking nature, although everything is done using a dataset of historical returns. For this reason, the ideal requirements set out above can be considered satisfied. As a final outcome, managing to simulate realistic conditions for an asset manager, the procedure will return overall $T - M$ out-of-sample returns for a given

[88]DeMiguel et al. (2009).

portfolio construction approach and will have allowed the generation of numerous asset allocation experiments. Secondly, for greater awareness of the procedure itself, it is good to recall the aspects that constitute the choices attributed to its user (although, as seen, general indications can be provided), namely:

- the constant length of the estimation window, i.e. the choice of M.
- the so-called rebalancing frequency implied by the choice of F; evidently if F was set equal to 6 months or 12 months, it would mean that, in applying the procedure, a rebalancing and/or revision of the portfolio is assumed, respectively, every 6 months and annually.

To tell the truth, this second aspect creates the need for new in-depth analysis. It is connected to the fact that, for each asset allocation experiment, the analysis horizon of the out-of-sample behaviour is multiple compared to the frequency that characterises the available historical returns.

For extreme clarity and, at the same time, in order to allow a quick summary, it must be said that if we consider the dataset used in this contribution and assume to use the procedure proposed by DeMiguel with the ultimate aim, by way of example, of evaluating which would have been the effects of relying, as an investment strategy, on the least risky portfolio suggested by the classic Mean-Variance Optimisation (as is known, this is the GMVP), it is easy to agree that, in the case of $M = 60$ and $F = 6$, 180 out-of-sample monthly returns would be obtained (subsequently useful for determining the cumulative wealth achieved by the strategy, the crash/drawdown that would have characterised it, etc.) and 31 asset allocation experiments[89] would be carried out, each of which (except obviously the last one) would allow, with the help of the historical monthly returns available in the database, to calculate the out-of-sample returns of the portfolio in the specific semester following its estimate. Well, as can be seen from these words, the future holding period (out of sample) for each asset allocation experiment corresponds to a semester, but the usable returns in the database are monthly!

Without any doubt, there is therefore the need to determine multiperiod portfolio returns starting from single-period returns for each asset class $R_{t,i}$. In this regard, the question that arises is whether, for this purpose, it is reasonable to adopt a methodology that involves applying the single-period returns of the asset classes to a constant weight vector in correspondence with each period t in the future holding period or if, otherwise, it makes sense to use a methodology which at each moment t recognises that the weights may have changed compared to the initial moment and to each intermediate step due to the depreciation/appreciation of the asset classes.

The first methodology, identifiable as constant or fixed weight approach, has received several criticisms in the context of scientific contributions. Liu and Strong, as well as Alexeev and Ignatieva[90] observe that such an approach assumes that the

[89] For the 31st asset allocation experiment, the available dataset would no longer have the returns of the asset classes for the construction of the out-of-sample returns of the portfolio; in fact, as written, the overall monthly out-of-sample returns will be 180 (not 186).

[90] Liu and Strong (2008) and Alexeev and Ignatieva (2021).

investor brings the portfolio back to the initial weight structure (corresponding to those suggested by the sample estimates) with a rebalancing action carried out in each period based on the sampling frequency of returns in the dataset used.[91] This way of proceeding, although not lacking in popularity due to its simplicity and algebraic tractability,[92] is understandably considered by the mentioned authors to be impracticable for a real investor, implying prohibitive transaction costs. Still on the subject of critical issues, the crucial problem that should discourage its use in the out-of-sample analysis procedure proposed by DeMiguel aimed at evaluating investment strategies is the inability to reflect/capture the real effects, in terms of wealth, recorded by an investor who holds the portfolio resulting from the in-sample estimates.

The second methodology, identified as the buy-and-hold approach, provides that the multiperiod return of a portfolio constructed in samples on the future horizon F is calculated starting from single-period returns which represent a weighted-average of the returns of the asset classes where the associated weight to each of them is changeable at each step t due to the fact that they are affected by what happened to them, in terms of appreciation/depreciation, in the previous periods (i.e. from $\tau = 0, \ldots, t - 1$). Put more rigorously, the correct weight vector to apply to the calculation of the portfolio return for period t is a time-varying function of the cumulative performance of each asset class since the start of the holding period out of sample that you want to consider until the beginning of the moment t in which we find ourselves. The above admits a more formal representation. In fact, the calculation of the portfolio return in period t on the basis of updated weights or time-varying weights can be, analytically, written as follows:

$$R_{t,P} = \sum_{i=1}^{N} w_{t-1,i} \cdot R_{t,i} \tag{4.59}$$

In the context of (4.59), the updated weights $w_{t-1,i}$, still in the form of monetary values, can be reformulated as follows:

[91] In essence, the portfolio should be rebalanced in each month of the out-of-sample holding period if monthly returns are used, every week if weekly returns are used, etc.. . .

[92] The calculation of the returns of a portfolio in the periods included in the out-of-sample holding period is simply given by the product of a matrix $F \times N$, containing the returns of each of the N asset classes at each moment of the interval F with the vector $N \times 1$ of the initial weights resulting from the sample estimates.

$$\sum_{i=1}^{N} w_{t-1,i} = \sum_{i=1}^{N} \left(\frac{w_{0,i} \prod_{\tau=1}^{t-1} (1 + R_{\tau,i})}{\sum_{j=1}^{N} w_{0,j} \prod_{\tau=1}^{t-1} (1 + R_{\tau,j})} \right) \cdot R_{t,i} \tag{4.60}$$

Although the notation of (4.60) may appear rather cumbersome and complex, in essence it proposes in the numerator of the quantity in round brackets the monetary value of the investment in the asset class i cumulative up to the beginning of period t (that is, up to the end of $t - 1$), while the denominator of the same quantity captures the monetary value of the entire portfolio at the same time. What, in conclusion, must be duly pointed out is the fact that only the $R_{t,P}$ obtained ensure that the compounded capitalisation of the single-period returns gives the multiperiod return of a buy-and-hold investor and correctly reflects the wealth effect of the investment made and held.

Table 4.14 proposes the application of the methodologies discussed in the hypothesis of a portfolio with an initial value of 100 euros equally distributed across asset classes Y and Z, for which the assumed single-period returns are indicated in the first 2 columns.

To confirm the above statement, only the compounding of the single-period returns calculated within the buy-and-hold approach is able to return the multiperiod performance of +6.39%, consistent with the effect of increasing wealth achieved by the "investor". In fact:

$$(1 + 5.500\%) \times (1 - 0.265\%) \times (1 + 0.530\%) \times (1 + 0.578\%) - 1 = +6.39\% \tag{4.61}$$

The replication of the calculation just presented using the single-period returns for the case of the fixed weight approach would instead produce a return bias, returning a +7.62% (instead of +6.39%) not consistent with the wealth that a buy-and-hold investor would have actually held.

In conclusion, it is therefore useful to reiterate the recommendation to conduct the out-of-sample analysis based on a rolling window mechanism as indicated by DeMiguel, resorting, within the future holding period constructed for each asset allocation experiment, to single-period returns resulting from the use of a buy-and-hold approach.

Table 4.14 Fixed weight approach and buy-and-hold approach for generating single-period returns

Periodic returns Y	Periodic returns Z	Portfolio return fixed weight approach	Cash investment in Y (buy-and-hold approach)	Cash investment in Z (buy-and-hold approach)	Portfolio value (buy-and-hold approach)	% Weights time-varying Y (buy-and-hold approach)	% Weights time-varying Z (buy-and-hold approach)	Portfolio return buy-and-hold approach
			50	50	100	50.00%	50.00%	
+15%	−4%	+5.500%	57.50	48	105.5	54.50%	45.50%	+5.500%
−8%	+9%	+0.500%	52.90	52.32	105.22	50.28%	49.72%	−0.265%
+6%	−5%	+0.500%	56.07	49.70	105.78	53.01%	46.99%	+0.530%
−6%	+8%	1.000%	52.71	53.68	106.39			+0.578%

References

Alexeev V, Ignatieva K (2021) To rebalance or not to rebalance: portfolio risk may be larger than you think. Int Rev Financ 21(4):1152

Allaj E (2013) The Black-Litterman model: a consistent estimation of the parameter tau. Financ Mark Portf Manag 27:217–251

Amenc N, Goltz F, Lioui A (2011) Practitioner portfolio construction and performance measurement: evidence from Europe. Financ Anal J 67(3):39–50

Barros Fernandes JL, Haas Ornelas JR, Martinez Cusicanqui OA (2012) Combining equilibrium, resampling, and analyst's views in portfolio optimisation. J Bank Financ 36:1354–1361

Best M, Grauer R (1991) On the sensitivity of mean variance efficient portfolios to changes in asset means. Some analytical and computational results. Rev Financ Stud 4(2):315–342

Bevan A, Winkelmann K (1998, June) Using the Black-Litterman global asset allocation model: three years of practical experience. Fixed Income Research, Goldman Sachs

Black F, Litterman R (1991, October) Global asset allocation with equities, bonds and currencies. Fixed Income Research, Goldman Sachs

Black F, Litterman R (1992) Global portfolio optimisation. Financ Anal J 48(5):28–43

Braga MD, Natale FP (2012) Active risk sensitivity to views using the Black-Litterman model. J Asset Manag 13(1):5–21

Ceria S, Stubbs RA (2006) Incorporating estimation errors into portfolio selection: robust portfolio construction. J Asset Manag 7(2):109–127

Chamberlain G (1993) A characterization of the distribution that imply mean-variance utility functions. J Econ Theory 29:185–201

Cheung W (2010) The Black-Litterman model explained. J Asset Manag 11(4):229–243

Chopra VK, Ziemba WT (1993) The effects of errors in means, variances, and covariances on optimal portfolio choice. J Portf Manag 19(2):6–11

Deguest R, Martellini L, Meucci A (2013) Risk parity and beyond: from asset allocation to risk allocation decisions. SSRN Working Paper

DeMiguel V, Garlappi L, Uppal R (2009) Optimal versus naive diversification: how inefficient is the 1/N portfolio strategy? Rev Financ Stud 22(5):1915–1953

Doornik JA, Hansen H (2008) An omnibus test for univariate and multivariate normality. Oxf Bull Econ Stat 70(s1):927–939

Drobetz W (2001) How to avoid the pitfalls in portfolio optimisation? Putting the Black-Litterman approach at work. Financ Mark Portf Manag 15(1):59–75

Eichhorn N, Gupta D, Stubbs F (1998) Using constraints to improve the robustness of asset allocation. J Portf Manag 24(3):41–48

Fabozzi FJ, Petter KN, Dessislava PA, Focardi SM (2007) Robust portfolio optimisation and management. Wiley, Hoboken, NJ

Fama E (1963) Mandelbrot and the stable paretian hypothesis. J Bus 36:420–429

Frankfurter GM, Phillips HE, Seagle JP (1971) Portfolio selection: the effects of uncertain means, variances and covariance's. J Financ Quant Anal 6(5):1251–1262

Frost PA, Savarino JE (1986a) Portfolio size and estimation risk. J Portf Manag 12(4):60–64

Frost PA, Savarino JE (1986b) An empirical Bayes approach to efficient portfolio selection. J Financ Quant Anal 21(3):293–305

Frost PA, Savarino JE (1988) For better performance: constrain portfolio weights. J Portf Manag 15(1):29–34

Fuhrer A, Hock T (2023) Uncertainty in the Black-Litterman mode: empirical estimation of the equilibrium. J Empir Financ 72:251–275

Garlappi L, Uppal R, Wang T (2007) Portfolio selection with parameter and model uncertainty: a multi-prior approach. Rev Financ Stud 20(1):41–81

Grauer R, Shen F (2000) Do constraints improve portfolio performance? J Bank Financ 24(8): 1253–1274

Harvey CR, Siddique A (2000) Conditional skewness in asset pricing tests. J Financ 55:1263–1245

He G, Litterman R (1999, December) The intuition behind Black-Litterman model portfolios. Investment Management Research, Goldman Sachs
Herold U, Maurer R (2003) Bayesian asset allocation and U.S. domestic bias. Financ Anal J 59(6): 54–65
Herold U, Maurer R (2006) Portfolio choice and estimation risk. A comparison of Bayesian to heuristic approaches. ASTIN Bull 36(1):135–160
Idzorek T (2002) A step by step guide to the Black-Litterman model. Working paper. Zephyr Associates
Jagannathan R, Ma T (2003) Risk reduction in large portfolios: why imposing the wrong constraints helps. J Financ 58(4):1651–1683
Jansen DW, de Vries CG (1991) On the frequency of large stock returns: putting booms and busts into perspective. Rev Econ Stat 13:18–24
Jobson JD, Korkie B (1980) Estimation for Markowitz efficient portfolios. J Am Stat Assoc 75(371):544–554
Jobson JD, Korkie B (1981) Putting Markowitz theory to work. J Portf Manag 7(4):70–74
Jondeau E, Rockinger M (2006) Optimal portfolio allocation under higher moments. Eur Financ Manag 12(1):29–55
Jorion P (1985) International portfolio diversification with estimation risk. J Bus 58(3):259–278
Jorion P (1986) Bayes-Stein estimation for portfolio analysis. J Financ Quant Anal 21(3):279–292
Jorion P (1991) Bayesian and CAPM estimators of the mean: implications for portfolio selection. J Bank Financ 15(3):717–727
Jorion P (1992) Portfolio optimisation in practice. Financ Anal J 15(1):68–74
Kan R, Zhou G (2007) Optimal portfolio choice with parameter uncertainty. J Financ Quant Anal 42(3):621–656
Kon SJ (1984) Models of stock returns—a comparison. J Financ 39(1):147–165
Kraus A, Litzenberger RH (1976) Skewness preference and the valuation of risk assets. J Financ 31(4):1085–1100
Kroll Y, Levy H, Markowitz HM (1984) Mean-variance versus direct utility maximization. J Financ 39(1):1177–1179
Lee W (2000) Theory and methodology of tactical asset allocation. Frank J. Fabozzi Associates, New Hope
Levy H, Markowitz HM (1979) Approximating expected utility by a function of mean and variance. Am Econ Rev 69(3):308–317
Liu W, Strong N (2008) Biases in decomposing holding-period portfolio returns. Rev Financ Stud 21:2243–2274
Longin FM (1996) The asymptotic distribution of extreme stock market returns. J Bus 69:383–408
Mardia KW (1970) Measures of multivariate skewness and kurtosis with applications. Biometrica 57(3):519–530
Markowitz HM (1952) Portfolio selection. J Financ 7(1):77–91
Markowitz HM (1959) Portfolio selection: efficient diversification of investments. Wiley, New York
Markowitz HM (1987) Mean-variance analysis in portfolio choice and capital markets. Blackwell, Oxford
Markowitz HM (1999) The early history of portfolio theory: 1600–1960. Financ Anal J 55(4):5–16
Martellini L, Ziemann V (2010) Improved estimates of higher-order comments and implications for portfolio selection. Rev Financ Stud 23(4):1467–1502
Meucci A (2005) Risk and asset allocation. Springer, New York
Meucci A (2010) The Black-Litterman approach: original model and extensions. In: The Encyclopedia of quantitative finance. Wiley, Hoboken
Michaud R (1989) The Markowitz optimisation enigma: is 'optimized' optimal? Financ Anal J 45(1):31–42
Michaud R, Michaud R (2008) Estimation error and portfolio optimization: a resampling solution. J Invest Manag 6:8–28

Mossin J (1968) Optimal multiperiod portfolio policies. J Bus 41(2):215–229
Peiró A (1999) Skewness in financial returns. J Bank Financ 23(6):843–862
Prakash AJ, Chang CH, Pactwa TE (2003) Selecting a portfolio with skewness: recent evidence from US, European and Latin America equity markets. J Bank Financ 27(7):1375–1390
Pratt JW (1964) Risk aversion in the small and in the large. Econometrica 32(1–2):83–98
Sarnat M (1974) A note on the implications of quadratic utility for portfolio theory. J Financ Quant Anal 9(4):687–689
Satchell S, Scowcroft A (2000) A demystification on the Black-Litterman model: managing quantitative and traditional portfolio construction. J Asset Manag 1(2):138–150
Scherer B (2002) Portfolio resampling: review and critique. Financ Anal J 58(6):98–109
Scherer B (2015) Portfolio construction and risk budgeting, 5th edn. Risk Books, London
Schmittmann JM (2010) Currency hedging for international portfolios. IMF Working Paper, WP/10/151
Scott RC, Horwath PA (1980) On the direction of preference for moments of higher order than the variance. J Financ 35(4):915–919
Sharpe WF (1974) Imputing expected security returns from portfolio composition. J Financ Quant Anal 9(3):463–472
Sharpe WF (2010) Adaptive asset allocation policies. Financ Anal J 66(3):45–59
Simaan Y (1993) Portfolio selection and asset pricing-three-parameter. Manag Sci 39:568–577
Simkowitz MA, Beedles WL (1980) Asymmetric stable distributed security returns. J Am Stat Assoc 75(370):306–312
Singleton JC, Wingender J (1986) Skewness persistence in common stock returns. J Financ Quant Anal 21(3):335–341

Chapter 5
Methods and Tools for Portfolio Selection

Maria Debora Braga

5.1 Portfolio Selection

Strategic asset allocation—combining Markowitz's optimisation algorithm and the techniques of estimation risk management—leads to the identification of a number of efficient portfolios in the risk-return space. It is this common feature that brings the topic of portfolio selection to the fore, since it is obvious that when presented with a variety of long-term investment options which cannot be ranked (being all optimal), we face a problem of choice and selection.

However, the identification of a portfolio within the opportunity set is inevitably influenced by priorities, preferences, and risk tolerance/risk aversion of the individual institutional investor who—it is important to bear in mind—is generally an "agent" for a number of subjects, i.e. delegating end-investors. In other words, portfolio selection is subjective. That said, a specific chapter devoted to the issue should not be seen as an attempt to replace the assessments of an institutional investor with an automatic device able to extract the "right" portfolio from a set of efficient alternatives. The aim is rather to support the decision-making process by exploring some tools and/or methods which can highlight certain aspects of optimal portfolios that do not emerge manifestly from the specification of their risk/expected returns combinations, but are still important to understand their consistency and congruity with the institutional investor's preferences. These tools and/or methodologies have been formulated in contributions (not only from academia) mainly related to the areas of risk budgeting and financial planning that appeared after

M. D. Braga (✉)
Department of Economics and Political Science, University of "Valle d'Aosta", Aosta, Italy

SDA Bocconi School of Management, Bocconi University, Milan, Italy
e-mail: m.braga@univda.it; mariadebora.braga@sdabocconi.it

I. Basile, P. Ferrari (eds.), *Asset Management and Institutional Investors*, Contributions to Finance and Accounting,
https://doi.org/10.1007/978-3-031-59819-7_5

Markowitz's Modern Portfolio Theory, with which they are neither incompatible nor in conflict, and indeed offer a complementary viewpoint. As said above, these tools and/or methods make possible a more conscious and informed selection among portfolios constructed using the mean-variance framework. A detailed analysis will clarify that the tools and/or methodologies being considered aim:

- To enhance our knowledge of the nature and intensity of the diversification offered by the portfolios in the opportunity set.
- To gain a more profound and/or alternative assessment of the risk of optimal portfolios.
- To verify, at least in probabilistic terms, the extent to which optimal portfolios satisfy the need to "protect" a financial result or defend against possible losses.
- To analyse possible future scenarios of cumulative wealth associated with investment in efficient portfolios.

The following sections will examine the tools used to realise these objectives. For ease of presentation, we will follow the order of the list above, without implying any inflexible categorisation of the methodologies and/or tools proposed to support portfolio selection.

5.2 Level of Diversification of Optimal Portfolios

When looking at the opportunity set, we are aware that these are the portfolios that optimise the trade-off between risk and expected return, i.e. portfolios that *ex ante* present the minimum possible risk for a given level of return.

What cannot be stated immediately, however, is the heterogeneous skill of the different portfolio allocations to counter risk by exploiting diversification as a "weapon" of defence. To put it another way, the graph of the efficient frontier does not immediately reveal the degree of risk reduction/risk saving offered by the portfolios on the frontier. This useful information can be obtained by calculating two closely interrelated values for each optimal portfolio:

- The benefit of diversification.
- The diversification ratio as proposed by Choueifaty and Coignard.[1]

The benefit of diversification (DIV_B) is a percentage expressing by how much overall portfolio risk is lower than the weighted average of the volatilities of the *N* asset classes in the portfolio. Analytically, this is defined as (5.1):

[1] Choueifaty and Coignard (2008).

$$\text{DIV\_B} = \frac{\sum_{i=1}^{N} w_i\sigma_i - \sigma_P}{\sum_{i=1}^{N} w_i\sigma_i} = \frac{\sum_{i=1}^{N} w_i\sigma_i - \sqrt{\sum_{i=1}^{N} w_i^2\sigma_i^2 + \sum_{i=1}^{N}\sum_{\substack{j=1\\ j\neq i}}^{N} w_i w_j \sigma_{ij}}}{\sum_{i=1}^{N} w_i\sigma_i} \tag{5.1}$$

where

DIV_B = benefit of diversification.
σ_P = portfolio risk.
w_i = weight of asset class i in the portfolio.
σ_{ij} = covariance between asset classes i and j.

The diversification ratio (DIV_R), on the other hand, is given by the ratio between the weighted average of the volatilities of the asset classes in a portfolio and the actual standard deviation of the same portfolio.

Both values (DIV_B, DIV_R), therefore, include the weighted average of the individual risks. The reason for this is simple: it is the maximum risk that an investor could face if every correlation were equal to +1.

The diversification ratio can also be shown in analytical form (5.2):

$$\text{DIV\_R} = \frac{\sum_{i=1}^{N} w_i\sigma_i}{\sigma_P} = \frac{\sum_{i=1}^{N} w_i\sigma_i}{\sqrt{\sum_{i=1}^{N} w_i^2\sigma_i^2 + \sum_{i=1}^{N}\sum_{\substack{j=1\\ j\neq i}}^{N} w_i w_j \sigma_{ij}}} \tag{5.2}$$

On the basis of this, the diversification ratio is a value which, when applied to long-only portfolios (such as the efficient portfolios resulting from the strategic asset allocation process described in Chap. 4), must always be strictly greater than 1, except in two cases: the first case is represented by a single-asset class portfolio, while the second case is represented by a portfolio composed of asset classes characterised by perfectly positive correlation for each possible pair.

Even though they are expressed differently, the benefit of diversification and the diversification ratio can be interpreted in similar ways since, in both cases, a higher/lower value indicates that a given portfolio allocation results in a more/less efficient diversification among asset classes. The different results for these measures across the optimal portfolios can be shown in a graph: the y-axis in Figs. 5.1 and 5.2 indicates the entity of the benefit of diversification and the diversification ratio, respectively, for each of the 100 portfolios on the Resampled Efficient Frontier (REF) shown in Fig. 4.16 in the previous chapter. The variation is significant for both measures: the benefit of diversification ranges from 9.603 to 46.656%, while the diversification ratio is between 1.106 and 1.875. Looking at these numbers, it is correct to state that, where the "risk savings" is more negligible, the real risk of the

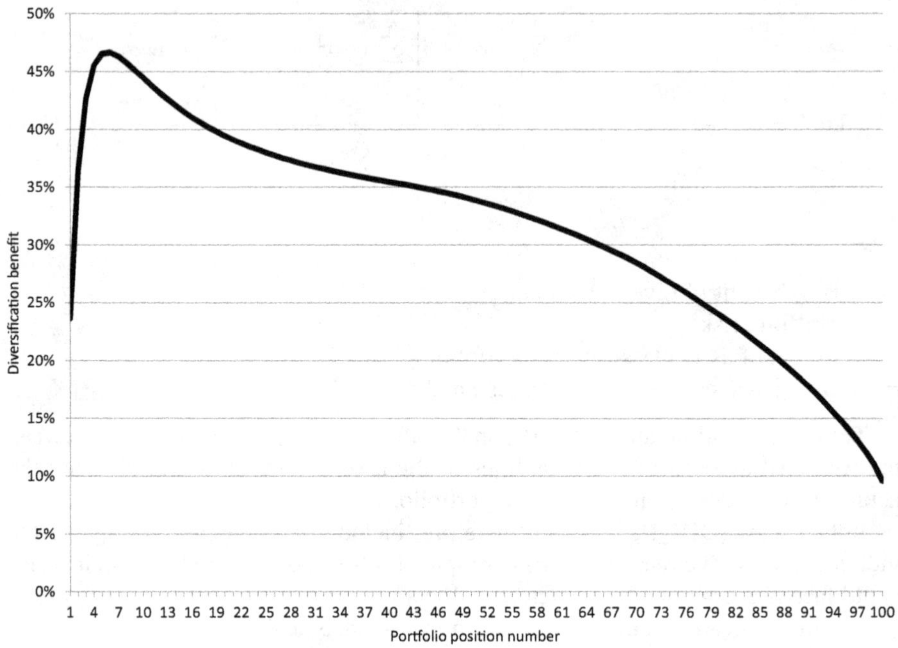

Fig. 5.1 The benefit of diversification for the portfolios on the resampled Efficient Frontier

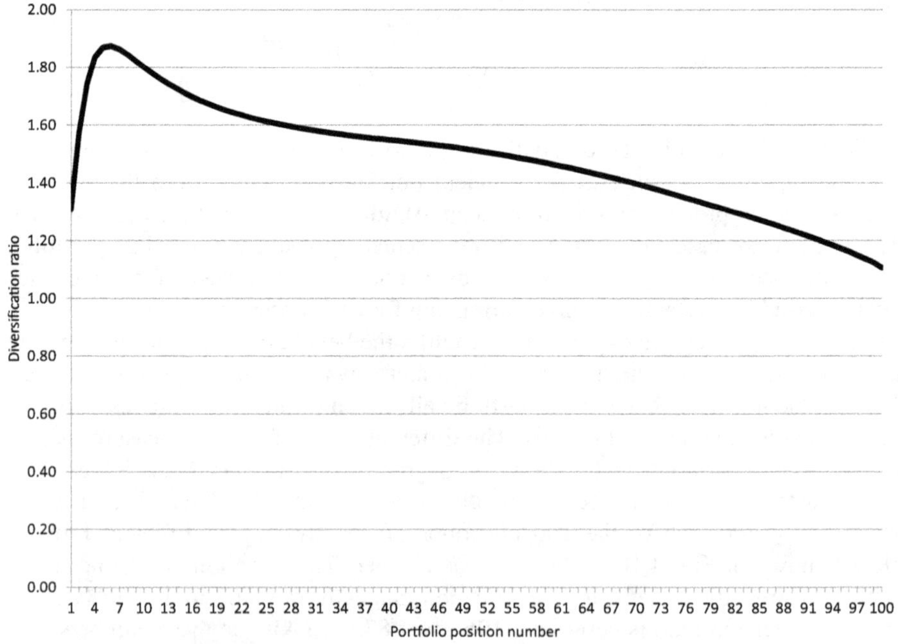

Fig. 5.2 The diversification ratio for the portfolios on the resampled Efficient Frontier

portfolio is lower than the weighted average of risks by only 9.603%, while where it is more massive, the actual risk of the portfolio is lower than the weighted average of risks by 46.656%. The same concepts can be reformulated by declaring that the portfolio whose composition is least able to ensure a "discount" on risk would have shown, if all the pairwise correlations had been equal to +1, a standard deviation equal to 1.106 times the one actually shown; while the portfolio whose asset allocation is more capable of safeguarding this "discount" would have shown, again under the hypothesis of perfectly positive correlations, a standard deviation equal to 1.875 times the actual one.

In the example, the portfolios offering the greatest reduction in risk (compared to the upper limit given by the weighted average of volatilities) are approximately located between position 2 and position 40, while those offering the smallest risk reduction are to the extreme right of the Resampled Efficient Frontier.

In this analysis of the benefit of diversification and the diversification ratio, we have focused on the total risk (standard deviation) of optimal portfolios, proposing some analytical tools which may help an institutional investor to compare heterogeneous degrees of risk reduction offered by these portfolios. However, the level of risk reduction may not be the only additional information which an institutional investor will find useful. Other interesting questions to ask may be: "What proportion of the total portfolio risk can be attributed to each of its components?" or "Is it true (for example) that 90 % of the risk of the portfolio under consideration derives from exposure to the asset class i?"

If we admit that such questions are valid and can be posed, and therefore we want to provide the answers to them, we must consider a different way of analysing optimal portfolios, i.e. the risk allocation perspective. In other words, the available and obvious knowledge of the subdivision of wealth among the different asset classes suggested by each optimal portfolio needs to be complemented with an awareness of the subdivision of total risk among the various asset class exposures. The latter has to be identified, however, since it is not immediately available: we cannot simply assume that the generic weight w_i represents the contribution of the asset class i to the portfolio risk. Intuitively, it is true that this contribution is influenced by the weight associated with asset class i in the portfolio, but it is also directly connected both to its volatility and to its correlation with other asset classes.

The step from a generic and intuitive assessment to a formal analysis of a portfolio from the point of view of risk allocation rather than asset allocation requires the use of the so-called risk decomposition tools, as proposed in Scherer and in Qian.[2] Among these, the first to be examined is the measurement of marginal risk, also known as marginal risk contribution, indicated here as MR_i. It expresses the variation in total portfolio risk caused by an infinitesimal change in the weight of asset class i. Mathematically, this value is obtained by calculating the first order partial derivative of the risk function with respect to the single w_i, as shown in (5.3):

[2] Scherer (2015) and Qian (2006).

$$\text{MR}_i = \frac{\partial \sigma_P}{\partial w_i} = \frac{w_i \sigma_i^2 + \sum\limits_{j \neq i}^{N} w_j \sigma_{ij}}{\sigma_P} \tag{5.3}$$

It is clear that the number of first order partial derivatives that can be calculated by (5.3) is equal to the number, N, of asset classes. In support of this argument, note that the formulation of the first order partial derivative with respect to w_i not only includes aspects that can be referred individually to i (its weight and volatility), but also the term incorporating the link between i and every other asset class (obviously this is the covariance σ_{ij}), weighted for the exposure indicated for each of the remaining asset classes.

Secondly, we have to consider the value of the risk contribution of asset class i, also called total risk contribution (TRC_i) or component risk, which expresses the portion of total portfolio risk that can be assigned to i. This value is obtained by multiplying the weight assigned to i by its marginal risk, as shown in (5.4):

$$\text{TRC}_i = w_i \cdot \text{MR}_i = w_i \cdot \frac{\partial \sigma_P}{\partial w_i} = w_i \cdot \frac{w_i \sigma_i^2 + \sum\limits_{j \neq i}^{N} w_j \sigma_{ij}}{\sigma_P} \tag{5.4}$$

Once the TRC_i has been determined, the sum shown in (5.5) gives portfolio risk:

$$\sigma_P = \sum_{i=1}^{N} \text{TRC}_i = \sum_{i=1}^{N} w_i \cdot \frac{\partial \sigma_P}{\partial w_i} \tag{5.5}$$

Formula (5.5) demonstrates how risk decomposition tools can quantify individual and separate contributions to risk from asset classes which can be added together.[3] This is a fundamental precondition for obtaining a clear representation of the level of diversification/concentration of risk in a portfolio. In this sense, though it may seem superfluous, it should be noted that total risk contributions can be represented as a fraction of 100%, made equal to the value σ_P. Briefly, the percentage total risk contribution (PTRC_i) associated with the asset class i is obtained as follows (5.6):

[3]We have considered risk decomposition from the point of view of portfolio volatility, because—when using the mean-variance approach—this is the risk measure which is associated by definition with each efficient portfolio. It can also, however, refer to other risk measures. Yet note that one condition must be respected: the metric chosen must be a first order homogenous function (e.g. standard deviation) or, equivalently, a linear homogenous function. Only by meeting this condition Euler's Theorem can be applied, and a similar function translated into the sum of its arguments (in the present case, the portfolio weights) multiplied by their first order partial derivative Braga (2008).

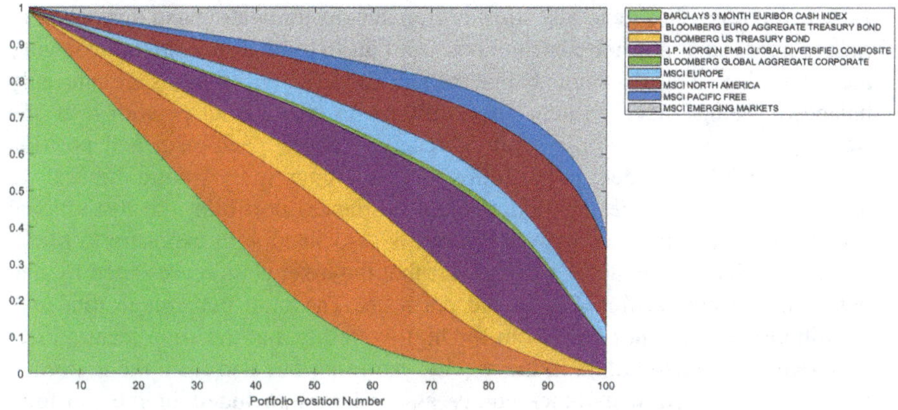

Fig. 5.3 Asset allocation of portfolios on the resampled efficient frontier

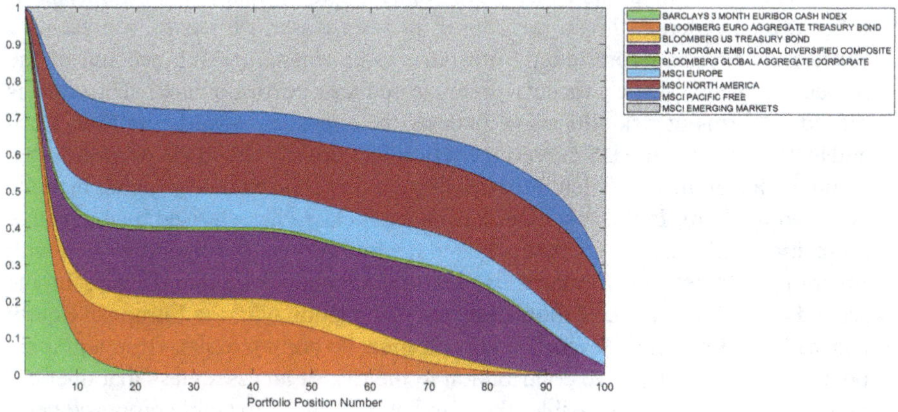

Fig. 5.4 Risk allocation of portfolios on the resampled efficient frontier

$$\mathrm{PTRC}_i = \frac{\mathrm{TRC}_i}{\sigma_P} \tag{5.6}$$

Also with respect to the issue of risk decomposition, it is useful to provide an empirical example. Figure 5.3 (taken from Chap. 4) shows the asset allocation of portfolios on the Resampled Efficient Frontier, while Fig. 5.4 illustrates their risk allocation in terms of $PTRC_i$.

The valuable insights gained from the comparison of Figs. 5.3 and 5.4 can be summarised as follows:

- Confirmation that the exposure to an asset class in terms of the proportion of wealth invested or to be invested in the asset class and the exposure to that same asset class in terms of the portfolio risk generated by the asset class are not identical.

- Resampled portfolios generally show reasonable diversification and dispersion of risk sources, as abnormal contributions from an individual investment category are not evident. This does not change the fact that risk contributions are unevenly balanced, being more evident in portfolios in the initial positions of the Resampled Efficient Frontier (approximately between position 5 and position 30) and much less evident in portfolios to the extreme right. Among these, if we take the example of the maximum return (variance) portfolio, the institutional investor will see that over 74% of the risk is associated with exposure to MSCI Emerging Markets, regardless of the fact that the asset class represented by that benchmark accounts for 62% in the portfolio. The high percentage total risk contribution derives not only from the high volatility and the significant allocation given to the asset class in the specific portfolio, but also from the important positive correlation with other equity asset classes included in the portfolio (in particular with MSCI Pacific Free and MSCI Europe and, to a slightly lesser extent, with MSCI North America).
- Admissibility of situations of negative contributions to risk. As long-only portfolios are considered, this is possible if an asset class in a portfolio presents a sufficiently negative correlation with other asset classes (taking account of the respective weight), such that its presence reduces portfolio risk.[4] This role is played, in terms of risk allocation, both by the exposure towards the Euro Cash and by that towards the US Government Bonds, respectively, in the portfolios that lie on the Resampled Efficient Frontier in positions approximately higher than the 30th and the 93rd. In the first case, it is an asset class characterised by a negative correlation with each equity asset class in the investment universe as well as with Emerging Markets Government Bonds and Global Corporate Bonds, in the second case, the asset class shows negative correlation with the European Equity and with the Emerging Markets equity. It must be admitted that, from a graphic point of view, the negative contribution to the risk of an asset class that operates as a hedger is not always visible (this also happens in the example proposed here) because, instead of appearing in a negative zone, it can appear overshadowed by the positive risk contribution of other components of the portfolio. However, from a mathematical point of view (Table 5.1), it is trivial in these cases to ascertain that the sum of the positive percentage contributions to portfolio risk would lead to a result greater than 100% and that only the consideration of the negative risk contributions also allows restoring the value of 100%.

We have compared the tools that can examine in greater detail the diversification of portfolios, considering their application in a real context, in line with the approach adopted in Chap. 4 where the focus was on portfolio construction. With the purpose of facilitating the understanding of the algebraic steps required by these tools, and of clarifying interpretation, Box 5.1 provides an intentionally didactic example to summarise our discussion so far.

[4]With long-only portfolios, a negative total risk contribution is possible only if marginal risk is negative. Taking the formula in (5.3), it is easy to deduce that this presupposes that the sum in the numerator is negative, meaning, in turn, that at least some of the values of σ_{ij} are negative.

Table 5.1 The representation in numerical terms of the risk allocation of the resampled efficient frontier portfolios

	BARCLAYS 3 MONTH EURIBOR CASH INDEX	BLOOMBERG EURO AGGREGATE TREASURY BOND	BLOOMBERG BARCLAYS US TREASURY BOND	J.P. MORGAN EMBI GLOBAL DIVERSIFIED COMPOSITE	BLOOMBERG BARCLAYS GLOBAL AGGREGATE CORPORATE	MSCI EUROPE	MSCI NORTH AMERICA	MSCI PACIFIC FREE	MSCI EMERGING MARKETS
PORT. 1	98.667%	0.213%	0.030%	0.011%	0.297%	0.147%	0.497%	0.126%	0.010%
PORT. 2	86.373%	1.967%	0.480%	1.210%	0.941%	1.341%	5.428%	0.985%	1.275%
PORT. 3	69.323%	4.559%	1.354%	4.100%	1.170%	3.092%	9.479%	2.105%	4.818%
PORT. 4	52.535%	7.122%	2.253%	7.204%	1.339%	4.736%	12.472%	3.187%	9.153%
PORT. 5	38.841%	9.228%	3.005%	9.851%	1.439%	6.025%	14.488%	4.042%	13.081%
PORT. 6	28.616%	10.803%	3.622%	11.880%	1.428%	6.979%	15.781%	4.638%	16.253%
PORT. 7	21.231%	11.947%	4.084%	13.382%	1.397%	7.663%	16.556%	5.049%	18.691%
PORT. 8	15.930%	12.773%	4.430%	14.487%	1.352%	8.151%	16.995%	5.335%	20.549%
PORT. 9	12.095%	13.372%	4.685%	15.299%	1.312%	8.489%	17.235%	5.537%	21.974%
PORT. 10	9.287%	13.810%	4.875%	15.914%	1.278%	8.736%	17.346%	5.679%	23.074%
PORT. 11	7.201%	14.137%	5.020%	16.383%	1.243%	8.915%	17.383%	5.780%	23.937%
PORT. 12	5.629%	14.382%	5.133%	16.743%	1.212%	9.047%	17.379%	5.853%	24.622%
PORT. 13	4.429%	14.570%	5.221%	17.024%	1.185%	9.144%	17.347%	5.908%	25.172%
PORT. 14	3.502%	14.716%	5.291%	17.245%	1.160%	9.215%	17.301%	5.951%	25.620%
PORT. 15	2.778%	14.830%	5.348%	17.423%	1.136%	9.268%	17.244%	5.982%	25.992%
PORT. 16	2.207%	14.920%	5.393%	17.569%	1.115%	9.308%	17.181%	6.005%	26.302%
PORT. 17	1.752%	14.991%	5.430%	17.689%	1.098%	9.338%	17.117%	6.022%	26.562%
PORT. 18	1.388%	15.049%	5.460%	17.789%	1.082%	9.359%	17.054%	6.036%	26.783%
PORT. 19	1.095%	15.096%	5.485%	17.871%	1.067%	9.377%	16.992%	6.045%	26.973%
PORT. 20	0.857%	15.134%	5.506%	17.941%	1.054%	9.390%	16.932%	6.052%	27.136%
PORT. 21	0.663%	15.165%	5.523%	17.999%	1.043%	9.399%	16.874%	6.057%	27.278%
PORT. 22	0.505%	15.190%	5.537%	18.048%	1.034%	9.405%	16.818%	6.061%	27.402%

(continued)

Table 5.1 (continued)

	BARCLAYS 3 MONTH EURIBOR CASH INDEX	BLOOMBERG EURO AGGREGATE TREASURY BOND	BLOOMBERG BARCLAYS US TREASURY BOND	J.P. MORGAN EMBI GLOBAL DIVERSIFIED COMPOSITE	BLOOMBERG BARCLAYS GLOBAL AGGREGATE CORPORATE	MSCI EUROPE	MSCI NORTH AMERICA	MSCI PACIFIC FREE	MSCI EMERGING MARKETS
PORT. 23	0.375%	15.211%	5.550%	18.090%	1.025%	9.410%	16.765%	6.063%	27.512%
PORT. 24	0.269%	15.228%	5.561%	18.126%	1.016%	9.413%	16.713%	6.064%	27.610%
PORT. 25	0.182%	15.243%	5.570%	18.156%	1.009%	9.415%	16.664%	6.064%	27.697%
PORT. 26	0.110%	15.255%	5.578%	18.183%	1.002%	9.415%	16.618%	6.064%	27.775%
PORT. 27	0.052%	15.265%	5.585%	18.206%	0.995%	9.415%	16.573%	6.064%	27.845%
PORT. 28	0.004%	15.273%	5.592%	18.227%	0.989%	9.414%	16.531%	6.062%	27.908%
PORT. 29	−0.034%	15.279%	5.597%	18.244%	0.983%	9.413%	16.491%	6.061%	27.966%
PORT. 30	−0.065%	15.285%	5.602%	18.260%	0.978%	9.411%	16.453%	6.059%	28.018%
PORT. 31	−0.090%	15.289%	5.607%	18.273%	0.972%	9.409%	16.416%	6.057%	28.065%
PORT. 32	−0.109%	15.292%	5.611%	18.286%	0.967%	9.407%	16.381%	6.055%	28.109%
PORT. 33	−0.123%	15.295%	5.614%	18.297%	0.963%	9.404%	16.348%	6.053%	28.149%
PORT. 34	−0.134%	15.297%	5.617%	18.306%	0.959%	9.401%	16.316%	6.051%	28.186%
PORT. 35	−0.142%	15.299%	5.620%	18.315%	0.955%	9.398%	16.286%	6.049%	28.220%
PORT. 36	−0.147%	15.300%	5.622%	18.322%	0.952%	9.396%	16.257%	6.047%	28.251%
PORT. 37	−0.149%	15.301%	5.624%	18.329%	0.948%	9.393%	16.229%	6.045%	28.280%
PORT. 38	−0.150%	15.302%	5.626%	18.335%	0.945%	9.390%	16.203%	6.042%	28.307%
PORT. 39	−0.148%	15.300%	5.627%	18.340%	0.943%	9.387%	16.177%	6.040%	28.334%
PORT. 40	−0.146%	15.284%	5.626%	18.351%	0.940%	9.386%	16.152%	6.038%	28.369%
PORT. 41	−0.142%	15.256%	5.623%	18.360%	0.937%	9.384%	16.128%	6.036%	28.418%
PORT. 42	−0.137%	15.175%	5.612%	18.371%	0.934%	9.383%	16.116%	6.037%	28.508%
PORT. 43	−0.132%	15.047%	5.595%	18.386%	0.931%	9.380%	16.118%	6.042%	28.633%
PORT. 44	−0.127%	14.883%	5.570%	18.415%	0.929%	9.373%	16.134%	6.047%	28.776%

PORT. 45	−0.121%	14.657%	5.534%	18.455%	0.933%	9.367%	16.158%	6.054%	28.963%
PORT. 46	−0.115%	14.382%	5.495%	18.502%	0.936%	9.366%	16.192%	6.061%	29.183%
PORT. 47	−0.109%	14.058%	5.448%	18.562%	0.938%	9.361%	16.232%	6.072%	29.438%
PORT. 48	−0.104%	13.688%	5.398%	18.634%	0.941%	9.356%	16.290%	6.088%	29.709%
PORT. 49	−0.098%	13.277%	5.343%	18.727%	0.942%	9.356%	16.360%	6.107%	29.986%
PORT. 50	−0.092%	12.819%	5.285%	18.828%	0.943%	9.360%	16.446%	6.136%	30.275%
PORT. 51	−0.087%	12.335%	5.229%	18.930%	0.944%	9.362%	16.549%	6.167%	30.570%
PORT. 52	−0.082%	11.831%	5.173%	19.047%	0.944%	9.365%	16.662%	6.199%	30.861%
PORT. 53	−0.076%	11.326%	5.116%	19.178%	0.939%	9.371%	16.776%	6.230%	31.140%
PORT. 54	−0.071%	10.801%	5.057%	19.325%	0.934%	9.379%	16.898%	6.265%	31.412%
PORT. 55	−0.066%	10.276%	4.995%	19.482%	0.929%	9.386%	17.024%	6.302%	31.672%
PORT. 56	−0.062%	9.738%	4.928%	19.646%	0.926%	9.387%	17.165%	6.340%	31.933%
PORT. 57	−0.057%	9.198%	4.861%	19.811%	0.926%	9.383%	17.309%	6.383%	32.187%
PORT. 58	−0.053%	8.667%	4.788%	19.985%	0.929%	9.382%	17.447%	6.426%	32.431%
PORT. 59	−0.049%	8.144%	4.711%	20.163%	0.934%	9.382%	17.582%	6.469%	32.664%
PORT. 60	−0.045%	7.629%	4.636%	20.341%	0.939%	9.382%	17.727%	6.510%	32.883%
PORT. 61	−0.042%	7.124%	4.559%	20.519%	0.945%	9.377%	17.872%	6.551%	33.095%
PORT. 62	−0.038%	6.641%	4.481%	20.695%	0.948%	9.368%	18.017%	6.590%	33.299%
PORT. 63	−0.035%	6.167%	4.399%	20.873%	0.949%	9.354%	18.159%	6.629%	33.504%
PORT. 64	−0.032%	5.710%	4.309%	21.060%	0.952%	9.334%	18.296%	6.669%	33.702%
PORT. 65	−0.029%	5.272%	4.215%	21.250%	0.956%	9.308%	18.424%	6.705%	33.899%
PORT. 66	−0.026%	4.853%	4.115%	21.435%	0.960%	9.282%	18.546%	6.742%	34.093%
PORT. 67	−0.024%	4.449%	4.012%	21.615%	0.964%	9.256%	18.663%	6.780%	34.285%
PORT. 68	−0.021%	4.067%	3.895%	21.774%	0.969%	9.216%	18.777%	6.816%	34.506%
PORT. 69	−0.019%	3.702%	3.766%	21.932%	0.975%	9.171%	18.891%	6.854%	34.729%
PORT. 70	−0.017%	3.358%	3.608%	22.055%	0.982%	9.123%	19.012%	6.896%	34.984%
PORT. 71	−0.015%	3.035%	3.432%	22.163%	0.987%	9.058%	19.154%	6.937%	35.249%

(continued)

Table 5.1 (continued)

	BARCLAYS 3 MONTH EURIBOR CASH INDEX	BLOOMBERG EURO AGGREGATE TREASURY BOND	BLOOMBERG BARCLAYS US TREASURY BOND	J.P. MORGAN EMBI GLOBAL DIVERSIFIED COMPOSITE	BLOOMBERG BARCLAYS GLOBAL AGGREGATE CORPORATE	MSCI EUROPE	MSCI NORTH AMERICA	MSCI PACIFIC FREE	MSCI EMERGING MARKETS
PORT. 72	−0.014%	2.736%	3.243%	22.243%	0.993%	8.973%	19.315%	6.977%	35.534%
PORT. 73	−0.012%	2.457%	3.049%	22.266%	0.994%	8.879%	19.480%	7.017%	35.871%
PORT. 74	−0.011%	2.195%	2.850%	22.241%	0.997%	8.777%	19.652%	7.054%	36.244%
PORT. 75	−0.010%	1.956%	2.646%	22.159%	1.001%	8.662%	19.839%	7.094%	36.653%
PORT. 76	−0.009%	1.739%	2.439%	22.013%	0.999%	8.531%	20.037%	7.136%	37.115%
PORT. 77	−0.008%	1.541%	2.226%	21.784%	0.990%	8.405%	20.241%	7.184%	37.637%
PORT. 78	−0.007%	1.359%	2.012%	21.471%	0.987%	8.289%	20.441%	7.233%	38.214%
PORT. 79	−0.006%	1.197%	1.802%	21.071%	0.971%	8.172%	20.661%	7.286%	38.846%
PORT. 80	−0.005%	1.048%	1.607%	20.596%	0.929%	8.054%	20.892%	7.343%	39.536%
PORT. 81	−0.005%	0.912%	1.422%	20.050%	0.884%	7.935%	21.128%	7.401%	40.272%
PORT. 82	−0.004%	0.790%	1.240%	19.402%	0.838%	7.833%	21.376%	7.454%	41.071%
PORT. 83	−0.004%	0.681%	1.065%	18.704%	0.777%	7.748%	21.642%	7.509%	41.877%
PORT. 84	−0.003%	0.582%	0.908%	17.963%	0.706%	7.654%	21.925%	7.566%	42.699%
PORT. 85	−0.003%	0.495%	0.764%	17.088%	0.640%	7.551%	22.235%	7.627%	43.602%
PORT. 86	−0.002%	0.418%	0.632%	16.150%	0.570%	7.446%	22.543%	7.692%	44.552%
PORT. 87	−0.002%	0.350%	0.511%	15.145%	0.504%	7.354%	22.862%	7.767%	45.509%
PORT. 88	−0.002%	0.290%	0.402%	14.113%	0.443%	7.265%	23.160%	7.829%	46.500%
PORT. 89	−0.001%	0.239%	0.302%	13.053%	0.392%	7.169%	23.464%	7.868%	47.515%
PORT. 90	−0.001%	0.195%	0.213%	11.964%	0.347%	7.048%	23.717%	7.866%	48.652%
PORT. 91	−0.001%	0.157%	0.136%	10.890%	0.298%	6.892%	23.920%	7.817%	49.891%
PORT. 92	−0.001%	0.125%	0.073%	9.812%	0.256%	6.732%	24.060%	7.708%	51.235%
PORT. 93	−0.001%	0.098%	0.020%	8.727%	0.220%	6.550%	24.093%	7.551%	52.742%

PORT. 94	0.000%	0.074%	−0.020%	7.684%	0.187%	6.359%	23.886%	7.295%	54.535%
PORT. 95	0.000%	0.056%	−0.049%	6.694%	0.149%	6.078%	23.437%	7.014%	56.622%
PORT. 96	0.000%	0.040%	−0.068%	5.725%	0.114%	5.744%	22.691%	6.741%	59.014%
PORT. 97	0.000%	0.028%	−0.079%	4.844%	0.081%	5.416%	21.713%	6.338%	61.659%
PORT. 98	0.000%	0.018%	−0.079%	4.050%	0.050%	5.012%	20.642%	5.704%	64.603%
PORT. 99	0.000%	0.010%	−0.065%	3.328%	0.036%	4.548%	19.014%	4.800%	68.329%
PORT. 100	0.000%	0.005%	−0.064%	2.443%	0.028%	3.800%	16.252%	3.354%	74.180%

Box 5.1 Mathematical Tools for the Analysis and Decomposition of Portfolio Risk

Consider a simplified portfolio with only two asset classes, A and B, whose characteristics are summarised as follows:

	Asset class A	*Asset class* B
Return	3.50%	6.80%
St. dev.	6.00%	16.00%
Weight	40.00%	60.00%
$\text{Correlation}_{A,B}$	0.28	

For the present purposes, we will look only at portfolio risk, calculated as follows:

$$\begin{aligned}\sigma_P &= \sqrt{(40.00\,\%\cdot 6.00\,\%)^2+(60.00\,\%\cdot 16.00\,\%)^2+2\cdot 40.00\,\%\cdot 60.00\,\%\cdot 6.00\,\%\cdot 16.00\,\%\cdot 0.28}\\ &= 10.53\,\%\end{aligned}$$

Obviously, the result is lower than the weighted average of standard deviations (12.00%).

The extent to which a portfolio can produce a reduction in risk can be measured firstly by calculating the benefit of diversification:

$$\text{DIV\_B}=\frac{12.00\ \%-10.53\ \%}{12.00\ \%}=12.27\ \%$$

It is correct to say, therefore, that the effective portfolio risk is lower than the weighted average of individual risks by 12.27%.

Secondly, we determine the diversification ratio:

$$\text{DIV\_R}=\frac{12.00\ \%}{10.53\ \%}=1.14$$

The next step involves assigning the total portfolio risk (10.53%) to the individual components. Following (5.3), we obtain the marginal risk for both asset classes:

$$\begin{aligned}\text{MR}_A &= \frac{40.00\ \%\cdot 6.00\ \%^2+60.00\ \%\cdot 6.00\ \%\cdot 16.00\ \%\cdot 0.28}{10.53\ \%}\\ &= 0.02900\end{aligned}$$

$$\begin{aligned}\text{MR}_B &= \frac{60.00\ \%\cdot 16.00\ \%^2+40.00\ \%\cdot 6.00\ \%\cdot 16.00\ \%\cdot 0.28}{10.53\ \%}\\ &= 0.15612\end{aligned}$$

(continued)

Box 5.1 (continued)

Once we have the marginal risk, it is easy to measure the component risk for the two asset classes with the following two steps:

$$\mathrm{TRC}_A = 0.02900 \cdot 40\ \% = 1.16\ \%$$

$$\mathrm{TRC}_B = 0.15612 \cdot 60\ \% = 9.37\ \%$$

The sum of the total risk contributions gives total portfolio risk, i.e. 10.53%.

It is evident that the contribution to portfolio risk of each asset class is very different. The percentage total risk contributions for A and B are:

$$\mathrm{PTRC}_A = \frac{1.16\ \%}{10.53\ \%} = 11.02\ \%$$

$$\mathrm{PTRC}_B = \frac{9.37\ \%}{10.53\ \%} = 88.98\ \%$$

These results prove that the examined portfolio is less balanced, in terms of risk allocation, than its composition in terms of asset allocation.

5.3 Analysis of Portfolio Risk Using Asymmetric Measures

In strategic asset allocation, standard deviation is the only risk measure associated with efficient portfolios. As is well-known, standard deviation measures the dispersion around a mean return and thus expresses conceptually the associated level of uncertainty.

If we shift the focus from the portfolio construction problem to the portfolio selection problem, it may be reasonable (and helpful for the decision-maker) to consider alternative ways of representing the risk of different investment options. Indeed, it seems plausible that, during this phase, institutional investors or their financial advisors should be interested in attributing a more restrictive and one-sided definition of risk to these options. In other words, estimating portfolio exposure to critical situations is an important part of obtaining a comprehensive, in-depth understanding of the opportunity set, and this step becomes feasible with the use of asymmetric, rather than symmetric, risk measures.

There are various asymmetric risk measures available in the literature, but for our present purposes, our choice can be guided if we ask those salient questions that decision-makers strive to answer. Examples of these are: "What is the worst loss this portfolio could cause, such that the risk of even more negative outcomes is extremely small?", "As regards the worst outcomes that an investment in a given portfolio could cause, what loss should be envisaged on average?", "If an investment had been made in this portfolio in the past, what is the maximum loss of wealth that would have been sustained?"

The wish to address these questions leads to the selection of the following asymmetric risk measures to be determined for each optimal portfolio: Value at Risk (VaR), Expected Shortfall (ES)—often also termed Conditional VaR (CVaR)—and Maximum Drawdown (MDD). The following sections contain an analysis of these measures.

5.3.1 Value at Risk

The VaR measure became established between the end of the 1980s and the 1990s.[5] It is generally defined as the maximum loss that can be expected in a given period of time at a given confidence level (close to 100%, e.g. 95% or 99%). To avoid any misunderstanding, we should add that the VaR expresses a potential loss whose probability of being exceeded (thus resulting in an even worse loss) in a given period of time is equal to $1 - \alpha$ (e.g. 5% or 1%). Therefore, if we take R_P as the random variable representing the returns of any optimal portfolio, it is evident that the VaR focuses on one tail of its probability distribution, i.e. the worst possible $(1 - \alpha)$ scenarios/cases. To be even more precise, the VaR corresponds to a threshold or cut-off point between the $(1 - \alpha)$ worst scenarios and the α best scenarios. Among the worst scenarios, VaR indicates the best value (i.e. the least negative).[6]

This description of the VaR can obviously be expressed formally. To begin with, we can point out its role as threshold, i.e. the possible performance of portfolio r_P that separates or splits a mass of probability (of the variable R_P) into two sections, as in (5.7):

$$\text{VaR}_\alpha(R_P) = \sup \; \{r_P \; |\text{Prob}(R_P \le r_P) \le 1 - \alpha\} \tag{5.7}$$

Consequently, we can also observe the equivalence between the VaR estimate and the estimate of the $1 - \alpha$ quantile of a generic probability distribution, as shown in (5.8):[7]

[5]The first, fundamental contributions on the VaR are attributed to J.P. Morgan (1996) with RiskMetrics™, Jorion (1997), and Duffie and Pan (1997).

[6]For greater clarity, note that although the VaR is described as a potential loss, it can also be positive. Indeed, it is possible that the cut-off-point, which isolates to its left a probability of $1 - \alpha$ of a certain distribution, indicates a profit. Consequently, the value below which the portfolio should most probably (probability α) not fall at the end of the time-frame incorporates in any case a positive result.

[7]Note that "quantile" is understood as a value of a random variable, such that a given percentage of its cumulative distribution function (cdf) lies below this value.

$$\text{VaR}_{\alpha}(R_P) = F_{R_P}^{-1}(1 - \alpha) \tag{5.8}$$

where $F_{R_P}^{-1}$ denotes the inverse of a generic probability distribution.

Therefore, the definition of VaR involves two parameters: the confidence level and the selected period of time (also called VaR holding period). All other conditions being equal, as the confidence level increases, so does the VaR[8] and, in general, the same is also true for an increase in the holding period. If we wish, then, to offer decision-makers the chance to assess optimal portfolios from a new perspective based on maximum potential loss, we need to make a choice regarding these parameters.

By way of example, we provide such a decision-making support option below, referring once again to optimal portfolios on the Resampled Efficient Frontier. As outlined above, we need to make decisions regarding two parameters in order to express the maximum potential loss associated with them. Specifically, we have decided to adopt a confidence level of 99% and to consider different holding periods of 1, 3, and 5 years, which are deliberately not too short, since the analysis intends to support the selection of strategic portfolios.

The step from the definition of the VaR to its practical application involves adopting a methodology to calculate the VaR itself. Note that in both scientific and operational environments, various methods of estimation have been used, but an analysis of their respective strengths and weaknesses (and the logic underpinning them) lies outside the scope of the present chapter, which seeks merely to highlight an alternative idea of the risk of optimal portfolios that is not apparent in the traditional representation using the efficient frontier.[9] As we already have the combinations of annualised standard deviation (σ_P) and annualised expected return (μ_P) for the resampled portfolios, we can, for the sake of simplicity, estimate the VaR using just the classic parametric method which assumes that R_P is normally distributed. For this reason, it is termed Normal VaR. Given that values of the available statistical parameters are annual, the VaR with a confidence level α and a holding period of 1 year is obtained by (5.9):

$$\text{VaR}_{\alpha}^{1\ \text{year}} = \mu_P - N_{\alpha}^{-1} \cdot \sigma_P \tag{5.9}$$

where N^{-1} is the inverse of a standard normal distribution, calculated here for a probability level α.

The calculation can be extended to multiple holding periods, compared to that to which the parameters describing R_P refer, by (5.10):

[8] This can be easily explained: all other conditions being equal, a higher confidence level means isolation of a smaller tail of the probability distribution. It is therefore easy to understand that in cases where the threshold that isolates a probability of $1 - \alpha$ to its left is positive, then as α increases, the estimate of the VaR becomes progressively less positive.

[9] The methods for estimating VaR are usually divided into the parametric, non-parametric, and hybrid approaches, and a good summary can be found in Dowd (2005) and in Jondeau et al. (2007).

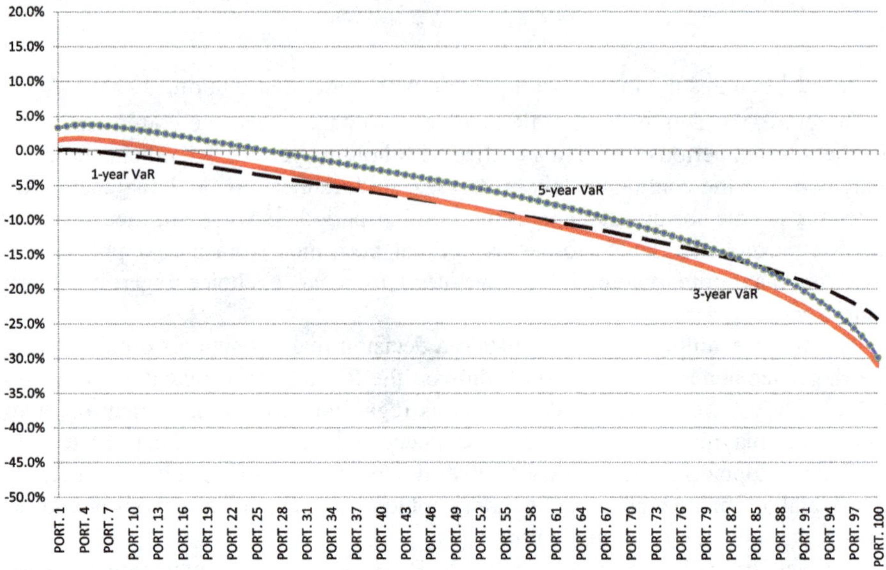

Fig. 5.5 VaR estimates (confidence level = 99%) over different holding periods for portfolios on the Resampled Efficient Frontier

$$\mathrm{VaR}_{\alpha}^{T\ \text{years}} = \mu_P \cdot T - N_{\alpha}^{-1} \cdot \sigma_P \cdot \sqrt{T} \tag{5.10}$$

that is, assuming that return is proportional to time and that risk follows the so-called square root of time rule.

For resampled portfolios, Eqs. (5.9) and (5.10) estimate the VaR shown in Fig. 5.5 with a confidence level of 99%. Going back to what was discussed previously, it highlights the fact that for a portion of the portfolios on the Resampled Efficient Frontier, the VaR appears more worrying for a holding period of 1 year, rather than for a holding period of 3 or 5 years. The figure offers useful elements for selection. For example, if the main financial goal is to protect or preserve capital, then, compared to standard deviation, the VaR suggests that attention should be concentrated on the first four portfolios on the Resampled Efficient Frontier or, in the case of a higher holding period, on the portfolios within the 14th position (for the 3-year time horizon) and within the 26th position (for the 5-year time horizon) at the same time it excludes from plausible preferences the portfolios located on the rightmost section of the aforementioned Resampled Efficient Frontier.

Though clearly an exaggerated comparison, we could assess the VaR for the first and the last portfolios, assuming an available sum to invest of €10 million in both cases. For the first portfolio (with $\mu_P = 1.13\%$ and $\sigma_P = 0.45\%$), the VaR predicts a probability of 99% that the investment will have a market value equal to or higher than €10,008,778 after 1 year, €10,158,869 after 3 years, and €10,332,796 after 5 years. For the examined portfolio, the extreme event remains positive, generating a potential profit of 0.088%, 1.589%, and 3.328%, respectively, with a probability of only 1% of not achieving these figures. For the 100th portfolio, on the other hand (with $\mu_P = 8.95\%$ and $\sigma_P = 14.35\%$), the VaR indicates a potential loss of 24.438% after 1 year, of 30.983% after 3 years, and 29.915% after 5 years. In this case, then, the decision-maker must reasonably take as a threshold (for which there is only a 1% probability that it could be even worse) a market value substantially below the original investment: €7,556,112 after 1 year, €6,901,671 after 3 years, and €7,008,510 after 5 years.

To conclude, the analyses and applications considered in this section demonstrate how, at the portfolio selection stage, the VaR can show the adverse event that may be hidden in the (σ_P; μ_P) combinations of each portfolio. The VaR, therefore, is a valid response to our first question. It is equally true, though, that the use of the VaR as a threshold reveals a kind of limitation: the VaR corresponds to the best of the worst-case scenarios isolated in the tail of the distribution[10] and completely disregards the severity of the losses (or of the less desirable outcomes) beyond the cut-off point indicated. In other words, the VaR says nothing about those worst cases and, as a result, does not give an adequate response to our second question: "With regard to the worst $(1-\alpha)$ outcomes that an investment in a given portfolio could cause, what loss/what return should be envisaged on average?".[11]

5.3.2 Expected Shortfall

The question at the end of Sect. 5.3.1 requires us to look at a different risk statistic/measure: the Expected Shortfall (ES) as proposed by Artzner et al., with which the expected value of the tail scenarios beyond the VaR can be measured.[12]

As with the VaR, it is clear that the ES, too, needs to be accompanied by the specification of a confidence level and a holding period, preferably the same ones used for the calculation of the VaR. The formal definition of ES, highlighting its feature of conditional mean, is given in (5.11):

$$\mathrm{ES}_\alpha(R_P) = E[R_P \ | R_P \leq \mathrm{VaR}_\alpha(R_P)] \tag{5.11}$$

Alternatively, ES corresponds to the computation of the following integral (5.12):

[10] Indeed, of the worst cases, the VaR represents the least severe loss.

[11] Though of no particular interest here, it should be noted that from a theoretical point of view the VaR has also been criticized for reasons other than those mentioned. There are claims in the literature that the VaR cannot be considered a coherent measure of risk as suggested by Artzner et al. (1999), where further details can be found. Here, it is sufficient to note that the reason is the possible violation of the subadditivity requirement.

[12] Artzner et al. (1999).

$$\mathrm{ES}_\alpha(R_P) = \frac{1}{1-\alpha}\int_0^{1-\alpha} F_{R_P}^{-1}(p)dp \text{ with } 0 \le p \le 1-\alpha \tag{5.12}$$

and thus represents, with respect to returns distribution, the value of the area to the left of the VaR.

Again, as with the VaR, there are various ways of estimating the ES with the aim of translating into concrete terms Eq. (5.12), which frequently does not admit a closed-form solution. The investigation of approaches for ES estimation, however, goes beyond the aims of the present study. For this reason, in proposing an application of the ES to portfolios on the Resampled Efficient Frontier we limit ourselves, as with the VaR, to a parametric method based on the assumed normality of the R_P returns. In this case, (5.12) is explicitly assessed through the calculation in (5.13):

$$\mathrm{ES}_\alpha = \mu_P - \frac{f(N^{-1}(1-\alpha))}{1-\alpha} \cdot \sigma_P \tag{5.13}$$

where $f(\cdot)$ represents the density function of a standard normal distribution.

By way of example, Fig. 5.6 gives an illustration of the dynamics of the 3-year ES, with a confidence level of 99%, together with the VaR calculated using the same parameters.

It is not surprising to see that the ES is smaller than the VaR (i.e. less positive or more strongly negative). In fact, it should be borne in mind that this measurement seeks to offer decision-makers a succinct idea of how negative the tail losses exceeding VaR may be on average. Returning, once again and by way of example, to the case of the 100th portfolio, the calculation of the ES indicates that, in the worst 1% of cases, that particular portfolio generates an expected loss of slightly more than 39% after 3 years. Previously, we were "only" able to establish that the cut-off point was a potential loss of 30.983%.

By virtue both of the theoretical logic and the examples given, it is evident that VaR and ES must be seen as *ex ante* or forward-looking measures of risk. Indeed, both measures express situations that investors in a given portfolio may face in the future, and therefore recommend that they be carefully considered so as to ensure informed investment choices. The formulation of both measures in probabilistic terms is proof of their nature.

5.3.3 Maximum Drawdown

Though it may appear contradictory, institutional investors have shown an interest, for some time, in a metric that has essentially opposite features of those recognised to the metrics examined so far: it is an *ex post* and purely empirical measure of risk,

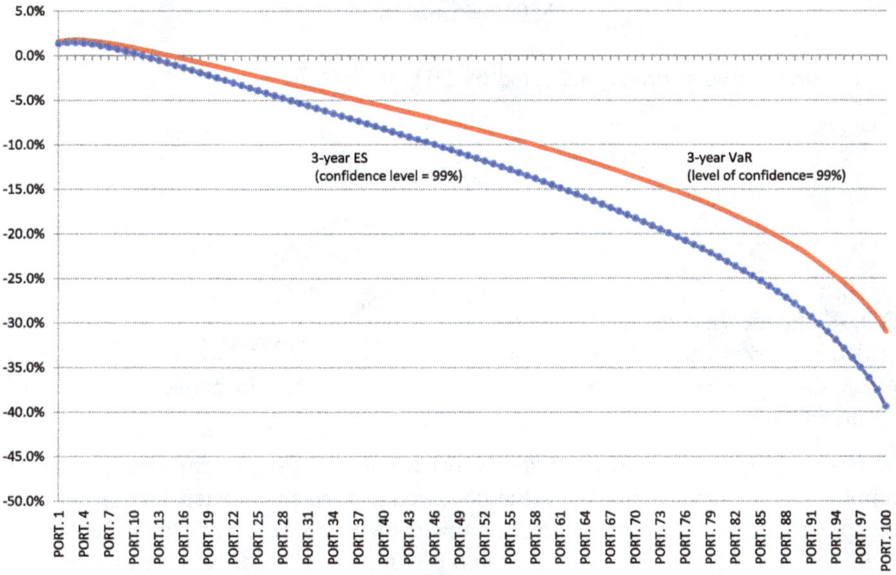

Fig. 5.6 Estimates of VaR and ES (confidence level = 99%) for portfolios on the Resampled Efficient Frontiers and a holding period of 3 years

i.e. definitely not concerned with the identification of an association between the phenomenon analysed and a reference probabilistic distribution. The metric being considered is Maximum Drawdown (MDD), deriving from the work of Zhou and Grossmann.[13]

Maximum Drawdown is calculated on a time series of the market values of a portfolio, i.e. on the dynamics of its cumulative wealth, which can also be expressed by means of a "rebased" series.[14] While drawdown is usually defined as a fall measurable (if present) by observing the current value of the portfolio compared to its previous maximum value, maximum drawdown is simply the largest drawdown and expresses, in percentage terms, a portfolio's maximum loss of value. Once again, it is important to note that there is a significant difference between maximum drawdown and the previous metrics: while VaR and ES identify adverse scenarios that may be faced in the future, MDD describes a real negative experience that has already occurred.

Formally, it is defined as in (5.14):

[13] Grossman and Zhou (1993).

[14] For example, starting with a series of market values for a portfolio, it is possible to construct a base 100 index series. Obviously, this involves eliminating those phenomena that represent external cash flows; see Chap. 8.

$$\mathrm{MDD} = \min(DD_t) \tag{5.14}$$

The drawdown at time t, indicated by DD_t, is:

$$\mathrm{DD}_t = \min\left(\frac{\mathrm{VM}_{P,t}}{\max(\mathrm{VM}_{P,0\ldots t})} - 1, 0\right) \tag{5.15}$$

where

$t = 0,\ldots,T$.
DD_t = drawdown calculated at time t.
$\mathrm{VM}_{P,t}$ = market value of the portfolio at time t.[15]
max $(\mathrm{VM}_{P,0\ldots t})$ = maximum market value of the portfolio between time 0 and time t.[16]

Given (5.15), the drawdown is always a value between zero and a negative number: zero when the market value observed at time t is greater than or equal to the previous maximum value reached up to then and negative when the current market value is lower than the maximum previously reached. As time passes, with (5.15) we obtain a historical series of drawdowns "from moment to moment" (DD_0, $\mathrm{DD}_1, \ldots, \mathrm{DD}_T$) of which, according to (5.14), the maximum drawdown is the lowest value of this sequence.

To facilitate understanding, the dynamics of an investment over time can be shown together with the MDD (Fig. 5.7).

Figure 5.7 also shows the sequence of drawdowns for a hypothetical investment, from $t = 0$ to $t = 15$. The most dramatic fall in the value of the portfolio (−50%), shown on the graph by the decline from 140 to 70, occurs between $t = 3$ and $t = 7$. On closer observation and considering (5.14), it is easy to note a further distinctive feature of maximum drawdown and drawdown in general: they are strongly path-dependent. It seems clear that the drawdowns given in the last column would certainly not remain unvaried if the values describing the path of the portfolio were "remixed". This becomes even more understandable if we underline that both the MDD and the DD_t are also multi-period performance measures that use compound returns in their calculation procedure; it is enough to observe that (5.15) can be re-written as follows:

[15] Here, we use the most common definition of drawdown. There is an alternative definition according to which calculation is based on the series of accumulated returns, rather than the values/prices series. In other words, the alternative definition requires that at each time-point t, the accumulated performance of the portfolio from the beginning up to that time-point is compared with the maximum potential performance of the portfolio accumulated at any time-point prior to t.

[16] Some authors consider the maximum market value up to $t - 1$ as the denominator. It is a variant coming from the mutual fund industry which recognizes this value as the high water mark useful for calculating performance fees. See D. Bradford and D. Siliski (2016).

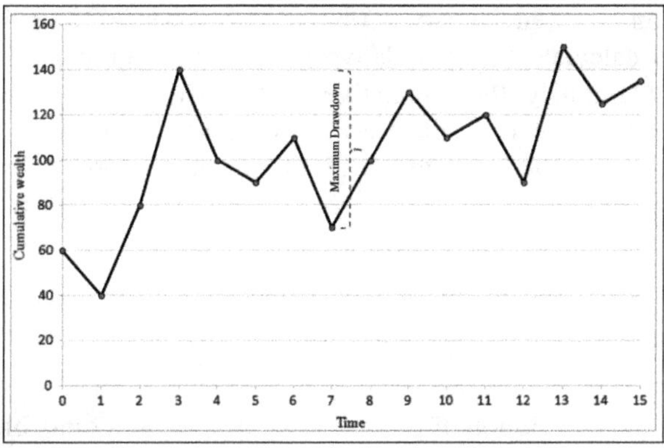

t	DRAWDOWNS
0	0.00%
1	-33.33%
2	0.00%
3	0.00%
4	-28.57%
5	-35.71%
6	-21.43%
7	**-50.00%**
8	-28.57%
9	-7.14%
10	-21.43%
11	-14.29%
12	-35.71%
13	0.00%
14	-16.67%
15	-10.00%

Fig. 5.7 An example of maximum drawdown calculation

$$\mathrm{DD}_t = \min\left(\frac{\prod_{i=0}^{t}(1+r_i)}{\max\limits_{t\in 0.....T-1}\left(\prod_{j=0}^{t}(1+r_j)\right)} - 1, 0\right) \tag{5.16}$$

According to (5.16), the periodic returns r_i must not be treated as independent events, but rather as sequential events. On an interpretative level, it is of fundamental importance not to confuse the drawdowns and the maximum drawdown with the calculation of a loss compared to the capital initially invested.

As shown in the example, the most significant fall in cumulative (but not yet realised) wealth, equal to -50%, does not cause the initial capital to be compromised. It is therefore clear that it is not correct to automatically assume the coincidence between the two measurements.

Wanting to be even more exhaustive in illustrating the drawdowns and maximum drawdown, it should also be noted that drawdown is sensitive both to the frequency of the data in the time series on which it is calculated and to the length of this series, a fact not to be overlooked if we are to avoid erroneous and misleading comparisons, in terms of drawdown, between portfolios: in fact, such comparisons are possible only if the frequency and length of the time series are homogeneous.

The application of maximum drawdown to efficient portfolios requires one further specification. As these portfolios originate from a process of strategic asset allocation, they obviously have not a past, a history. This means that we must first reconstruct the path that each optimal portfolio (given its composition) would have followed in the past, which enables institutional investors to explore their spontaneous and emotive reaction to the worst experience that these compositions would have generated.

The reconstruction was conducted using the strategic asset allocations derived from the application of Resampling. First, we calculated the monthly performances of each strategic asset allocation between January 2003 and December 2022, and then we used these figures to calculate the cumulative wealth that would have resulted over that time period. Obviously, the cumulative wealth then becomes the basis for the calculation of the drawdowns. Table 5.2 shows the maximum drawdown to which each of the 100 portfolios on the REF would have exposed an investor.

5.4 Constraints on Risk-Taking

One further step in completing the description of portfolio risk from other points of view besides that of volatility involves the definition and implementation of tools that enable portfolio selection to respect constraints on risk-taking. These tools can "filter" down the number of efficient portfolios in the opportunity set by eliminating those that do not respect a risk-taking constraint. Depending on the specific constraint being considered, the "filtering" criterion could be a portfolio's probability of reaching a minimum return threshold or of avoiding a loss greater that a given amount.

5.4.1 The Shortfall Constraint

The best known and most widely used risk-taking constraint is the shortfall constraint proposed by Leibowitz et al.[17] With this constraint, we can exclude from the selection those portfolios of the opportunity set which show a shortfall probability—that is a probability of not attaining and/or exceeding a desired minimum level of return—higher than the tolerated value.

In order to define the shortfall constraint formally, we need to be clear about the initial information required and the result sought. Regarding the former, we need to specify or gain knowledge of two elements: the shortfall probability and the minimum return to protect. The result sought, on the other hand, is the expected return that a portfolio, given a level of volatility and with returns following a normal distribution, should exhibit such that the cumulative probability of performances below the minimum return threshold does not exceed the shortfall probability. This expected return is compared to the actual μ_P of an efficient portfolio with the same standard deviation.

We can bring all the "ingredients" of shortfall constraint together with an example based on the following question: assuming that we can tolerate a maximum probability of 10% of not reaching an annual return of 1%, what is the expected return that would be required to a portfolio with a standard deviation of 6%? This is the same as

[17] Leibowitz et al. (1996).

Table 5.2 Maximum drawdown of portfolios on the resampled efficient frontier

Portfolio position number	Maximum drawdown (%)	Portfolio position number	Maximum drawdown (%)
PORTF. 1	−1.819	PORTF. 51	−12.707
PORTF. 2	−1.358	PORTF. 52	−13.168
PORTF. 3	−0.961	PORTF. 53	−13.632
PORTF. 4	−0.677	PORTF. 54	−14.104
PORTF. 5	−0.827	PORTF. 55	−14.580
PORTF. 6	−0.977	PORTF. 56	−15.066
PORTF. 7	−1.204	PORTF. 57	−15.556
PORTF. 8	−1.679	PORTF. 58	−16.050
PORTF. 9	−2.122	PORTF. 59	−16.547
PORTF. 10	−2.344	PORTF. 60	−17.048
PORTF. 11	−2.570	PORTF. 61	−17.553
PORTF. 12	−2.805	PORTF. 62	−18.060
PORTF. 13	−3.039	PORTF. 63	−18.573
PORTF. 14	−3.272	PORTF. 64	−19.088
PORTF. 15	−3.506	PORTF. 65	−19.605
PORTF. 16	−3.761	PORTF. 66	−20.127
PORTF. 17	−4.024	PORTF. 67	−20.653
PORTF. 18	−4.287	PORTF. 68	−21.189
PORTF. 19	−4.549	PORTF. 69	−21.733
PORTF. 20	−4.811	PORTF. 70	−22.301
PORTF. 21	−5.072	PORTF. 71	−22.883
PORTF. 22	−5.334	PORTF. 72	−23.478
PORTF. 23	−5.594	PORTF. 73	−24.089
PORTF. 24	−5.855	PORTF. 74	−24.716
PORTF. 25	−6.115	PORTF. 75	−25.359
PORTF. 26	−6.375	PORTF. 76	−26.021
PORTF. 27	−6.634	PORTF. 77	−26.713
PORTF. 28	−6.893	PORTF. 78	−27.433
PORTF. 29	−7.152	PORTF. 79	−28.180
PORTF. 30	−7.410	PORTF. 80	−28.951
PORTF. 31	−7.668	PORTF. 81	−29.746
PORTF. 32	−7.925	PORTF. 82	−30.583
PORTF. 33	−8.182	PORTF. 83	−31.449
PORTF. 34	−8.439	PORTF. 84	−32.332
PORTF. 35	−8.695	PORTF. 85	−33.254
PORTF. 36	−8.951	PORTF. 86	−34.208
PORTF. 37	−9.207	PORTF. 87	−35.192
PORTF. 38	−9.462	PORTF. 88	−36.197
PORTF. 39	−9.716	PORTF. 89	−37.223
PORTF. 40	−9.968	PORTF. 90	−38.281
PORTF. 41	−10.217	PORTF. 91	−39.355
PORTF. 42	−10.455	PORTF. 92	−40.448

(continued)

Table 5.2 (continued)

Portfolio position number	Maximum drawdown (%)	Portfolio position number	Maximum drawdown (%)
PORTF. 43	−10.683	PORTF. 93	−41.570
PORTF. 44	−10.903	PORTF. 94	−42.702
PORTF. 45	−11.108	PORTF. 95	−43.836
PORTF. 46	−11.301	PORTF. 96	−44.988
PORTF. 47	−11.481	PORTF. 97	−46.132
PORTF. 48	−11.646	PORTF. 98	−47.277
PORTF. 49	−11.809	PORTF. 99	−48.491
PORTF. 50	−12.254	PORTF. 100	−49.916

identifying the "centre" (mean) of a Gaussian distribution such that if we move to the left 1.28155 times the standard deviation, we reach 1%.[18] If we translate this into algebraic form (5.17), we get the result 8.69%:

$$\mu^* - 1.2815 \cdot 6\ \% = 1\ \% \tag{5.17}$$

from which we obtain:

$\mu^* = 1\% + 1.2815 \cdot 6\%$ and, therefore, $\mu^* = 8.69\%$.

If an efficient portfolio with a volatility of 6% really does have an expected return not lower than 8.69%, then the shortfall constraint is respected; otherwise, the constraint is not satisfied and, consequently, the level of protection associated with the minimum return is below the required value.

At first sight, the calculation of this example should be repeated for the entire range of standard deviations of the portfolios on the efficient frontier. However, it is clearly more logical and practical to specify a generalised version of the shortfall constraint that determines the "required" expected return for any level of risk, such that the probability of not reaching the minimum return target does not exceed the shortfall probability. If R_{MIN} is the minimum return to protect, with μ^* being the "required" expected return (the unknown quantity) and π is the shortfall probability, considering the calculations in (5.15), we can obtain the shortfall constraint equation using (5.18):

$$\mu^* + N_\pi^{-1} \cdot \sigma_P \geq R_{MIN} \tag{5.18}$$

from which, given that R_{MIN} is fixed, we get:

$$\mu^* \geq R_{MIN} - N_\pi^{-1} \cdot \sigma_P$$

[18] 1.28155 is the volatility multiplier that isolates a probability density of 10% in the tail of a Gaussian distribution.

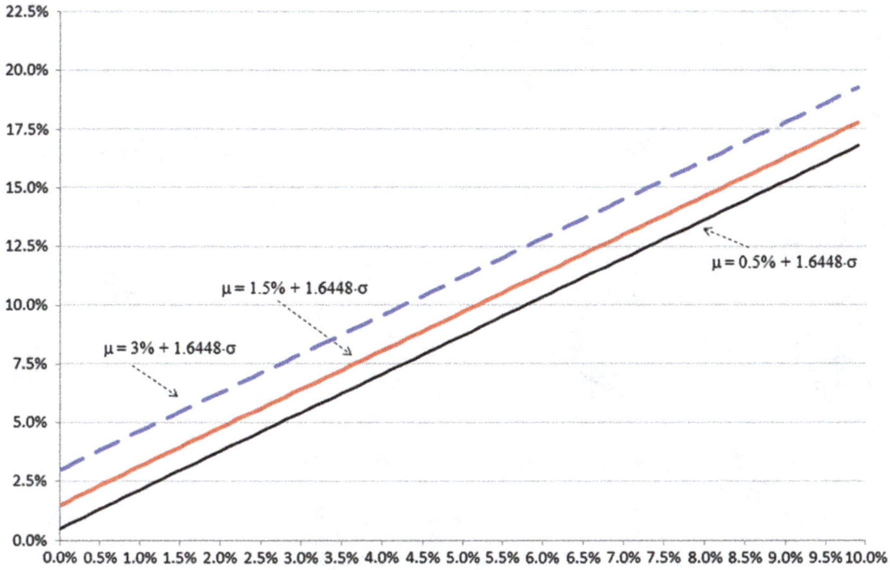

Fig. 5.8 Sensitivity of the shortfall constraint to changes in the minimum return

Once we have the algebraic formulation of the shortfall constraint, it is easy to obtain its graphical representation, which can be plotted in a classical risk-return space as a straight line (the shortfall line) using the equation $\mu^* = R_{\text{MIN}} - N_{\pi}^{-1} \cdot \sigma_p$. As a result, the area of the graph will be divided into two sectors, one above (and including) the shortfall line, which contains the risk-return combinations that do not violate the shortfall constraint, and one below the shortfall line with the combinations which instead do not meet the constraint.

It is also useful to examine the sensitivity of the shortfall constraint to changes in the parameters of the minimum return threshold and the shortfall probability.

In the first case, since R_{MIN} identifies geometrically the intercept of the shortfall line, any modification leads to a parallel shift of the shortfall line in the same direction in which R_{MIN} varies. This is confirmed in Fig. 5.8, which shows the movements of a shortfall line with a shortfall probability of 5% (so with $N_{5\%}^{-1} = -1.6448$) and values for R_{MIN} of 0.5%, 1.5%, and 3%.

Differently, if the shortfall probability is varied, the shortfall line changes as a result of modifications in the value of the inverse of the normal N_{π}^{-1} which conditions its slope coefficient. Specifically, the slope of the shortfall line increases if the probability of not reaching the tolerated threshold return is reduced, whereas the line gets flatter if the acceptable shortfall probability is increased. In other words, a decrease/increase in shortfall probability is associated with a decrease/increase in the risk-return combinations that satisfy the constraint. This is not surprising, since a variation in π is symptomatic of a greater/lesser tolerance of shortfall risk, i.e. of the possibility of not being able to protect R_{MIN}. Again, it is useful to have a graphical

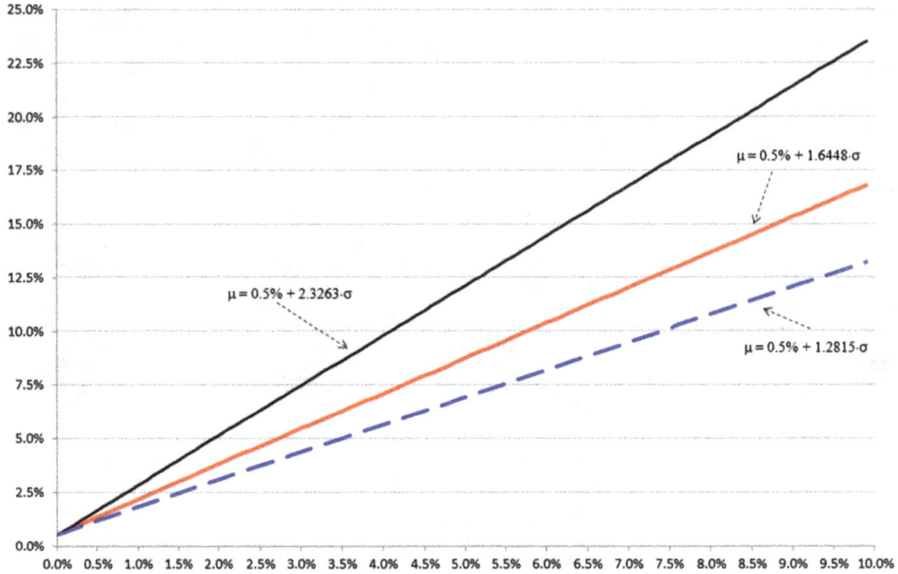

Fig. 5.9 Sensitivity of the shortfall constraint to variations in the shortfall probability

representation of the impact of changes in shortfall probability: Fig. 5.9 shows the shortfall line for an annual R_{MIN} of 0.5%, with three different levels of shortfall probability (1%, 5%, 10%).

Finally, having seen the sensitivity of the shortfall constraint to the parameters on which it is based, we can represent this risk-taking constraint together with the set of efficient portfolios. Figure 5.10 shows the annualised risk-return combinations of the portfolios on the Resampled Efficient Frontier and the shortfall constraint with an annual R_{MIN} of 0.5% and π of 10%.

In terms of portfolio selection, Fig. 5.10 aims to show institutional investors who need to be sure (with a good degree of certainty) that a chosen investment will produce a minimum level of return, that only efficient portfolios which lie above the shortfall line can legitimately guarantee that this objective will be fulfilled. For the portfolios below the shortfall line, the threat of not achieving the minimum return has a probability of occurrence greater than the tolerated one. In the specific case, only the selection of one of the resampled portfolios that occupy the first 7 positions of the Resampled Efficient Frontier would justify high confidence in the safeguarding of the minimum annual return of 0.5%.

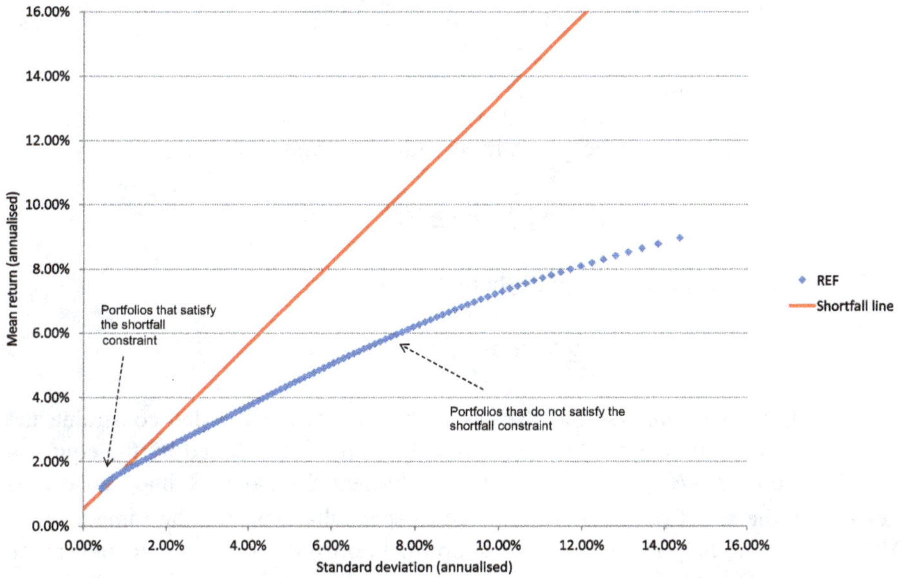

Fig. 5.10 Shortfall line and portfolio selection

5.4.2 *Constraints Based on Value at Risk*

At the beginning of Sect. 5.4, we briefly noted the possibility that the constraint on risk-taking may take a slightly different form from that of the shortfall constraint. Actually, it is also possible to address the need to protect (with a high level of probability) against potential losses larger than a given sum set by the investor, as an alternative to the previous objective of achieving (with a high level of probability) a minimum return. In this case, we use a VaR constraint rather than a shortfall constraint. Despite the different name, it is easy to show that the procedure to implement the new constraint does not differ substantially from that for the shortfall constraint. As in that case, there are two initial parameters to specify: for VaR constraint, these are the maximum potential loss that can be tolerated, defined as VaR_bound, and the accepted low level of probability that actual loss may be larger, indicated as π_{VaR}. Again, the value sought is the mean that should characterise the distribution of portfolio returns, such that the cumulative probability of results worse than that loss is not greater than π_{VaR}.

To summarise, the question behind the use of a VaR constraint is, for example: given a potential annual loss of 2% (which in very adverse conditions would be tolerated), what the expected return from a portfolio with a volatility of 6% should be in order to ensure that the probability of sustaining or exceeding that loss is 10% at most? The answer can be derived by using (5.19), which in many respects is very similar to (5.17):

$$\mu^* - 1.2815 \cdot 6\ \% = -2\ \% \tag{5.19}$$

from which we obtain:

$\mu^* = -2\% + 1.2815 \cdot 6\%$ and, therefore, $\mu^* = 5.69\%$.

The VaR constraint can be generalised with the formula in (5.20):

$$\mu^* + N_{\pi_{\text{VaR}}}^{-1} \cdot \sigma_P \geq \text{VaR\_bound} \tag{5.20}$$

From which, given a stated VaR_bound, we obtain:

$$\mu^* \geq \text{VaR\_bound} - N_{\pi_{\text{VaR}}}^{-1} \cdot \sigma_P$$

Formula (5.20) clearly shows that, as in the case of the shortfall constraint, the VaR constraint can also be shown graphically with a line defined by the equation $\mu^* = \text{VaR\_bound} - N_{\pi_{\text{VaR}}}^{-1} \cdot \sigma_p$. This line is termed the iso-VaR line, since it is defined by the set of points in the risk-return space that generate the same level of VaR. In order to help in the selection of optimal portfolios, the iso-VaR line can be shown together with the efficient frontier: Fig. 5.11 superimposes the iso-VaR line onto the Resampled Efficient Frontier assuming a maximum tolerable loss of 2% per year and a willingness to accept a probability of 10% that losses are even higher.

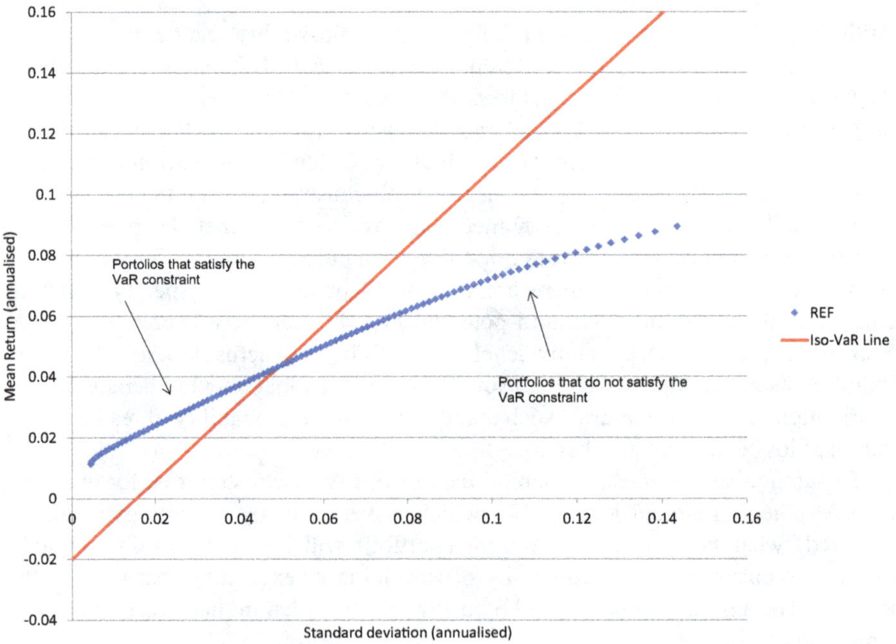

Fig. 5.11 Iso-VaR line and portfolio selection

As can be seen, the interpretation of the graph, too, is the same as for the shortfall constraint. The risk-taking constraint is satisfied by portfolios above the iso-VaR line, but not by those below the line. It is also interesting to observe that having expressed a VaR constraint, rather than a shortfall constraint using the same probability level, has led to an expansion of the portfolios that can be selected.

5.5 Analysis of Future Scenarios of the Cumulative Wealth

The constraints on risk-taking (shortfall constraint and VaR constraint) discussed in the previous section are based on a single-period setting. If we bear in mind that the parameters describing optimal portfolios (σ_P; μ_P) are relevant to the definition and/or interpretation of these constraints, this implies that by using the constraints, we can identify which condition we have to respect[19] so that, at the end of the given period, the probability that an institutional investor will face a return lower than R_{MIN}, or a loss exceeding the VaR_bound, does not exceed the tolerated limit. In general (σ_P; μ_P) are expressed as annualised figures. The understandable convenience or need to "ensure" the protection of a minimum performance, or to "shield" against the accumulation of a loss greater than a given threshold over a time period that is a multiple of the one defined by (σ_P; μ_P), is addressed by extrapolating the multi-period risk and return using the square root of time rule for the former and the proportionality of time for the latter, and not by calculating a fully-fledged multi-period shortfall or VaR constraint.

To understand the significance of this last claim, it is useful to recall that, in a multi-period model, time is not a rigid parameter (T), but a variable t in the range [0, …, T]. Therefore, a model of this type is able to monitor the threatening phenomenon not only at the end of the time period, but also continuously throughout this same period. This is certainly of interest, because the maximum tolerable loss can be exceeded over the lifetime of a given investment in a portfolio, and therefore it is worth considering from the outset whether or how this would be tolerated before the effects of time diversification can be appreciated.[20]

One interesting multi-period model is the Ibbotson-Sinquefield simulation,[21] whose characteristic graphical representation is similar to a cone. More precisely, the Ibbotson-Sinquefield simulation is a continuous-time model which aims to predict in probabilistic terms the evolution of the cumulative wealth derived from an investment, for example from a portfolio. Lewis et al. and Kaplan have provided

[19] As we have seen, this involves an expected return that a portfolio should provide, together with a given standard deviation.

[20] For a brief analysis of the issue of time diversification, see Box 5.2.

[21] Ibbotson and Sinquefield (1976).

the formal description for the model, which was initially absent as the authors had originally adopted an empirical/simulation approach.[22]

According to the Ibbotson-Sinquefield simulation, the future evolution of the cumulative wealth follows a log-normal distribution, as do, obviously, the factors of wealth accumulation from one period to the next. These two values can be related as shown in (5.21):[23]

$$W_t = (1 + R_t) \cdot W_{t-1} \tag{5.21}$$

where W_t is cumulative wealth up to time t, and R_t the rate of return between $(t - 1)$ and t.

The properties of the log-normal distribution reveal that the continuous rate of return defined in (5.22):

$$r_t = \ln(1 + R_t) \tag{5.22}$$

is normally distributed. It is, therefore, correct to write $r_t \sim N(m, s^2)$. Consequently, the logarithm of cumulative wealth as given in (5.23) also has a Gaussian distribution.

$$\ln(W_T) = r_1 + r_2 + \ldots + r_T \tag{5.23}$$

In particular, the value $\ln\left(\frac{W_T}{W_0}\right)$ is normally distributed, with expected value mT and standard deviation $s\sqrt{T}$.

The parameters still needed to estimate the future evolution of the cumulative wealth over time are m and s. As indicated by Kaplan, both must be derived from the values of σ_P and μ_P associated, in a discrete context, with efficient portfolios using the conversion rules in (5.24) and (5.25), respectively:

$$s = \sqrt{\ln\left(1 + \left(\frac{\sigma_P}{1 + \mu_P}\right)^2\right)} \tag{5.24}$$

$$m = \ln(1 + \mu_P) - \frac{1}{2}s^2 \tag{5.25}$$

At this point, if we consider an initial investment W_0 in an efficient portfolio with an expected return μ_P and standard deviation σ_P, both converted into their continuous equivalents using m and s, then the expected cumulative wealth, in probabilistic terms, at a generic time t is:

[22] Lewis et al. (1980) and Kaplan (1998).

[23] Note that a random variable with a log-normal distribution can take values between 0 and $+\infty$.

$$W_{t,\mathrm{prob\_0}\alpha} = W_0 \cdot \exp\left(m \cdot t + N^{-1}_{\mathrm{prob\_}\alpha} \cdot s \cdot \sqrt{t}\right) \tag{5.26}$$

This can be re-written in extended form as shown in (5.27):

$$W_{t,\mathrm{prob\_}\alpha} = \mathrm{W}_0 \cdot \exp\left\{ \left[\ln(1+\mu_P) - \frac{1}{2} \cdot \ln\left(1 + \left(\frac{\sigma_P}{1+\mu_P}\right)^2\right)\right] \cdot t \right.$$
$$\left. + N^{-1}_{\mathrm{prob\_}\alpha} \cdot \sqrt{\ln\left(1 + \left(\frac{\sigma_P^2}{1+\mu_P}\right)\right) \cdot t} \right\} \tag{5.27}$$

The fact that both (5.26) and (5.27) include $N^{-1}_{\mathrm{prob\_}\alpha}$, i.e. the quantile of a normal distribution, proves that the estimate of the evolution of the future cumulative wealth is given in a continuous-time setting, specifying the level of cumulative probability that remains below the simulated path. In practical applications of the Ibbotson-Sinquefield simulation, the evolution of the cumulative wealth in a worst-case scenario, a best case scenario and the expected scenario are usually shown together in one graph. In defining the quantile $N^{-1}_{\mathrm{prob\_}\alpha}$ to use, symmetrical levels of probability are used (e.g. 5% and 95%) to represent the worst and the best cases, while the expected scenario is given a probability of 50%.

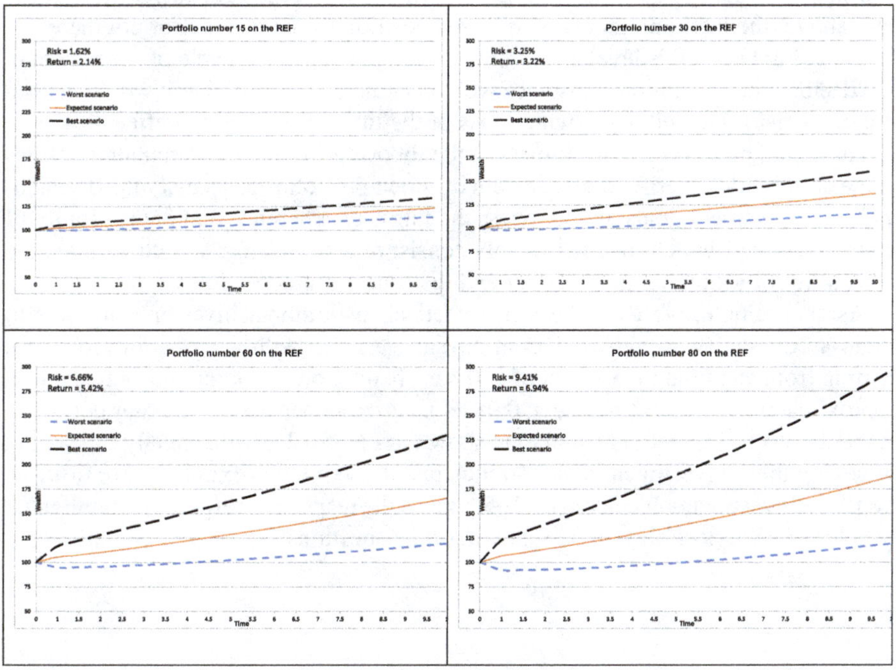

Fig. 5.12 Graphs of the Ibbotson-Sinquefield simulation for values of prob_α of 5%, 50%, and 95%

Using these values, for which the respective quantile is $N_{5\%}^{-1} = -1.6449$, $N_{95\%}^{-1} = +1.6449$ and $N_{50\%}^{-1} = 0$, Fig. 5.12 shows a graphical representation of the Ibbotson-Sinquefield simulation for 4 portfolios on the Resampled Efficient Frontier with risk-return combinations in a discrete context.[24] With reference to this illustration of Ibbotson-Sinquefield simulation, we can say that:

- The probability of obtaining amounts of cumulative wealth below those on the worst scenario path is at most 5%.
- The probability of obtaining amounts of cumulative wealth above those on the best scenario path is at most 5%.
- The probability of obtaining amounts of cumulative wealth between the worst scenario path and the best scenario path is 90%.

So as not to omit anything regarding the terms used with the Ibbotson-Sinquefield simulation, note that it is common practice to provide a confidence level along with the graph. The confidence level specifies the probability associated with those amounts of cumulative wealth that are not below those on the worst scenario path. Considering the choices used for the graphs in Fig. 5.12, a 95% confidence level was selected.

The Ibbotson-Sinquefield simulation can support portfolio selection in two ways. First, it can verify if and/or which portfolios are able to protect against a fall in the value of the cumulative wealth below a safety level that an investor may consider crucial with regard to a deadline in the reference time-frame. Actually, unlike in the case above, the institutional investor is in a position to see whether or not there is a risk of going below this level at any time during the entire life cycle of the portfolio, and decide whether this circumstance would be tolerated or whether it would instead lead to decisions to sell, even with a loss of the time diversification effect.

Above all, however, the usefulness of the Ibbotson-Sinquefield simulation resides in the possibility to verify to what extent different portfolios can satisfy the decision-maker's investment goals and to assess the intrinsic coherence and compatibility of these objectives, which may reasonably consist of a cumulative wealth to attain the probability of achieving that amount and the time required to do so.

As an example, we can take a hypothetical institutional investor who aims to attain a cumulative wealth of 105 in 5 years (T) and with a probability of 95%, starting from an initial investment of 100, i.e. "betting on" a compound annual return approximately of 1.00% (precisely, 0.981%). As explained above, the four portfolios in the Ibbotson-Sinquefield simulation shown in Fig. 5.12 are not equally efficient in achieving these investment goals. To illustrate the varying proximity to (or distance from) the investment objective, Table 5.3 gives the respective responses (highlighted in bold) of each portfolio to the following four questions:

[24] Consequently, the parameters shown in the graphs in Fig. 5.12 have been transformed to show the Ibbotson-Sinquefield simulation as in (5.22) and (5.23).

Table 5.3 Consistency check of financial investment goals

	W_0	W_T	T	Prob_α (%)
PORTFOLIO 15	**100.22**	105	5	95
	100	**104.78**	5	95
	100	105	**5.14**	95
	100	105	5	**94.34**
PORTFOLIO 30	**100.84**	105	5	95
	100	**104.12**	5	95
	100	105	**5.42**	95
	100	105	5	**93.65**
PORTFOLIO 60	**102.76**	105	5	95
	100	**102.18**	5	95
	100	105	**5.95**	95
	100	105	5	**92.68**
PORTFOLIO 80	**105.71**	105	5	95
	100	**99.33**	5	95
	100	105	**6.67**	95
	100	105	5	**91.34**

- Given a target cumulative wealth of 105, a confidence level of 95%, and a time period of 5 years, what is the initial wealth W_0 required to achieve the target?
- Given an initial wealth of 100 and a time period of 5 years, what is the final attainable cumulative wealth W_T with a confidence level of 95%?
- Given an initial wealth of 100, how long is the time period T required to achieve the target cumulative wealth of 105 with a confidence level of 95%?
- Given an initial wealth of 100, a time period of 5 years and a target cumulative wealth of 105, what is the associated confidence level, prob_α?

If we look at the results in Table 5.3, it is easy to see that portfolios 60 and 80 are poorly suited to satisfy the investor's goal. The other two portfolios appear more capable of approaching the investment goals of the hypothetical institutional investor.

The description of the mathematical steps involved in obtaining the values given in Table 5.3 goes beyond the scope of the present study; note, however, that they require the formulation of (5.26) or (5.27) with a change each time of the dependent variable to be calculated as a function of the other three.

However, for the mere fact that it is often a priority to grasp the probability of achieving the investment goal that can be associated to the portfolios, it is opportune to enunciate the formal way of deducing the quantile implicit in (5.26) or in (5.27), knowing: the capital that can be placed by the institutional investor; the final amount that the investor would not like to see compromised at the end of the time horizon T; the continuous equivalents (s and m) of risk and return of the portfolios in the opportunity set. The above quantile can be derived according to (5.28):

$$N_*^{-1} = \frac{\ln\left(\frac{W_T}{W_0}\right) - m \cdot T}{s \cdot \sqrt{T}} = \frac{\ln(W_T) - \ln(W_0) - m \cdot T}{s \cdot \sqrt{T}} \tag{5.28}$$

Once obtained N_*^{-1}, the probability of reaching a final wealth equal to or greater than W_T is estimated, in a Gaussian framework, as follows (5.29):

$$1 - \text{Norm\_cdf}\left(N_*^{-1}\right) = \text{prob\_}\alpha \tag{5.29}$$

where Norm_cdf indicates the cumulative distribution function of a normal distribution.

By way of example, the application of (5.28) to the portfolio in the 80th position of the Resampled Efficient Frontier, already considered in Table 5.3, returns the following value for the implicit quantile:

$$\frac{\ln\left(\frac{105}{100}\right) - 6.3271\% \cdot 5}{8.7837\% \cdot \sqrt{5}} = -1.3623$$

And, subsequently, (5.29) leads to the probability of success in pursuing the investment goal as indicated below:

$$1 - \text{Norm\_cdf}(-1.3623) = 91.34\%$$

Box 5.2 The Time Diversification Issue

Time diversification is an expression that refers to the relationship between risk and the holding period of an investment. Typically, it is summarised in a statement such as: the returns of equity asset classes are less risky over long time horizons than over short time horizons. A plausible corollary of this general principle is to suggest the selection of a more aggressive asset allocation if, for various reasons (age, financial situation, etc. of the end investor), it is realistic to "grant" the investment an extended holding period.

In the financial literature, the time diversification matter is at the centre of a rather heated and animated debate which began in 1969 with the help of Samuelson (1969) and represents a real puzzle since different schools of thought opposed to each other in the questioning whether diversification works across time, as well as between asset classes and securities, have reached strongly conflicting conclusions: the different schools of thought in the financial literature have discussed the issue of time diversification by referring, respectively, to expected utility theory, option pricing theory, and behavioural finance (Bianchi et al., 2016). The reason can perhaps be found in the fact that each of them wants to classify the issue within its own theoretical

(continued)

Box 5.2 (continued)
construct equipped with its own assumptions, often not realistic, about the behaviour of returns and/or the behaviour or attitude of investors, to which over time a growing number of risk definitions have been added as possible references. Here, however, we do not want to make any effort to highlight the merits or fragility of one or the other approach.

With an approach closer to that of practitioners, we want to formulate some considerations on the relationship between risk and investment horizon, addressing it in an extremely empirical way.

In fact, a realistic example is proposed, starting from the monthly returns provided to an European investor by an undoubtedly aggressive asset class, represented by the MSCI Emerging Markets, in the 22 years between January 2001 and December 2022. Therefore, using 264 monthly returns, we estimated the outcomes that could have been achieved by an investor with a holding period of 6 months, 1 year, 3 years, 5 years, 7 years, and 10 years. The simulations of investments in an asset class start in correspondence of each months in the time series available for each specific holding period considered. In doing so, it is possible to count 259 investment cases lasting 6 months, 253 investment cases lasting 1 year, 229 investment cases lasting 3 years, 205 investment cases lasting 5 years, 181 investment cases lasting 7 years, and 145 cases of 10-year long investment. It would have been methodologically preferable to conduct an analysis of this type using time horizons with a specific extension that are independent of each other and not using, even in part, redundant and overlapping returns; however, this would have required a much larger set of initial information than the 264 monthly returns actually available. The next step consists in quantifying, for each starting point of the investment and for each time horizon, the annual geometric mean return so as to be able to express, with a common and comparable language, at which constant rate (positive or negative) the initial sum is implicitly invested. Finally, employing the same data used to calculate the geometric mean return, also the standard deviation of such data, the best value and the worst value as well as the distance between the latter two are measured for each investment horizon. These values are shown in the table below with reference to the different holding periods (HP) examined.

MSCI Emerging Markets	6 months	1 year	3 years	5 years	7 years	10 years
MAX GEOMETRIC MEAN RETURN	129.34%	79.15	41.04%	29.64%	19.75%	15.17%
MIN GEOMETRIC MEAN RETURN	−70.81%	−51.23%	−13.79%	−1.34%	0.53%	2.77%
RANGE	200.15%	130.37%	54.83%	30.97%	19.22%	12.40%
STANDARD DEVIATION	32.81%	22.79%	10.50%	6.44%	4.28%	3.20%

It reveals a narrowing of the variability of returns, i.e. of constant compound rates, as a function of the duration of the investment. According to the measurement of the geometric mean return, therefore, as the time horizon extends, the potential outcomes of the more and less prosperous investment becomes less extreme. In order not to draw misleading conclusions and have a more complete knowledge, it is also important to observe the behaviour of the final wealth achieved by the investor in the same situations considered in the previous table. The graph proposed here provides this.

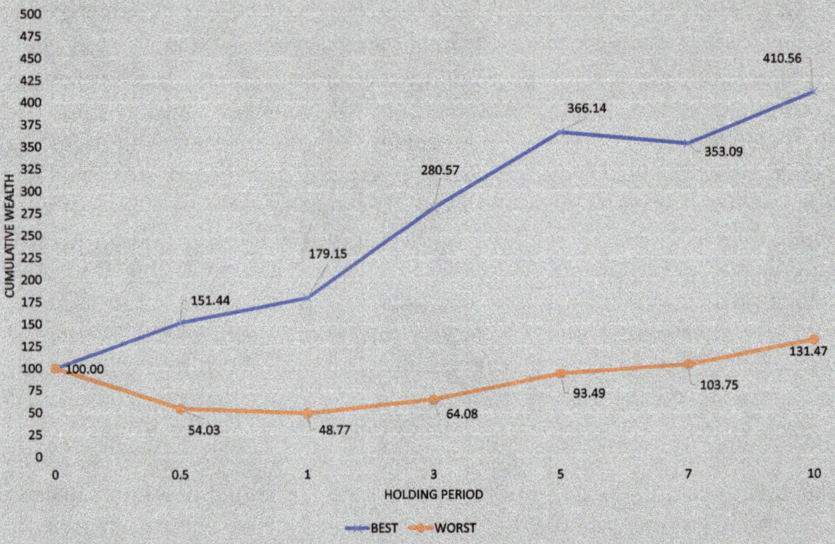

Based on it, the range of variation of the so-called terminal wealth expands as the time horizon extends and the compounding effect associated with the returns reported in the table becomes visible. The results seem contradictory which makes it difficult to answer the question: what is the relationship between risk and time horizon?

With the awareness of not advancing radical conclusions, looking at the behaviour found for the MSCI Emerging Markets (which is reported to be similar to what a similar investigation would reveal for other equity asset classes), it is legitimate to state that time seems capable of smoothing the risk since the standard deviation of the single-period geometric average return is effectively reduced.

It can then be added that if the table had reported, in conjunction with more extended holding periods, maximum and minimum geometric mean returns not dissimilar to those indicated for the shorter horizons (6 months, 1 year), the proposed figure would have shown a greater gap between the line of the "best scenario" and the "worst scenario" and made visible a much more worrying

(continued)

Box 5.2 (continued)

dynamic for the latter. It is also reasonable to argue that time seems capable of making risk "more friendly" or "less antagonistic" to the investor in the sense that it supports the recovery of losses; however, it is erroneous or at least forced to assert that the lengthening of the time horizon constitutes a guarantee or automatic realisation of a positive cumulative performance. However, the above does not make it reasonable to assume that the volatility of independent periodic returns underlying different calculation periods (days, months, years) is meant to reduce with the increase in the duration of the investment; in truth, it increases but in a less than proportional way and, as commonly accepted given the multi-period conversion rule of a risk already used in this chapter, in relation to the square root of time.

Coming to an evaluation of what emerged from the perspective of portfolio selection, the idea is that a non-trivial presence of asset classes similar to the one on which attention has been focused here in asset allocation solutions may be appropriate for a "patient" investor, i.e. an investor equipped with an extended investment time horizon and, possibly, also available to consider the implementation of the investment itself in gradual steps.

In truth, without giving up an empirical approach, for the sake of completeness it should be noted that a more rigorous investigation of time diversification is feasible through a verification of the behaviour of returns which allows one to ascertain, alternatively, the existence of the random walk or rather the existence of some form of dependence or serial correlation which, obviously, constitutes a trace of non-randomness. If, for example, a mean-reversion phenomenon (return to the mean) is found through negative serial correlation values (at certain significantly deferred time lags), the idea of risk reduction due to a greater temporal extension of the investment appears validated. As suggested by Poterba and Summers (1988) and Lo and MacKinley (1988, 1989), the use of the variance ratio test can be considered a rigorous approach to conduct such a verification. The variance ratio test is defined as the ratio between the variance of the returns of a k-period compared to k times the variance of the returns of the individual periods included in the k-period. It can be assumed that there are indications for a mean-reversion phenomenon if such a test takes on a value lower than 1.

References

Artzner P, Delbaen F, Eber JM, Heath D (1999) Coherent measures of risk. Math Financ 9(3)

Bianchi RJ, Drew ME, Walk AN (2016) The time diversification puzzle: A survey. Financ Plan Res J. 2(2)

Bradford D, Siliski D (2016, Fall) Performance drawdowns in asset management: extending drawdown analysis to active returns. J Perform Meas 2016:34–48

Braga MD (2008) Il risk budgeting nell'asset management. Strumenti e tecniche per la misurazione, scomposizione e allocazione del rischio. Bancaria Editrice, Rome
Choueifaty Y, Coignard Y (2008) Toward maximum diversification. J Portf Manag 35(1)
Dowd D (2005) Measuring market risk. Wiley, Chichester
Duffie D, Pan J (1997) An overview of value at risk. J Deriv 4(3)
Grossman S, Zhou Z (1993) Optimal investment strategies for controlling drawdowns. Math Financ 3(3)
Ibbotson R, Sinquefield R (1976) Stocks, bonds, bills, and inflation: Simulation of the future. J Bus 49(3)
Jondeau E, Poon SH, Rockinger M (2007) Financial modeling under non-Gaussian distribution. Springer, London
Jorion P (1997) Value at risk: The new benchmark for controlling market risk. Irwin, Chicago
Kaplan P (1998) "The Ibbotson-Sinquefield simulation made easy" made easy. Ibbotson Associates, Chicago
Leibowitz ML, Bader LN, Kogelman S (1996) Return targets and shortfall risks: studies in strategic asset allocation. Irwin Professional Publishing, Chicago
Lewis A, Kassouf S, Brehm D, Johnston J (1980) The Ibbotson-Sinquefield simulation made easy. J Bus 53(2)
Lo A, MacKinley A (1988) Stock market prices do not follow random walks: evidence from a simple specification test. Rev Financ Stud 1:41–66
Lo A, MacKinley A (1989) The size and power of variance ratio test in finite samples: a Monte Carlo investigation. J Econ 40:203–238
Morgan JP (1996) RiskMetrics™—technical document. J. P. Morgan, New York
Poterba J, Summers L (1988) Mean reversion in stock prices: evidence and implications. J Financ Econ 22:27–59
Qian E (2006) On the financial interpretation of risk contributions: risk budgets do add up. J Invest Manag 4(4)
Samuelson PA (1969, August) Lifetime portfolio selection by dynamic stochastic programming. Rev Econ Stat 51
Scherer B (2015) Portfolio construction and risk budgeting, 5th edn. Risk Books, London

Chapter 6
Alternative Approaches to Traditional Mean-Variance Optimisation

Maria Debora Braga

6.1 Alternative Asset Allocation Approaches to Traditional Mean-Variance Optimisation

Chapter 4 took an in-depth look at the practical problems affecting the Markowitz mean-variance framework. In particular, we pointed to the difficulty involved in obtaining accurate estimates of expected returns and the extreme sensitivity of Markowitz portfolios to estimation errors. In recent years, having become aware of these problems, academics, practitioners, and institutional investors have re-evaluated existing strategies or promoted new approaches to portfolio construction that no longer strive to optimise the trade-off between expected return and risk, and consequently no longer resort to the application of Mean-Variance Optimisation (MVO).

The common feature of these strategies is that they remove expected returns from the set of inputs required for portfolio construction, which explains their being called μ-free strategies.[1] Obviously, their development only involves estimating and modelling risk parameters, and their focus is exclusively on the risk dimension of portfolios. For this reason, they are commonly known as risk-based asset allocation approaches or strategies. An obvious practical implication of accepting the new portfolio construction framework is that different optimisation algorithms have to be set up, and these will be discussed later as each strategy is presented.

[1] For further details, refer to Braga (2015, 2016).

M. D. Braga (✉)
Department of Economics and Political Science, University of "Valle d'Aosta", Aosta, Italy

SDA Bocconi School of Management, Bocconi University, Milan, Italy
e-mail: m.braga@univda.it; mariadebora.braga@sdabocconi.it

I. Basile, P. Ferrari (eds.), *Asset Management and Institutional Investors*, Contributions to Finance and Accounting,
https://doi.org/10.1007/978-3-031-59819-7_6

This chapter illustrates two examples which are representative of this class of asset allocation methodology:

- The global minimum-variance strategy/portfolio.
- The optimal risk parity strategy/portfolio.

The presentation will be carried out in a pragmatic way; for a detailed discussion of the technical aspects and for an out-of-sample evaluation of risk-based strategies, the interested reader should refer to Braga and Roncalli.[2]

Despite the central role played by risk in both the aforementioned strategies, one significant difference between the two will become evident in the following discussion, and this difference has to do with the way in which risk is managed in order to identify a strategic asset allocation proposal. To put it simply, while the former is concerned with knocking down risk, the latter is more interested in spreading it as fairly as possible among the asset classes in the investment universe.

Traditionally, it is erroneously believed that the goal pursued by the optimal risk parity portfolio could be achieved more simply by a naive asset allocation strategy, which is also commonly included among μ-free strategies: the equally weighted strategy/portfolio. According to this, the asset allocation problem is solved by assigning the same weight to each asset class, corresponding to the ratio of 1 over N. That said, it is clear that this strategy not only disregards expected returns, but any input generally required for portfolio construction, and so the idea that equally weighted strategy can perfectly diversify risk is questionable.

To demonstrate this, thanks to the availability of the quantitative tools for risk decomposition presented in Chap. 5, it is appropriate to provide an easy example of risk allocation for a portfolio equally invested in two-asset classes. Let us consider a Eurozone investor who has opted for a domestic portfolio which he/she believes to be well diversified between domestic bonds and equities consisting of 50% Bloomberg Euro Aggregate Treasury Bond and 50% MSCI Europe. The portfolio's standard deviation is 7.72%, using the risk and covariance/correlation parameters estimated in Chap. 4 ($\sigma_{\text{Bloomberg euro aggregate treasury bond}} = 4.65\%$; $\sigma_{\text{MSCI Europe}} = 14.44\%$; and $\rho_{\text{Bloomberg euro aggregate treasury bond, MSCI Europe}} = +0.06$). With reference to this portfolio, Table 6.1 provides the calculation of marginal risks, total risk contributions, and percentage total risk contributions.[3]

It is clear from this simple example that the 50%/50% portfolio is really a 10.769%/89.231% portfolio in terms of risk allocation. Therefore, it is also clear that the equally weighted strategy cannot be considered equivalent to the optimal

[2]Braga (2016) and Roncalli (2014).

[3]The reader will remember from Chap. 5 that the marginal risk, the total risk contribution, and the percentage total risk contribution denote, respectively:

- The change in portfolio risk (volatility) induced by a small increase in the weight of one asset class;
- The amount of risk for which the exposure in an asset class is deemed responsible;
- The proportion of individual risk contribution from an asset class to overall portfolio risk.

Table 6.1 Risk allocation for an equally weighted portfolio

Marginal risk Bloomberg Euro Aggregate Treasury Bond	Marginal risk MSCI Europe
$\frac{4.65\%^2 \cdot 50\% + 4.65\% \cdot 14.44\% \cdot 50\% \cdot 0.06}{7.72\%}$ $= 0.016620$	$\frac{14.44\%^2 \cdot 50\% + 4.65\% \cdot 14.44\% \cdot 50\% \cdot 0.06}{7.72\%}$ $= 0.137715$
Total risk contribution Bloomberg Euro Aggregate Treasury Bond	Total risk contribution MSCI Europe
$50\ \% \cdot 0.016620 = 0.008310$	$50\ \% \cdot 0.137715 = 0.068857$
Percentage total risk contribution Bloomberg Euro Aggregate Treasury Bond	Percentage total risk contribution MSCI Europe
$0.008310/7.72\ \% = 10.769\%$	$0.068857/7.72\ \% = 89.231\%$

risk parity strategy. The acceptance of the equally weighted strategy as a risk-based strategy has a psychological rather than concrete, documented explanation.[4] This validates the choice of paying attention to strategies that really are risk-focused. The interest in risk-based strategies, which dates back to the global financial crisis of 2008, reveals investors' increased sensitivity to drawdowns and attention to capital preservation. In other words, one might say that risk-based strategies have achieved greater popularity because investors have become "agnostic" about expected returns but want to feel confident that their capital is invested safely or more prudently. To tell the truth, this attitude became the driving force behind the methodological innovations that affected risk-based strategies many years after the global financial crisis of 2008. In fact, recently, some authors have formally developed both the strategy aimed at minimising portfolio risk and the strategy aimed at its homogeneous distribution among the portfolio components, opting for the replacement of the portfolio standard deviation, as a reference risk measure, with a metric representing an approximation of the portfolio kurtosis that pays more attention to the extreme dispersion of returns around its mean. The efforts made have led to the identification of the minimum-kurtosis portfolio and the kurtosis-based risk parity.[5]

6.2 The Global Minimum-Variance Strategy

Under the global minimum-variance strategy, the asset manager recommends a portfolio already known by the reader. In fact, it is the portfolio at the leftmost end of the efficient frontier shown in Fig. 4.2. Given this location, it is obviously the portfolio with the smallest attainable *ex ante* standard deviation or variance.

[4]This is confirmed by the fact that it is behavioral finance that has primarily offered explanations for why the equally weighted portfolio is considered a risk-based strategy. See Fisher and Statman (1997a, b), Benartzi and Thaler (2001), and Windcliff and Boyle (2004).

[5]Braga et al. (2023a, b).

An important aspect to note here is that it is not necessary to implement Mean-Variance Optimisation in order to identify the global minimum-variance portfolio (GMVP): it can be replaced with a simplified optimisation algorithm with the formula for portfolio variance as an objective function to be minimised and the inclusion of the traditional long-only and budget constraints. Therefore, the optimisation problem to be solved in order to identify the GMVP can be written formally as follows:

$$\begin{aligned} &\min_{w} \sum_{i=1}^{N} \sum_{j=1}^{N} w_i w_j \sigma_{ij} \\ &\text{subject to} \\ &\sum_{i=1}^{N} w_i = 1 \\ &w_i \geq 0 \end{aligned} \tag{6.1}$$

or in matrix notation:

$$\begin{aligned} &\min_{\mathbf{w}} \mathbf{w}'\boldsymbol{\Sigma}\mathbf{w} \\ &\text{subject to} \\ &\mathbf{w}'\mathbf{e} = 1 \\ &[\mathbf{w}] \geq 0 \end{aligned} \tag{6.2}$$

On the basis of (6.1) and (6.2), this strategy falls into the group of risk-based strategies for two reasons:

- The financial goal of the strategy is solely concerned with the risk of the portfolio.
- Only the estimate of the covariance matrix of asset class returns is required.

In order to understand the characteristics of the global minimum-variance portfolio, we first apply (6.1) and (6.2) by numerical techniques using the same investment universe considered in Chap. 4, and then we examine the obtained GMVP from a risk allocation perspective. The results of these activities are shown in Table 6.2.

On the basis of Table 6.2, the global minimum-variance portfolio can be accurately described. In terms of composition, it is characterised by a partial inclusion of the asset classes available in the selected investment universe. Specifically, the portfolio shows an extreme concentration on the Eurozone money market (the asset class, among those included in the investment universe, with the lowest volatility) and negligible, almost imperceptible exposures to global corporate bonds and towards North American equity. It is easy to verify that this composition for the GMVP is the same obtained in Chap. 4 by implementing (4.3a) or (4.3b) and, obviously, entails the same 0.4486% standard deviation.

Table 6.2 Identification and analysis of the global minimum-variance portfolio

	Weights GMVP (%)	Marginal risk	Total risk contribution	Percentage total risk contribution (%)
BARCLAYS 3 MONTH EURIBOR CASH INDEX	99.09	0.00449	0.00445	99.09
BLOOMBERG EURO AGGREGATE TREASURY BOND	0.00	0.00540	0.00000	0.00
BLOOMBERG US TREASURY BOND	0.00	0.00751	0.00000	0.00
J.P. MORGAN EMBI GLOBAL DIVERSIFIED COMPOSITE	0.00	0.01023	0.00000	0.00
BLOOMBERG GLOBAL AGGREGATE CORPORATE	0.22	0.00449	0.00001	0.22
MSCI EUROPE	0.00	0.00596	0.00000	0.00
MSCI NORTH AMERICA	0.70	0.00449	0.00003	0.70
MSCI PACIFIC FREE	0.00	0.00580	0.00000	0.00
MSCI EMERGING MARKETS	0.00	0.01529	0.00000	0.00

In terms of risk allocation, two aspects deserve particular attention. The first one is that we observe essentially homogeneous marginal risks among asset classes included in the portfolio allocation. This indicates that the effect on overall portfolio risk of an infinitesimal change in the weight of asset class i is no different from the effect of a change resulting from an infinitesimal variation of weight j. According to Scherer (2014), this is necessary because: "if this were not so, we could always find a pair of assets where slightly increasing one holding while at the same time reducing the other would result in lowered risk".[6] The second aspect is the perfect matching between asset class weights and percentage total risk contributions, which means that exposure to an asset class also expresses the portion of risk for which it is responsible.

After the distinguishing features of the global minimum-variance portfolio/strategy have been studied, we can conclude by noting that the asset allocation pursuing the most powerful risk reduction in absolute terms[7] corresponds to a portfolio which cannot really be considered diversified both in terms of "euro allocation" and in terms of risk allocation. It is therefore a portfolio that takes significant bets on certain asset classes, but it should be borne in mind that these tend to be the most prudent ones.

[6] Scherer (2014, p. 21).

[7] Although it goes beyond the aim of this chapter, it should be noted that a strategy focused on risk reduction in relative terms has recently been proposed, called the most diversified portfolio strategy/approach. For more details, see Choueifaty and Coignard (2008) and Braga (2016).

6.3 The Optimal Risk Parity Strategy

As we mentioned in the introduction, the second risk-based strategy we discuss, i.e. the optimal risk parity strategy, aims to overcome the problem of risk concentration. The alternative designation of optimal risk parity as equally weighted risk contribution strategy or portfolio (abbreviated, ERC portfolio), provided by Maillard et al.,[8] suggests the criteria used to structure the portfolio coherently with this goal: to give each asset class a weight so that the amount of risk it contributes to overall portfolio risk is equal to the amount contributed by any other asset class in the portfolio. In practice, the optimal risk parity strategy requires the application of the "1 over *N*" rule no longer to asset allocation, but to risk allocation. Using the tools already known related to risk decomposition,[9] the above condition can be written formally as equality among component risks, defined as the simple product of asset class weight in the portfolio and its marginal contribution to risk. This is shown in (6.3):

$$w_i \cdot \frac{\partial \sigma_P}{\partial w_i} = w_j \cdot \frac{\partial \sigma_P}{\partial w_j} \quad \forall i,j \tag{6.3}$$

Intuitively, the identification of optimal asset class weights for a risk parity portfolio involves solving an optimisation problem which is different from MVO and is formulated by Maillard et al.[10] More precisely, it includes a new objective function to be minimised together with the traditional constraints on portfolio weights. Ultimately, the optimisation problem can be fully stated as follows:

$$\begin{aligned} &\min_{w} \sum_{i=1}^{N} \sum_{j=1}^{N} \left(w_i \cdot \frac{\partial \sigma_P}{\partial w_i} - w_j \frac{\partial \sigma_P}{\partial w_j} \right)^2 \\ &\text{subject to} \\ &\sum_{i=1}^{N} w_i = 1 \\ &0 \leq w_i \leq 1 \end{aligned} \tag{6.4}$$

Looking at the formulation of the objective function, one sees that its minimisation is achieved for weights such that it is equal to zero. It goes without saying that this condition holds when (6.3) is verified. The solution of the optimisation problem is not available in a closed form and can be obtained by numerical

[8] Maillard et al. (2010).

[9] See Sect. 5.2.

[10] Maillard et al. (2010).

Table 6.3 Identification and analysis of optimal risk parity portfolio

	Weights optimal risk parity portfolio (%)	Marginal risk	Total risk contribution	Percentage total risk contribution (%)
BARCLAYS 3 MONTH EURIBOR CASH INDEX	84.48	0.00134	0.00114	11.11
BLOOMBERG EURO AGGREGATE TREASURY BOND	4.93	0.02300	0.00114	11.11
BLOOMBERG US TREASURY BOND	2.44	0.04657	0.00114	11.11
J.P. MORGAN EMBI GLOBAL DIVERSIFIED COMPOSITE	1.46	0.07753	0.00114	11.11
BLOOMBERG GLOBAL AGGREGATE CORPORATE	1.77	0.06415	0.00114	11.11
MSCI EUROPE	1.41	0.08077	0.00114	11.11
MSCI NORTH AMERICA	1.20	0.09446	0.00114	11.11
MSCI PACIFIC FREE	1.19	0.09528	0.00114	11.11
MSCI EMERGING MARKETS	1.11	0.10212	0.00114	11.11

techniques.[11] For the sake of completeness, it is worth mentioning the existence of an alternative way of writing the optimisation algorithm for the optimal risk parity strategy, which is considered less demanding from a computational point of view while clearly remaining fixable only with numerical procedures.[12] In fact, leaving the constraints in (6.4) unchanged, it is possible to replace the objective function as indicated in (6.5):

$$\min_{w} \sum_{i=1}^{N} \left(w_i \cdot \frac{\partial \sigma_P}{\partial w_i} - \theta \right)^2 \tag{6.5}$$

where

$$\theta = \frac{1}{N} \sum_{i=1}^{N} w_i \cdot \frac{\partial \sigma_P}{\partial w_i}$$

Similarly to what has been done in Sect. 6.2, below we present the solution for the optimal risk parity portfolio considering the same investment universe. The ERC

[11] Maillard et al. (2010) suggest that this constrained non-linear programming problem can be tackled using the sequential quadratic programming algorithm (SQP). For a brief general description, see Fabozzi et al. (2007).

[12] Bai et al. (2016)

portfolio has a higher volatility than the global minimum-variance portfolio equal to 1.0215% (Table 6.3).

For the optimal risk parity strategy too, attention is paid to the capital allocation and risk allocation characterising the portfolio. From the point of view of capital allocation, the most distinctive feature when compared to the global minimum-variance strategy is that all the asset classes in the investment universe are included in the suggested portfolio. However, given the strategy's primary goal, not surprisingly asset classes exhibiting simultaneously low volatility and low or negative correlation with others tend to be over-weighted. To put it another way, preference is given to asset classes with low marginal risk, which leads us to observe the risk allocation of the portfolio. As expected, the total risk contribution and, consequently, the percentage total risk contribution prove that no asset class has a dominant role in driving portfolio volatility, and that risk is perfectly distributed.

To take a more concrete approach in using the optimal risk parity strategy, it has to be recognised that an institutional investor who were to opt for an ERC portfolio would lean towards defining its structure starting from an investment universe that excludes the monetary asset class, unlike what happened in the previous example. Since the optimal risk parity strategy requires that each asset class contribute equally to portfolio risk, it is inevitable that an excessive concentration of asset allocation towards cash is recommended so that the latter, with its low volatility, can bring the same portion of risk as the other components of the portfolio. However, the investor who selects the optimal risk parity strategy is not necessarily looking for a low volatility portfolio and the performance associated with it; its undoubted interest is that of the homogeneous distribution of risk between the components of the portfolio. This suggested solving the optimisation problem illustrated by (6.4) and (6.5) with regard to an investment universe revised in the way previously described. The characteristics of the risk parity portfolio obtained in this case are shown in Table 6.4. Its volatility is equal to 6.2096%.

In the previous section we stated implicitly that the global minimum-variance portfolio is *ex ante* optimal or efficient under the Markowitz framework, even if it has been computed without considering expected returns. The literature on risk parity has often questioned the possibility that the ERC portfolio too could be optimal according to mean-variance analysis. The general conclusion is that there are two necessary conditions for this to be possible:

- Pair-wise correlations among asset classes must be the same.
- All asset classes must have the same Sharpe ratio.[13]

It should be fairly obvious that satisfying both of these conditions is arduous, because they basically take for granted the existence of a world where asset classes

[13] To be precise, if these conditions occur (and a risk-free asset is available), the risk parity portfolio would prove to be the *ex ante* tangency portfolio or the maximum Sharpe ratio portfolio. In other words, it would fall *ex ante* on the Capital Market Line. See Maillard et al. (2010), Roncalli (2014), and Braga (2016).

Table 6.4 Identification and analysis of optimal risk parity portfolio when cash is excluded as an asset class

	Weights optimal risk parity portfolio (%)	Marginal risk	Total risk contribution	Percentage total risk contribution (%)
BLOOMBERG EURO AGGREGATE TREASURY BOND	33.33	0.023285	0.007762	12.50
BLOOMBERG US TREASURY BOND	15.88	0.048893	0.007762	12.50
J.P. MORGAN EMBI GLOBAL DIVERSIFIED COMPOSITE	9.38	0.082760	0.007762	12.50
BLOOMBERG GLOBAL AGGREGATE CORPORATE	11.04	0.070309	0.007762	12.50
MSCI EUROPE	8.64	0.089859	0.007762	12.50
MSCI NORTH AMERICA	7.21	0.107719	0.007762	12.50
MSCI PACIFIC FREE	7.38	0.105172	0.007762	12.50
MSCI EMERGING MARKETS	7.15	0.108597	0.007762	12.50

are extremely redundant. Therefore, anyone who is interested in risk parity strategy should not presume to select a portfolio that is *ex ante* mean-variance efficient, while ignoring expected returns.

6.4 Portfolio Selection Under Alternative Asset Allocation Approaches

In the Mean-Variance Optimisation framework, regardless of the extensions and/or amendments introduced to address the issue of estimation risk, the asset manager identifies, on the basis of the chosen portfolio construction method, a set of efficient portfolios such that for any target return the portfolio with the lowest variance (standard deviation) is displayed or, alternatively, such that for any risk level the portfolio with the highest expected return is shown. It is obvious that once optimal portfolios are constructed, the next step is up to the investor and is that of choosing the optimal risky portfolio from the efficient frontier most suited to his/her preferences.[14]

[14]The reader will remember that, in this phase, investors can rely on the support of financial advisors, who can use the tools illustrated in Chap. 5 to make investors' choices more informed.

When risk-based strategies are considered, the portfolio selection phase inevitably undergoes some changes for a simple reason: as previously demonstrated, these strategies come up to a single risky portfolio proposal, therefore an investor has not to choose an optimal point on the efficient frontier according to his/her investment objective and his/her risk tolerance/aversion.

This circumstance justifies the idea that the investor's decision-making process has to be similar to the one recognisable in the context of the Capital Asset Pricing Model (CAPM). In this framework, it is well known that an asset management company offers the same risky portfolio to all potential investors. Then, it is the individual characteristics of the investor that come into play in the selection of the desired point on the Capital Market Line. This is the line, in the risk-return space, that originates from the risk-free rate and is tangent to the efficient frontier; the tangency portfolio is also the portfolio with the maximum Sharpe ratio. It must be emphasised that the inclusion of the risk-free rate in the analysis reflects the possibility for the investor to combine borrowing or lending at this rate with the tangent portfolio, in order to move along the Capital Market Line.

To address the problem of portfolio selection under alternative asset allocation approaches to MVO, the idea that can be drawn from the CAPM is that it is also possible to combine the only portfolio recommended by a risk-based strategy with borrowing or lending at the risk-free rate. This idea is particularly acceptable if the risk-based strategy adopted is the optimal risk parity, whereas it is a bit less acceptable for the global minimum-variance strategy. In fact, in this situation, it makes little sense to imagine that GMVP is combined with an investment at the risk-free rate because, considering a problem of allocation among multiple asset classes, it is likely that GMVP is already very close to the vertical axis on which the risk-free rate lies.[15] Therefore, we prefer to refer below to the optimal risk parity strategy.

After this clarification, and drawing on CAPM logic, we can say that for potential investors in the optimal risk parity portfolio, addressing the issue of portfolio selection means making a decision on where to stay or move on the Risk Parity Line, i.e. the line connecting the risk-free rate to the optimal risk parity portfolio. In this way, investors can choose the level of volatility depending on their varying degrees of risk aversion. Thus, a more or extreme risk-averse investor will invest a portion of his/her capital in the risk-free (cash) asset and the remaining capital (obviously less than 100%) in the ERC Portfolio, whereas a less risk-averse investor will borrow additional funds to invest in the ERC portfolio, achieving an exposure higher than 100%.

As a starting point for a practical treatment of the subject just discussed, Table 6.5 provides the optimal risk parity portfolio for the case with two-asset classes previously considered in Sect. 6.1. According to our original sample parameters, it presents a 3.885% expected return and a 5.126% standard deviation. Its structure is characterised, as expected, by an overexposure to the bond component. This composition, in the extremely simplified context of the investment universe assumed

[15] The combination of GMVP with positive leverage is more acceptable.

Table 6.5 Optimal risk parity portfolio for the two-asset cases

	Weights (%)	Marginal risk (%)	Total risk contribution (%)	Percentage total risk contribution (%)
BLOOMBERG EURO AGGREGATE TREASURY BOND	75.63	0.0339	2.563	50
MSCI EUROPE	24.37	0.1052	2.563	50
Total	100		5.126	100

Table 6.6 Combination of optimal risk parity portfolios with lending or borrowing

	Overall portfolio description	Overall portfolio risk	Overall portfolio return
More risk-averse investor	36.88% Bloomberg euro aggregate Treasury bond (48.77% × 75.63%) 11.88% MSCI Europe (48.77% × 24.37%) 51.23% investment at the risk-free rate	48.77% × 5.126% = 2.50%	(2.80% × 36.88%) + (7.24% × 11.88%) + (1.50% × 51.23%) = 2.66%
Less risk-averse investor	110.65% Bloomberg euro aggregate Treasury bond (146.30% × 75.63%) 35.65% MSCI Europe (146.30% × 24.37%) 46.30% borrowing at the risk-free rate	146.30% × 5.126% = 7.50%	(2.80% × 110.65%) + (7.24% × 35.65%) + (−1.50% × 46.30%) = 4.99%

here, is immediate to derive since there is no need to use the information relating to the correlation. In a two-asset class context, the optimal weight of each within the risk parity portfolio is, in fact, simply proportional to the inverse of the respective volatility, taking care to provide for normalisation to ensure that the identified weights comply with traditional full investment constraint. Therefore, in this specific case, the 75.63% exposure recommended for the Bloomberg Euro Aggregate Treasury Bond index arises from the calculation: (1/4.65%)/ ((1/4.65%) + (1/14, 44%)); that, however, of 24.37% indicated for the MSCI Europe can be obtained from the calculation: (1/14.44%)/ ((1/4.65%) + (1/14.44%)).

We then assume that a prudent investor is going to bear a 2.50% annualised standard deviation and that a more risk tolerant investor will accept a 7.50% volatility. According to the above arguments, the crucial point to stress is that an asset manager can serve both clients with the optimal risk parity portfolio shown in Table 6.5, so long as the first investor can invest a 51.23% portion (2.50%/ 5.126% − 1) of his/her overall portfolio at the risk-free rate, and the second investor can borrow 46.40% (7.50%/5.126% − 1) of additional funds at the risk-free rate to invest more than his/her capital in the ERC portfolio (146.30% to be precise).

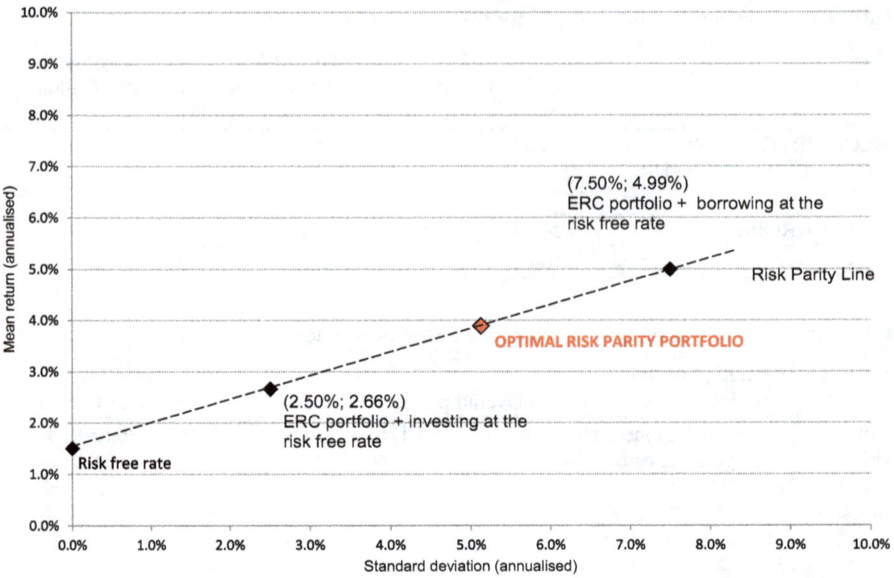

Fig. 6.1 The risk parity line

Table 6.6 summarises the characteristics of the overall portfolio proposed to each investor in terms of composition, risk, and return. We assume a 1.50% annualised risk-free rate, necessary to calculate the impact of negative and positive leverage on the overall portfolio return.

In Fig. 6.1 the reader can verify the position of the two portfolios corresponding to a combination of the risk-free and the optimal risk parity portfolio along the Risk Parity Line. In some ways, Fig. 6.1 leads to an obvious conclusion: the position on the line is the means by which each investor reaches a target volatility which inevitably influences the potential return target. It has been shown that the use of leverage (negative or positive) is essential and we specify that the approach followed here is to lever (either negatively or positively) the entire ERC portfolio in order to achieve the volatility/return objective.[16]

References

Bai X, Scheinberg K, Tutuncu R (2016) Least-squares approach to risk parity in portfolio selection. Quant Financ 16(3):357

Benartzi S, Thaler R (2001) Naive diversification strategies in defined contribution saving plans. Am Econ Rev 91(1):79–98

[16]Ruban and Melas (2011) suggest taking a levered or unlevered position exclusively in the lower-risk asset classes in the optimal risk parity portfolio.

Braga MD (2015) Risk parity versus other μ-free strategies: a comparison in a triple view. Invest Manag Financ Innov 12(2):277–289
Braga MD (2016) Risk-based approaches to asset allocation. Concepts and practical applications, Springer briefs in finance. Springer, London
Braga MD, Nava CR, Zoia MG (2023a, June) Kurtosis-based vs volatility-based asset allocation strategies: do they share the same properties? A first empirical investigation. Financ Res Lett 54
Braga MD, Nava CR, Zoia MG (2023b, June) Kurtosis-based risk parity: methodology and portfolio effects. Quant Financ 23(3)
Choueifaty Y, Coignard T (2008) Toward maximum diversification. J Portf Manag 35(1):40–51
Fabozzi FJ, Kolm PN, Pachamanova DA, Focardi SM (2007) Robust portfolio optimization and management. Wiley, Hoboken, NJ
Fisher KL, Statman M (1997a) The mean-variance-optimization puzzle: security portfolios and food portfolios. Financ Anal J 53(4):41–50
Fisher KL, Statman M (1997b) Advice from mutual fund companies. Closer to the Talmud than to Markowitz. J Portf Manag 24(1):9–25
Maillard S, Roncalli T, Teïletche J (2010) The properties of equally weighted risk contribution portfolios. J Portf Manag 36(4):60–70
Roncalli T (2014) Introduction to risk parity and budgeting. Chapman & Hall, Boca Raton, FL
Ruban O, Melas D (2011) Constructing risk parity portfolios: rebalance, leverage, or both? J Invest 20(1):99–107
Scherer B (2014) Portfolio construction and risk budgeting, 4th edn. Risk Books, London
Windcliff H, Boyle PB (2004) The 1/n pension investment puzzle. N Am Actuar J 8(3):32–45

Chapter 7
Integrating Sustainability Criteria into the Investment Process

Ignazio Basile and Pierpaolo Ferrari

7.1 Sustainable Investing

Sustainable investing encompasses a set of investment strategies that, when evaluating companies and institutions, integrate financial analysis with environmental, social, and governance criteria (ESG), aiming to create value for investors and society as a whole.[1] Sustainable investments involve a holistic approach that considers not only financial performance but also the broader impact of companies on the environment, society, and governance structures. Specifically, environmental

[1] The realm of sustainable finance needs a concise elucidation of terminological nuances that have accumulated over the years. "Ethical investments" typically involve selection procedures applied to assets considered for inclusion in a portfolio. These procedures primarily focus on themes aligned with specific sets of values, often leading to the exclusion of economic sectors that contradict these values. "Socially responsible investments" (SRI) can be viewed as an evolution of ethical investments, encompassing strategies that consistently and explicitly incorporate factors related to social development into the portfolio allocation process. Consequently, they surpass a perspective based solely on excluding specific economic sectors or companies, leaning more towards a positive and proactive stance on investments as a means of fostering social development. Notably, the same acronym SRI is employed to denote "Sustainable and Responsible Investments", which boast a more extensive scope. These investments are distinguished by portfolio policies geared towards compliance with ESG criteria, expanding their purview compared to the preceding categories. See: Fung et al. (2010), Ballestero et al. (2015), and Boubaker et al. (2018).

I. Basile (✉)
Department of Economics and Management, University of Brescia, Brescia, Italy
e-mail: ignazio.basile@unibs.it

P. Ferrari (✉)
Department of Economics and Management, University of Brescia, Brescia, Italy

SDA Bocconi School of Management, Bocconi University, Milan, Italy
e-mail: pierpaolo.ferrari@unibs.it; pierpaolo.ferrari@unibocconi.it

I. Basile, P. Ferrari (eds.), *Asset Management and Institutional Investors*, Contributions to Finance and Accounting,
https://doi.org/10.1007/978-3-031-59819-7_7

considerations pertain to the challenges of climate change, ecosystem impacts, and related risks; social considerations address issues of inequality, inclusiveness, labour relations, investment in human capital, and improving relations within communities; and governance relates to the evolution of the corporate governance model and the issues faced by workers and their trade union representatives. Moreover, ESG considerations are closely intertwined, and, therefore, require appropriate rebalancing over time.

In recent years, awareness among investors regarding ESG issues has experienced exponential growth globally, albeit with significant variations in diffusion across countries. Consequently, asset managers' investment policies on a global scale have increasingly prioritised factors promoting sustainable growth, with a keen focus on social considerations, environmental impact, and respect for human rights. This trend has led to the establishment of specific ESG constraints in investment allocation for many portfolios and investment products.[2] These constraints aim to channel resources towards companies adopting virtuous practices, emphasising environmentally friendly production methods, ensuring inclusive and human rights-conscious working conditions, and adhering to the highest standards of corporate governance.

The integration of ESG analysis into traditional financial analysis enhances the overall assessment by allowing the consideration of risks that, though financially relevant, are typically overlooked in traditional analyses. Inappropriate business conduct can generate costs and risks not only for individual companies but also for the broader economic system, impacting financial stability and economic growth, sometimes even in the short term. Conversely, as supported by a substantial body of literature and numerous empirical studies, companies that prioritise ESG factors tend to be less exposed to operational, legal, and reputational risks. They also exhibit a greater orientation towards innovation, efficiency in resource allocation, and benefit from lower performance volatility, resulting in lower risk premia demanded by the market.

7.2 The European Union Framework

Since 2001, the European Union (EU) has been actively working towards incorporating the principles of sustainable development into its short, medium, and long-term operational programmes through the EU Sustainable Development Strategy. However, it was only with the signing of the 2030 Agenda for Sustainable Development and the Paris Agreement on Climate Change that the EU committed to a comprehensive and coordinated strategy for sustainable development.

The 2030 Agenda for Sustainable Development, initiated by the United Nations (UN), is a programme of action comprising 17 goals to be achieved by 2030, known

[2] Abate et al. (2023).

Source: United Nations (2018)

Fig. 7.1 The seventeen global sustainable investment goals of the United Nations 2030 Agenda. Source: United Nations (2018)

as sustainable development goals.[3] Adopted in September 2015 by the governments of the 193 UN member states, it became operational in January 2016. The goals, depicted in Fig. 7.1, span interconnected areas such as the protection of natural resources, promotion of responsible consumption, the fight against poverty, and gender equality.[4]

The Paris Agreement, resulting from negotiations during the 2015 Paris Climate Conference (COP21), stands as the first universal and legally binding global climate agreement. Ratified by the EU on 5 October 2016 and in force since 4 November 2016, its primary objective is to intensify the global response to climate change. This involves limiting the average global temperature increase to below 2 °C compared to pre-industrial levels, with an aspiration to further restrict this increase to 1.5 °C. The agreement also seeks to enhance countries' capabilities to manage climate change impacts and channel financial resources towards activities aligned with low greenhouse gas emissions development. The EU's commitment to these international agreements underscores its dedication to sustainable development and environmental responsibility.

In March 2018, the European Commission unveiled the Action Plan for Sustainable Finance, a strategic initiative crafted based on recommendations from a dedicated group of experts.[5] This plan delineated a comprehensive strategy and measures intended to cultivate a financial system capable of fostering sustainable development in economic, social, and environmental dimensions. The overarching goal was to

---

[3] United Nations (2018).

[4] United Nations (2015).

[5] The EU High-Level Expert Group on Sustainable Finance, established in 2018, played a pivotal role in advancing sustainable finance initiatives. EU High Level Expert Group on Sustainable Finance (2018).

contribute actively to the realisation of the UN 2030 Agenda for Sustainable Development and the Paris Agreement on Climate Change. The EU acknowledged the specific responsibilities of the financial system in transitioning towards a sustainable development model by directing capital towards activities aligned with this paradigm. The Action Plan outlined key objectives:

- Redirect capital flows towards investments promoting sustainable and inclusive growth.
- Manage financial risks associated with climate change, resource depletion, environmental degradation, and social disputes.
- Promote transparency and instil a long-term vision in economic and financial activities.

To achieve these goals, the plan proposed ten actionable items (Box 7.1):

1. Establish an EU classification system for sustainable activities.
2. Create standards and labels for green financial products.
3. Foster investment in sustainable projects.
4. Incorporate sustainability into financial advice.
5. Develop sustainability benchmarks.
6. Improve the integration of sustainability in ratings and market research.
7. Clarify the duties of institutional investors and asset managers' duties.
8. Incorporate sustainability into prudential requirements.
9. Strengthen sustainability disclosure and accounting rule-making.
10. Foster sustainable corporate governance and mitigate short-termism in capital markets.

This strategic approach underscores the EU's unwavering commitment to cultivating a financial landscape that not only supports sustainable growth but also seamlessly integrates ESG considerations into the core of investment decision-making processes.

Box 7.1 The European Union Action Plan for Sustainable Finance

The objectives of the EU Action Plan for sustainable finance

1. Establishing an EU classification system for sustainable activities

To develop progressively an EU Taxonomy for climate change, environmentally and socially sustainable activities. The aim is to embed the EU sustainability Taxonomy in EU law and provide the basis for using such a classification system in different areas.

2. Creating standards and labels for green financial products

To prepare a report on an EU green bond standard, based on current best practices.

To define the content of the prospectus for green bond issuances to provide potential investors with additional information.

To rule the use of the EU Ecolabel framework for certain financial products.

3. Fostering investment in sustainable projects

To reinforce advisory capacity for sustainable projects, including infrastructure projects.

(continued)

Box 7.1 (continued)

To take further measures that will improve the efficiency and impact of instruments aiming at sustainable investment support in the EU and in partner countries.

4. Incorporating sustainability when providing financial advice

To amend the Markets in Financial Instruments Directive II (MiFID II) and the Insurance Distribution Directive (IDD) delegated acts to ensure that sustainability preferences are taken into account in the suitability assessment.

5. Developing sustainability benchmarks

To adopt delegated acts, within the framework of the Benchmark Regulation, on the transparency of the methodologies and features of benchmarks to allow users to better assess the quality of sustainability benchmarks; and to put forward, subject to the outcome of its impact assessment, an initiative for harmonising benchmarks comprising low-carbon issuers, based on a sound methodology to calculate their carbon impact.

6. Better integrating sustainability in ratings and market research

To explore the merits of amending the Credit Rating Agency Regulation to mandate credit rating agencies to explicitly integrate sustainability factors into their assessments in a proportionate way to preserve market access for smaller players.

To assess current practices in the credit rating market, analysing the extent to which environmental, social, and governance considerations are taken into account.

To analyse methodologies and explore aspects like the market structure of sustainability ratings and market research services, the depth and breadth of sustainability research assessments and scoring, and the independence of those research/scoring providers. To encourage sustainability ratings and market research.

7. Clarifying institutional investors' and asset managers' duties

To require explicitly that institutional investors and asset managers integrate sustainability considerations in the investment decision-making process and to increase transparency towards end-investors on how they integrate such sustainability factors in their investment decisions, in particular as concerns their exposure to sustainability risks.

8. Incorporating sustainability in prudential requirements

To explore the feasibility of the inclusion of risks associated with climate and other environmental factors in institutions' risk management policies and the potential calibration of capital requirements of banks and insurance companies. The aim is to take into account such factors, where this is justified from a risk perspective, to safeguard the coherence and effectiveness of the prudential framework and financial stability.

9. Strengthening sustainability disclosure and accounting rule-making

To launch a fitness check of EU legislation on public corporate reporting to assess whether public reporting requirements for listed and non-listed companies are fit for the purpose to disclose their sustainability commitment.

To request asset managers and institutional investors to disclose how they consider sustainability factors in their strategy and investment decision-making process, in particular for their exposures to climate change-related risks.

To assess the impact of new or revised International Financial Reporting Standards (IFRSs) on sustainable investments.

To explore how the adoption process of IFRSs can allow for specific adjustments to standards where they could pose an obstacle to long-term investment objectives.

10. Fostering sustainable corporate governance and attenuating short-termism in capital

To promote a corporate governance that is more conducive to sustainable investments, by assessing: the possible need to require corporate boards to develop and disclose a sustainability strategy, including appropriate due diligence throughout the supply chain, and measurable sustainability targets; the possible need to clarify the rules according to which directors are expected to act in the company's long-term interest.

Source: European Commission (2018a)

Building upon the Action Plan unveiled in March 2018, the European Union advocated for the financial system to strike a delicate balance between addressing two imperatives:

- Enhancing the financial system's contribution to sustainable growth by catering to society's long-term needs.
- Integrating ESG factors into the decision-making processes of investments.

Aligned with the outlined intentions in the Action Plan, the European Commission took significant steps in May 2018 by introducing a comprehensive package of legislative proposals geared towards fostering "sustainable finance". Subsequently, the legislative process reached fruition, resulting in the issuance of specific regulations aligned with several objectives of the Action Plan.[6]

To fortify the foundations of the initially defined framework, the European Commission, in June 2023, formulated a new package of measures (Box 7.2) focusing on:[7]

- The taxonomy for categorising sustainable activities.
- Regulating the activities of ESG rating providers.
- Enhancing the usability of the new regulations by key stakeholders.
- Introducing the concept of "transition finance".

In the rest of this chapter our emphasis is directed exclusively at issues considered most crucial for intermediaries and investors.

Box 7.2 The European Union's Second Package of Sustainable Investment Measures

The objectives of the second package of sustainable investment measures

1. EU Taxonomy: The aim is to approve a new set of criteria to define the sustainability of economic activities with reference to environmental but not strictly climate-related objectives such as:

- The sustainable use and protection of water and marine resources.
- The transition to a circular economy.
- The prevention and control of pollution.
- The protection and regeneration of biodiversity and ecosystems.

2. ESG rating agency: The aims are: to address the lack of transparency typical of the sector; to increase the reliability of ESG rating processes; to establish clear rules to prevent conflict of interest situations; and to ensure market integrity. The new framework, which

(continued)

[6] European Commission (2018b).

[7] European Commission (2023).

Box 7.2 (continued)

requires ESG rating agencies to be subject to an authorisation regime by the European Securities and Markets Authority (ESMA), should ensure that investors make more informed sustainable investment decisions.

3. Use of information by stakeholders: The objective is to help companies correctly interpret the principles underlying the new taxonomy and correctly translate them into the required public reporting.

4. Transition finance: The aim is to help companies and intermediaries to manage the difficult transition to more sustainable finance more effectively by defining reporting standards for companies subject to the Corporate Sustainability Reporting Directive (CSRD).

Source: European Commission (2023)

7.2.1 The EU Taxonomy

Concerning the first facet of the EU Action Plan, Regulation (EU) 2020/852, enacted on 18 June 2020 (Taxonomy Regulation), laid the groundwork for embedding ESG criteria at the core of the financial system. The objective is to orient investments towards activities that not only ensure a positive social and environmental impact but also adhere to minimum governance standards. The Taxonomy Regulation precisely defines the criteria for determining the environmental sustainability of an economic activity. It establishes a process, facilitated by a multi-stakeholder platform, to create a unified classification system across the EU. This system relies on specific criteria to ascertain which economic activities qualify as sustainable.

In delineating what constitutes "green" and "sustainable", the EU Taxonomy establishes a shared language for sustainability, imposing stringent standards to prevent greenwashing.[8] This is achieved through the establishment of minimum thresholds, detailed in "technical screening criteria", aiding in the identification of environmentally friendly activities.

[8]The term "greenwashing" lacks a universally accepted definition; however, in the EU, the European Supervisory Authorities have presented a shared understanding of greenwashing (ESMA, 2023a). They characterise it as a practice where sustainability-related statements, declarations, actions, or communications fail to transparently and fairly represent the true sustainability profile of an entity, financial product, or financial services. This deceptive practice has the potential to mislead consumers, investors, or other market participants. In a comprehensive study conducted in 2023, ESMA utilised a dataset encompassing historical information on 36,000 funds managing EUR 16 trillion in assets. The findings suggest a concerning risk: the robust demand for ESG products might incentivise greenwashing behaviour among asset managers. This implies that fund managers could be tempted to attract larger net inflows by employing misleading, confusing, or inaccurate claims about the ESG profile of their funds, aiming to enhance their appeal to end-investors (ESMA, 2023b).

As part of the new measures introduced in June 2023, the European Commission has both expanded the spectrum of sectors contributing to climate change mitigation and adaptation, particularly in manufacturing, transport, and services, and enhanced and supplemented criteria to encompass environmental objectives beyond strictly climate-related aspects.

The EU Taxonomy is structured around four environmental objectives:

- Sustainable use and protection of water and marine resources.
- Transition to a circular economy optimising the use of scarce resources.
- Prevention and reduction of pollution.
- Protection and restoration of biodiversity and ecosystems.

Screened economic activities are assessed based on three fundamental principles to ascertain their alignment with the EU Taxonomy's objectives. For an activity to be deemed sustainable, it must:

- Substantially contribute to at least one environmental objective outlined in the Regulation.
- Not significantly harm any of the other environmental objectives (Do No Significant Harm criteria).
- Adhere to minimum safeguards preventing adverse impacts on social stakeholders.

The resulting delegated acts are designed to promote investments in sectors capable of tangibly contributing to EU-level environmental, climate, and other objectives without imposing additional reporting obligations on the involved companies.

7.2.2 The EU Sustainability Disclosure

The European Commission's Regulation 2019/2988, known as the Sustainable Finance Disclosure Regulation (SFDR), plays a pivotal role in fortifying and promoting sustainable investment practices within the EU's financial services sector. Adopted in 2019 and implemented on 10 March 2021, the SFDR stands as a key component of the EU Action Plan for Sustainable Finance, aimed at redirecting capital flows towards a more sustainable economy. The regulation seeks to integrate sustainability into financial decision-making processes and enhance transparency regarding sustainability practices.

Under the SFDR, investment companies, portfolio managers, and financial sector operators are obligated to disclose information related to sustainability risks and potential negative impacts on sustainability concerning their investment decisions. This disclosure requirement extends to both the financial company level and individual product levels.

Financial market participants are mandated to fulfil several obligations:

- Publish information on their policies regarding the integration of sustainability risks into decision-making processes and the negative effects of such decisions.
- Include information in remuneration policies to showcase their consistency with the integration of sustainability risks.
- Indicate in pre-contractual information how sustainability risks are incorporated into investment decisions and the assessment results of the potential impacts on the financial products' performance, despite having no obligation to consider ESG factors.

The SFDR classifies financial products into three categories:

- Products that do not promote environmental or social characteristics and lack specific sustainable investment objectives but are monitored for sustainability risks (Article 6), termed as "brown products" or products pursuant to Article 6.
- Products that generally promote environmental or social characteristics or a combination thereof (Article 8), referred to as "light green products" or products pursuant to Article 8.
- Products with the specific objective of creating sustainable investments that yield positive impacts at the ESG level (Article 9), known as "dark green products" or products pursuant to Article 9.

The products pursuant to Article 6 have no ESG objectives, the products pursuant to Article 8 generically promote ESG objectives, and the products pursuant to Article 9 have specific objectives to produce a positive ESG impact.[9] This systematic approach within the SFDR aims to elevate transparency, furnishing investors with clear and comparable insights into sustainable financial products. Financial intermediaries are mandated to publicly disclose comprehensive information pertaining to the integration of ESG factors into their investment policies.

These disclosures are expected to be detailed, transparent, and exhaustive, encompassing the level of exposure to ESG risks, the tangible impact of investment decisions on sustainability, and the methodologies for selecting and monitoring financial instruments adopted to integrate the analysis and management of sustainability risks as part of the investment process. These methodologies involve the application of specific processes for selecting financial instruments, appropriately graduated according to the characteristics and objectives of the individual products managed and taking into account ESG factors. For products governed by Articles 8 and 9, financial market participants must publish in-depth information on environmental or social characteristics, sustainable investment objectives, and the

[9] In September 2023, the European Commission initiated a wide-ranging consultation on the SFDR, marking the first phase of a review of the regulation. The review seeks to explore the potential introduction of new product categories based on sustainability objectives and precise criteria. Two proposals include the introduction of new categories based on a product's investment strategy, leading to the potential dismantling of the existing approach based on Articles 8 and 9, or the conversion of Articles 8 and 9 into formal product categories with adding criteria.

methodologies employed for assessment, measurement, and monitoring, including data sources, screening criteria for underlying assets, and relevant sustainability indicators.[10]

7.2.3 The Creation of an EU Green Bond Market

In advancing sustainable finance, the EU has established green bond standards that articulate consistent guidelines for issuers to designate a bond as green, addressing concerns related to "greenwashing".[11] Delegated Regulation (EU) 2023/2632, enacted on 22 November 2023, imposes uniform requirements for bond issuers seeking to use the designation of European Union Green Bond (EuGB) in alignment with the EU Taxonomy. This regulation not only establishes a registration and supervisory system but also incorporates voluntary disclosure requirements for issuers, aiming to mitigate the risk of greenwashing in the green bond market.

EU green bonds emerge as pivotal instruments for financing investments in environmentally friendly technologies, energy efficiency, sustainable transport, and research infrastructures. According to the regulation, the proceeds from all EuGBs must be invested in economic activities aligned with the EU Taxonomy, provided the sectors are already covered. To ensure the comprehensive usability of the EU green bond standards from their inception, a 15% flexibility margin is envisaged for sectors not yet covered by the EU Taxonomy and specific activities.

In terms of supervision, national competent authorities designated by the home member state–in line with the Prospectus Regulation–take on the responsibility of monitoring issuers' compliance with their obligations under this new framework. This strategic approach–combining clear standards, voluntary disclosure requirements, and robust supervision–underlines the EU's commitment to fostering a green bond market that genuinely supports sustainable investments while guarding against deceptive practices.

---

[10] The SFDR should be read in conjunction with the Corporate Sustainability Reporting Directive (CSRD) 2022/2464/EU, as both regulations play pivotal roles in the EU's strategy to promote sustainability in the financial and corporate sectors. Although these regulations contribute on different levels to improve the transparency and standardisation of information relating to sustainability, they focus on different aspects. The CSRD focuses mainly on non-financial companies and replaces the Non-Financial Reporting Directive 2014/95/EU, a directive on the disclosure of non-financial information. The CSRD requires public interest bodies, large companies and, in future, small and medium-sized listed companies to provide clear, detailed, and comparable information on how they address sustainability issues.

[11] For a definition of "greenwashing", see footnote 8.

7.2.4 The Integration of Sustainability Criteria in Financial Advice

In steering the financial system towards sustainability, investment companies, through their advisory role, hold a fundamental position. Throughout the advisory process, investment firms are mandated to evaluate clients' investment objectives and risk tolerance to recommend suitable financial instruments. However, sustainability preferences of investors are often insufficiently considered in the advisory landscape. Addressing this gap, Delegated Regulation (EU) 2021/1253, amending Delegated Regulation (EU) 2017/565, underscores the European Commission's commitment to ensuring that investment firms inquire about their clients' preferences, including environmental, social, and governance factors. These preferences must then be taken into account during the assessment of the range of recommended financial instruments, encompassing the product selection process and suitability assessment.

The new Regulation introduces a comprehensive definition of customer preferences regarding sustainability, encapsulating choices to integrate financial instruments into investments that:

- Include sustainable investments in a specified minimum quantity, as per the EU Taxonomy.
- Include sustainable investments in a specified minimum quantity, as outlined in the SFDR.
- Consider principal adverse impacts, encompassing harm stemming from investment decisions and referencing negative indicators such as carbon emissions, fossil fuel exposure, waste levels, gender pay gap, and more.

This regulatory framework compels intermediaries to incorporate sustainability risks at the organisational level within risk management policies and procedures. In analysing and identifying conflicts of interest inherent in providing investment services to clients, the clients' sustainability preferences must be duly considered.

Further, the regulation mandates that investment firms, when disseminating information on financial instruments, inform clients about the sustainability factors analysed into the selection process of financial instruments. Moreover, it ensures that investment recommendations, in addition to aligning with the client's objectives and risk tolerance, adhere to their sustainability preferences. This signifies the integration of ESG factors into questionnaires utilised for suitability assessments, fostering a more holistic and client-centric approach to sustainable financial advice.

7.2.5 *The Integration of Sustainability Factors in Benchmark Construction*

In the realm of sustainability, providers of indices serving as benchmarks fall under the purview of the Benchmark Regulation (EU Regulation 2016/1011 of the European Parliament and the European Council), entailing significant disclosure obligations. Firstly, all indices employed as investment benchmarks are mandated to declare whether and how they integrate ESG criteria into their processes. Secondly, expanding beyond generic ESG benchmarks, the Benchmark Regulation introduces two distinct categories:

- EU Climate Transition Benchmarks (EU CTB).
- EU Paris Aligned Benchmarks (EU PAB).

Both categories share a focus on decarbonisation, with differing threshold parameters. EU CTB indices must adhere to the minimum decarbonisation standards set by EU authorities, while EU PAB indices align with the Paris Agreement's goals to limit global temperature increases. These new benchmarks at the EU level are designed to channel private capital towards a climate transition marked by low greenhouse gas emissions.

EU climate benchmarks provide valuable assistance to investors seeking climate-friendly investment strategies, enabling more informed decision-making. An index provider can only claim compliance with these new benchmarks if the construction technique includes securities from issuers and sectors meeting parameters stipulated by the EU itself. Both benchmark categories are geared towards progressively reducing emissions and excluding activities that undermine ESG objectives, with the Paris-aligned category adhering to stricter standards. This strategic integration of sustainability factors in benchmark construction underscores the commitment to aligning investment practices with climate goals and ESG objectives, fostering a more responsible and environmentally conscious financial landscape.[12]

7.3 Relation Between Investment Sustainability and Performance

Research on ESG investment performance is extensive and can be divided into two main strands. The first strand of research examines the relationship between corporate sustainability policies and the financial performance of adopting companies. The second strand of research seeks potential positive deviations between the performance of sustainable investment and that of more traditional investment strategies.

[12] The rules established by the EU have the potential to encompass any benchmark type related to debt securities, equity, or alternative investments, except for benchmarks representing interest rates and exchange rates.

The first strand, examining the relationship between social and financial performance, combines theoretical and empirical methodologies. Neoclassical theory suggests that actions taken by companies to enhance sustainability may result in uncompensated costs, leading to a competitive disadvantage.[13] In contrast, stakeholder theory, acknowledging the intricacies of companies within their social context, places emphasis on the reputational dimension.[14] According to this perspective, a company's enhanced positioning with external stakeholders, such as investors, consumers, and authorities, translates into improved market performance, both in financial terms and returns on securities markets. Empirical studies within this realm mostly indicate a positive relationship between social and financial performance, although a unanimous consensus is lacking.[15] A notable systematic review conducted by Friede et al. (2015) analysed approximately 2200 empirical studies spanning four decades. The findings revealed non-negative, though not always statistically significant, relationships between ESG corporate policies and financial performance in around 90% of the studies reviewed. The occasional presence of contradictory empirical evidence can be attributed to biases in the data sample, such as sample representativeness, selected time series, or underlying assumptions in the analytical models employed.[16] Despite these nuances, the prevailing trend suggests a positive association between ESG practices and financial performance, reinforcing the discourse on the integral role of sustainability in investment strategies.

The second strand, assessing the connection between ESG strategies and the performance of investment vehicles and instruments utilising them, can be gleaned from two comprehensive reviews of relevant literature and a recent analysis by the ESMA.[17] Over the past decade, analyses conducted at both academic and operational levels have undergone evolution. Earlier studies in the early 2000s leaned towards a predominantly neutral stance, whereas more recent investigations tend to acknowledge a positive and sometimes highly significant correlation between sustainable strategies and superior performance compared to traditional allocation methodologies.

Examining the impact of individual ESG factors on share prices reveals distinctive patterns:

- Environmental: Empirical evidence establishes a direct relationship between companies' environmental performance and their share performance in stock markets. News about a company's environmental impact triggers corresponding stock price movements, with notable declines observed in the chemical sector post-environmental disasters. Moreover, highly polluting companies exhibit

[13] Refer to Levitt (1958).

[14] See Donaldson and Preston (1995).

[15] Among the numerous contributions, particular attention should be given to: Waddock and Graves (1997) and Margolis and Walsh (2003).

[16] Refer to Ballestero et al. (2015).

[17] See the United Nations Environment Programme Finance Initiative and Mercer (2007), Clark et al. (2015), and ESMA (2022).

lower stock market valuations, while environmentally efficient companies outperform their less efficient counterparts.
- Social: Research by Edmans (2012) focusing on companies in the annual reports of the 100 Best Companies to Work for in America identifies a robust positive relationship between employee satisfaction and share price performance. Further studies on the social dimension of ESG strategies indicate that companies actively engaged in charitable giving report better stock market performance, particularly over the long term.
- Governance: The influence of corporate governance quality on equity performance has been extensively analysed. Gompers et al. (2003) empirically determined that a portfolio combining long investments in shares of companies with optimal governance models and short investments in shares of companies with governance shortcomings generated an alpha of 8.5% per annum over the study period.

The ESMA's analysis aligns with these findings, highlighting factors explaining the outperformance of ESG funds.[18] Beyond different sectoral allocations and lower management costs, the distribution of emphasis across the three ESG pillars emerges as crucial. Funds more focused on governance (G) and social (S) issues tend to exhibit superior performance compared to those emphasising environmental factors (E). This nuanced perspective underscores the multifaceted nature of ESG considerations and their varying impacts on investment performance.

7.4 ESG Rating

The evolution of sustainable investments has spurred the demand for concise indicators assessing the sustainability of issuers and portfolios of financial instruments, notably investment funds and other asset management products. ESG ratings, also known as sustainability ratings, emerge as a solution available to investors to integrate ESG criteria into the investment process.[19]

An ESG rating serves as a synthetic judgment, reflecting the level of attention and sensitivity exhibited by an issuer towards its ESG performance. Unlike traditional credit ratings that primarily focus on financial variables, ESG rating focuses on sustainability practices of a given issuer, enhancing assessments and subsequent investment decisions.[20]

---

[18] See ESMA (2022).

[19] For a comprehensive examination of the ESG rating methodologies employed globally by major companies in the sector and the challenges associated with identifying a common rating metric in the industry, consult the following works: Escrig-Olmedo et al. (2010), Stubbs and Rogers (2013), and Mooij (2017).

[20] To streamline the incorporation of ESG criteria into the decision-making process for end-investors, specialised firms calculate sustainability ratings for investment funds and other

It is crucial to note that the ESG rating does not replace the traditional credit rating but acts as a complementary tool, offering a holistic evaluation.

While traditional credit ratings provide a synthetic judgment on the creditworthiness of an issuer or issue, predominantly considering the financial dimension, ESG ratings stem from an extra-financial analysis focused on the sustainability aspects of the issuer.[21] Therefore, when making investment decisions that incorporate ESG considerations, both measures should be considered concurrently to obtain a comprehensive understanding.

The primary goal of ESG ratings is to assist investors in selecting financial instruments that balance economic and financial performance while aligning with sustainability requirements.[22]

The foundation for ESG ratings is broad, encompassing a range of quantitative and, notably, qualitative information. Table 7.1 serves as a visual representation, offering a breakdown of the topics covered within each area of analysis.

Notably, sovereign states, as bond issuers, are also subject to ESG rating assessments. In such instances, data considered for the ESG rating extend to encompass diverse factors such as healthcare spending levels, freedom of the press, the extent of corruption in public administration, and potential violations of human, civil, and political rights. This inclusive approach ensures a comprehensive evaluation, acknowledging the multifaceted nature of ESG considerations, even within the realm of sovereign bond issuers.

7.4.1 Prospects for Integration of Credit Rating and ESG Rating

Major credit rating agencies are currently engaged on dual fronts concerning the integration of ESG considerations into their processes:[23]

– Integrating ESG analyses into credit rating assignment processes. Challenges in this domain primarily arise from environmental and social responsibility risks, with a notable discrepancy in the assessment time horizons for financial and non-financial risks. Although an issuer's ability to repay is conventionally

---

asset management products. These ratings represent the weighted averages of the ESG ratings assigned to the securities within the portfolio. It is important to note that these ratings can be computed effectively only when a substantial percentage of the securities in the portfolio possess an ESG rating.

[21] Unlike credit ratings, which encompass both issuer and issue ratings, ESG ratings specifically pertain to issuer ratings and do not extend to individual issue ratings. These ESG ratings cover all types of securities, including debt, equity, and hybrids, issued by the respective issuers.

[22] Amel-Zadeh and Serafeim (2018).

[23] Refer to Fitch Ratings (2022), Moody's Investors Service (2021), and S&P Global Ratings (2022).

Table 7.1 The information base typically used for ESG rating assignment

Environment	Environmental approach	Environmental strategy
		Environmental policy
		Quality of the environmental policy
		Internal control bodies on environmental issues
		Level of emissions
		Actions to anticipate future environmental trends and standards
	Environmental management policy	Environmental management system
		Legislative compliance
		Inspections and audits
		Information transparency
		Supply chain environmental management
		Communication and training
		Risk management
	Products	Products life-cycle impact
	Production process	Environmental impact and eco-efficiency
		Pollution prevention
		Alternative energy sources
		Attention to water consumption
		Intensity of greenhouse gas emissions
		Circular economy and attention to waste
Social responsibility	Community relations	Respect for workers' rights
		Corporate citizenship strategy
		Social dialogue and community initiatives
		Contribution to social and economic development
		Philanthropy and donations
		Social accountability
		Disputes
	Employees and human capital	Respect for workers' rights
		Respect for minorities
		Remuneration scheme
		Training activities
		Health and safety
		Disputes with employees
	Relations with markets	Relations with suppliers
		Relations with customers
		Corruption and abuse of dominant positions
		Relations with competitors
Governance	Governance model	Principles of corporate governance
		Transparency on corporate governance
		Shareholders' rights and duties
		Composition and functioning of the board of directors

(continued)

Table 7.1 (continued)

		Transparency of transactions with related parties
		Transposition of international accounting standards
		Management experience and effectiveness

evaluated concerning the maturity of various debt instruments, ESG assessments extend their significance over a long-term horizon, unrelated to the issuer's immediate financial commitments.[24] However, there is consensus that governance, distinct from the other pillars, significantly influences both financial and non-financial (ESG) rating processes due to its impact on corporate decisions and strategies at various levels.

– Ensuring transparent and objective integration. The integration process aims to be transparent and objective, acknowledging the inherent challenges in estimating ESG risk factors with less precision and relying on qualitative assessment elements. Effective integration hinges on acquiring the necessary human and technological resources to develop the rating process, regardless of the integration method.

Difficulties encountered in the implementation of this integration process include:[25]

– Duration of trends influencing potential ESG risks.
– Challenge of predicting the effects of ESG risk factors on debt repayment capacity.
– Borrowers' need to adjust their business model in response to the potential negative effects of ESG risks.
– Uncertainty regarding the construction of a shared framework for collecting and processing relevant information in ESG screening.
– Limited efforts by rating agencies to independently guarantee completeness and transparency in information flow, especially in qualitative analyses.

Moreover, two alternative approaches have emerged in managing these critical issues:[26]

– Integration strategy: It effectively integrates ESG assessments into credit ratings, revising the weights of qualitative and quantitative analyses, with a priority focus on governance issues.

[24]The occurrence of CO2 emissions fraud within the automotive sector in 2017 has heightened the sensitivity of credit rating agencies to the necessity of appropriately assessing environmental risks in evaluating issuers' repayment capacity. This has led to an increased collaboration with ESG analyses, reflecting a growing awareness of the interconnectedness between environmental factors and creditworthiness.

[25]Refer to Cheng et al. (2014).

[26]For an in-depth analysis of this specific issue, refer to PRI (2017, 2018).

- Add-on strategy: It considers ESG and creditworthiness assessments in parallel, maintaining independence to facilitate flexibility in the final scoring assignment phase and enhance comparability among issuers with distinct profiles in each area.

Currently, the prevailing approach favours the second option, aligning more coherently with the underlying philosophy and purposes of both credit rating and ESG rating providers. This approach ensures an integrated assessment of the risk-return profile of various investment alternatives while acknowledging the inherent difficulties in merging these two distinct philosophies.[27]

The challenges in this integration process have led to the coexistence of ESG ratings offered by major credit rating agencies and specialised operators exclusively dedicated to ESG services. These operators, equipped with proprietary research models and historically independent or loosely associated with index providers, have operated without direct forms of supervision or discipline. This historical perspective perceived their ratings as external or marginal to investors' risk assessments.[28]

The significance of sustainability in investors' decision-making processes, coupled with the lack of transparency in ESG rating models and the variability in ratings from different providers, has prompted EU authorities to consider regulating this sector. This regulation would be akin to the established frameworks for credit rating agencies. Underlying this decision are the market findings summarised in two studies conducted by the European Commission (2020) and IOSCO (2021). Both studies highlight critical grey areas in ESG rating assignment process that need swift resolution to meet investors' real needs:

- Lack of transparency: ESG rating methodologies and data sources vary among providers, hindering the comparability of analyses.
- Absence of industry standards: The sector lacks industry standards validated by certification bodies or forms of self-regulation/supervision.
- Reliability and timeliness concerns: ESG assessments are criticised for reduced reliability, limited accuracy, and poor timeliness.
- Potential conflicts of interest: Providers involved in engagement activities may face conflicts of interest, particularly in dialogues between rated companies and ESG rating providers.
- Widespread mistrust: Rated companies often express mistrust in the ability of ESG ratings to accurately reflect their true sustainability performance.

[27] Refer to Windolph (2011).

[28] In response to the rapid development of the European market spurred by regulatory demands from EU authorities, major US rating agencies and prominent information providers (such as S&P, Moody's, and Morningstar) have opted for a faster market entry by acquiring established European research and rating companies. This strategic choice allows them to expedite their presence and leverage the expertise and market understanding of the acquired entities rather than develop internal know-how.

Empirical verifications on ESG ratings reveal critical issues, as demonstrated by Berg et al. (2022), who conducted a thorough analysis of divergences among major ESG rating providers.[29] The identified divergences fall into three main categories:

- Use of metrics: Differences arise from employing metrics collected with different methods within ESG criteria.
- Scope of application: Providers exhibit variations in the fields of application of the ESG criteria analysed.
- Weighting ambiguity: Ambiguity in evaluating the weighting of individual ESG criteria is observed.

The study found that these differences in metrics, scopes, and weights accounted for 56%, 38%, and 6% of rating divergences, respectively. While convergence may occur over time, particularly with increased standardisation of ESG data published by companies, the agencies are likely to treat the same information differently due to variations in their ESG criteria approach. Divergence in opinions may be acceptable given the heterogeneous preferences of ESG rating users. However, measurement divergence becomes problematic if ESG ratings are expected to be based on objective observations.

In June 2023, the European Commission proposed rules to achieve the dual objective of enhancing comparability and reliability of ESG ratings and ensuring credit ratings consider relevant ESG risks. The suggested rules include:

- Authorisation regime: Establishing an authorisation regime for ESG rating providers.
- Qualification and methodologies: Precisely defining the activities performed and methodologies adopted by ESG rating providers.
- Conflict of interest management: Prevention and management of potential conflicts of interest arising from advisory activities.
- Supervision mechanisms: Introduction of sanction mechanisms and forms of supervision at the EU level.
- Integration of ESG analysis: Implementing more effective ways of integrating ESG analysis into the allocation of credit risk for companies offering both credit ratings and ESG ratings.

The anticipated positive impact of these regulations is expected to benefit all stakeholders, providing improved information to investors, issuers, index providers, investment management companies, and regulators alike. This regulatory intervention aims to address the identified challenges and create a more transparent, standardised, and trustworthy ESG rating landscape.

[29] Refer to Berg et al. (2022) and Brandon et al. (2021).

7.4.2 *The Main ESG Rating Providers*

In contrast to the credit rating landscape, where a few agencies dominate with established standards, the ESG rating arena is characterised by a multitude of companies, each employing distinct methodologies. Although this diversity allows for comprehensive assessments, it also introduces challenges, primarily stemming from the subjectivity inherent in the absence of an internationally shared standard.

One significant limitation is the subjectivity of ESG ratings, influenced by the lack of a universal standard. Various variables impact rating determination, often in qualitative ways. The evaluator's preferences and interpretation of "sustainability" play a pivotal role, with some emphasising environmental concerns, while others prioritise an issuer's commitment to employees.

The information nature adds another layer of variability. Some ESG rating providers rely solely on publicly available data, while others engage directly with rated issuers through questionnaires, enhancing accuracy but introducing the risk of biased responses. The assessment output varies across companies, with different scales and symbolism. For example, some companies define 0 as the best level and 100 as the worst, whereas others assign 0 as the worst value and 100 as the best in terms of ESG practices. The absence of universal standards for the variables, calculation methodologies, and common and well recognised ranking criteria limit the comparability of ratings, undermining their credibility and potential.[30]

Table 7.2 highlights ten globally recognised ESG rating providers, showcasing their sustainability rating scales, the number of ESG indicators employed, and the volume of companies rated. This diversity reflects the evolving landscape of ESG assessments and underscores the need for standardised practices to enhance transparency and credibility in this critical domain.[31]

7.5 Approaches to Sustainable Investing

Sustainable investing embraces various approaches for integrating ESG criteria into the investment process, each contributing to the overarching goal of aligning investments with environmental, social, and governance considerations. Here are the possible approaches:

- Positive/best in class: In this case, the issuers are among the best in a given sector or peer group in terms of ESG criteria. Thus, issuers with the highest ESG ratings within a given universe, category, or asset class are included or overweighted.

---

[30]The correlation levels among ESG ratings provided by six rating agencies, as analysed by Berg et al. (2022), range from 0.38 to 0.71. In contrast, the credit ratings assigned by Moody's and Standard & Poor's exhibit a notably higher correlation of 0.99.

[31]The information, as of the end of 2023, is sourced from the companies' websites and their methodological guides.

Table 7.2 A comparison of leading ESG rating providers

ESG rating providers	Scale used for ESG rating	Number of indicators used	Number of companies rated
Bloomberg ESG disclosure score	The Bloomberg ESG disclosure score comprises E, S, and G scores, and the scale ranges from 0 to 100, where a score near to 100 is the best score a company can achieve.	Around 800 ESG indicators	Over 15,000 companies
CDP	The scoring methodology assesses the level of detail and comprehensiveness in a response, as well as the company's awareness of environmental issues, its management methods and progress towards environmental stewardship. The scoring system is divided into four levels: Disclosure (D-, D), awareness (C-, C), management (B-, B), and leadership (A-, A), with companies assessed and awarded points at each level.	Industry-specific questionnaire updated annually	Over 14,000 companies
ISS ESG	The rating is alphabetically coded from A+ to D-, with A +, A, A- (excellent), B+, B, B- (good), C+, C, C- (medium), D+, D, D- (poor).	ESG indicators vary according to the production sector	Over 8000 companies
LSEG Data & Analytics	From 0 to 100 with 0 identifying the worst level and 100 indicating the best level.	630 ESG indicators	Over 15,500 companies
Moody's ESG solutions	From 0 to 100 with 0 identifying the worst level and 100 indicating the best level.	30 generic ESG indicators plus sector-specific indicators	Over 5000 companies
MSCI ESG Ratings	The rating has an alphabetical code ranging from AAA to CCC, with: AAA, AA (best cases); A, BBB, BB (intermediate cases); B, CCC (worst cases).	35 ESG indicators	Over 14,000 companies, as well as states and investment funds
Refinitiv	ESG score ranges from 0 to 100, where closer to 100 is the better score a company can achieve. It also includes a controversy score generated on controversial ESG news events happening	186 ESG indicators	Over 6000 companies

(continued)

Table 7.2 (continued)

ESG rating providers	Scale used for ESG rating	Number of indicators used	Number of companies rated
	across the globe that relate to the specific company.		
RepRiskRating	The rating is alphabetically coded from AAA to D, with: AAA, AA, and A indicating low ESG risk; BBB, BB, B indicating medium ESG risk; CCC, CC, C indicating high ESG risk; D indicating very high ESG risk.	102 ESG indicators	Over 250,000 companies
S&P Global	ESG scores are measured on a scale from 0 to 100, where 100 represents the maximum score.	Points are awarded at the question-level, with an average of 130 per company, based on an assessment of up to 1000 underlying data points	Over 10,000 companies
Sustainalytics	The risk score ranges from 100 (highest risk) to 0 (lowest risk). In practice, company ratings are categorised across five risk levels: negligible risk (overall score of 0–9.99 points), low risk (overall score of 10–19.99 points), medium risk (overall score of 20–29.99 points), high risk (overall score of 30–39.99 points), and severe risk (overall score of 40 points and above). Almost all companies score below 50.	ESG indicators vary according to the production sector (at least 250)	Over 16,000 companies

- Sustainability themes: Under this approach, investments are concentrated in specific thematic areas or activities closely related to sustainability development, resulting in a focus on one or more ESG themes. General areas of focus are climate change, renewable energy, energy efficiency, sustainable transport, water emergencies, waste management, land use, and forests.
- Norms-based screening: In this approach, investment selection is based on issuers' level of compliance with international standards and norms; therefore, threshold values, which are necessary for the application of selection mechanisms, are defined in the scoring process. International norms and standards of reference are those related to the ESG dimensions defined on a global scale by the UN or other international institutions.
- Engagement and voting: This approach involves dialogue between the investor (typically an institutional investor) and the company under evaluation. The

objective of interaction with management is to encourage the company to take actions that can improve its sensitivity to ESG issues and increase its degree of transparency on these issues.

- Impact investing: This approach involves implementing investments deemed to have a positive impact on sustainable development. Examples include microfinance, community investment, and social venture capital funds, in which the objective of generating an environmental and/or social impact must not, however, be detrimental to economic performance or the ability to generate financial returns for investors.
- ESG integration: In this approach, ESG rating factors are explicitly and systematically included in traditional financial analysis. Owing to the significant lack of clarity in the scope of ESG factor integration, the extent to which the strategies in this approach can be compared is extremely difficult to assess. The concept of ESG integration is yet to be unambiguously defined, as its content varies between countries and investment managers.
- Exclusion: This approach systematically excludes from the investment universe issuers, sectors, and countries that do not meet the minimum standards for ESG issues. Generally, the excluded companies operate in controversial sectors that are far from adopting minimum ESG standards, such as the production and marketing of tobacco, chemical or biological weapons, and so on. In the case of countries, exclusion stems from human rights violations or restrictions on personal freedom.

Operationally, investment funds and other asset management products that adopt ESG investment policies may apply one or more of the previous approaches. On the one hand, some funds have an internal ethics committee, which is responsible for overseeing the fund's activity and identifying the investable universe. On the other hand, some funds leverage analyses from external sustainability-focused companies, which, using predefined criteria, select a set of securities that could be the object of investment by the fund.[32]

Indexed investment funds integrate their ESG policies into the investment process by replicating an ESG index provided by an external index provider, which is used as a benchmark.

Table 7.3 provides a summary of the possible approaches to sustainable investing, distinguishing between positive and negative screening approaches.

These multifaceted approaches contribute to the evolving landscape of sustainable investing, reflecting the growing importance of ESG considerations in investment decision-making.

[32] Staub-Bisang (2012) and Berry and Junkus (2013).

Table 7.3 A summary of approaches to sustainable investing

Positive screening	
Best in class	This approach refers to using ESG information to identify companies that are sustainability leaders, often by industry or sector or those that are using sustainability to establish or enhance a competitive advantage.
Sustainability themes	Under this approach, investments may include environmental-related themes like renewable energy, clean tech, and clean water, or those related to social themes, such as gender equity, managing population growth, or health and well-being.
Norms-based screening	In this investment style, investment selection is based on the level of issuer compliance with international standards and norms.
Engagement and active ownership	This style refers to seeking positive ESG outcomes via active ownership activities, primarily made possible because asset managers are shareholders in public companies.
Impact investing	In this approach, investors seek, in addition to financial returns, a measurable positive impact on society.
ESG integration	This style refers to including explicitly and systematically ESG information and ESG rating in the traditional financial analysis to assess the investment universe and to act security selection.
Negative screening	
Exclusion	This approach refers to excluding stocks or sectors that do harm to society. Common sector exclusions include tobacco, alcohol, controversial weapons, etc.

References

Abate G, Basile I, Ferrari P (2023) The integration of environmental, social and governance criteria in portfolio optimization: an empirical analysis. Corp Soc Respon Environ Manag:1–12

Amel-Zadeh A, Serafeim G (2018) Why and how investors use ESG information: evidence from a global survey. Financ Anal J 74(3)

Ballestero E, Pérez-Gladish B, Garcia-Bernabeu A (eds) (2015) Socially responsible investment, A multi-criteria decision making approach. Springer, Heidelberg

Berg F, Koelbel JF, Rigobon R (2022) Aggregate confusion: the divergence of ESG ratings. Rev Financ 26(6):1315

Berry TC, Junkus JC (2013) Socially responsible investing: an investor perspective. J Bus Ethics 112(4)

Boubaker S, Cumming D, Nguyen DK (eds) (2018) Research handbook of finance and sustainability. Edward Elgar, Cheltenham

Brandon RG, Krueger P, Schmidt PS (2021) ESG rating disagreement and stock returns. Financ Anal J 77(4)

Cheng B, Ioannou I, Serafeim G (2014) Corporate social responsibility and access to finance. Strateg Manag J 35(1)

Clark GL, Feiner A, Viehs M (2015) From the stockholder to the stakeholder. How sustainability can drive financial outperformance, working paper, University of Oxford/Arabesque Partners, Oxford-London

Donaldson T, Preston LE (1995) The stakeholder theory of the corporation: concepts, evidence, and implications. Acad Manag Rev 20(1)

Edmans A (2012) The link between job satisfaction and firm value, with implications for corporate social responsibility. Acad Manag Perspect 26(4)

Escrig-Olmedo E, Muñoz-Torres MJ, Fernández-Izquierdo MA (2010) Socially responsible investing: sustainability indices, ESG rating and information provider agencies. Int J Sustain Econ 2(4)

ESMA (2022) The drivers of the costs and performance of ESG funds, "ESMA document number" 50

ESMA (2023a) "Progress report on greenwashing", ESMA30-1668416927-249831, 31 May 2023

ESMA (2023b) ESG names and claims in the EU fund industry, ESMA50-524821-2931, 2 October 2023

EU High Level Expert Group on Sustainable Finance (2018) *Financing a sustainable European economy. Final report 2018*, Bruxelles

European Commission (2018a) *Action plan: financing: sustainable growth*, COM(2018) 97 final, Bruxelles

European Commission (2018b) *Proposal for a Regulation of the European Parliament and of the Council on the establishment of a framework to facilitate sustainable investment*, COM(2018) 353 final, Bruxelles

European Commission (2020) Study on sustainability-related ratings, data and research, Bruxelles.

European Commission (2023) Sustainable finance. Investing in a sustainable future, Bruxelles

Fitch Ratings (2022) ESG ratings, data and analysis, New York

Friede G, Busch T, Bassen A (2015) ESG and financial performance. Aggregated evidence from more than 2000 empirical studies. J Sustain Financ Invest 5(4)

Fung HG, Law SA, Yau J (2010) Socially responsible investment in a global environment. Edward Elgar, Cheltenham

Gompers PA, Ishii JL, Metrick A (2003) Corporate governance and equity prices. Q J Econ 118(1)

IOSCO (2021) *Environmental, social and governance (ESG) ratings and data products providers final report*, Madrid.

Levitt T (1958) The dangers of social responsibility. Harv Bus Rev 36(5)

Margolis JD, Walsh JP (2003) Misery loves companies: rethinking social initiatives by business. Adm Sci Q 48(2)

Moody's Investors Service (2021) *General principles for assessing environmental, social and governance risks methodology*, New York

Mooij S (2017) *The ESG rating and ranking industry; vice or virtue in the adoption of responsible investment?*, working paper, University of Oxford – Smith School of Enterprise and the Environment

PRI (2017) Shifting perceptions: ESG, credit risk and ratings. Part 1: the state of play, London

PRI (2018) Shifting perceptions: ESG, credit risk and ratings. Part 2: exploring the disconnects, London

S&P Global Ratings (2022) S&P global ESG scores methodology

Staub-Bisang, M. (2012). Sustainable investing for institutional investors. Risks, Regulations and strategies, Wiley, Singapore.

Stubbs W, Rogers P (2013) Lifting the veil on environment-social-governance rating methods. Soc Responsib J 9(4)

United Nations (2015) United Nations framework convention on climate change. Paris Agreement, Paris

United Nations (2018) The sustainable development goals report 2018, New York.

United Nations Environment Programme Finance Initiative, Mercer (2007) *Demystifying responsible investment performance.* A review of key academic and broker research on ESG factors, UNEP FI/Mercer, New York

Waddock SA, Graves SB (1997) The corporate social performance-financial performance link. Strateg Manag J 18(4)

Windolph SE (2011) Assessing corporate sustainability through ratings: challenges and their causes. J Environ Sustain 1(1)

Part III
Performance Evaluation for Traditional Investment Portfolios

Chapter 8
Performance Evaluation

Pierpaolo Ferrari

8.1 Performance Evaluation Stages

The purpose of evaluating the performance of an investment portfolio is to assess the quality of the asset manager's work by analysing the results achieved and the manner in which these results were achieved. This activity involves the identification of:

- The return on the investment portfolio, taking into account alternative calculation methods and selecting the most appropriate one according to the objective pursued.
- The risk of the portfolio, in its various absolute, asymmetric, and relative forms, and the related risk-adjusted performance indicators.
- The asset manager's ability to perform effective stock picking by selecting the best stocks in the reference market.
- The asset manager's ability to perform profitable market timing through effective tactical asset allocation by temporarily overweighting sectors and asset classes that will outperform the market, while underweighting those that will underperform.
- Performance persistence, understood as an assessment of the continuity and frequency with which the asset manager ranks in the top percentile of his/her peer group.

P. Ferrari (✉)
Department of Economics and Management, University of Brescia, Brescia, Italy

SDA Bocconi School of Management, Bocconi University, Milan, Italy
e-mail: pierpaolo.ferrari@unibs.it; pierpaolo.ferrari@unibocconi.it

I. Basile, P. Ferrari (eds.), *Asset Management and Institutional Investors*, Contributions to Finance and Accounting,
https://doi.org/10.1007/978-3-031-59819-7_8

8.2 Return on an Investment Portfolio

The measurement of return is the basis for evaluating the performance of any investment portfolio. Two premises are useful in this respect:

- Measuring return is merely the starting point for performance evaluation and not, as one might imagine, the objective.
- There are many forms of return, all of which are potentially proper and whose suitability should be assessed according to the objectives being pursued through its measurement.

The return on a portfolio can be considered as the expression of the performance of the investment activity or, in other words, a measure of the profitability registered by the same in a given time period. To measure it, the final value has to be compared with the initial portfolio value within a given time period:

$$r_{0-T} = \frac{V_T}{V_0} - 1$$

However, the return obtained in this way overlooks a series of key factors which, depending on the objective pursued, may be essential both for the correct and complete measurement of the result actually recorded and for having an indicator of absolute and relative concrete use. A thorough measurement of the return on a financial portfolio should therefore consider all the potentially relevant factors, which may be summarised as follows:

- The capital initially invested.
- The final value at the time of evaluation.
- The periodic income distributed, if any.
- Any periodic cash flows in terms of new capital injections and/or withdrawal of previously invested resources.
- Commissions.
- Tax charges.
- The length of the reference time period.

A host of alternative returns is thus derived, all of which are potentially correct and able to meet different needs of interpretation and comparison. The following concepts will now be analysed:

- Simple, compound, and continuous return.
- Arithmetic and geometric return.
- Money-weighted and time-weighted return.
- Gross and net return.

8.2.1 Simple, Compound, and Continuous Return

The length of the reference time period for performance evaluation is relevant both with regard to the accumulation system adopted for income accrued periodically and to the calculation of average returns suitable for enabling comparison of investments with different maturities.

Without attempting an exhaustive description of interest accumulation systems, it may be useful to describe the basic features of simple, compound, and continuous interest and briefly explain them.

In simple interest, the periodic income accrued in the investment life does not generate new interest and it is always and only the initial capital that is considered. Consequently, the final or accumulated value at a given date grows linearly according to the time elapsed:

$$V_t = V_0 + I = V_0 + V_0 \cdot r_s \cdot t = V_0 \cdot (1 + r_s \cdot t)$$

from which we obtain the simple return:

$$r_s = (V_t - V_0)/(V_0 \cdot t)$$

where:

V_t	= final value
V_0	= initial capital
t	= days/365
r_s	= simple return on a yearly basis, expressed as a percentage

In compound interest, the periodic income accrued in the life of the investment generates new interest by virtue of its reinvestment at the end of each accrual period. In this case, the final value at a given date increases exponentially with time:

$$V_t = V_0 \cdot (1 + r_c)^t$$

from where the compound return is equal to:

$$r_c = \sqrt[t]{\frac{V_t}{V_0}} - 1$$

In continuously compounded interest (or continuous interest), we assume that the periodic income accrues continuously and generates new interest instantly, unlike the case of compound interest, where periodic income accrues at pre-established dates, and is reinvested and generates new interest only once it has accrued. On the basis of this assumption, the initial capital still undergoes exponential growth over time, despite following a different path with respect to compound interest, by virtue of the different frequency of reinvestment:[1]

$$V_t = V_0 \cdot e^{r_{co} \cdot t}$$

from which:

$$r_{co} = \frac{1}{t} \cdot \ln(V_t/V_0)$$

The choice of which accumulation system to use is purely conventional: no one system is better in absolute terms, even though a market convention exists in favour of the simple interest system for maturities of less than 1 year and the compound interest system for longer ones. In financial models, and especially in derivative pricing models, the continuous interest system is the most popular, thanks to the useful analytical properties of the exponential function and its opposite, the logarithmic function.[2]

Table 8.1 shows the effects of the different capitalisation conventions in the case of an initial capital of 100 and a nominal annual rate of 7%, highlighting the development of the final value over subsequent time periods.

First of all, it should be noted how, with the same nominal annual rate, the final values differ depending on the accumulation system adopted. With the same nominal annual rate, the final value with simple interest is lower than the one with compound interest, which is in turn lower than the one with continuous interest. It is interesting to point out that with a nominal annual rate of 7%, interest on interest represents the following portions of the final value: after 10 years, 0% in simple interest, 13.6% in compound interest, and 15.6% in continuous interest; after 30 years, 0% in simple interest, 59.3% in compound interest, and 62% in continuous interest. It is therefore clear that, with the same nominal annual rate, the effective annual return differs according to the type of interest and corresponds:

(a) In the case of simple interest, to an effective rate which is always 7% per year, whatever the frequency of interest calculation, as the periodic income accrued does not generate new interest.

---

[1]Continuous interest is a special case of compound interest, whereby interest is instantly accrued. The term e, known as Napier's or Euler's number, is about 2.7183 and represents the final value of 1 euro invested per 1 year at a yearly rate of 100%. Starting with $V_t = V_0 \cdot (1 + 1/k)^{k \cdot 1}$, we can show that the limit of 1 $(1 + 1/k)^k$ as k approaches positive infinity is in fact equal to e. As a matter of generalisation, for periods longer or shorter than 1 year and for rates different from 100% per year, it is possible to calculate the limit of $(1 + r/k)^{k \cdot t}$ as k approaches positive infinity, equal to $e^{r \cdot t}$.

[2]De Fusco et al. (2007).

Table 8.1 The development of final values over time, depending on the system of interest accumulation adopted (nominal annual rate: 7%)

Years	0	1	2	3	4	5	10	15	20	30
Simple interest	100	107.00	114.00	121.00	128.00	135.00	170.00	205.00	240.00	310.00
Compound interest	100	107.00	114.49	122.50	131.08	140.26	196.72	275.90	386.97	761.23
Continuous interest	100	107.25	115.03	123.37	132.31	141.91	201.38	285.77	405.52	816.62

Table 8.2 The relation between nominal annual rate and effective annual return

Interest calculation frequency	k	Effective return with nominal annual rate of 7%
Annual	1	7.000%
Six-monthly	2	7.122%
Quarterly	4	7.186%
Monthly	12	7.229%
Weekly	52	7.246%
Daily	365	7.250%
Continuous	Instantly	7.251%

(b) In the case of compound interest, to an effective rate which varies from 7%, for annual compounding, to 7.25% for daily compounding.[3]

(c) In the case of continuous interest, to an effective rate of 7.251%, very similar to compound interest with interest paid daily (Table 8.2).[4]

In order to understand this reasoning, we must analyse the convertibility relations between rates belonging to the same accumulation convention but expressed with different time periods. The convertibility relation of the rates in the different periods varies depending on the accumulation system.

In the simple-interest system, the annual rate (r_{ann}) may be converted into a periodic rate r_k, where k is the number of periods in a year for which interest is paid, or, conversely, the periodic rate r_k may be converted into an annual rate with the following rule:

$$(1 + r_{\text{ann}}) = (1 + r_k \cdot k)$$

from which:

$$r_{\text{ann}} = r_k \cdot k$$

and conversely:

$$r_k = r_{\text{ann}}/k$$

[3] As we will learn shortly, the effective annual rate in compound interest can be calculated as follows:

$$r_{\text{ann}} = (1 + r_k)^k - 1$$

with k equal to the frequency with which interest is paid in a year.

[4] In continuous interest, the effective annual rate, as we will see, can be calculated as follows:

$$r_{\text{ann}} = r_k \cdot k$$

In the compound interest system, the annual rate r may be converted into a rate r_k or, conversely, the rate r_k may be converted into an annual rate with the formula:

$$(1 + r_{\text{ann}}) = (1 + r_k)^k$$

from which:

$$r_{\text{ann}} = (1 + r_k)^k - 1$$

and conversely:

$$r_k = (1 + r_{\text{ann}})^{1/k} - 1$$

Lastly, in the continuous interest system, the relation connecting the annual rate r and the periodic rate r_k is the following:

$$e^{r_{\text{ann}}} = e^{r_k \cdot k}$$

from which:

$$r_{\text{ann}} = r_k \cdot k$$

and conversely:

$$r_k = r_{\text{ann}}/k$$

Turning this reasoning around, if, instead of starting from the same nominal annual rate of 7%, we sought a rate which could ensure an identical final value regardless of the interest accumulation system, it would be possible to resort to what are called financially equivalent rates. In order to guarantee an identical final value, these rates will obviously have to be different from one another: higher in simple interest, lower in compound interest, and even lower in continuous interest.

To ensure the same final value in compound interest, a simple annual rate of 7% should be equal to:

$$(1 + r_s \cdot t) = (1 + r_c)^t$$

which, for example, with $t = 3$, produces a compound rate equal to:[5]

$$r_c = \sqrt[t]{(1 + r_s \cdot t)} - 1 = \sqrt[3]{(1 + 0.07 \cdot 3)} - 1 = 6.56\ \%$$

[5] If one wishes to convert the compound rate into its financially equivalent simple rate, it is enough simply to reverse the logic: $r_s = [(1 + r_c)^t - 1]/t$.

To ensure the same final value in continuous interest after t years, an annual compound rate of 6.56% should instead be equal to:

$$e^{r_{co} \cdot t} = (1 + r_c)^t$$

which, for example, with $t = 3$, would produce a continuous rate of:[6]

$$\begin{aligned} e^{r_{co} \cdot 3} &= (1 + r_c)^3 \\ 3 \cdot r_{co} &= 3 \cdot \ln(1 + r_c) \\ r_{co} &= \ln(1 + 0.0656) = 6.354\ \% \end{aligned}$$

The logic of financially equivalent annual rates is especially useful as it enables the transformation of rates obtained following a given accumulation system into rates corresponding to another system, thus making it possible to reverse the choice one has made while still ensuring comparability between returns, even though these have been calculated with a different method.

8.2.2 *Arithmetic and Geometric Return*

In addition to the accumulation system, the length of the reference time period for evaluating the performance of a given portfolio is significant for determining the mean return applicable to a predefined unit of time, required in order to compare returns on investments with different maturities.

In this regard, two types of mean return can be calculated:

- An arithmetic mean return, obtained by the sum of the returns over the sub-periods, divided by the number of sub-periods under consideration:

$$r_{\text{arithmetic } 0-n} = \frac{\sum_{t=1}^{n} r_t}{n}$$

- A geometric mean return, obtained from the outcome of the chain linking of the returns over the sub-periods under the nth root. We adopt the geometric mean return calculation method through the product of the growth factors, instead of directly from the returns, because the presence of a negative result would not allow extraction of the nth root. In this way, if the assumption of negative returns over the sub-periods exceeding -100% is excluded, the mean return can still be

[6]Here too, if the conversion desired was from continuous rate to financially equivalent compound rate, it is enough to reverse the logic: $r_c = e^{r_{co}} - 1$.

calculated, but in order to achieve a result in the form of a return, and not as a growth factor, one unit must be subtracted:

$$r_{\text{geometric } 0\text{-}n} = \sqrt[n]{\prod_{t=1}^{n}(1+r_t)} - 1$$

Here is an example which can illustrate the difference between arithmetic and geometric mean. Let us assume that portfolio *P* had a value of 10 at the time of initial investment and developed as described by the two following alternative scenarios:

Scenario 1		Scenario 2	
$P_0 = 10$		$P_0 = 10$	
$P_1 = 12$	$r_{0-1} = (P_1 - P_0)/P_0 = +20\%$	$P_1 = 10.5$	$r_{0-1} = (P_1 - P_0)/P_0 = +5\%$
$P_2 = 13.2$	$r_{1-2} = (P_2 - P_1)/P_1 = +10\%$	$P_2 = 13.2$	$r_{1-2} = (P_2 - P_1)/P_1 = +25.71\%$

In both cases, the final value of the investment is equal to 13.2. In scenario 1, however, the returns over the sub-periods are 20% and 10%, respectively, whereas in scenario 2, the returns over the sub-periods are 5% and 25.71%. In the first case, the arithmetic mean of the returns is 15%, while in the second case, it is 15.36%:

	Scenario 1	Scenario 2
$r_{\text{arith } 0-n} = \frac{\sum_{t=1}^{n} r_t}{n}$	$r_{\text{arith } 0-2} = \frac{+20\%+10\%}{2} = 15\%$	$r_{\text{arith } 0-2} = \frac{+5\%+25.71\%}{2} = 15.36\%$

The geometric mean, on the other hand, is 14.89% in both scenarios:

	Scenario 1	Scenario 2
$r_{\text{geo } 0-n} = = \sqrt[n]{\prod_{t=1}^{n}(1+r_t)} - 1$	$r_{\text{geo } 0-2} = \sqrt[2]{(1+20\ \%)\cdot(1+10\%)} - 1 = +14.89\%$	$r_{\text{geo } 0-2} = = \sqrt[2]{(1+5\ \%)\cdot(1+25.71\%)} - 1 = +14.89\%$

This example can be used to draw some general conclusions:

- The arithmetic mean of the returns for each sub-period (15% and 15.36%) is always higher than the geometric mean (14.89% in both cases). The difference between the two means widens with the increasing historical volatility of the returns over the sub-periods. The two means can be the same only in the (rare) event that returns over the sub-periods are all identical.[7]

[7] The relation connecting the two means can be approximated as follows:

$$r_{\text{geometric}} \approx r_{\text{arithmetic}} - \sigma^2/2$$

If the distribution of returns is normal, then the relation will change from approximate to exact. Bodie et al. (2024).

- The arithmetic mean of returns involves a systematic overestimation of the return actually achieved, unless all the returns for each sub-period are identical. In the example above, the return actually obtained in both scenarios is equal to 14.89% and is demonstrated by the fact that, by multiplying the initial portfolio value for two periods at the rate of 14.89%, the final result is always 13.2.[8]
- The arithmetic mean of returns leads to different results depending on the path followed by the investment value. Although the final values are the same at the end of the second period, the arithmetic mean return in scenario 1 is different from the arithmetic mean return in scenario 2.
- The geometric mean provides returns which are not affected by the path followed by the investment value: however long the time period taken into consideration, it is only the initial and final values that count.[9] With the same final values, the geometric mean returns are in fact identical in the two scenarios.[10]

From these considerations we may conclude that:

- The calculation of actual *ex post* returns is measured by the geometric mean and not by the arithmetic mean, which would lead to an overestimation of the actual return. The geometric mean takes into account the variation of the initial capital at the end of each sub-period due to gains or losses, thus assuming that the return over the sub-periods is reinvested in the same portfolio at the end of each sub-period. Geometric return is the average growth rate that causes an initial investment to grow to its final value. In fact, when the aim is to measure a portfolio's past performance, the solution is to calculate the geometric mean return.
- The arithmetic mean of returns reasons in terms of invested capital, without considering the impact of the returns for each sub-period on the size of initial capital. In *ex post* terms, this can therefore be interpreted as the average value of withdrawals and contributions, expressed as a percentage of the initial capital, which it would have been possible to carry out at the end of each sub-period to

[8]In both scenarios: $10 \times (1 + 14.89\%)^2 = 13.2$

[9]If we consider the definition of geometric return provided above:

$$r_{\text{geometric } 0-n} = \sqrt[n]{\prod_{t=1}^{n} (1 + r_t)} - 1$$

it is possible to rewrite the returns in this way, thus showing how the geometric mean return depends only on the initial and final values and not on intermediate values:

$$r_{\text{geometric } 0-n} = \sqrt[n]{\left[1 + \left(\frac{P_1}{P_0} - 1\right)\right] \cdot \left[1 + \left(\frac{P_2}{P_1} - 1\right)\right] \cdot \ \dots \ \cdot \left[1 + \left(\frac{P_n}{P_{n-1}} - 1\right)\right]} - 1$$
$$= \sqrt[n]{\frac{P_1}{P_0} \cdot \frac{P_2}{P_1} \cdot \ \dots \ \cdot \frac{P_n}{P_{n-1}}} - 1 = \sqrt[n]{\frac{P_n}{P_0}} - 1$$

[10]Christopherson et al. (2009).

keep the initial capital value intact and constant. As investors rarely withdraw the positive return for the period at the end of each sub-period, or pay in negative return, thereby reinstating the initial capital value, the arithmetic mean is hardly an effective measure of past performance.

– When a time series is used to extrapolate forward-looking data, it undoubtedly seems more correct to use arithmetic returns, as they are better able to show the average return for the generic reference sub-period. To a good approximation, we can say that if the geometric mean is the average rate of growth, constant for each sub-period, which leads a given initial capital to grow until its final value, the arithmetic mean, on the other hand, expresses the average return for the generic sub-period. Therefore, when the aim is to obtain generic information from past data able to represent forward-looking data, the solution lies in the arithmetic average return.[11]

To make this aspect clearer, let us consider an extreme scenario of a portfolio with an initial value of 10, characterised by a change of +100% in the first period and –50% in the second period.

Extreme scenario	
$P_0 = 10$	
$P_1 = 20$	$r_{0-1} = (P_1 - P_0)/P_0 = +100\%$
$P_2 = 10$	$r_{1-2} = (P_2 - P_1)/P_1 = -50\%$

If we wish to know the actual *ex post* return on the portfolio, we need to calculate the geometric mean, which in this particular case is equal to 0%:

$$r_{\text{geo } 0-2} = \sqrt[2]{(1 + 100\ \%) \cdot (1 - 50\ \%)} - 1 = 0\%$$

Moreover, this return is confirmed by the fact that final and initial portfolio values coincide. Whereas if we wish to know the average return for the generic sub-period and accept it as expected return (provided that the past returns are an effective measure for estimating future returns), we need to calculate the arithmetic average, which in this particular case is equal to 25%:

$$r_{\text{arith } 0-2} = \frac{+100\% - 50\%}{2} = +25\%$$

This return expresses the average result expected for a generic sub-period. Considering a starting value of 10, the investment in the first sub-period would have generated a return of 10, while the investment in the second sub-period would have generated a loss of 5. The average value expected would therefore be 2.5 = (10 – 5)/2, which corresponds to exactly 25% of the initial capital. In this case, the arithmetic mean provides a more accurate measurement of the returns expected in the future.

[11] Bailey et al. (2007).

8.2.3 Money-Weighted and Time-Weighted Return

Whenever additional contributions and/or withdrawals are made within the measurement period with respect to the initial investment, two complementary measures for returns need to be calculated:

- The money-weighted rate of return (MWRR), which expresses the actual return earned on the assets invested during the measurement period.
- The time-weighted rate of return (TWRR), which expresses the return that could have been earned if no additional contributions and/or withdrawals had been made.

It is essential that both measures for returns are available, in order to meet various information requirements.[12]

The first measure (MWRR), since it also accounts for the impact generated by new contributions and withdrawals, makes it possible to understand what return was actually achieved by a portfolio characterised by cash flows within a given time period. It is therefore an actual *ex post* return, incorporating both the return generated by the asset manager and the return generated by cash flows in the portfolio through subsequent contributions and withdrawals.

The second measure (TWRR), disregarding the impact generated by new contributions and withdrawals, makes it possible to isolate only that portion of return attributable to the manager's activity in the reference time period. It is therefore a hypothetical *ex post* return, expressing the return which would have been obtained if no new contributions and withdrawals had been made in that given period. As it is unaffected by the timing of cash flows in the portfolio, this measure is the only one which can be compared to a benchmark indicator, in order to give an initial verdict of the actual quality of the asset manager's work (Box 8.1).

Box 8.1 The Difference Between Money-Weighted and Time-Weighted Return

Assume that portfolio P is fully and perfectly indexed to a given benchmark B, to the extent that the portfolio return (r_p) and the index return (r_b) are equal ($r_p = r_b$) in every period. Take a 1-year period and assume an investment of 100,000 euros at the start of the period. Assume also that:

- The benchmark achieved a return of +50% in the first 6-month period, at which point another 100,000 euros were invested.
- In the subsequent 6-month period, the benchmark recorded a return of −30% and there was no further movement.

(continued)

[12] Feibel (2003).

Box 8.1 (continued)

The initial investment, the subsequent contribution, and the assumption of perfectly index-linked management result in a year-end portfolio value of 175,000 euros: the 100,000 euros initially invested provided a 50% return in the first 6 months, becoming 150,000 euros; at that point, another 100,000 euros were contributed and, after this contribution, the portfolio value rose to 250,000 euros; in the second 6-month period, the benchmark and therefore the portfolio returned −30%, bringing the total value down to 175,000 euros, with a final loss of 25,000 euros compared to the overall capital invested of 200,000 euros (100,000 from initial investment and 100,000 invested at the end of the first 6 months).

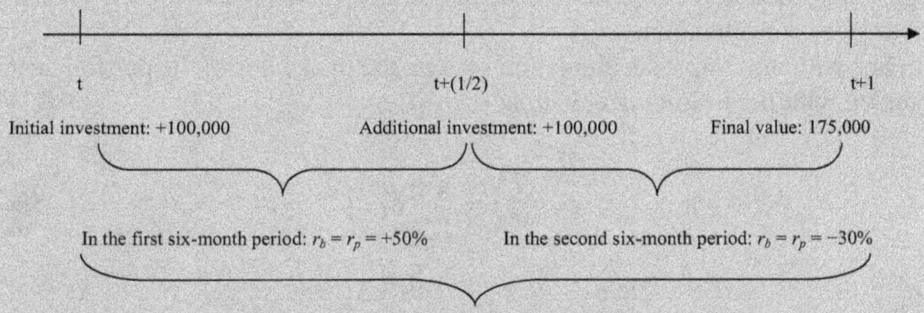

The return can be correctly represented by means of the geometric return on the portfolio over the entire year, which, given the assumptions made above, is consistent with the geometric return of the benchmark:

$r_p = r_b = (1 + 0.5) \cdot (1 - 0.3) - 1 = +0.05 = +5\%$

This return was therefore equal to +5%. So with a positive return of +5%, why did the final value of the portfolio fall to 175,000 euros in the space of a year, with a total investment of 200,000 euros?

The answer lies in the fact that two different measures for returns had to be calculated every time there were cash flows within the reference time period:

– A money-weighted rate of return, which shows what the actual return on the portfolio was, taking into account contributions and withdrawals.

– A time-weighted rate of return, which expresses the potential return that the portfolio would have generated if no contributions and withdrawals had been made.

Specifically:

– The actual return of the portfolio is measured by a money-weighted rate of return equal to −16.67%, as we will learn shortly, which corresponds to the ratio between the investment gain/loss of the portfolio in the measurement period (−25,000 euros) and the average amount of capital invested, equal to 150,000 euros, obtained by considering that the first 100,000 euros were invested for the entire year and the following 100,000 euros for only 6 months $(150{,}000 = 100{,}000 \cdot 1 + 100{,}000 \cdot 0.5)$.

– The potential return over the entire period is obtained by geometrically linking the returns over the sub-periods and is equal to +5%. Only this return (and not the previous one) can be compared to the benchmark for an initial evaluation of the relative return on the portfolio with respect to the progress of the same reference index.

8.2.3.1 Calculating Money-Weighted Return

The MWRR accounts for the cash flows occurring in the reference period and the time at which they took place. In other words, the MWRR is affected by contribution and withdrawal decisions made independently by the client during the period of measurement. A new contribution before a rise, just like a withdrawal before a decline, leads to an increase in return as a result of effective timing. Conversely, a new contribution before a decline, just like a withdrawal before a rise, leads to a reduction in the return as a result of incorrect timing.[13]

To calculate the money-weighted return, it is necessary to find the interest rate that makes the final value of the portfolio at the end of the period equal to its initial value and to all the other contributions and/or withdrawals made during the period of measurement, by ensuring that each contribution and/or each withdrawal is given a weight proportional to the period between the time it was made and the end of the period under consideration.

In line with the simple accumulation system, the final value of the portfolio can be expressed with the following equation:

$$V_T = V_0 \cdot \left(1 + r_{\mathrm{mw}} \cdot \frac{T}{T}\right) + \sum_{t=1}^{T} F_t \cdot \left(1 + r_{\mathrm{mw}} \cdot \frac{T-t}{T}\right)$$

with:

V_T	= final value of the portfolio
V_0	= initial value of the portfolio
F_t	= inflows (contributions) and outflows (withdrawals) concerning the portfolio
T	= duration of the period expressed in days
T - t	= number of days between the moment in which cash flows occur and the last day of the reference period for performance measurement
r_{mw}	= rate of return sought

If compound interest is adopted, on the other hand, the final value of portfolio is obtained using the following equation:

$$V_T = V_0 \cdot (1 + r_{\mathrm{mw}})^{\frac{T}{T}} + \sum_{t=1}^{T} F_t \cdot (1 + r_{\mathrm{mw}})^{\frac{T-t}{T}}$$

If continuous interest is used instead, the final value of portfolio is obtained as follows:

$$V_T = V_0 \cdot e^{r_{\mathrm{mw}} \cdot \frac{T}{T}} + \sum_{t=1}^{T} F_t \cdot e^{r_{\mathrm{mw}} \cdot \frac{T-t}{T}}$$

Recourse to simple interest is implicit when calculating a return measure known as the modified Dietz money-weighted rate of return (MDMWRR), while compound

[13] De Fusco et al. (2007).

interest (and more rarely, continuous interest) is implicit in the calculation of the money-weighted internal rate of return (MWIRR).[14]

At an international level, simple interest—and therefore the MDMWRR—is preferred, since calculating the MWIRR needs a function with a degree higher than the first order, thereby involving an iterative process in its solution which can give multiple results in the presence of cash flows with dissimilar terms of sign.

The MDMWRR is calculated in the following three different stages:

– The first stage involves calculating the investment result for the period (IR), equal to the difference between the final value and the initial value of the portfolio, taking into account cash flows (F_t) occurring during the measurement period, adding withdrawals (R_t) and deducting contributions (C_t) made in that period:

$$\mathrm{IR}_{0-T} = V_T - V_0 - \sum_{t=1}^{T} F_t = V_T - V_0 + \sum_{t=1}^{T} W_t - \sum_{t=1}^{T} C_t$$

– The second stage involves calculating the average capital employed (ACE), or rather the average invested balance:

$$\mathrm{ACE}_{0-T} = V_0 \cdot \frac{T}{T} + \sum_{t=1}^{T} F_t \cdot \frac{T-t}{T} = V_0 - \sum_{t=1}^{T} W_t \cdot \frac{T-t}{T} + \sum_{t=1}^{T} C_t \cdot \frac{T-t}{T}$$

– The third stage produces a percentage return measure by dividing the investment result by the average capital employed in the period referred to:[15]

[14] This is a modified version of the calculation approach proposed originally by Peter Dietz, in which, for simplicity's sake, all contributions and withdrawals are assumed as having occurred halfway through the period analysed. Dietz (1966) and Bacon (2023).

[15] Starting from the final value of the portfolio:

$$V_T = V_0 \cdot \left(1 + r_{\mathrm{mw}} \cdot \frac{T}{T}\right) + \sum_{t=1}^{T} F_t \cdot \left(1 + r_{\mathrm{mw}} \cdot \frac{T-t}{T}\right)$$

and developing the equation, we find that:

$$V_T - V_0 - \sum_{t=1}^{T} F_t = V_0 \cdot r_{\mathrm{mw}} \cdot \frac{T}{T} + \sum_{t=1}^{T} F_t \cdot r_{\mathrm{mw}} \cdot \frac{T-t}{T}$$

$$V_T - V_0 - \sum_{t=1}^{T} C_t + \sum_{t=1}^{T} W_t = V_0 \cdot r_{\mathrm{mw}} \cdot \frac{T}{T} + \sum_{t=1}^{T} C_t \cdot r_{\mathrm{mw}} \cdot \frac{T-t}{T} - \sum_{t=1}^{T} W_t \cdot r_{\mathrm{mw}} \cdot \frac{T-t}{T}$$

therefore:

$$r_{\mathrm{mw}} = \frac{V_T - V_0 - \sum_{t=1}^{T} C_t + \sum_{t=1}^{T} W_t}{V_0 + \sum_{t=1}^{T} C_t \cdot \frac{T-t}{T} - \sum_{t=1}^{T} W_t \cdot \frac{T-t}{T}} = \frac{\mathrm{IR}_{0-T}}{\mathrm{ACE}_{0-T}}$$

$$\text{MWRR} = \frac{\text{IR}_{0-T}}{\text{ACE}_{0-T}}$$

The return obtained in this way represents a periodic rate, which can be converted on an annual basis according to the conversion rules examined above.

8.2.3.2 Calculating Time-Weighted Return

The time-weighted rate of return makes it possible to obtain a return indicator which is unaffected by the impact of withdrawals and contributions which cannot be attributed to the asset manager (and therefore to the investment policy). By comparing the time-weighted portfolio return and the benchmark return, we can begin to assess the quality of the asset manager's work, however imprecise this may be, due to the lack of consideration of the level of risk.

The TWRR is calculated in three subsequent steps:

- The first step breaks the period under examination into the various sub-periods, ranging from the investment of the initial assets, to all new contributions and/or withdrawals, until the end of the period under consideration.[16]
- The second step involves calculating the return (r_t) for each of the sub-periods identified in the previous step, obtained by comparing the value of the portfolio at the end of the sub-period (FV_t) with the value of that portfolio at the beginning (IV_t):

$$r_t = \frac{\text{FV}_t}{\text{FV}_{t-1} + F_t} - 1 = \frac{\text{FV}_t}{\text{IV}_t} - 1$$

where FV_t and FV_{t-1} represent the value of the portfolio at the end of each sub-period and of the previous one, respectively, i.e. before each withdrawal or contribution (F_t) is made.

- The third step gives us the TWRR by geometrically linking the returns for the individual sub-periods:

$$\text{TWRR} = (1 + r_1) \cdot (1 + r_2) \cdot \ldots \cdot (1 + r_T) - 1$$

The time-weighted return thus obtained also represents a periodic rate, which can be converted on an annual basis according to the conversion rules examined previously.

[16] The number of significant sub-periods is equal to the number of withdrawals and contributions occurring during the period, increased by one to account for the initial value of the portfolio. By way of example, if two contributions and one withdrawal took place during the period under analysis, there would be four significant sub-periods.

When calculating the TWRR, the mark-to-market value of the portfolio must be available a moment before each cash flow; this input is needed to calculate returns for each sub-period. If this input is not available—typical of private equity and real estate investments—it is necessary to adopt alternative procedures allowing you to identify a TWRR proxy.[17]

Box 8.2 provides a money-weighted and time-weighted return calculation following the steps described above.

Box 8.2 Calculating Money-Weighted and Time-Weighted Returns

Consider a discretionary mandate which began on 1st July 2024 with an investment of 100,000 euros and which was subjected to two additional cash flows before the end of the 6-month period: a contribution of 80,000 euros, reported on 1st September 2024, and a withdrawal of 50,000 euros, occurring on 16th October 2024.

On 1st September 2024, the value of the portfolio before the contribution was 122,000 euros, while on 16th October, before the withdrawal, it was 180,000 euros.

On 31st December 2024, the date on which the client received the management report, the assets under management amounted to 125,000 euros.

Below is a table summarising cash flows:

Date	Number of days until the end of the period	Value of the portfolio before contributions or withdrawals	Contributions (+) and withdrawals (−)	Value of the portfolio after contributions or withdrawals
01/07/2024	184	100,000		100,000
01/09/2024	122	122,000	+80,000	202,000
16/10/2024	77	180,000	−50,000	130,000
31/12/2024		125,000		125,000

What was the money-weighted rate of return? And what was the time-weighted rate of return?

Money-weighted return is calculated in three steps:

1. Calculation of the investment result of the period:

$\text{IR} = \text{FV} - \text{IV} + W - C = 125{,}000 - 100{,}000 + 50{,}000 - 80{,}000 = -5{,}000$

2. Calculation of the average capital employed:

$\text{ACE} = 100{,}000 \cdot (184/184) + 80{,}000 \cdot (122/184) - 50{,}000 \cdot (77/184) = 132{,}119.57$

3. Calculation of the money-weighted return:

$\text{MWRR} = \text{IR}/\text{ACE} = -5{,}000/132{,}119.57 = -3.78\%$

Time-weighted return is calculated in three steps, as set out below:

(continued)

[17] One possibility is to calculate money-weighted returns for each sub-period—using the modified Dietz approach MWRR or, alternatively, the MWIRR approach—and then geometrically link the returns obtained in that way over the sub-periods. Feibel (2003).

Box 8.2 (continued)

1. We need to identify the sub-periods, which in this case are three and range:
   - From the start of the 6-month period up to the first contribution.
   - From the first contribution up to the first withdrawal.
   - From the first withdrawal to the end of the 6-month period.
2. Returns over the sub-periods are calculated by comparing the value of the portfolio at the end with the value of the portfolio at the start of each sub-period:
   - First period: $r_1 = (FV_t/IV_t) - 1 = (122{,}000/100{,}000) - 1 = 22.00\%$.
   - Second period: $r_2 = (FV_t/IV_t) - 1 = (180{,}000/(202{,}000) - 1 = -10.89\%$.
   - Third period: $r_3 = (FV_t/IV_t) - 1 = (125{,}000/130{,}000) - 1 = -3.85\%$.
3. Lastly, it is necessary to geometrically link the returns for each sub-period in order to calculate the time-weighted rate of return for the measurement period:

$\text{TWRR} = (1 + r_1) \cdot (1 + r_2) \cdot \ldots \cdot (1 + r_T) - 1 = (1 + 22\%) \cdot (1 - 10.89\%) \cdot (1 - 3.85\%) - 1 = +4.53\%$

8.2.3.3 A Specific Time-Weighted Return Calculation Technique: The Unit Price Method

In the case of collective investment vehicles (CIVs) and other asset management products characterised by the value of the individual unit or share, time-weighted rate of return may be calculated more simply, using the so-called unit price method or shares method. This allows you to obtain a time-weighted rate of return in a simplified way, which is not affected by the subscription of new units and/or by the redemption of units subscribed previously. This is possible because, when new contributions and/or redemptions take place in the reference time period, it is not the individual value of the units or shares that changes, but the number of the units or shares. In fact, the value of each unit or share is unaffected by the cash flows occurring in the period and depends exclusively on the value of the assets under management at the valuation dates. On subscription, we need to calculate the number of units or shares corresponding to the initial investment, dividing the invested capital (net of any entry load, fees, and expenses) by the initial unit or share value (V_0/NAV_0). Then every time a further contribution and/or withdrawal takes place, the number of units or shares held by the investor changes, but with no change in the individual value of each unit or share.[18]

The unit price method is fairly simple to implement. First of all, the initial unit or share value must be defined, a value corresponding to the net asset value of the CIV at the initial subscription date (NAV_0). The return for the period is calculated simply by dividing the unit or share value at the end of the period under evaluation by the unit or share value at the start of the period: $\text{NAV}_T/\text{NAV}_0 - 1$. The rate thus obtained

[18] Bacon (2023).

can be annualised using the well-known conversion rules: an example of the method is given in Box 8.3.

Box 8.3 An Example of Application of the Unit Price Method

Consider an investment fund with the following characteristics:

– It is a no-load fund.

– 100,000 euros was invested on 1st July 2024 and the fund had a unit value of 10 euros on that date.

– A further 80,000 euros was invested on 1st September 2024, on which date each unit was worth 12.2 euros.

– 50,000 euros was withdrawn on 16th October 2024, when the unit value was 10.871 euros.

– The unit value on 31st December 2024 is 10.453 euro and this is the date when the investor decides to work out the return over the period.

The table below summarises the investment activity during the measurement period:

Date	Value of the portfolio before contributions and withdrawals	Contributions (+) and withdrawals (−)	Value of the portfolio after contributions and withdrawals	Value of each unit	Number of units
01/07/2024	100,000		100,000	10.000	10,000.00
01/09/2024	122,000	+80,000	202,000	12.200	16,557.38
16/10/2024	180,000	−50,000	130,000	10.871	11,957.98
31/12/2024	125,000		125,000	10.453	11,957.98

How can we calculate time-weighted rate of return in a simplified way?

In a CIV, given that the value of the unit or share exists, calculating time-weighted return is greatly simplified, because new contributions and withdrawals do not alter the net asset value (NAV) per unit or share, but merely change the overall number of units or shares held.

10,000 units were allocated against an initial investment of 100,000 euros. The number of units increases or decreases for each cash flow, according to the type of movement, but these operations do not impact the value of each unit.

Therefore, to calculate the TWRR, all we have to do is compare the unit value at the end of the period under evaluation (NAV_T) with the initial unit value (NAV_0):

$TWRR = NAV_T/NAV_0 - 1 = (10{,}453 - 10) - 1 = +4.53\%$

In this case, too, the return obtained expresses a rate over the period, which can be annualised, as necessary, using the well-known conversion rules.

8.2.4 *Gross and Net Return*

The inclusion or not of commissions and tax charges leads us to distinguish between:

- Gross return.
- Net return.

With regard to commissions, we need to distinguish between costs which represent direct burdens on the portfolio under management (performance and management fees, costs related to portfolio turnover, and all other charges borne by the asset management product) and costs borne directly by the investor (entry load, exit load, or switch load). While the former are already included in the measures of return calculated previously, since the portfolio value is already net of them, the latter must be expressly considered with a view to identifying an accurate and reliable performance measure, net of all commissions.

With reference to fiscal burden, the investment portfolios in the various countries are normally taxed on the income generated (or more rarely accrued) in a given time period and, in some cases, also on the value of the portfolio.[19]

Income taxation involves differentiated criteria depending on the earner category (natural person or company), the type of income, and the instrument category. This results in a certain room for optimisation of the fiscal burden by taking into account earner category, the corresponding tax position, and the type of instrument chosen. Taxation of asset value involves periodic charging of a tax linked to the size of the capital through rates which may be proportional, progressive, or regressive. In the case of tax burden, too, we must distinguish between those charges borne directly by the portfolio and drawn periodically out of the same, and those borne directly by the investor. While the former are already included in the measures for returns presented above, the latter must be expressly considered by the investor to identify a measure for after-tax return.

The issue of commissions and fiscal burden becomes clear when we compare the return on a given portfolio with the return of a reference market benchmark index. As the benchmark is a theoretical portfolio, it includes neither commissions of any type nor fiscal charges which may be envisaged for the different categories of investment instruments. There are two alternatives for making an economically rational comparison:

- Grossing up the performance of the portfolio under evaluation, in order to reinstate the return component which was reduced by commissions directly burdening the portfolio being managed and by any fiscal charges taken out of the same portfolio.
- Considering the net value of the benchmark performance, by attributing implicit costs, representative of commissions and any tax burden on the portfolio being managed.[20]

In the absence of such an adjustment, any comparisons would certainly be inappropriate and the conclusions reached undoubtedly misleading.

---

[19] More recently, fiscal charges have been introduced in some countries on the total value of transactions made on certain financial instrument categories, through the introduction of different forms of the so-called Tobin tax.

[20] While both solutions are possible in theory, considering the net value is much simpler from the practical point of view and is in many cases actually more rational than the grossing up approach.

8.3 Investment Portfolio Risk Measures

Having regard to the rational investor's risk aversion, a simple analysis of return is not sufficient to evaluate the performance of a given investment portfolio. What is needed is to supplement the evaluation with a measure able to express the level of risk, which can be used for comparison both with the risk of alternative investments and with the risk of a reference market benchmark index.

A large number of risk measures have been proposed over time, which can be divided into three different categories:

- Absolute risk measures.
- Asymmetric risk measures.
- Relative risk measures.

As we will see, these are not alternative but rather complementary measures that provide us with differing information, according to the objective pursued.

8.3.1 *Absolute Risk*

Traditionally, the risk of an investment is taken to be the variability of its result, which can be measured through the range of dispersion of its returns around the average return. If this is the meaning of risk we wish to espouse, the statistical indicators used to measure what is termed absolute risk are variance and standard deviation, which measure the variability of the returns of a portfolio around the average return. In doing so, these indicators deal with positive and negative deviations from the mean in the same way, meaning that uncertainty of result coincides with the total risk, whether it be upside or downside.

With an *ex post* approach and when the dispersion of portfolio returns is compared within the same time period, variance (σ^2) and standard deviation (σ) are calculated by dividing the sum of the squared deviations from the mean by the number of observations considered:

$$\sigma^2 = \frac{1}{n}\sum_{t=1}^{n}\left(r_{p,t} - \bar{r}_p\right)^2$$

$$\sigma = \sqrt{\sigma^2} = \sqrt{\frac{1}{n}\sum_{t=1}^{n}\left(r_{p,t} - \bar{r}_p\right)^2}$$

where:

$r_{p,t}$	= portfolio return in the sub-period t
$\overline{r}_p$	= average portfolio return in the period under analysis
n	= number of sub-periods considered

Note that by raising deviations from the mean to their squared values, the plus and minus signs are removed from deviations above or below the mean, thereby preventing any positive or negative deviations from the mean from cancelling each other out, and higher deviations are more than proportionally increased. However, this means that variance is expressed as a unit of measurement different from the mean and, in practice, makes standard deviation a more favourable risk indicator, since (being the square root of the variance) it is expressed in the same unit of measurement as the returns.

When the time period of the portfolios under evaluation is not the same or, in any case, when we wish to account for the fact that the period analysed represents a mere sample of the whole set of returns, we can use sample standard deviation, obtained by multiplying the standard deviation of the population by $n/(n - 1)$, even though the effect of such an adjustment obviously becomes increasingly irrelevant as the number of observations increases:[21]

$$\widehat{\sigma} = \sqrt{\widehat{\sigma}^2} = \sqrt{\frac{1}{n-1}\sum_{t=1}^{n}\left(r_{p,t} - \overline{r}_p\right)^2}$$

The main advantages of standard deviation lie in the fact that it is relatively easy to calculate and completely objective. Unlike other risk measures, standard deviation does not require clarification of a market portfolio or a proxy and therefore does not provide different valuations according to the reference passive portfolio (market portfolio or benchmark) with respect to which it is calculated.

On the other hand, standard deviation is adversely affected by two significant restrictions. First and foremost, it deals with upside and downside risk in the same way. In fact, upside risk is just as important as downside risk in determining deviations from the mean which are significant for its calculation, despite the fact that, in practical terms, it is only downside risk that is bad for the investor, while upside risk is good. From this point of view, the significance of standard deviation as a risk measure of a portfolio is based on the assumption of normal return distribution: in this case, since distribution is symmetrical (as well as mesokurtic), the classification of portfolios depending on total risk or downside risk is unaltered.[22]

---

[21] Note that the Excel function "=STDEV.S" calculates the sample standard deviation, while the population standard deviation is obtained with the function "=STDEV.P".

[22] Normal distribution is symmetrical, since the mean and median coincide, and mesokurtic, as it has zero excess kurtosis. For the concepts of skewness and kurtosis and for the verification of the assumption of normality of the distribution of returns, see the comments in Chap. 4 on strategic asset allocation, and especially Chap. 13 on the performance evaluation of hedge funds.

Standard deviation is also limited to calculating the dispersion of the returns around the average return, but, in certain situations, investors may be interested in the variability of the returns around a subjective return threshold in line with the objectives of their own investment policy or, alternatively, around a maximum expected loss.[23]

8.3.2 Asymmetric Risk

When the distribution of *ex post* returns is not symmetrical, standard deviation does not correctly measure the actual risk of a portfolio, as it does not distinguish between deviations above or below the mean. In the case of asymmetric distributions, the probability of having values below the mean is in fact higher or lower according to the type of asymmetry.[24]

The so-called downside variance and the corresponding square root, known as the downside deviation, are useful measures in these contexts, focusing solely on the downside risk.

Analytically, downside variance and downside deviation consider only the negative deviation values calculated with respect to a target return assumed by investors at their discretion based on the objectives of their own investment policy:[25]

$$\text{Downside variance} = \frac{1}{n-1} \sum_{\substack{t=1 \\ r_{p,t} < r_{\text{target}}}}^{n} \left(r_{p,t} - r_{\text{target}}\right)^2$$

$$\text{Downside deviation} = \sqrt{\frac{1}{n-1} \sum_{\substack{t=1 \\ r_{p,t} < r_{\text{target}}}}^{n} \left(r_{p,t} - r_{\text{target}}\right)^2}$$

In this case too, in practice, downside deviation is usually preferred to downside variance as it is expressed in the same unit of measurement as the returns.

[23] Lawton and Jankowski (2009) and Bodie et al. (2024).

[24] In the case of asymmetry towards the right (or positively skewed distribution), there are more returns on the right-hand side of distribution than there are in normal distribution. In this case, the mean is greater than the median, which in turn is greater than the mode. In this case of skewness, standard deviation overestimates the portion of returns below the mean and underestimates those above. In the case of skewness towards the left (or negatively skewed distribution) there are more returns on the left-hand side of distribution than in normal distribution. In this case, therefore, the mean is lower than the median, which is in turn lower than the mode. In this case, standard deviation underestimates the portion of returns below the mean and overestimates those above.

[25] When target return coincides with the mean of returns, the two indicators are called semi-variance and semi-standard deviation.

Downside deviation can be used to find a solution for the two shortcomings of standard deviation described above, accounting for any asymmetry in the distribution of the returns and calculating the dispersion of the same around a threshold return identified at the investor's discretion.

However, despite the potential advantages of the asymmetric risk indicators, the following considerations should be borne in mind:

- Standard deviation is simple to calculate and combine for a number of portfolios.
- Certain portfolios have a roughly symmetrical return distribution, which justifies the use of standard deviation.
- Even if returns distributions are asymmetrical, a portfolio ranking based on standard deviation might be similar to the one obtained with downside risk measures if the portfolios under evaluation have the same degree of skewness.
- Forward-looking risk, i.e., the expected risk of a portfolio is the main risk to measure, and not backward-looking risk, whether it be total or downside risk. Historical risk statistics become relevant only if they are useful for forecasting future risk. It should therefore come as no surprise that *ex post* total risk measures, such as the classic standard deviation, may be preferable to *ex post* downside measures when the former is a better estimation of the *ex ante* downside risk, which is after all the measure that really counts for the investor.[26]

8.3.3 Relative Risk

Relative risk measures compare the trend of a given investment portfolio with market performance, represented by an appropriate benchmark index, thereby making it possible to make a comparative assessment of the portfolio examined and a passive reference portfolio.

The following two main measures can be attributed to this category:

- Beta (β).
- Tracking error volatility (TEV).

Beta represents the quantification of systematic risk in a given portfolio. By identifying the behaviour of the stocks in the portfolio defined as aggressive ($\beta > 1$), neutral (β close to 1) or defensive ($\beta < 1$), it measures the portfolio's sensitivity to variations of a given benchmark index adopted as a proxy of the reference market. Having originated from the Capital Asset Pricing Model (CAPM), this risk measure is vulnerable to all the criticisms concerning validity which have been levelled at the CAPM theory and its capacity to represent an effective, realistic equilibrium and pricing model for the capital market. As is known, the most important concerns are:

[26] Rasmussen (2003) and Schulmerich et al. (2013).

- The nature of the market portfolio, understood as both the difficulty in concretely identifying its nature and the variability of results obtained, depending on which benchmark index is adopted to represent it in practice.
- The capacity of a single factor to explain the overall non-diversifiable risk component of a portfolio.
- The instability of beta over time, which makes it impossible to use past values to perform *ex ante* evaluation.[27]

In the case of a portfolio, the beta can either be calculated directly (by reconstructing the time series of the portfolio returns and calculating the covariance of that variable with market returns) or, if the betas of the individual stocks are already known, simply as the weighted average of the betas of the individual portfolio securities.

In the first case, the portfolio beta can be obtained through one of the alternative formulas set forth below:

$$\beta_p = \frac{\sigma_{p,m}}{\sigma_m^2} = \frac{\rho_{p,m}\ \sigma_m \sigma_p}{\sigma_m^2} = \rho_{p,m}\frac{\sigma_p}{\sigma_m}$$

where:

$\sigma_{p,m}$	= covariance between the returns of the portfolio and the returns of the market portfolio benchmark
σ^2_m	= variance of the returns of the market portfolio benchmark
σ_m	= standard deviation of the returns of the market portfolio benchmark
σ_p	= standard deviation of the returns of the portfolio examined
$\rho_{p,m}$	= correlation coefficient between the returns of the portfolio and the returns of the market portfolio benchmark

In the second case, the portfolio beta is estimated as a weighted average:[28]

$$\beta_p = \sum_{i=1}^{n} \beta_i \cdot x_i$$

where:

[27] Roll (1992, 1977), Ross (1977).

[28] Starting with the definition of the return of a portfolio, which, as we know, coincides with the weighted average of the returns of the individual stocks:

$$r_p = \sum_{i=1}^{n} r_i \cdot x_i$$

and continuing with the definition of portfolio beta as the weighted average of the betas of the individual stocks, taking into account the amount of each:

$$\beta_p = \sum_{i=1}^{n} \beta_i \cdot x_i$$

we can demonstrate the following relation:

β_i	= beta of the generic i-th security
x_i	= weight of the i-th security within the portfolio examined, expressed as a percentage

Below, using a combined analysis of beta and the stock picking evaluation measures, we shall be returning to the estimate of beta as the slope or gradient of the line of interpolation obtained by the regression of portfolio returns with respect to benchmark returns.

A further useful indicator to understand how the relative return on the portfolio with respect to the benchmark was obtained is the so-called tracking error volatility (TEV).[29] In order to make an initial assessment of the manager's level of activism, a comparison can be made between the historical returns of the portfolio and of the benchmark, and the relative return (also known as active return) calculated, a measure also known as tracking error (TE).

However, merely analysing the relative return does not explain how the performance was actually obtained, since a portfolio with zero active return may not be passive when compared to the benchmark, but may instead have performed better or worse than the benchmark in the reference period, so that the sum of the periodic active returns is zero. By way of an example, Table 8.3 below lists the monthly returns of the two portfolios A and B.[30] We can infer from these that while the trend of portfolio A was very similar to that of the benchmark, portfolio B deviated significantly from it. Despite this, the average monthly active returns (tracking error) and the relative annualised figure are identical. This is because in the case of portfolio B, positive and negative deviations from the benchmark cancelled each other out.

In order to give an accurate assessment of the manager's degree of loyalty to the reference benchmark, it is essential to calculate the volatility of the tracking error, i.e. the standard deviation of the difference between the portfolio return and the benchmark return: known as tracking error volatility (or TEV), this measure accounts for deviations from the benchmark, whether these be positive or negative,

---

$$\sum_{i=1}^{n}\beta_i \cdot x_i = \sum_{i=1}^{n}\frac{\sigma_{i,m}\cdot x_i}{\sigma_m{}^2} = \frac{\sum_{i=1}^{n} x_i \sum_{t=1}^{T}(r_{i,t}-\bar{r}_i)(r_{m,t}-\bar{r}_m)\frac{1}{T}}{\sigma_m{}^2} = \frac{\sum_{i=1}^{n}\sum_{t=1}^{T} x_i(r_{i,t}-\bar{r}_i)(r_{m,t}-\bar{r}_m)\frac{1}{T}}{\sigma_m{}^2} =$$

$$= \frac{\sum_{t=1}^{T}\left(\sum_{i=1}^{n} x_i r_i - \sum_{i=1}^{n} x_i \bar{r}_i\right)(r_{m,t}-\bar{r}_m)\frac{1}{T}}{\sigma_m{}^2} = \frac{\sum_{t=1}^{T}(r_{p,t}-\bar{r}_p)(r_{m,t}-\bar{r}_m)\frac{1}{T}}{\sigma_m{}^2} = \frac{\sigma_{p,m}}{\sigma_m{}^2} = \beta_p$$

[29]The terminology proposed by Roll (1992) is used here, so that the difference between portfolio return and benchmark return is defined as tracking error and the standard deviation of tracking error is defined as tracking error volatility. It should be noted, however, that in practice, tracking error is often used as a synonym of tracking error volatility.

[30]For the sake of simplicity, the example assumes use of monthly data and a time period of just 12 months. Actually, with monthly data, a time period of at least 36 months should be used.

Table 8.3 A TE and TEV calculation example

	Portfolio A monthly return	Portfolio B monthly return	Benchmark monthly return	Portfolio A monthly tracking error	Portfolio B monthly tracking error
	a (%)	b (%)	c (%)	d = a – c (%)	e = b – c (%)
Month 1	5.36	6.37	5.25	0.11	1.12
Month 2	3.19	0.70	3.24	−0.05	−2.54
Month 3	−2.38	−3.08	−2.57	0.19	−0.51
Month 4	2.21	4.53	2.24	−0.03	2.29
Month 5	−3.64	−3.66	−3.45	−0.19	−0.21
Month 6	1.62	6.68	1.49	0.13	5.19
Month 7	4.48	3.11	4.56	−0.08	−1.45
Month 8	3.21	0.20	3.25	−0.04	−3.05
Month 9	−1.82	−3.41	−1.96	0.14	−1.45
Month 10	−2.21	−6.12	−2.14	−0.07	−3.98
Month 11	3.48	5.56	3.61	−0.13	1.95
Month 12	1.61	4.23	1.69	−0.08	2.54
		Monthly TE		−0.0083	−0.0083
		Annualised TE		−0.10	−0.10
		Monthly TEV		0.12	2.69
		Annualised TEV		0.42	9.33

and enables us to make an exhaustive, reliable assessment of the loyalty of the manager with respect to the selected benchmark.

In analytical terms, TEV is equal to:

$$\text{TEV} = \sigma(\text{TE}) = \sqrt{\frac{1}{n-1}\sum_{t=1}^{n}\left[\left(r_{p,t} - r_{b,t}\right) - \overline{\text{TE}}\right]^2}$$

where:

$(r_{p,\,t} - r_{b,\,t})$ = active return, i.e. the difference between the returns of the portfolio and those of the benchmark in each sub-period:

$$\overline{\text{TE}} = \text{tracking error} = \text{average active return over the entire period} = \\ = \frac{1}{n}\sum_{t=1}^{n}\left(r_{p,t} - r_{b,t}\right).$$

If tracking error volatility is high, then the asset manager has adopted an active strategy: his/her investment choices have differentiated the managed portfolio from the composition of the benchmark. If tracking error volatility is low, then the manager has adopted a passive strategy: he/she has invested in a portfolio very similar to the one making up the index, or one which fairly closely replicates its performance over time, even though it may be composed of different stocks. For this reason, the TEV of an index fund is extremely low when calculated with respect to the index to which the portfolio is connected.[31]

A relatively "high" TEV does not necessarily imply a high level of risk-taking by the asset manager, but rather a lower level of fidelity with respect to the chosen benchmark. In order to understand this, we have only to take the example of an index portfolio with a rather low TEV due to its accurate replication of a global equity market index. If fifty per cent of that portfolio were to be converted into cash at any given time, thereby reducing the proportion of the global equity investment, the total risk represented by the standard deviation would decrease, but, at the same time, the relative risk represented by TEV would increase. To sum up, TEV is an indicator of relative risk and may be used to assess the manager's degree of conformity to the chosen benchmark, but it should not be used to measure either absolute risk or downside risk.

TEV is the main relative risk indicator used for management purposes and is often one of the constraints imposed *ex ante* on asset managers for the fulfilment of their managerial mandate. In this case, the manager's objective becomes that of optimising the return difference with respect to the benchmark—the so-called tracking error or active return—in compliance with the restrictions imposed in terms of tracking error volatility.[32]

8.4 Risk-Adjusted Performance Measures

For each of the risk measures described above, there is at least one corresponding risk-adjusted performance indicator which is able to combine the risk-return on the portfolio under evaluation into one single value. In reality, the last 50 years have seen increasing proliferation of these indicators, which are often mere variants of the original measures.[33] Below, we examine the main risk-adjusted performance indicators for traditional investment portfolios, while the measures used for alternative investments will be discussed in Chap. 13. Where possible, a common model will be used to analyse the indicators which covers the following points:

- Formal definition of the indicator.
- Economic meaning.

---

[31] TEV is never zero, even when management is perfectly index-linked, due to various disturbances that make it impossible to replicate the index perfectly, such as: periodic cash flows allocated from securities included in the index and not reinvested immediately, portfolio liquidity, commissions and transaction costs, fiscal charges, etc.

[32] It is worth pointing out that if the reference benchmark were not mean-variance efficient, implementing this type of approach would lead to the systematic construction of inefficient portfolios dominated by other feasible portfolios. In this regard, see Roll (1992).

[33] Le Sourd (2007) and Fischer and Wermers (2013).

- Algebraic interpretation.
- Geometric meaning.
- Problems of interpretation.

Depending on the particular situation, the various risk-adjusted performance indicators refer either to the risk-free rate, or to a reference market benchmark index, or to a target rate of return for the investor. The risk-free rate is actually estimated using a money market index for short-term government bonds or an interbank rate on short-term deposits. Depending on the particular case, the benchmark index may be the one declared by the asset manager, the one which is better able to explain the past performance of the portfolio and attained through regression analysis (best fit benchmark) or the one obtained through a style model, i.e. the analysis of the investment style of the portfolio under evaluation (customised style benchmark), a topic examined in more detail in Chap. 9. Target return, on the other hand, is the objective, at the investor's discretion, around which the risk of not meeting expectations is calculated.

The length of the time series used is a significant factor in the calculation of the various risk-adjusted performance measures, and while there is general agreement about the need to use a series of returns which covers all market phases, there is less agreement on the type of data to use (weekly or monthly returns) and on the time period to adopt (number of years). In this respect, it is useful to note how companies specialised in the evaluation of collective investment vehicles generally use monthly data over time periods of 3, 5, or 10 years, which may vary according to the current life of the portfolio under evaluation.

The building-up of a uniform peer group composed of portfolios all managed in the same way is also significant for the comparative use of the various indicators.

A first approach could be to build peer groups based on the benchmark declared by the manager, or with reference to the official classification for certain asset management product categories. In this case, however, there is the risk that the group of portfolios analysed may not actually present uniform investment policies, due to both potential ambiguity in the institutional classification itself, which groups products with potentially different investment policies together in the same category, and to potential opportunistic behaviour by the asset manager who may adopt an investment policy in certain market situations that does not comply with the declared benchmark or with the appropriate institutional category. In actual fact, therefore, this first approach poses the risk of creating irregular peer groups due to an initial misclassification causing portfolios to be considered similar even when they are actually very different from each other in terms of actual investment policies and resulting risk-return combinations.[34]

A second approach—followed by companies specialised in the evaluation of asset management products—involves creating peer groups based on the "investment style" actually adopted through a deductive method, based on Sharpe's

---

[34] DiBartolomeo and Witkowski (1997).

regression style model.[35] Based on this reasoning, the actual investment policy of the portfolio is inferred by a regression analysis of the returns of the fund and the returns of a series of benchmark indices representative of the entire market, setting the weights of the benchmarks within the portfolio as unknown and limiting regression to ensure that the sum of each asset class is equal to one (budget constraint), and that each asset class has an eligible value between zero and one (no short-selling constraint). This approach is discussed in Chap. 9, but we now need to consider its importance in building homogenous peer groups so that the risk-adjusted performance of portfolios with effectively similar risk-return combinations can be compared.

In line with the aim of evaluating past performance, the analysis below will refer mainly to the calculation and use of risk-adjusted performance measures in an *ex post* approach.

8.4.1 The Sharpe Ratio

The best-known and most widely used risk-adjusted performance measure is the Sharpe ratio, which measures the excess return of a portfolio per unit of total risk over a given time period.[36] The Sharpe ratio—originally known as the reward-to-variability ratio—is obtained from the ratio between the excess return of a portfolio and the standard deviation of its returns, where excess return is understood as the difference between the average return on the portfolio and the risk-free rate. In analytical terms, the Sharpe ratio is defined thus:

$$\text{Sharpe ratio}_p = \frac{\overline{r}_p - r_f}{\sigma_p}$$

where:

$\overline{r}_p$	= average portfolio return
r_f	= risk-free rate[37]
σ_p	= standard deviation of the portfolio returns

[35] Sharpe (1992).

[36] Sharpe (1966, 1994).

[37] The risk-free rate is not calculated as a mean return, because the assumed absence of risk suggests that the standard deviation is zero and that the rate is therefore constant. Actually, since the risk-free rate is estimated using money market indices with non-zero standard deviation, it is worth thinking in terms of mean return of this proxy of the risk-free rate, but it is just as correct in this case to use the standard deviation of the excess returns of the portfolio calculated with respect to the same risk-free rate as the denominator in the Sharpe ratio.

The Sharpe ratio makes it possible to compare different investments, defining their relative degree of efficiency. The portfolio with the highest Sharpe ratio is the one able to produce the greatest excess return per unit of total risk: it is therefore the most efficient portfolio and the best option with respect to the others.

Let us consider two portfolios A and B, both with the same characteristics:

Portfolio	Return (%)	Std. dev. (%)	Sharpe ratio
A	12	10	1
B	16	20	0.70
Benchmark	14	15	0.80
Risk-free	2	0	–

The Sharpe ratio of portfolio A is 1 = (12% – 2%)/10%, while that of portfolio B is 0.70 = (16% – 2%)/20%: despite having a lower return than portfolio B, portfolio A is the more efficient, since it achieves a higher return per unit of total risk.

In geometric terms, the Sharpe ratio may be interpreted as the slope of a straight line which, in a two-dimensional *ex post* return-standard deviation space, has the risk-free rate as the y-intercept and connects the return of the risk-free asset to the average return-standard deviation combination of the portfolio under evaluation (Fig. 8.1).

Each point on the line thus traced may be interpreted as a combination, in different proportions, of risk-free and risky assets. Each portfolio along that line is available to the investor who, according to his/her own risk appetite, will decide how to combine risk-free and risky assets. In analytical terms, the Sharpe ratio is derived from the *ex post* Capital Market Line (CML) of the CAPM, or rather from all the optimal risk-return combinations which can be obtained when it is possible to invest or run into debt at the risk-free rate.[38]

When selecting the optimal portfolio, the investor has no need to be concerned about the level of absolute risk characterising each portfolio, or to eliminate it from the set of the possible choices in case of mismatch with his/her own preferences, since this level can be amended accordingly and brought back to the desired one by

[38] The *ex post* CML equation is set out below:

$$r_p = r_f + \frac{r_m - r_f}{\sigma_m} \cdot \sigma_p$$

Its slope is the Sharpe ratio of the market portfolio, which is indeed represented by an appropriate benchmark index. The portfolios lying above the *ex post* CML—and which as such have a higher Sharpe ratio—are more efficient in terms of risk-adjusted performance compared to the market portfolio. The portfolios lying below the *ex post* CML—which therefore have a lower Sharpe ratio—are less efficient. Note however that the ordinal capacity of the Sharpe ratio is independent of the clarification of the market portfolio: the most efficient portfolio is the one with the highest Sharpe ratio, regardless of the identification of a proxy of the so-called market portfolio.

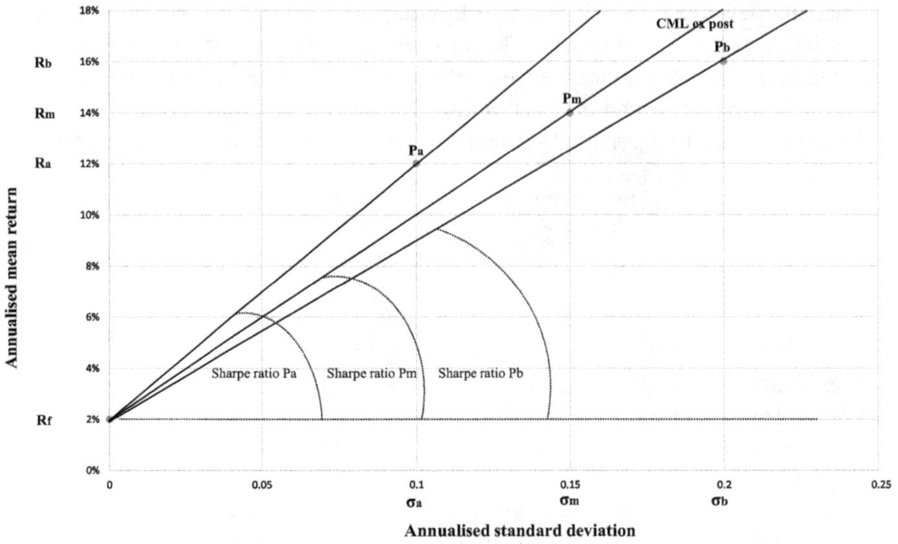

Fig. 8.1 Illustration of the Sharpe ratio

means of the so-called zero-investment strategy, which makes it possible to combine a risky asset with a risk-free one in order to obtain the desired level of risk.[39]

It should be noted that, because the Sharpe ratio does not require the identification of a market portfolio, it is not vulnerable to Roll's critique regarding the unobservable nature of the market portfolio, according to which the results of the equilibrium model are rather unstable when the benchmark adopted as the proxy of the market portfolio of the CAPM is changed.[40]

From a selection point of view, the rationale behind the use of the Sharpe ratio to evaluate the best portfolio is the combination of one single risky portfolio with the risk-free asset only. If more than one risky portfolio is owned overall, the Sharpe ratio would be inadequate, since it does not take into consideration the linear correlation coefficients (or the covariance) among the different portfolios, which

[39] The Sharpe ratio is used based on the assumption that it is possible to separate the identification of the most efficient portfolio available on the market from the influence exerted by the risk appetite of the investor. Use of financial leverage, borrowing or lending at the risk-free rate, makes it possible to reach the level of risk that is consistent with the degree of tolerance of the individual investor.

In the example above, if the investor had a maximum risk tolerance level of 5% in terms of standard deviation, portfolio A—with a standard deviation of 10%—would in any case be the preferable portfolio. In order to reduce the risk, it would be enough to allocate 50% of the portfolio to A and 50% to the risk-free asset. In this way, the resulting risk would be 5% and the return would be equal to $7\% = 2\% \cdot 0.5 + 12\% \cdot 0.5$. This choice is represented by the so-called zero-investment strategy, which allows combination of risky and risk-free assets in order to obtain the desired level of risk.

[40] Roll (1977).

would be essential in order to identify the overall risk of a portfolio made up of two or more risky assets.

The Sharpe ratio takes on a negative sign when the return obtained on the risky portfolio is lower than the risk-free rate. In this case, according to some, calculation and use of the index become insignificant, as it is no longer univocally correct to state that the portfolio with the higher (in this particular case, less negative) Sharpe ratio is preferable to an alternative portfolio with a lower (i.e. more negative) Sharpe ratio. This is because a less negative ratio may result both from a greater proximity of the return on the risky investment to the risk-free rate (and in this case the selection would be correct) or from a higher standard deviation (and in this case the selection would not be correct).

However, even in the case of a negative value, the Sharpe ratio remains valid as it identifies the portfolio which, by means of a suitable zero-investment strategy, delivers the highest return per unit of total risk faced. As proof of this, let us take two portfolios, C and D, with the following characteristics:

Portfolio	Return (%)	Std. dev. (%)	Sharpe ratio
C	1.5	10	−0.05
D	0.5	40	−0.0375
Risk-free	2.0	0	–

It may seem obvious that portfolio C dominates portfolio D, since it has a higher return and lower risk, but the Sharpe ratio of portfolio D is higher (−0.0375) than that of C (−0.05). However, if we carry over the standard deviation of portfolio D to that of portfolio C by combining D with a risk-free asset, we find that the return on the portfolio is equal to $1.625\% = 0.25 \times 0.5\% + 0.75 \times 2.0\%$, and, therefore, higher than the return of C.

Although negative, in an *ex post* CML context from which it originates, the Sharpe ratio therefore maintains its validity and capacity to identify the most efficient asset manager, who is able to generate the highest excess return, even if less negative, per unit of total risk.[41]

8.4.2 The Modigliani RAP

In 1997, Franco and Leah Modigliani created an additional risk-adjusted return measure, which they named the Risk-Adjusted Performance (RAP) index, but which was later renamed M^2 by the market, in honour of the two authors.

[41] The role played by the risk-free rate level in defining the ranking based on this indicator is also evident, especially where financial assets subject to analysis have average returns that are close to the return of the risk-free asset.

The Modigliani RAP index expresses the return that the portfolio would have achieved if its level of absolute risk had been identical to that of the market, represented by an appropriate benchmark index.[42]

In analytical terms, RAP is formulated as follows:

$$\mathrm{RAP}_p = (\sigma_m/\sigma_p) \cdot (r_p - r_f) + r_f$$

Through the RAP index, it is possible to compare different investments and define their degree of relative efficiency. The portfolio with the highest RAP is the most efficient and is preferable to all the others. The idea underlying this measure is to use the zero-investment strategy presented in the section above to calculate the return that each risky portfolio would have attained if its risk level had been identical to that of the market (σ_m), represented through an appropriate benchmark index.

Just like in the Sharpe ratio, the decision as to which risky portfolio to hold is seen as separable from the decision as to how much risk to assume in the RAP index. The operation allowing us to establish this equivalence is leverage, which can be positive or negative.[43] It takes the positive sign if the risk of the portfolio examined is lower than the market risk; in this case, in order to establish an equivalency, it is necessary to get into debt at the risk-free rate and allocate the funding received in the investment to the risky assets. Conversely, it takes the negative sign if the portfolio being examined presents a risk that is higher than that of the market; in this case, a portion of the assets invested in the risky portfolio must be sold and the proceeds used to invest in the risk-free asset.[44]

Once the return on each portfolio has been adjusted (RAP_p) according to the risk recorded on the market, it is possible to compare it with the benchmark return, which coincides with the RAP of the benchmark ($\mathrm{RAP}_m = r_m$).[45] If the RAP of the portfolio is higher than the market RAP, i.e. ($\mathrm{RAP}_p - r_m) > 0$, then we can infer that the portfolio was more efficient than the benchmark in terms of risk-adjusted performance. If the RAP of the risky portfolio is lower than the market RAP, i.e. ($\mathrm{RAP}_p - r_m) < 0$, then we can infer that the portfolio is less efficient than the benchmark in terms of risk-adjusted performance.

By way of example, let us consider two portfolios, A and B, with the same characteristics:

[42] Modigliani and Modigliani (1997).

[43] The degree of financial leverage (L_p) required to attain this equivalency can be identified using the following equation:

$$(1 + L_p) \cdot \sigma_p = \sigma_m$$

where L_p is the unknown. Solving the equation, we obtain $L_p = \sigma_m/\sigma_p - 1$.

[44] Sharpe et al. (1999).

[45] As we can easily demonstrate, $\mathrm{RAP}_m = (\sigma_m/\sigma_m) \cdot (r_m - r_f) + r_f = r_m$

Portfolio	Return (%)	Std. dev. (%)
A	12	10
B	16	20
Benchmark	14	15
Risk-free	2	0

In the case of portfolios A and B, the respective RAP indices are equal to:

$$\text{RAP}_a = \sigma_m/\sigma_a \cdot (r_a - r_f) + r_f = (15\%/10\%) \cdot (12\% - 2\%) + 2\% = 17\%$$
$$\text{RAP}_b = \sigma_m/\sigma_b \cdot (r_b - r_f) + r_f = (15\%/20\%) \cdot (16\% - 2\%) + 2\% = 12.5\%$$

In Fig. 8.2, by measuring standard deviation on the horizontal axis and the *ex post* return on the vertical axis, any portfolio can be represented with coordinates (σ_p; r_p). Point P_m identifies the market portfolio with standard deviation σ_m and return r_m.

The lines passing through point r_f and any other portfolio P identify the point where portfolio P itself can move using financial leverage. Calculation of the Modigliani RAP index assumes that, with a portfolio P characterised by return r_p and degree of variability σ_p, it is possible to build a version of this portfolio with a level of risk equivalent to that of the benchmark, in such a way, therefore, as to ascertain the equivalence $\sigma_p = \sigma_m$.

Each point on the line to the right of the starting portfolios P_a and P_b implies positive leverage, while each point to the left of the starting portfolios P_a and P_b implies negative leverage: in order to calculate the Modigliani RAP index, we need only follow each of the lines joining the risk-free rate to each portfolio until they intersect with the vertical line stemming from σ_m and then read the corresponding return on the vertical axis of the graph.

The difference between the RAP of the portfolio in question and the return on the market portfolio (which is the same as the RAP of the market portfolio) allows us to see how the portfolio would have over- or under-performed the market benchmark if it had assumed the same level of absolute risk as the reference market:

$$\text{RAP}_a - r_m = 17\% - 14\% = 3\%$$
$$\text{RAP}_b - r_m = 12.5\% - 14\% = -1.5\%$$

In the illustration provided (Fig. 8.2), while the return on portfolio A is lower than that of the market, its risk-adjusted performance (RAP_a) is higher ($\text{RAP}_a > \text{RAP}_m \equiv r_m$). This means that, with respect to benchmark risk, the investor is more than adequately compensated for the risk taken with portfolio A. Conversely, the return on portfolio B is higher than that of the market, but its risk-adjusted performance (RAP_b) is lower ($\text{RAP}_b < \text{RAP}_m \equiv r_m$).

The strength of the Modigliani RAP lies in its ease of interpretation: while the Sharpe ratio represents the slope of the line which joins the risk-free rate to the risk-return combination of the portfolio under evaluation, the RAP shows a result as a percentage, expressed in terms of risk-adjusted performance, highlighting the

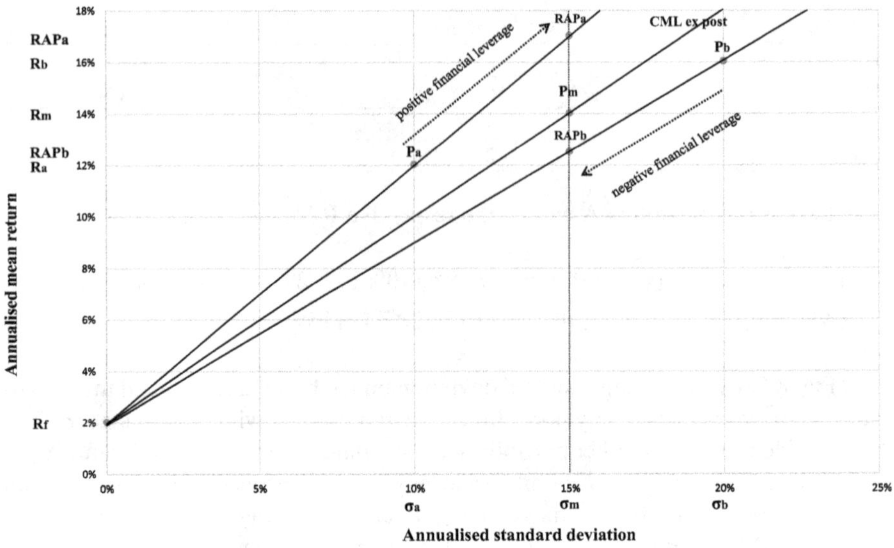

Fig. 8.2 Illustration of the Modigliani RAP

difference between the return on the portfolio evaluated and the performance of the benchmark, where the risk taken is the same.

If the portfolios subject to evaluation, as is logical, make up a common peer group and share the same benchmark, then the classifications obtained with the Sharpe ratio and with the Modigliani RAP are identical, making the second indicator—having established the assumption of a common benchmark—a linear transformation of the first, through addition of a constant (r_f) and a common multiplier (σ_m):

$$\text{RAP}_p = r_f + \left[(r_p - r_f)/\sigma_\text{p}\right] \cdot \sigma_m$$

$$\text{RAP}_p = r_f + \text{Sharpe ratio}_p \cdot \sigma_m$$

From a selection point of view, the assumption rationalising the use of the Modigliani RAP to evaluate the best portfolio is the combination of a single risky portfolio with the risk-free asset only. If more than one risky portfolio is owned overall, the RAP would also be inadequate (just like the Sharpe ratio), since the correlation coefficients or covariances among the various portfolios are not taken into consideration.

Lastly, by way of example, it is useful to reconsider the two portfolios C and D, and also add their benchmark data:

Portfolio	Return (%)	Std. dev. (%)	Sharpe ratio (%)
C	1.5	10	−0.05
D	0.5	40	−0.0375
Benchmark	1.1	20	−0.045
Risk-free	2.0	0	–

The RAP of the two portfolios C and D, the benchmark risk being known, is equal to:

$$\mathrm{RAP}_c = (\sigma_m/\sigma_c) \cdot (r_c - r_f) + r_f = (20\%/10\%) \cdot (1.5\% - 2\%) + 2\% = +1\%$$

$$\mathrm{RAP}_d = (\sigma_m/\sigma_d) \cdot (r_d - r_f) + r_f = (20\%/40\%) \cdot (0.5\% - 2\%) + 2\% = +1.25\%$$

If we carry over the standard deviation of the two portfolios C and D to that of the benchmark and compare the RAP of both with the return of the same index, we can infer that portfolio C is less efficient than the benchmark in terms of risk-adjusted performance, unlike portfolio D, which outperforms the benchmark in terms of risk-adjusted performance:

$$\mathrm{RAP}_c - r_m = 1\% - 1.10\% = -0.10\%$$

$$\mathrm{RAP}_d - r_m = 1.25\% - 1.10\% = +0.15\%$$

This example is also useful to highlight and confirm how negative Sharpe ratios maintain their validity in reporting the level of efficiency of the various portfolios: these conclusions are indeed identical to those obtained in Sect. 8.4.1, considering the relative Sharpe ratios for the same two portfolios.

8.4.3 The Sortino Ratio

With asymmetric return distribution, the mean-variance principle is not sufficient to catch the downside risk, and the classic risk-adjusted performance indicators can no longer represent the essential characteristics of the distribution. In 1991, Sortino and Van der Meer developed a risk-adjusted performance measure, known as the Sortino ratio, which can account for this situation.[46]

The Sortino ratio has an analytical formula very similar to that of the Sharpe ratio, but is based on a definition of asymmetric risk. Its numerator is the difference between the average return on the portfolio and, instead of the risk-free rate, a target return identified according to the expectations and financial requirements of the investor. At the denominator, the Sortino ratio replaces the standard deviation with

[46] Sortino and Van der Meer (1991) and Sortino (2010)

the downside deviation of portfolio returns, calculated by considering only the returns below the target. In analytical terms, the Sortino ratio is obtained by:

$$\text{Sortino ratio}_p = \frac{\overline{r}_p - r_{\text{target}}}{\text{Down dev}_p} = \frac{\overline{r}_p - r_{\text{target}}}{\sqrt{\sum_{\substack{t=1 \\ r_{p,t} < r_{\text{target}}}}^{n} \frac{\left(r_{p,t} - r_{\text{target}}\right)^2}{n-1}}}$$

where:

$\overline{r}_p$	= average portfolio return
r_{target}	= target return desired by the investor[47]
Down dev_p	= downside deviation of portfolio returns compared to the target return

Even in a context of asymmetrical risk, if we assume a target return equal to the risk-free rate, then the Sortino ratio would end up with exactly the same numerator as in the Sharpe ratio:

$$\text{Sortino ratio}_p = \frac{\overline{r}_p - r_f}{\text{Down dev}_p} = \frac{\overline{r}_p - r_f}{\sqrt{\sum_{\substack{t=1 \\ r_{p,t} < r_f}}^{n} \frac{\left(r_{p,t} - r_f\right)^2}{n-1}}}$$

where:

Down dev_p	= downside deviation portfolio returns with respect to the risk-free rate

In any case, whatever the target return assumed, it is possible to use the Sortino ratio to compare different investments and define their degree of relative efficiency in a context of downside risk. The portfolio with the highest Sortino ratio is the one able to optimise the return difference with respect to the target per unit of downside risk. By way of example, let us consider two portfolios, A and B, with the following characteristics:

[47] This target return, which Sortino initially defined as the minimum acceptable return (MAR), was later renamed desired target return (DTR®) because of potential legal problems stemming from information transparency and possible misunderstandings on minimum return guarantees for the investor.

Portfolio	Return (%)	Down.dev. with respect to target return (%)	Sortino ratio
A	12	6.25	1.600
B	16	12.90	1.085
Target return	2	0	–

The Sortino ratio of portfolio A is equal to 1.6 = (12% − 2%)/6.25%, whereas that of portfolio B is equal to 1.085 = (16% − 2%)/12.90%; portfolio A is therefore more efficient than portfolio B, as it made it possible to attain greater return per unit of downside risk.

The graphic representation and the geometric interpretation of the Sortino ratio do not seem logical, as it does not derive from a capital market equilibrium model (unlike the Sharpe ratio, the Modigliani RAP and, as we shall see, the Treynor measure) and it does not allow recourse to the zero-investment strategy, but the Sortino ratio has two important qualities. On the one hand, it is able to adapt the risk-adjusted performance measure to the expectations of return of the investor, who, according to his/her own financial requirements, may assume a target return at least equal to zero, to the inflation rate, to the risk-free rate, to a given benchmark, or to any other measure consistent with the objectives of the investment policy. In other words, it becomes possible to include a minimum return threshold in the measurement of risk-adjusted performance, which must be met in order not to compromise the objective of the investment. Table 8.4 summarises some examples of target return and gives the objectives related to each one. The second important quality of the Sortino ratio is that of being able to catch downside risk only, through use of the downside deviation, thus assuming that the most appropriate variability measure should account only for negative deviations from the target return pursued, and not for positive deviations. This indicator is especially useful to evaluate those portfolios which pursue total and absolute return objectives, where the risk is the uncertainty of reaching a given target return, and in all those contexts where the distribution of returns is characterised by a more or less pronounced asymmetry in which standard deviation is not able to infer the actual exposure to downside risk. The Sortino ratio may therefore be considered as an effective auxiliary to the other risk-adjusted performance indicators for traditional investments. In this sense, for very close levels of evaluation of two or more portfolios, the Sortino ratio will select the most efficient

Table 8.4 Choosing the target return

Alternative target returns	Objectives
Zero return	To protect the capital in nominal terms
Inflation	To protect the capital in real terms
Risk-free rate	Not to go below the risk-free investment return assumed as an opportunity cost
Target return	To reach a given sum of capital at a predefined date in the future
Benchmark	Not to go below the indexed management return assumed as an opportunity cost

Source: Adapted from Scherer (2015)

one per unit of downside risk, in this case playing the role of supplementary indicator alongside the others.

Just like the Sharpe ratio and the Modigliani RAP, the Sortino ratio does not consider the linear correlation coefficients among the various portfolios, which are essential in order to identify the overall risk of a portfolio made up of two or more risky assets.

8.4.4 The Treynor Measure

The Treynor measure—originally known as the reward-to-volatility ratio[48]—calculates the excess return over the risk-free rate per unit of systematic risk over a given time period.[49] In fact, the Treynor measure is obtained from the ratio between the excess return on a portfolio—the difference between the average portfolio return and the risk-free rate—and its beta, calculated with respect to an appropriate benchmark index. In analytical terms, the Treynor measure is equal to:

$$\text{Treynor measure}_p = \frac{\overline{r}_p - r_f}{\beta_p}$$

where:

$\overline{r}_p$	= average portfolio return
r_f	= risk-free rate
β_p	= portfolio beta

The financial portfolio with the highest Treynor measure is the one that can give the greatest excess return per unit of systematic risk; it is the most efficient portfolio and therefore preferable with respect to all the others.

By way of example, let us consider two portfolios, A and B, with the following characteristics:

Portfolio	Return (%)	Beta	Treynor measure
A	12	0.75	0.133
B	16	1.25	0.112
Benchmark	14	1	0.120
Risk-free	2	0	–

[48] Originally, the term "volatility" was synonymous with systematic risk, while "variability" with total risk. However, this original meaning has now changed, and the two terms are both used to mean total risk, while the expression "systematic risk" now indicates the non-eliminable component of total risk.

[49] Treynor (1965).

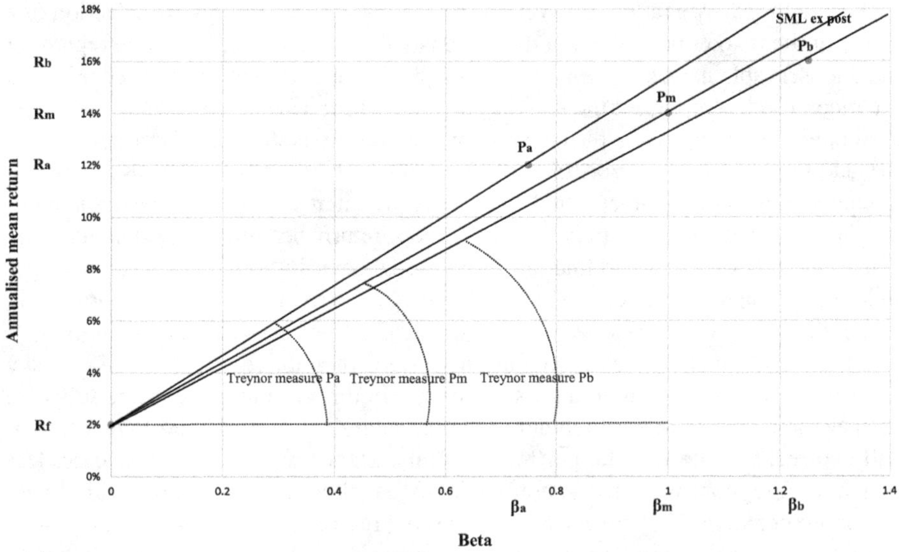

Fig. 8.3 Illustration of the Treynor measure

The Treynor measure of portfolio A is equal to 13.33% = (12% − 2%)/0.75, while that of portfolio B is equal to 11.2% = (16% − 2%)/1.25; despite having a lower return than portfolio B, portfolio A is more efficient, as it has delivered a higher return per unit of systematic risk taken.

In geometric terms, the Treynor measure may be interpreted as the slope of a straight line which, in a two-dimensional *ex post* return-beta space, has the risk-free rate as intercept on the vertical axis and connects the risk-free rate with the average return-beta combination of the portfolio under evaluation (Fig. 8.3).

In this case, too, every point on the line thus traced may be interpreted as a combination, in different proportions, of risk-free and risky assets. Each portfolio along that line is available to the investor who, depending on his/her own risk appetite, will decide how to combine risk-free and risky assets. In analytical terms, the Treynor measure is derived from the *ex post* Security Market Line (SML) of the CAPM, or rather from all the combinations of return and systematic risk that can be obtained when it is possible to invest or borrow at the risk-free rate.[50]

[50] In this case, too, the investor can follow the line starting from the risk-free rate and passing through the *ex post* return-beta combination of the portfolio. Using β_p to represent the beta of a portfolio that includes the risk-free asset and risky portfolio A, this coincides with the weighted average of the betas of the individual assets of which it is composed. Therefore:

$$\beta_p = x \cdot \beta_a + (1 - x) \cdot \beta_{rf}$$

As the beta of the risk-free asset is zero, it follows that: $\beta_p = x \cdot \beta_a$, from which $x = \beta_p / \beta_a$. The return of portfolio P—made up of risk-free asset and risky portfolio A—is also equal to the weighted average of the returns on the individual assets:

Unlike the Sharpe ratio, however, the Treynor measure requires the selection of a proxy of the market portfolio and therefore is vulnerable to Roll's critique regarding the unobservable nature of the market portfolio. In fact, its result varies according to the index used to represent the market portfolio. From a selection point of view, the assumption enabling users of the Treynor measure to evaluate the best portfolio is the addition of the new portfolio within a broader basket of risky assets which already contains many others. In this case, we are interested not in the excess return per unit of total risk, but rather in the excess return per unit of systematic risk. Therefore, it is significant to find out the marginal contribution in terms of return per unit of systematic risk that the new portfolio provides to the pre-existing portfolio in which the idiosyncratic risk was eliminated thanks to effective diversification.

The Treynor measure takes on a negative sign when the return obtained on risky asset portfolios is lower than the risk-free rate. In this case too, according to some, the calculation and use of the index become insignificant, as it is no longer univocally correct to state that the portfolio with the higher (in this particular case, less negative) Treynor measure is preferable to an alternative portfolio with lower (i.e. more negative) Treynor measure. The same reasoning we applied to the Sharpe ratio earlier is valid here too, but with some provisos. By way of example, let us consider two portfolios, C and D, with the following characteristics:

Portfolio	Return (%)	Beta	Treynor measure (%)
C	1.5	0.6	−0.83
D	0.5	1.3	−1.15
Benchmark	1.1	1	−0.90
Risk-free	2.0	0	–

The Treynor measure of portfolio C is equal to −0.83% = (1.5% − 2%)/0.6, while that of portfolio D is equal to −1.15% = (0.5% − 2%)/1.3; portfolio C is more efficient as it delivers a higher return per unit of systematic risk taken.

Although negative, in the *ex post* SML context from which it originates, the Treynor measure therefore maintains its validity and capacity to identify the most efficient manager in terms of ability to generate the highest excess return (even if this

---

$$r_p = x \cdot r_a + (1 - x) \cdot r_f$$

from which, by replacing x with β_p/β_a, we obtain:

$$r_p = (\beta_p/\beta_a) \cdot r_a + [1 - (\beta_p/\beta_a)] \cdot r_f$$
$$r_p = r_f + [(r_a - r_f)/\beta_a] \cdot \beta_p$$

The above equation, with a slope equal to the Treynor measure of portfolio A, expresses the return on the portfolio group that can be created by combining the risky asset A with the risk-free asset. These portfolios should be compared with the *ex post* SML, the equation of which is the following:

$$r_p = r_f + [(r_m - r_f)/\beta_m] \cdot \beta_p = r_f + (r_m - r_f) \cdot \beta_p$$

The portfolios lying below the *ex post* SML are less efficient and those above the *ex post* SML are more efficient. Note however that the ordinal capacity of the Treynor measure depends on the index adopted as the proxy of the market portfolio.

means least negative return), per unit of systematic risk. For this purpose, it might be useful, in a return-systematic risk context, rather than in a return-total risk context, to use the same line of reasoning as Modigliani's RAP and calculate the so-called Market RAP (M-RAP), which, in line with the RAP, shows the return on the portfolio examined if its systematic risk had been exactly the same as the market one.[51] To that end, by establishing the same equivalence through the use of financial leverage, as seen above, the risk-adjusted return would be equal to:

$$\mathrm{M}-\mathrm{RAP}_p = r_f + \left[\left(r_p - r_f\right)/\beta_p\right] \cdot \beta_m$$

$$\mathrm{M}-\mathrm{RAP}_p = r_f + \text{Treynor measure}_p \cdot \beta_m$$

$$\mathrm{M}-\mathrm{RAP}_p = r_f + \text{Treynor measure}_p \cdot 1$$

In this particular case, the risk-adjusted performance indicators for the two portfolios C and D would be equal to:

$$\mathrm{M}-\mathrm{RAP}_c = 2\% + (1.5\% - 2\%)/0.6 = 1.17\%$$
$$\mathrm{M}-\mathrm{RAP}_d = 2\% + (0.5\% - 2\%)/1.3 = 0.85\%$$

The risk-adjusted performances calculated in this way can be compared to the market return in order to evaluate their relative efficiency. In this particular case, given a market return of 1.1%, it turns out that portfolio C, with equal systematic risk, has outperformed the market, while portfolio D has underperformed it.

In the presence of negative betas, a special case of Treynor measure calculation occurs. Unlike all the other risk-adjusted performance indicators examined in this chapter, which imply positive denominators (being the average of squared deviations), in the case of the Treynor measure, it is theoretically possible to have negative betas for portfolios with a systematic risk in direct contrast to that of the market portfolio. Of course, it should be emphasised that this is a highly unlikely scenario in the case of traditional investments and over sufficiently long time periods, but in this case, the Treynor measure loses its significance and portfolio evaluation can no longer be based on this risk-adjusted performance indicator, for obvious motives of potential confusion between the signs of the numerator and of the denominator.

8.4.5 The Information Ratio

The information ratio is the ratio between tracking error and tracking error volatility. This indicator is another variant of the Sharpe ratio: the numerator considers the difference between the portfolio return and, instead of the risk-free rate, the

[51] Bacon (2023).

benchmark return; the denominator considers the standard deviation of the difference between the portfolio return and the benchmark return, the measure of relative risk identified as tracking error volatility.[52]

The information ratio is used to gauge the manager's skill to generate value added through active management and it measures the additional return over the benchmark (active return) per unit of additional risk taken with respect to the benchmark (active risk). From an analytical point of view, the information ratio is defined as follows:

$$\text{Information ratio}_p = \frac{\text{TE}_p}{\text{TEV}_p}$$

where:

$$\text{TE}_p = \text{average active return over a given time period} = \frac{1}{n}\sum_{t=1}^{n}\left(r_{p,t} - r_{b,t}\right)$$

$$\text{TEV}_p = \sqrt{\frac{\sum_{t=1}^{n}\left(\text{TE}_t - \overline{\text{TE}}\right)^2}{n-1}}$$

The information ratio is able to synthesise and supplement both the active return of the portfolio compared to an appropriate benchmark and the active risk with respect to the same benchmark. With specific regard to the latter, a number of alternatives are available for calculating the information ratio depending on the benchmark considered, which may be any one of the following: any benchmark declared by the asset manager, if mandatory; a discretional benchmark linked to the pertinent institutional category or to the asset allocation declared by the asset manager; a benchmark derived from a regression analysis; a benchmark calculated using a style analysis.[53]

As stated above, the information ratio is an indicator of the value added by the asset manager in relation to indexed portfolios: it represents the active return per unit of additional active risk compared to the benchmark. A positive information ratio shows that the asset manager has generated value added in relation to a passive strategy. A negative information ratio, on the other hand, is an indication of the manager's inability to generate value added in relation to the reference benchmark.

A passive portfolio managed in relation to an index will obviously tend to have an information ratio close to zero, whereas an actively managed portfolio will be efficient when it is able to generate additional return with respect to the market, under controlled risk conditions compared to those of the benchmark.

---

[52] Sharpe (1994) and Goodwin (1998).

[53] When the information ratio is calculated using a benchmark obtained through style analysis, it is commonly defined as selection ratio.

8.5 Evaluation of Stock Picking Ability

Jensen's alpha is the most popular measure of stock picking ability, defined as the skill of an active manager to identify financial instruments whose market price differs from their fair value. Just like the Treynor measure, this indicator is also derived, in analytical terms, from the *ex post* SML of the CAPM.[54]

Jensen's alpha is measured by the difference between actual portfolio return and the portfolio return predicted by the *ex post* SML equation:

$$\alpha_p = r_p - r_{\text{CAPM},p} = r_p - \left[r_f + \beta_p \cdot \left(r_m - r_f\right)\right]$$

The asset manager with the best stock picking ability is the one who maximises the alpha. If the Jensen's alpha has a positive value, the asset manager demonstrates a good selection skill, which leads to an over-performance with respect to the return justified by the systematic risk of the portfolio, represented by its beta. In other words, the asset manager has created value added through an effective stock selection activity, which led him/her to pick undervalued stocks and avoid overvalued stocks.[55] If the Jensen's alpha has a negative value, then the manager demonstrates poor selection activity that has resulted in a portfolio return worse than the return predicted by that same portfolio's systematic risk. In other words, the asset manager has destroyed value due to ineffective stock picking activity.

It is therefore clear that, in the case of an indexed strategy, we expect an alpha close to zero, while for active management the alpha may be either positive or negative.

From a graphical point of view, in a two-dimensional return-beta space, the Jensen's alpha represents the vertical distance between the actual return attained by the portfolio and the return based on SML at the level of beta of the portfolio (Fig. 8.4).

Alternatively, Jensen's alpha may also be obtained by a regression analysis where excess portfolio returns are the dependent variable and the excess market returns are the independent variable, based on the following equation, of which all the terms are known apart from $\varepsilon_{p,t}$, which is the error term of the regression:

$$r_{p,t} - r_{f,t} = \alpha_p + \beta_p \cdot \left(r_{m,t} - r_{f,t}\right) + \varepsilon_{p,t}$$

In this case, the Jensen's alpha is the intercept on the vertical axis of the interpolation line obtained by means of regression, while the beta is the slope of the same line of interpolation (Fig. 8.5). The reliability of the measures calculated in this way can be assessed through the analysis of the statistical significance of the regression parameters.[56]

---

[54] Jensen (1968, 1969).

[55] It is implicitly assumed that we are dealing with a long-only management policy.

[56] Specifically, standard error, *t*-value, and *p*-value.

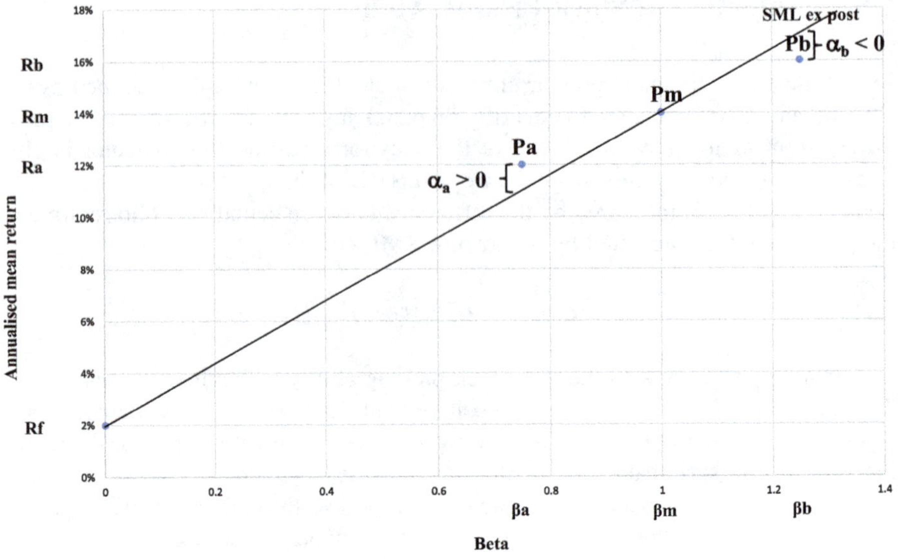

Fig. 8.4 Jensen's alpha: a first representation

The alpha calculated using the two methods described above implicitly assumes that the beta is constant throughout the whole period of observation. This hypothesis, therefore, excludes market timing activity, which is based on changing the portfolio's exposure to systematic risk in accordance with short-term market trend forecasts, and only allows stock picking activity, based on the selection of undervalued and overvalued stocks. In practice, this assumption is not always verified and the asset manager may perform more or less intense market timing activity, by changing the portfolio's exposure to systematic risk based on his/her expectations in terms of future market trend. If, in addition to stock picking, we also assume the presence of market timing, then both management abilities should be measured jointly—as we will do in Sect. 8.6—so as not to commit errors, which might confuse the first ability with the second one.

The use of Jensen's alpha and the Treynor measure points to similar conclusions in terms of the portfolio manager's abilities to outperform the market: a positive alpha indicates greater actual *ex post* return than the return predicted by the *ex post* SML equation; in this case, even the Treynor measure of the portfolio, which is the gradient of the straight line connecting the risk-free rate to the return-beta combination of the portfolio under evaluation, is undoubtedly greater than the Treynor measure of the market portfolio. Exactly the same relation applies to the opposite case with negative alphas.[57] However, if the two indicators were used to classify a

[57] Sharpe et al. (1999).

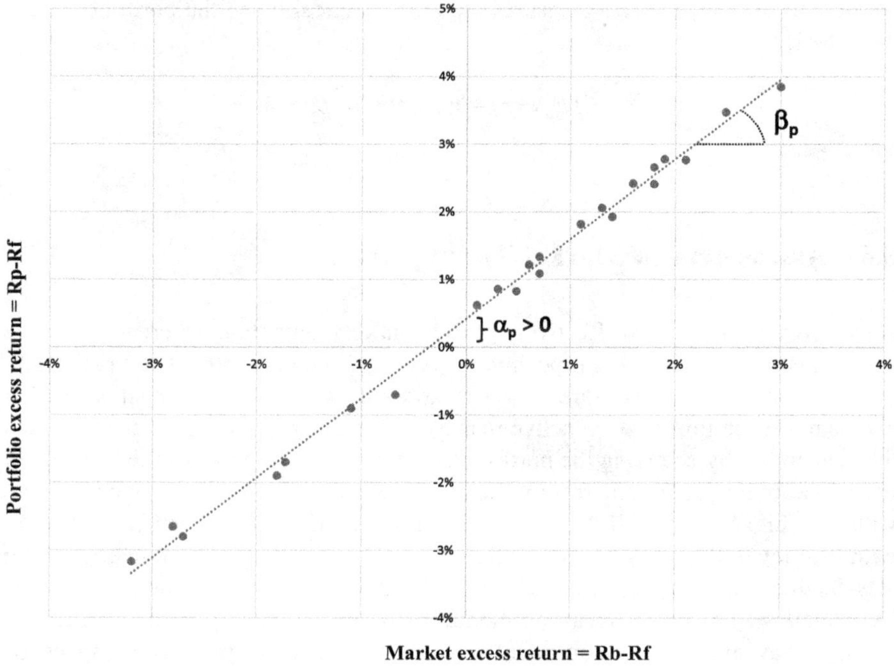

Fig. 8.5 Jensen's alpha: a second representation

number of portfolios making up a homogenous peer group, the resulting ranking would not necessarily be the same. This is because, as can be easily shown, the ordinal capacity of the two indicators also depends on the beta of each portfolio considered, which can bring about a different relative positioning:[58]

$$\alpha_p = \left(\text{Treynor}_p - \text{Treynor}_m\right) \cdot \beta_p$$

Jensen's alpha also requires clarification of a market portfolio and is therefore vulnerable to Roll's critique regarding the unobservable nature of the market portfolio. In practice, its result varies depending on the index used to represent the market portfolio.

If we examine the ratio between alpha and tracking error, which are often used as synonyms by the scientific literature, it can be seen that the difference between the two measures lies in whether or not the level of systematic risk of the portfolio is

[58] As can be easily shown:

$$\alpha_p = \left(\text{Treynor}_p - \text{Treynor}_m\right) \cdot \beta_p = \left(\frac{r_p - r_f}{\beta_p} - \frac{r_m - r_f}{1}\right) \cdot \beta_p = r_p - \left[r_f + \beta_p \cdot \left(r_m - r_f\right)\right]$$

considered. In fact, the two measures are only identical when the portfolio beta is equal to 1:

$$\alpha_p = TE_p = r_p - r_m \text{ only if } \beta_p = 1$$

8.6 Assessment of Market Timing Ability

An active manager able to effectively predict market trends should increase exposure to systematic risk when there is positive excess market return with respect to the risk-free rate, and reduce exposure to systematic risk when there is negative market excess return. In this case, an active manager attempts to obtain a better result than the benchmark by changing the portfolio beta with respect to that of the benchmark over time. In particular, an asset manager will aim to build: an aggressive portfolio, with beta higher than 1, if he/she predicts a bullish market; a defensive portfolio, with beta lower than 1, if he/she predicts a bearish market. An active manager who acts in this way aims to achieve a positive differential return compared to the benchmark by properly forecasting the timing of ups and downs in the market.[59]

As we saw above, the independent evaluation of stock picking ability implies the absence of market timing by the asset manager. In the case of portfolios managed actively through the use of market timing strategies, stock picking activity would be misrepresented if not evaluated together with the market timing ability.[60]

In effect, admitting the possibility of market timing means acknowledging that systematic portfolio risk is made up of two elements, one constant over time and the other linked to the performance of excess market return with respect to the risk-free rate. Thus, the portfolio beta, in a generic time period *t*, becomes equal to:

$$\beta_{p,t} = \beta_p + \gamma_p \cdot \left(r_{m,t} - r_{f,t}\right)$$

from which, by replacing it in the following equation:

$$r_{p,t} - r_{f,t} = \alpha_p + \beta_{p,t} \cdot \left(r_{m,t} - r_{f,t}\right) + \varepsilon_{p,t}$$

we obtain:

$$r_{p,t} - r_{f,t} = \alpha_{p,\mathrm{TM}} + \beta_p \cdot \left(r_{m,t} - r_{f,t}\right) + \gamma_p \cdot \left(r_{m,t} - r_{f,t}\right)^2 + \varepsilon_{p,t}$$

---

[59] Reilly and Brown (2012).

[60] Francis and Dongcheol (2013).

In the new relation, which reinterprets Treynor and Mazuy's CAPM version, the measures aimed at expressing a judgement on the ability of the active manager are:[61]

- The Treynor–Mazuy alpha, which is a stock picking indicator adjusted for market timing. Unlike Jensen's alpha, which we examined above, the indicator in this case accepts that it is possible for active managers to simultaneously perform stock selection activity (causing them to move away from the benchmark composition) and market timing activity (causing them to have a variable beta over time in accordance with forecasted market developments).
- The gamma, which is a market timing indicator, or rather an indicator of the manager's skill in varying the systematic risk of the portfolio in accordance with expected market developments. There are alternatives for performing this strategy, starting from the simplest method, whereby the proportional amount of risk-free asset is varied, to the use of derivatives as part of the most advanced techniques.

Expressed in terms of a graph, the assumption of a changeable beta causes the risk-return combination previously expressed as a line to be shown as a curve, whose changing slope depends essentially on whether or not the asset manager is able to perform effective market timing. Figures 8.6 and 8.7 show both hypotheses. Figure 8.6 highlights a positive gamma position, where the asset manager is able to create value added by varying the level of systematic risk: when the excess market return has been positive, the asset manager has increased the systematic risk of the portfolio by creating extra-performance; when the excess market return has been negative, the manager has reduced the systematic risk of the portfolio, mitigating the loss with respect to the market and thus once again achieving extra-performance. Figure 8.7 shows the opposite situation, where the asset manager has destroyed value through incorrect market timing: when the market excess return has been positive, the manager has reduced the systematic risk of the portfolio, creating under-performance; when the market excess return has been negative, the manager has increased the systematic risk of the portfolio, by enhancing the loss with respect to the market.

For the combined evaluation of stock picking and market timing abilities, Henriksson and Merton introduced a simplification into Treynor and Mazuy's evaluation model, incorporating a dummy variable (D) into the regression model which takes the value of zero when excess market return is negative and a value of 1 when excess market return is positive.[62] In this case, the equation for calculation of alpha, beta, and gamma is:

$$r_{p,t} - r_{f,t} = \alpha_{p,\mathrm{HM}} + \beta_p \cdot \left(r_{m,t} - r_{f,t}\right) + \gamma_p \cdot \left[D \cdot \left(r_{m,t} - r_{f,t}\right)\right] + \varepsilon_{p,t}$$

where:

[61] Treynor and Mazuy (1966).

[62] Henriksson and Merton (1981).

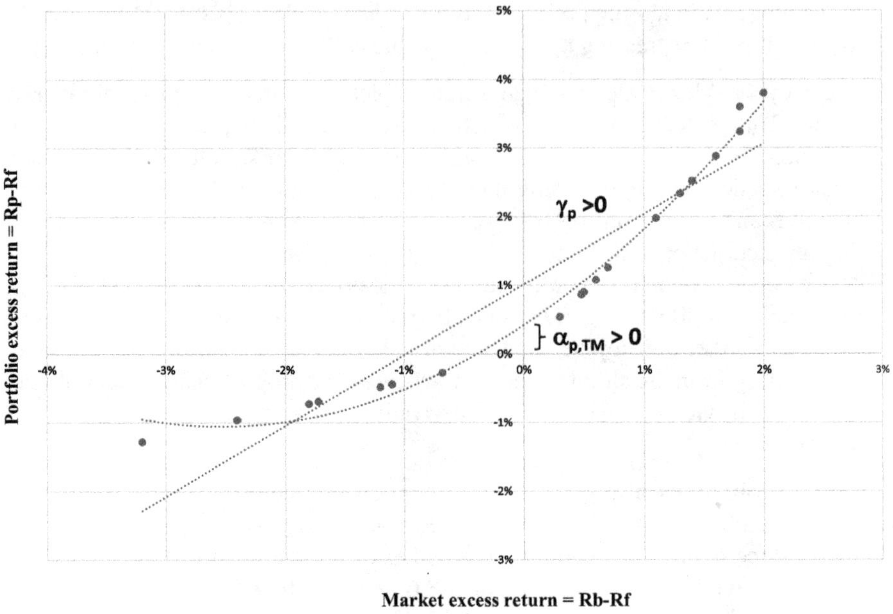

Fig. 8.6 Evaluation of market timing: an example of effective timing

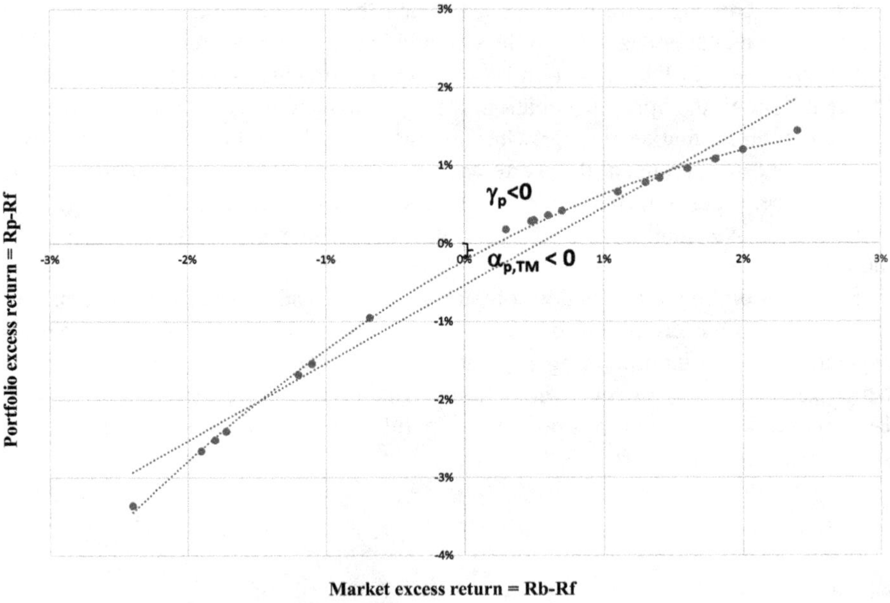

Fig. 8.7 Evaluation of market timing: an example of ineffective timing

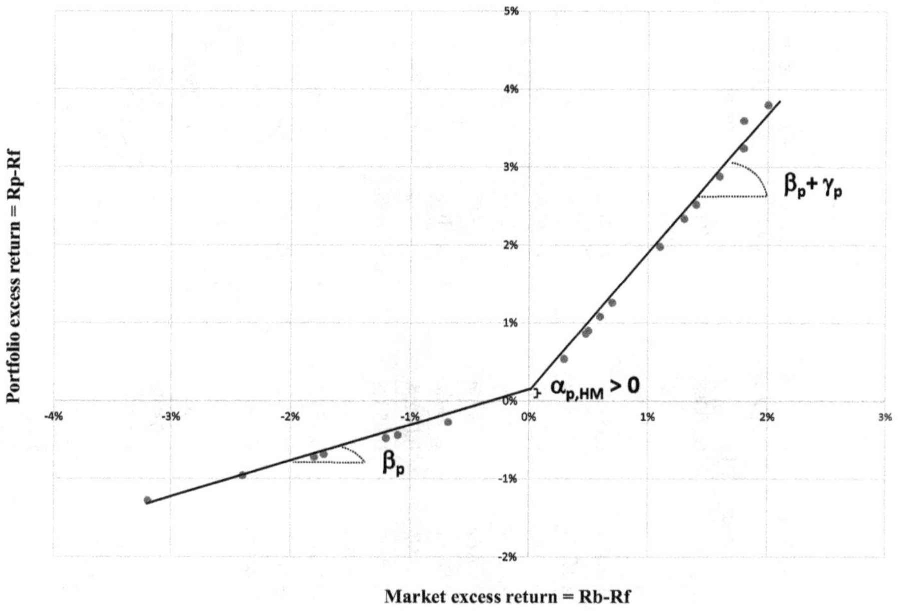

Fig. 8.8 Evaluation of market timing: Henriksson and Merton's simplified approach

$$D = 0 \text{ if } r_{m,t} < r_{f,t};$$
$$D = 1 \text{ if } r_{m,t} > r_{f,t}.$$

Here, the portfolio beta can take two different values: in bearish phases, it is equal to β_p and in bullish phases to ($\beta_p + \gamma_p$). Figure 8.8 illustrates what we have just described.

Tables 8.5 and 8.6 show how to set up the calculation of the various return, risk, risk-adjusted return, and ability measures in Excel.[63]

[63] In the case of the Treynor–Mazuy gamma, the quickest way of determining the regression parameters is to use Excel to trace a trend line on a scatter plot, in which the excess returns of the benchmark are shown on the horizontal axis and the excess returns of the portfolio on the vertical axis, expressing that the relation is not linear, but polynomial, and requiring the equation to be viewed on the graph. The gamma, beta, and alpha of the portfolio will be so depicted.

In the case of the Henriksson–Merton model, it is a question of calculating, at the same time—given the same point of interception on the vertical axis—the slopes of the two separate lines of interpolation: the one connecting the positive excess returns of the benchmark to the excess returns of the portfolio and the other one connecting the negative excess returns of the benchmark to the excess returns of the portfolio.

Table 8.5 An example of Excel calculation of returns, risk, risk-adjusted performance, and ability indicators—Part I

	A	B	C	D	E	F	G	H	I
1		**Input data**							
2	Months	Fund return	Risk-free rate	Benchmark return	Growth factor	Excess return (ER) of the fund	(ER of the fund)^2 < 0	Tracking error	Excess return (ER) of the benchmark
3	1	18.45%	0.12%	17.05%	1.185	18.335%	0.000%	1.400%	16.935%
4	2	−6.84%	0.13%	−5.82%	0.932	−6.965%	0.485%	−1.020%	−5.945%
5	3	−2.70%	0.08%	−1.59%	0.973	−2.780%	0.077%	−1.110%	−1.670%
6	4	1.57%	0.09%	1.15%	1.016	1.480%	0.000%	0.420%	1.060%
7	5	2.30%	0.15%	0.53%	1.023	2.155%	0.000%	1.770%	0.385%
8	6	0.35%	0.12%	−0.14%	1.004	0.230%	0.000%	0.490%	−0.260%
9	7	4.98%	0.16%	3.40%	1.050	4.820%	0.000%	1.580%	3.240%
10	8	−4.78%	0.19%	−3.64%	0.952	−4.970%	0.247%	−1.140%	−3.830%
11	9	2.71%	0.17%	3.32%	1.027	2.540%	0.000%	−0.610%	3.150%
12	10	−2.28%	0.21%	−0.68%	0.977	−2.485%	0.062%	−1.600%	−0.885%
13	11	−4.80%	0.24%	−4.98%	0.952	−5.035%	0.254%	0.180%	−5.215%
14	12	2.56%	0.22%	1.46%	1.026	2.345%	0.000%	1.100%	1.245%
15	13	−9.62%	0.14%	−7.77%	0.904	−9.760%	0.953%	−1.850%	−7.910%
16	14	−3.39%	0.20%	−2.83%	0.966	−3.585%	0.129%	−0.560%	−3.025%
17	15	4.51%	0.08%	2.26%	1.045	4.430%	0.000%	2.250%	2.180%
18	16	−4.23%	0.24%	−2.62%	0.958	−4.465%	0.199%	−1.610%	−2.855%
19	17	−4.28%	0.19%	−3.94%	0.957	−4.465%	0.199%	−0.340%	−4.125%
20	18	−1.80%	0.16%	−1.74%	0.982	−1.955%	0.038%	−0.060%	−1.895%
21	19	−3.69%	0.20%	−3.93%	0.963	−3.885%	0.151%	0.240%	−4.125%
22	20	15.89%	0.23%	14.77%	1.159	15.660%	0.000%	1.120%	14.540%
23	21	3.52%	0.19%	4.83%	1.035	3.335%	0.000%	−1.310%	4.645%
24	22	4.39%	0.15%	5.14%	1.044	4.245%	0.000%	−0.750%	4.995%

25	23	0.61%	0.10%	0.87%	1.006	0.510%	0.000%	−0.260%	0.770%
26	24	0.08%	0.09%	0.18%	1.001	−0.005%	0.000%	−0.100%	0.095%
27	25	−0.67%	0.12%	−1.11%	0.993	−0.790%	0.006%	0.440%	−1.230%
28	26	5.10%	0.10%	5.21%	1.051	5.005%	0.000%	−0.110%	5.115%
29	27	−2.48%	0.16%	−2.37%	0.975	−2.635%	0.069%	−0.110%	−2.525%
30	28	−4.19%	0.12%	−5.03%	0.958	−4.310%	0.186%	0.840%	−5.150%
31	29	−5.82%	0.15%	−7.27%	0.942	−5.970%	0.356%	1.450%	−7.420%
32	30	−7.54%	0.16%	−6.20%	0.925	−7.695%	0.592%	−1.340%	−6.355%
33	31	1.15%	0.13%	0.23%	1.012	1.020%	0.000%	0.920%	0.100%
34	32	14.11%	0.15%	12.47%	1.141	13.965%	0.000%	1.640%	12.325%
35	33	7.33%	0.12%	7.20%	1.073	7.210%	0.000%	0.130%	7.080%
36	34	7.03%	0.15%	8.15%	1.070	6.885%	0.000%	−1.120%	8.005%
37	35	−6.90%	0.13%	−7.48%	0.931	−7.030%	0.494%	0.580%	−7.610%
38	36	−3.47%	0.11%	−3.80%	0.965	−3.580%	0.128%	0.330%	−3.910%
39	**Annualised arithmetic return**	5.72%	1.79%	5.09%					
40	**Annualised geometric return**				3.46%				
41	**Standard deviation**	22.30%		20.81%		22.31%			
42	**Downside deviation**						12.59%		
43	**Beta**	1.058	or	1.058	or	1.058			
44	**Tracking error volatility**							3.75%	
45	**Sharpe ratio**	0.176	or	0.176	or	0.176	or	0.176	
46	**Sortino index**	0.31							
47	**Information ratio**	0.17	or	0.17					
48	**Treynor measure**	0.04							
49	**Alpha**	0.004	or	0.004					

Table 8.6 An example of Excel calculation of returns, risk, risk-adjusted performance, and ability indicators—Part II

	A	B	C	D	E	F	G	H	I
1		**Input data**							
2	Months	Fund return	Risk-free rate	Benchmark return	Growth factor	Excess return (ER) of the fund	(ER of the fund)^2 < 0	Tracking error	Excess return (ER) of the benchmark
3	1	18.45%	0.12%	17.05%	=1 + B3	=(B3 – C3)	=IF(F3 > 0;0;F3^2)	=B3 – D3	=D3 – C3
4	2	−6.84%	0.13%	−5.82%	=1 + B4	=(B4 – C4)	=IF(F4 > 0;0;F4^2)	=B4 – D4	=D4 – C4
5	3	−2.70%	0.08%	−1.59%	=1 + B5	=(B5 – C5)	=IF(F5 > 0;0;F5^2)	=B5 – D5	=D5 – C5
6	4	1.57%	0.09%	1.15%	=1 + B6	=(B6 – C6)	=IF(F6 > 0;0;F6^2)	=B6 – D6	=D6 – C6
7	5	2.30%	0.15%	0.53%	=1 + B7	=(B7 – C7)	=IF(F7 > 0;0;F7^2)	=B7 – D7	=D7 – C7
8	6	0.35%	0.12%	−0.14%	=1 + B8	=(B8 – C8)	=IF(F8 > 0;0;F8^2)	=B8 – D8	=D8 – C8
9	7	4.98%	0.16%	3.40%	=1 + B9	=(B9 – C9)	=IF(F9 > 0;0;F9A^2)	=B9 – D9	=D9 – C9
10	8	−4.78%	0.19%	−3.64%	=1 + B10	=(B10 – C10)	=IF(F10 > 0;0;F10^2)	=B10 – D10	=D10 – C10
11	9	2.71%	0.17%	3.32%	=1 + B11	=(B11 – C11)	=IF(F11 > 0;0;F11^2)	=B11 – D11	=D11 – C11
12	10	−2.28%	0.21%	−0.68%	=1 + B12	=(B12 – C12)	=IF(F12 > 0;0;F12^2)	=B12 – D12	=D12 – C12
13	11	−4.80%	0.24%	−4.98%	=1 + B13	=(B13 – C13)	=IF(F13 > 0;0;F13^2)	=B13 – D13	=D13 – C13
14	12	2.56%	0.22%	1.46%	=1 + B14	=(B14 – C14)	=IF(F14 > 0;0;F14^2)	=B14 – D14	=D14 – C14
15	13	−9.62%	0.14%	−7.77%	=1 + B15	=(B15 – C15)	=IF(F15 > 0;0;F15^2)	=B15 – D15	=D15 – C15
16	14	−3.39%	0.20%	−2.83%	=1 + B16	=(B16 – C16)	=IF(F16 > 0;0;F16^2)	=B16 – D16	=D16 – C16
17	15	4.51%	0.08%	2.26%	=1 + B17	=(B17 – C17)	=IF(F17 > 0;0;F17^2)	=B17 – D17	=D17 – C17
18	16	−4.23%	0.24%	−2.62%	=1 + B18	=(B18 – C18)	=IF(F18 > 0;0;F18^2)	=B18 – D18	=D18 – C18
19	17	−4.28%	0.19%	−3.94%	=1 + B19	=(B19 – C19)	=IF(F19 > 0;0;F19^2)	=B19 – D19	=D19 – C19
20	18	−1.80%	0.16%	−1.74%	=1 + B20	=(B20 – C20)	=IF(F20 > 0;0;F20^2)	=B20 – D20	=D20 – C20
21	19	−3.69%	0.20%	−3.93%	=1 + B21	=(B21 – C21)	=IF(F21 > 0;0;F21^2)	=B21 – D21	=D21 – C21
22	20	15.89%	0.23%	14.77%	=1 + B22	=(B22 – C22)	=IF(F22 > 0;0;F22^2)	=B22 – D22	=D22 – C22
23	21	3.52%	0.19%	4.83%	=1 + B23	=(B23 – C23)	=IF(F23 > 0;0;F23^2)	=B23 – D23	=D23 – C23
24	22	4.39%	0.15%	5.14%	=1 + B24	=(B24 – C24)	=IF(F24 > 0;0;F24^2)	=B24 – D24	=D24 – C24
25	23	0.61%	0.10%	0.87%	=1 + B25	=(B25 – C25)	=IF(F25 > 0;0;F25^2)	=B25 – D25	=D25 – C25
26	24	0.08%	0.09%	0.18%	=1 + B26	=(B26 – C26)	=IF(F26 > 0;0;F26^2)	=B26 – D26	=D26 – C26

27	25	−0.67%	0.12%	−1.11%	=1 + B27	=(B27 − C27)	=IF(F27 > 0;0;F27^2)	=B27 − D27	=D27 − C27
28	26	5.10%	0.10%	5.21%	=1 + B28	=(B28 − C28)	=IF(F28 > 0;0;F28^2)	=B28 − D28	=D28 − C28
29	27	−2.48%	0.16%	−2.37%	=1 + B29	=(B29 − C29)	=IF(F29 > 0;0;F29^2)	=B29 − D29	=D29 − C29
30	28	−4.19%	0.12%	−5.03%	=1 + B30	=(B30 − C30)	=IF(F30 > 0;0;F30^2)	=B30 − D30	=D30 − C30
31	29	−5.82%	0.15%	−7.27%	=1 + B31	=(B31 − C31)	=IF(F31 > 0;0;F31^2)	=B31 − D31	=D31 − C31
32	30	−7.54%	0.16%	−6.20%	=1 + B32	=(B32 − C32)	=IF(F32 > 0;0;F32^2)	=B32 − D32	=D32 − C32
33	31	1.15%	0.13%	0.23%	=1 + B33	=(B33 − C33)	=IF(F33 > 0;0;F33^2)	=B33 − D33	=D33 − C33
34	32	14.11%	0.15%	12.47%	=1 + B34	=(B34 − C34)	=IF(F34 > 0;0;F34^2)	=B34 − D34	=D34 − C34
35	33	7.33%	0.12%	7.20%	=1 + B35	=(B35 − C35)	=IF(F35 > 0;0;F35^2)	=B35 − D35	=D35 − C35
36	34	7.03%	0.15%	8.15%	=1 + B36	=(B36 − C36)	=IF(F36 > 0;0;F36^2)	=B36 − D36	=D36 − C36
37	35	−6.90%	0.13%	−7.48%	=1 + B37	=(B37 − C37)	=IF(F37 > 0;0;F37^2)	=B37 − D37	=D37 − C37
38	36	−3.47%	0.11%	−3.80%	=1 + B38	=(B38 − C38)	=IF(F38 > 0;0;F38^2)	=B38 − D38	=D38 − C38
39	**Annualised arithmetic return**	=AVERAGE (B3:B38)*12	=AVERAGE (C3:C38)*12	=AVERAGE(D3: D38)*12					
40	**Annualised geometric return**				=PRODUCT (E3:E38)^ (1/3) − 1				
41	**Standard deviation**	=STDEV.S (B3:B38) *12^(0,5)		=STDEV.S(D3: D38)*12^(0,5)		=STDEV.S(F3: F38)*12^(0,5)			
42	**Downside deviation**						=(SUM(G3:G38)/ (COUNTA(G3:G38) − 1)) ^0,5*(12^0,5)		
43	**Beta**	=CORREL (B3:B38;D3: D38)*B41/ D41	or	=COVARIANCE. S(B3:B38;D3: D38)/VAR.S(D3: D38)	or	=SLOPE(F3: F38;I3:I38)			
44	**Tracking error volatility**							=STDEV.S(H3: H38)*12^0,5	

(continued)

Table 8.6 (continued)

	A	B	C	D	E	F	G	H	I
45	**Sharpe ratio**	=(B39 – C39)/B41	or	=(AVERAGE (B3:B38)*12 – AVERAGE(C3: C38)*12)/ (STDEV.S(B3: B38)*12^(0,5))	or	=(B39 – C39)/ F41	or	=(AVERAGE (B3:B38)*12 – AVERAGE(C3: C38)*12)/ (STDEV.S(F3: F38)*12^(0,5))	
46	**Sortino index**	=(B39 – C39)/G42							
47	**Information ratio**	=(B39 – D39)/H44	or	=AVERAGE(H3: H38)*12/H44					
48	**Treynor measure**	=(B39 – C39)/B43							
49	**Alpha**	=B39 – (C39 + B43* (D39 – C39))	or	=INTERCEPT (F3:F38;I3:I38) *12					

8.7 Performance Persistence

Some concluding remarks need to be made regarding performance persistence in terms of return, risk-adjusted return, stock picking, and market timing abilities. In other words, it is obviously of some interest to us to ascertain whether or not past performance measured over a certain time period—intended in relative terms or in terms of positioning within a reference peer group—will repeat itself in the future. This is of considerable importance, because in the case of persistence, performance evaluation does not simply provide us with an *ex post* analysis of returns, risks, and the various ability measures, but also makes it possible to use these measures (with all due caution) in an *ex ante* approach, so as to identify the most efficient and/or capable asset managers on a given market, based on their past track record.

Companies specialised in the evaluation of funds and other asset management products analyse performance persistence based on the frequency with which asset managers position themselves in the various quantiles as well as on the consistency of their positioning. With historic performance measures as their starting point, then, these companies break performance over the periods down into a series of homogeneous sub-periods and identify the positioning of each asset manager in a predefined number of quantiles to which they belong (often the four quartiles), assigning a score to the asset manager positioned in the best quantiles for each sub-period, and another score to the manager who displays consistency by remaining in the same quantiles from sub-period to sub-period. This makes it possible to draw up a persistence index, able to classify managers on the basis of the frequency with which they are positioned in the best quantiles and the consistency of their past performances. On other occasions, these same companies specialised in the evaluation of investment funds only show as a graph the positioning of each asset manager in the various quantiles for each sub-period, highlighting the progress over time of the quantile to which each manager belongs in relation to the reference peer group.

There is a large body of academic literature on the persistence of investment funds, and the conclusions reached are not always univocal. Some agreement does seem to exist, however, about the following:

- When measured on the basis of past returns only, performance rarely persists in time, since the asset managers belonging to the best quantiles in a given period do not necessarily belong to the best quantiles in subsequent periods. The performance of the worst managers, on the other hand, who often belong to the last decile with continuity over the course of time, seems more persistent.
- When measured in terms of risk-adjusted performance and ability, the level of performance persistence increases slightly, at least in relative terms, and past performance is repeated more.
- When persistence measurement is based on the creation of homogeneous peer groups, deduced from the asset managers' style analysis, instead of being based on the benchmark declared, or on the category to which the fund belongs, then persistence improves even more, at least in relative terms.

- In all cases, however, evaluations of past performances are invalidated by a distortion related to the survival of the funds, known as survivorship bias. This derives from the fact that evaluations of past performance are generally based solely on the analysis of data from investment funds which survived during the time period under evaluation, and disregard all those funds which have ceased to exist in the meantime. There is solid empirical evidence which indicates that this distortion results in the overestimation of actual past performances, due to the greater frequency with which management companies close funds that have performed below average compared to those which have performed better than the reference peer group.[64]

References

Bacon C (2023) Practical portfolio performance measurement and attribution, 3rd edn. Wiley, Chichester

Bailey J, Richards T, Tierney D (2007) Evaluating portfolio performance. In: Maginn L, Tuttle DL, McLeavey DW, Pinto JE (eds) Managing investment portfolios. A dynamic process, 3rd edn. Wiley, Hoboken

Bodie Z, Kane A, Marcus A (2024) Investments, 13th edn. McGraw-Hill, New York

Brown SJ, Goetzmann WN (1995) Performance persistence. J Financ 50(2):679–698

Carhart MM (1997) On persistence in mutual fund performance. J Financ 52(1):57–82

Christopherson JA, Carino DR, Ferson WE (2009) Portfolio performance measurement and benchmarking. McGraw-Hill, New York

De Fusco R, McLeavey D, Pinto J, Runkle D (2007) Quantitative methods for investment analysis, 2nd edn. CFA Institute, Charlottesville

DiBartolomeo D, Witkowski E (1997) Mutual fund misclassification: evidence based on style analysis. Financ Anal J 53(5):32–43

Dietz P (1966) Pension funds. Measuring investment performance. Free Press, New York

Feibel B (2003) Investment performance measurement. Wiley, Hoboken

Fischer B, Wermers R (2013) Performance evaluation and attribution of security portfolios. Academic, Oxford

Francis JC, Dongcheol K (2013) Modern portfolio theory. Wiley, Hoboken

Goetzmann WN, Ibbotson RG (1994) Do winners repeat? J Portfolio Manag 20(2):9–18

Goodwin T (1998) The information ratio. Financ Anal J 54(4):34–43

Henriksson RD, Merton RC (1981) On market timing and investment performance II. Statistical procedures for evaluating forecasting skills. J Bus 54(4):513–533

Ibbotson R, Patel A (2002) Do winners repeat with style? Yale school of management working papers, No 00-70

Jensen MC (1968) The performance of mutual funds in the period 1945–1964. J Financ 23(2): 389–416

Jensen MC (1969) Risk, the price of capital assets, and the evaluation of investment portfolios. J Financ 24(5):167–247

Lawton P, Jankowski T (eds) (2009) Investment performance measurement: evaluating and presenting results. CFA Institute, Charlottesville

[64] Goetzmann and Ibbotson (1994), Brown and Goetzmann (1995), Carhart (1997), and Ibbotson and Patel (2002).

Le Sourd V (2007) Performance measurement for traditional investment. Literature survey. EDHEC Business School, Lille-Nice
Modigliani F, Modigliani L (1997) Risk-adjusted performance. J Portfolio Manag 23(2):45–54
Rasmussen M (2003) Quantitative portfolio optimisation, asset allocation and risk management. Palgrave MacMillan, Houndmills
Reilly FK, Brown KC (2012) Investment analysis & portfolio management, 10th edn. South-Western-Cengage Learning, Mason
Roll R (1977) A critique of the asset pricing theory's tests. J Financ Econ 4(2):129–176
Roll R (1992) A mean/variance analysis of tracking error. Journal of Portfolio Management 18(4): 13–22
Ross S (1977) Return, risk and arbitrage. In: Friend I, Bicksler J (eds) Risk and return in finance. Ballinger, Cambridge
Scherer B (2015) Portfolio construction and risk budgeting, 5th edn. Risk Books, London
Schulmerich M, Leporcher YM, Eu CH (2013) Applied asset and risk management. Springer, Heidelberg
Sharpe W (1966) Mutual fund performance. J Bus 39(1):119–138
Sharpe W (1992) Asset allocation: management style and performance measurement. J Portfolio Manag 18(2):7–19
Sharpe W (1994) The Sharpe ratio. J Portfolio Manag 21(1):49–58
Sharpe W, Alexander G, Bailey J (1999) Investments, 6th edn. Prentice Hall, Upper Saddle River
Sortino F (2010) The Sortino framework for constructing portfolios. Elsevier, New York
Sortino F, van der Meer R (1991) Downside risk. J Portfolio Manag 17(4):27–31
Treynor J (1965) How to rate management of investment funds. Harv Bus Rev 43(1):63–75
Treynor J, Mazuy K (1966) Can mutual funds outguess the market? Harv Bus Rev 44(4):131–136

Chapter 9
Returns-Based Style Analysis

Maria Debora Braga

9.1 The Role of Returns-Based Style Analysis

Returns-based style analysis (RBSA), introduced by William F. Sharpe in 1992, became well-known among both scholars and practitioners following the publication of the article "Asset Allocation: Management Style and Performance Measurement",[1] which highlighted its usefulness in situations where there was a need and/or desire to understand the investment style of a fund.[2] This means that RBSA aims to support those who cannot be considered insiders of an asset management company in identifying the asset allocation of a fund and understanding how the assets under its management are divided among different investment categories.

Besides RBSA, there is another approach that seeks to provide the same information: holdings-based style analysis (HBSA), also called characteristics-based style analysis. The names of the respective approaches indicate the key element that distinguishes them: while HBSA needs the actual portfolio constituents, RBSA only requires the time series of returns of the relevant fund. Understandably, HBSA may encounter issues regarding accessibility to up-to-date information, forcing the user to accept a fixed time-point representation of the allocation of the fund portfolio when the asset management company publishes the periodic reports required by

[1] Sharpe (1992).

[2] A kind of anticipation of what is commonly called RBSA can actually be found in a previous study by the same author: see Sharpe (1988).

M. D. Braga (✉)
Department of Economics and Political Science, University of "Valle d'Aosta", Aosta, Italy

SDA Bocconi School of Management, Bocconi University, Milan, Italy
e-mail: m.braga@univda.it; mariadebora.braga@sdabocconi.it

I. Basile, P. Ferrari (eds.), *Asset Management and Institutional Investors*, Contributions to Finance and Accounting,
https://doi.org/10.1007/978-3-031-59819-7_9

regulations; by the same token, HBSA can be significant and objective only for the time-point to which it refers.

Unlike the analytical approach of HBSA, Sharpe's RBSA offers a deductive method with which to analyse investment styles. The nature of the method is also evident in the following, extremely brief description: RBSA is a statistical methodology that uses exclusively the past returns of a fund and of the asset class benchmarks to identify the combination of long positions in these indices that would have replicated (more closely than other combinations) the effective performance of the fund over a specified time period. Essentially, RBSA makes it possible to assess the asset allocation of the fund without any direct knowledge of the portfolio holdings, since this is implicit in the tangible elements available, i.e. the fund's performance as resulting from the audited values of its units or shares and the performance of verifiable and widely used benchmarks.

The present chapter will seek to provide an in-depth examination of the methodological profiles of RBSA, as well as a few practical examples of its application.

9.2 Methodological Aspects of Returns-Based Style Analysis

The basic idea underlying RBSA is extremely simple: there must be a relation between the returns of a fund and those of a certain number of indices/benchmarks representing asset classes that are acceptable investment options for the fund, given the fund's investment policy as disclosed in the communications required by the supervisory authorities and/or in regulations governing investment proposals. From this there follows a useful operational consideration: asset allocation can be inferred by comparing the returns of the fund with those of the identified asset class benchmarks.

During the modelling phase, the logical conclusion is thus that the best path to follow to explain the performance of a fund is the estimates obtained from a multifactor model. Correspondingly, in statistical terms, returns-based style analysis (RBSA) takes the form of a multivariate linear regression in which the independent variables are the explicative factors identified in the asset class benchmarks (or to be more precise, the historical series of their returns), while the dependent variable is the performance over time of the fund under consideration. This is generally indicated as the Sharpe style regression and is written as in (9.1a) or (9.1b):

$$\begin{aligned} R_{\text{fund},t} = {} & w_1 \times R_{\text{benchmark}_1,t} + w_2 \times R_{\text{benchmark}_2,t} + \ldots + \\ & + w_i \times R_{\text{benchmark}_i,t} + \ldots + w_N \times R_{\text{benchmark}_N,t} + \varepsilon_t \end{aligned} \tag{9.1a}$$

$$R_{\text{fund},t} = \sum_{i=1}^{N} w_i \times R_{\text{benchmark}_i,t} + \varepsilon_t \tag{9.1b}$$

where:

$t =$	$1,\ldots,T$
$R_{\text{fund},t} =$	Return of the fund over the time period t (i.e. from $t - 1$ to t)
$R_{\text{benchmark}_i,t} =$	Return of the generic benchmark i (with $i = 1,\ldots, N$) over the time period t
$w_i =$	Regression coefficient for each independent variable $R_{\text{benchmark}_i,t}$
$\varepsilon_t =$	Residual return, the portion of the return of the fund not explained by or related to the set of independent variables (i.e. the selected benchmarks)

Obviously, the multivariate model to be estimated can also be formulated in matrix algebra. In this case, we use a compact form, as shown in (9.2a):

$$\mathbf{Y}_{\text{FUND}} = \mathbf{R}_{\text{BENCH}} \cdot \mathbf{w} + \mathbf{e} \tag{9.2a}$$

where:

$\mathbf{Y}_{\text{FUND}} =$	Column vector $(T \times 1)$ of the returns of the analysed fund
$\mathbf{R}_{\text{BENCH}} =$	Matrix $(T \times N)$ of the returns for $t = 1,\ldots,T$ of the N selected indices
$\mathbf{w} =$	Column vector $(N \times 1)$ of the exposures to the asset class benchmarks
$\mathbf{e} =$	Column vector $(T \times 1)$ of the error terms

In the expanded form, the model is described as in (9.2b):

$$\begin{bmatrix} R_{\text{fund},1} \\ R_{\text{fund},2} \\ \cdots \\ R_{\text{fund},T} \end{bmatrix} = \begin{bmatrix} R_{\text{bench}.1,1} & R_{\text{bench}.1,2} & \cdots & R_{\text{bench}.1,N} \\ R_{\text{bench}.2,1} & R_{\text{bench}.2,2} & \cdots & R_{\text{bench}.2,N} \\ \cdots & \cdots & \cdots & \cdots \\ R_{\text{bench}.T,1} & R_{\text{bench}.T,2} & \cdots & R_{\text{bench}.T,N} \end{bmatrix} \begin{bmatrix} w_1 \\ w_2 \\ \cdots \\ w_N \end{bmatrix} + \begin{bmatrix} \varepsilon_1 \\ \varepsilon_2 \\ \cdots \\ \varepsilon_T \end{bmatrix} \tag{9.2b}$$

The main objective of style regression estimation is clearly to determine w_i, i.e. the unknown values of the regression coefficients. In order to make possible an intuitive interpretation of the estimated coefficients in terms of the weights of the asset class benchmarks in the fund portfolio and, thus, of the fund's exposures (implicit in its performance) to the investment categories represented by those benchmarks, the range of values that w_i can take has to be constrained.

The first constraint, known as portfolio constraint or restriction, requires that the total assets under management of the fund are exposed to the dynamics of the selected benchmarks. Consequently, this constraint is imposed with the Eq. (9.3):

$$\sum_{i=1}^{N} w_i = 1 \quad \text{that equals to } w_1 + w_2 + \ldots + w_i + \ldots + w_N = 1 \tag{9.3}$$

The second constraint, called short-selling restriction or non-negativity constraint, seeks to exclude the possibility of short exposures to the asset class benchmarks. It takes the form of the weak inequality found in (9.4):

$$w_i \geq 0 \text{ with } i = 1, \ldots, N \tag{9.4}$$

The inclusion (and respect) of these constraints has a dual effect. First, the model for the style analysis proposed by Sharpe must be reformulated as follows in (9.5):[3]

$$\begin{cases} R_{\text{fund},t} = \sum_{i=1}^{N} w_i \times R_{\text{benchmark}_i,t} + \varepsilon_t \\ \sum_{i=1}^{N} w_i = 1 \\ w_i \geq 0 \end{cases} \tag{9.5}$$

Second, once the w_i have been estimated, it is easy to show the fund's investment style "as slices in a pie chart".[4]

Although the addition of constraints to Sharpe's RBSA is natural and logical from an economic point of view, it does have an impact on the way in which the regression coefficients are estimated. In fact, the presence of the constraints means that the traditional method of ordinary least squares (OLS) can no longer be used, as in a classical multivariate linear regression. Instead, a quadratic programming algorithm has to be applied to solve for the style weights w_i. It is clear that in order to identify, among the many possibilities, the linear combination of benchmark exposures that can be considered the best set of asset class exposures, we need an appropriate criterion, and this criterion is the minimisation of variance in error terms, or the so-called residual returns, ε_t, each of which expresses the difference between the actual return of the fund in a given time t and the return that a portfolio would have registered over the same period when represented by the linear combination of the selected benchmarks that seeks to interpret the fund's investment style. In practice, a generic residual return is defined as follows:

---

[3]DeRoon et al. (2004) define Sharpe's model for analysing the investment style of funds as "strong style analysis". This expression allows the authors to distinguish "Sharpe style analysis" from "weak style analysis" (which does not impose any constraints), which is also called "generalized style analysis" by Agarwal and Naik (2000), and from "semi-strong style analysis", which imposes only one of the two constraints described in the text. Strictly speaking, it should be noted that there is also a small difference between "Sharpe style analysis" and "strong style analysis": the former, as shown in (9.1a), (9.1b) and (9.5), does not include the intercept, while the latter does.

[4]Lobosco and DiBartolomeo (1997), p. 81.

$$\varepsilon_t = \begin{array}{l} R_{\text{fund},\,t} - \left(w_1 \times R_{\text{benchmark}_{1,t}} + w_2 \times R_{\text{benchmark}_{2,t}} + \ldots + \right. \\ \left. w_i \times R_{\text{benchmark}_{i,t}} + \ldots + w_N \times R_{\text{benchmark}_{N,t}}\right) \end{array} \tag{9.6}$$

If we look carefully at (9.6), we can see that aiming to minimise the variance of the ε_t implies minimising the portion of total fund's returns variability across time not explained by the dynamics (the variations) of the returns of the asset classes considered in the analysis.

Given this criterion, the estimates of the coefficients of the constrained multivariate linear regression representing the RBSA are obtained by resolving the following problem (9.7):

$$\begin{gathered} \min_{w_i} \operatorname{var}(\varepsilon_t) = \\ \min_{w_i} \operatorname{var}(R_{\text{fund},t} - (w_1 \times R_{\text{benchmark}_1,t} + w_2 \times R_{\text{benchmark}_2,t} + \ldots + \\ w_i \times R_{\text{benchmark}_i,t} + \ldots + w_N \times R_{\text{benchmark}_N,t})) \\ \sum_{i=1}^{N} w_i = 1 \\ w_i \geq 0 \end{gathered} \tag{9.7}$$

We should also bear in mind that the output of RBSA is always accompanied by a parameter, R-squared, which serves to indicate the quality of fit and effectiveness of the output in explaining the variability of $R_{\text{fund},t}$ and is defined as (9.8):

$$R^2 = 1 - \frac{\sigma^2(\varepsilon_t)}{\sigma^2(R_{\text{fund},t})} \tag{9.8}$$

It is not difficult, then, to see that minimising the variance of the residual returns and maximising the R^2 amount to substantially the same thing. Indeed, the right side of (9.8) is the difference between 100% and the proportion of the variance in the fund's performance that is not explained by the returns variability of the asset classes considered in the analysis. It goes without saying that the value of R^2 can be between 0% and 100% (or between 0 and 1).

The set of w_i which minimises the variance of ε_t or maximises R^2—obtained, as stated above, by resolving the problem in (9.7)—identifies the so-called style weights, which in turn describe the mimicking portfolio, i.e. the portfolio of indices that would most closely replicate the performance of the fund.

To further clarify the interpretation of the result of RBSA, we can quote Sharpe himself, who speaks of "an average of potentially changing styles over the period covered",[5] i.e. average exposures of the fund to selected asset class benchmarks over the specified time period. We can see, then, that RBSA does not necessarily identify

[5] Sharpe (1992), p. 11.

what was actually in the fund (in terms of asset class). It indicates instead that, over the period used to estimate the coefficients, the fund performed "as if" it had invested using those weights.[6]

Up to this point, our attention has quite rightly been focused on the way the style weights are estimated. However, equal consideration should also be given to another methodological issue that is crucial for a correct, reliable implementation of RBSA: the choice of benchmarks. Following Sharpe's suggestions, it may be stated in qualitative terms, that there are three requirements (or conditions) that the selected benchmarks must satisfy. They should:

- Be exhaustive.
- Be mutually exclusive.
- Have low reciprocal correlation.

The first condition (exhaustiveness) reflects the need for the selected benchmarks to ensure total coverage of the investment universe accessible to the analysed fund. If important asset classes are omitted, it should come as no surprise if there is an unconvincing fitting between the returns of the fund and those of the asset class benchmarks, resulting in a poor R^2: given the portfolio constraint in (9.3), the quadratic programming algorithm shown above in (9.7) is forced to associate 100% of the fund's assets to the selected benchmarks, even if the latter do not take account of alternative investments that are available to the fund manager.

The second desirable feature for the independent variables (exclusivity) reflects the need to avoid any overlapping of benchmarks in the event of a not-so-careful choice of the latter. In other words, it is important to check beforehand that the potential benchmarks are not redundant, i.e. that they do not include the same constituents. Indeed, in case of redundancy, the algorithm would obviously have difficulty in distinguishing and individually assessing the effects of exposure to these asset classes on the fund's performance.

The third requirement (low correlation) can be formulated more broadly. On this issue, Sharpe stated in 1992 that: "[...] asset class returns should have either low correlations with one another or, in cases in which correlations are high, different standard deviations".[7] Essentially, this means that we should not use benchmarks that can be interpreted as linear combinations of any other benchmarks. In technical terms, Sharpe is suggesting that we avoid the problem of multicollinearity.[8] To

---

[6]To emphasise this point, the financial community usually quotes the metaphor used by Sharpe: "If it walks like a duck and talks like a duck, for all intents and purposes it is a duck". Sharpe also observed: "Returns-based style analysis is not going to dissect the creature to determine if its DNA belongs to that of a duck, but it will tell you if it has enough duck-like characteristics to qualify" Sharpe (1988), p. 60.

[7]Sharpe (1992), p. 8.

[8]Multicollinearity describes a condition of strong reciprocal linear dependence of the regressors. This is "labelled" as a problem because it means that the individual regression coefficients are not very significant, even if the explicative value (the fitting) of their joint action or their combined effect is not compromised. Domian and Reichenstein (2008).

conclude this illustration of the guidelines for the choice of explicative benchmarks, it is useful to point out a feature common to both the second and third requirements: if they are not respected, the reliability of the estimated w_i coefficients can be jeopardised, making them unstable.[9]

Having completed the examination of the methodological profiles of RBSA, we can now draw some conclusions as regards the interpretation of a fund's performance based on its output; obviously, such considerations assume that the analysis has been implemented according to the method presented in (9.7) and in conformity with the ideal requirements for the best choice of the asset class benchmarks. In this respect, it can be said that RBSA "breaks down" a fund's performance into two components:

- The "style return", corresponding to the (periodic) performance of the mimicking portfolio. Considering that this is a combination of passive positions in asset class indices, it should be seen as a sort of ad hoc benchmark (i.e. a customised benchmark for the analysed fund) known as the style benchmark. It can then be automatically deduced that the style return reflects the performance of the style benchmark and, consequently, expresses the returns which would have been produced by that type of asset allocation in the time period under consideration.
- The "selection return", corresponding to an idiosyncratic component in the fund's performance. This is measured as the difference for each period t between the fund's performance and that of the style benchmark. In practice, it coincides with ε_t at the point in which the best set of w_i has already been identified and, consequently, the variance of ε_t has already been minimised. In logical terms, this identifies a "part" of the fund's performance that cannot be attributed to its implicit asset allocation.[10] In some cases, it may be termed alpha return because it is a measure of value added relative to the fund's customised benchmark resulting from active behaviour.[11]

[9]On the topic of the reliability of the estimated style weights, see Sect. 9.2.1 below.

[10]For the sake of completeness, note that when RBSA is used as a performance measurement tool, the mean value of the residual returns ε_t (which is still called selection return) as well as their standard deviation (called style tracking error volatility) are often considered. The ratio between these two values gives the so-called selection ratio. It is easy to see that this latter value differs from the information ratio seen in Chap. 8, as the series of residual returns ε_t used for the calculations of mean and volatility is determined with reference to a customised benchmark (the style benchmark), rather than to a standardised benchmark such as a generic market index.

[11]DiBartolomeo and Witkowski (1997).

9.2.1 Analysis of the Accuracy of the Style Weights

The user of RBSA obtains point estimates of the exposures of a fund to the asset class benchmarks that serve as independent variables. As known, the overall output that represents the style benchmark is accompanied by a measurement of the goodness of fit of the estimated regression model, the R^2 statistic. Generally, the user of RBSA does not ask spontaneously for anything else, nor do the application programs provide any other details. However, if we look carefully, stopping style analysis implementation at this stage omits an important issue. Indeed, the accuracy and reliability of the estimated parameters of the regression, and thus of the style weights, are not verified in any way. Therefore, on completing the RBSA, we have a result which measures the overall value of the model, but no indications of the quality of fit of each estimated coefficient. At the very least, this fact must be taken into consideration, because it is in contrast with what happens with the estimates in a classical model of multivariate linear regression. In that case, it is common practice to highlight the level of uncertainty surrounding each estimated regression coefficient by indicating its standard error, and to subject each coefficient to an inferential test with a null hypothesis that its true value be zero.

In RBSA, however, the absence of a test of the statistical significance of the style weights is no coincidence. The calculation of the standard error associated with the point estimates and, therefore, also of the confidence interval, is hampered or complicated by the use of a constrained regression model that includes inequality constraints or restrictions. If the latter are binding, then we have to decline the asymptotic distribution of the estimators stipulated by the standard theory, on the basis of which the confidence intervals are defined. Despite such awareness, the objective importance of realising whether the individual style weights can (or cannot) express the actual relationship between the performance of the fund and that of the selected benchmark has encouraged efforts to find solutions, other than the standard procedures, to approximate both the standard errors and the confidence intervals of Sharpe style weights. In particular, the solution described in the paper "Approximating the Confidence Intervals for Sharpe Style Weights", published in the *Financial Analysts Journal*,[12] has been quite well accepted and used. The procedure involves a number of steps that can be described as follows:

- Estimate the style weights by applying a regular style analysis as described in (9.5) to a fund, using T returns for both the fund and for the N asset class benchmarks selected.
- Calculate the standard deviation of the residual returns ε_t of the fund (σ_{ε_t}) unexplained by the N style weights estimated in the previous step.
- Perform RBSA using the i-th of the N benchmarks originally selected for the RBSA of the fund as a dependent variable and the remaining $N-1$ benchmarks

[12]Lobosco and DiBartolomeo (1997).

as explanatory variables, without introducing the inequality constraint requiring that the estimated regression coefficients be not negative.

- Calculate what is known as the unexplained Sharpe style index volatility for index i. This is the standard deviation of that "portion" of the returns of the benchmark i that is not attributable to sensitivity to the dynamics of the other market indices. We denote this volatility by $\sigma_{\text{Bench.}_i}$.
- Calculate the standard error of the style weight of the i-th asset class benchmark (σ_{w_i}) specified for the RBSA of the fund, combining the volatility of the residual (selection) return of the RBSA of the fund with the unexplained Sharpe style index volatility for index i. In practice, this last step leads to formula (9.9) for the calculation of σ_{w_i}:

$$\sigma_{w_i} = \frac{\sigma_{\varepsilon_t}}{\sigma_{\text{Bench.}_i} \times \sqrt{T - N^* - 1}} \tag{9.9}$$

where N^* is the number of non-zero style weights obtained from the RBSA on the index i performed using the remaining benchmarks.

From (9.9), we can see that the volatility associated with the fund's exposures to benchmarks, σ_{w_i}, resulting from RBSA increases as more similar indices are included in the analysis. In this case, the "original contribution" of a benchmark turns out to be particularly limited; in other words, it is difficult to identify a part of its dynamics over time that is independent of the dynamics of the other benchmarks; therefore, the quantity $\sigma_{\text{Bench.}_i}$, that is the denominator in (9.9), is small, and this obviously increases the result of the fraction. As a result, we obviously find ourselves in complete agreement with the observation: "Style analysis can only reliably attribute portfolio returns to the portions of the market index returns that are themselves not attributable to the returns of the other indexes".[13] From the procedure illustrated to estimate the standard error associated with each style weight, a general recommendation can therefore be extrapolated which is essential whenever this model is applied: the number of explanatory benchmarks included in the RBSA should be carefully counterbalanced with the level of reliability sought in the analysis. The addition of independent variables can make the model more complete, but this does not necessarily produce greater reliability in the relationship identified between the behaviour of the fund and that of the individual benchmark.

After this in-depth analysis of the steps and logic involved in the procedure for calculating the standard errors of the style weights, the reasons why these are useful can be summarised. Calculation of the standard errors allows practitioners:

- To assess (as repeatedly stated) the accuracy of the estimated style coefficients and construct, as will be shown below, the respective confidence interval.
- To test whether the coefficients are significantly different from zero.
- To verify whether the coefficients are significantly different from each other.

[13] Lobosco and DiBartolomeo (1997), p. 81.

With regard to the calculation of the confidence interval within which, with a given probability, the "true" (but unknown) value of the style weight should be located, it is enough to apply to the estimated value of the parameter a multiplier of the standard error identified using the hypothesis of a normal distribution and respecting the chosen level of probability. By way of example, if we wish to have a confidence interval of 90% or 95% for a given style weight, we need to proceed as indicated in (9.10) and (9.11), respectively:

$$w_i \pm 1.645 \times \sigma_{w_i} \tag{9.10}$$

$$w_i \pm 1.960 \times \sigma_{w_i} \tag{9.11}$$

Observation of (9.9), (9.10), and (9.11) reveals that the confidence interval:

- Increases as the standard deviation of ε_t increases.
- Decreases the greater the number of past returns for the fund and for the explanatory benchmarks used in the RBSA.
- Decreases the more the behaviour of the *i*-th benchmark is independent of the dynamics of the other benchmarks.

Once the confidence interval has been estimated for each of the *N* benchmarks selected, if zero is included in this interval, it is fair to conclude that the estimated style weight is not significantly different from zero.

However, this is not the only way to assess whether to accept the null hypothesis that an estimated style weight is equal to zero, or rather the alternative hypothesis that it differs from zero. In fact, a two-tail *t*-test can be performed which presumes the calculation of a *t*-ratio defined by (9.12):

$$t_{w_i} = \frac{w_i}{\sigma_{w_i}} \tag{9.12}$$

that can be compared with a critical value obtained from Student's *t* distribution with degrees of freedom equal to $T - N^* - 1$, and obviously taking account of the confidence level that we wish to adopt in the analysis of the significance of the individual estimated parameters. If we find that $t_{w_i} >$ critical value, it is reasonable to conclude that the parameter w_i is statistically significant and that the null hypothesis can be rejected.

Both the procedure to determine the confidence intervals and the *t*-test can be reasonably considered essential supports in identifying the indices to include in the final version of RBSA.[14]

[14] A different procedure for identifying the confidence interval for each of the coefficients expressing the style weights which does not need to assume their normal distribution is described in Kim et al. (2005).

9.3 An Empirical Application of Returns-Based Style Analysis

Having given a detailed description of the theory underlying Sharpe style analysis, it is now useful to provide an example of a practical application of RBSA. Let us consider a fund classified by Morningstar in the Euro Area Large Cap Equity category, whose net asset value (NAV) is expressed in local currency, i.e. euro. The goal is to determine the implicit asset allocation of the fund from a geographic perspective starting from its historical performances, meaning that we want to deduce exposure to indices representing countries in the Eurozone as implied by the actual performance of the fund. To this end, it seemed proper to select indices of countries constituting the MSCI EMU, namely: MSCI Austria, MSCI Belgium, MSCI Finland, MSCI France, MSCI Germany, MSCI Ireland, MSCI Italy, MSCI Netherlands, MSCI Portugal, and MSCI Spain. Use of these indices should ensure that we meet the exhaustiveness and non-redundancy constraints. More difficult, in the specific case, is being compliant with the third condition: as can be seen from the correlation estimates in Table 9.1 for the period of 60 months from January 2018 to December 2022, they are rather high, oscillating between +0.59 and +0.93. However, in the last column, there is, at least, some differentiation between the indices in terms of annualised standard deviation (the minimum and maximum calculated values are, respectively, 15.20% and 29.05%).

Given the choice of asset class benchmarks, the estimated style regression respecting the restrictions described in (9.3) and (9.4) can be expressed as follows:

$$
\begin{aligned}
R_{\text{Euro area large cap equity fund},t} = {} & w_{\text{MSCI Austria}} \cdot R_{\text{MSCI Austria},t} \\
& + w_{\text{MSCI Belgium}} \cdot R_{\text{MSCI Belgium},t} + w_{\text{MSCI Finland}} \cdot R_{\text{MSCI Finland},t} \\
& + w_{\text{MSCI France}} \cdot R_{\text{MSCI France},t} + w_{\text{MSCI Germany}} \cdot R_{\text{MSCI Germany},t} \\
& + w_{\text{MSCI Ireland}} \cdot R_{\text{MSCI Ireland},t} + w_{\text{MSCI Italy}} \cdot R_{\text{MSCI Italy},t} \\
& + w_{\text{MSCI Netherlands}} \cdot R_{\text{MSCI Netherlands},t} + w_{\text{MSCI Portugal}} \cdot R_{\text{MSCI Portugal},t} \\
& + w_{\text{MSCI Spain}} \cdot R_{\text{MSCI Spain},t} + \varepsilon_t
\end{aligned}
\tag{9.13}
$$

Obviously, the time series of the monthly returns of the fund and of the selected explicative benchmarks are the input to the RBSA. These are taken from an Excel spreadsheet and shown in Fig. 9.1.

With RBSA, the estimate of the regression coefficients is obtained by resolving the quadratic programming problem that seeks to minimise the variance of residual returns. In order to initialise their values, we can first attribute arbitrary values to factor loadings (providing that they are positive and they add up to 1). In Fig. 9.2, we have assumed that the fund is equally exposed to the ten selected benchmarks and have calculated the performance that it would have registered monthly in that case. In each month, the arithmetic difference between the fund's actual performance and the artificial figure obtained gives the residual returns shown in column P. These show a variance of 0.000405382 with a standard deviation of 2.0134%. Applying the formula in (9.8), we can conclude that the investment style randomly attributed to the

Table 9.1 Correlations and standard deviations for the explanatory benchmarks

	MSCI AUSTRIA	MSCI BELGIUM	MSCI FINLAND	MSCI FRANCE	MSCI GERMANY	MSCI IRELAND	MSCI ITALY	MSCI NETHER-LANDS	MSCI PORTUGAL	MSCI SPAIN	Standard deviation (annualised)
MSCI AUSTRIA	1.00	0.76	0.70	0.87	0.86	0.67	0.86	0.73	0.63	0.84	29.05%
MSCI BELGIUM	0.76	1.00	0.71	0.83	0.77	0.60	0.79	0.75	0.64	0.80	19.80%
MSCI FINLAND	0.70	0.71	1.00	0.80	0.80	0.68	0.70	0.73	0.62	0.65	16.65%
MSCI FRANCE	0.87	0.83	0.80	1.00	0.93	0.77	0.93	0.84	0.69	0.90	18.75%
MSCI GERMANY	0.86	0.77	0.80	0.93	1.00	0.83	0.88	0.86	0.69	0.82	19.27%
MSCI IRELAND	0.67	0.60	0.68	0.77	0.83	1.00	0.69	0.76	0.60	0.62	20.34%
MSCI ITALY	0.86	0.79	0.70	0.93	0.88	0.69	1.00	0.75	0.65	0.92	22.46%
MSCI NETHERLANDS	0.73	0.75	0.73	0.84	0.86	0.76	0.75	1.00	0.68	0.67	19.23%
MSCI PORTUGAL	0.63	0.64	0.62	0.69	0.69	0.60	0.65	0.68	1.00	0.66	15.20%
MSCI SPAIN	0.84	0.80	0.65	0.90	0.82	0.62	0.92	0.67	0.66	1.00	20.64%

	A	B	C	D	E	F	G	H	I	J	K	L	M
1					STYLE ANALYSIS INPUTS								
2													
3			DEPENDENT VARIABLE	INDEPENDENT VARIABLES: ASSET CLASS BENCHMARKS									
4			EURO AREA LARGE CAP EQUITY FUND	MSCI AUSTRIA	MSCI BELGIUM	MSCI FINLAND	MSCI FRANCE	MSCI GERMANY	MSCI IRELAND	MSCI ITALY	MSCI NETHER-LANDS	MSCI PORTUGAL	MSCI SPAIN
5		Jan-18	3.84%	6.86%	1.61%	3.00%	3.16%	2.11%	-1.11%	7.62%	2.50%	0.70%	5.00%
6		Feb-18	-3.93%	-3.65%	-2.10%	4.87%	-2.82%	-5.18%	-5.02%	-3.90%	-3.02%	-2.43%	-6.04%
7		Mar-18	-2.55%	-3.10%	-1.35%	-2.18%	-2.28%	-2.70%	-2.15%	-0.54%	-0.81%	2.44%	-2.73%
8		Apr-18	3.16%	0.31%	-1.69%	6.17%	6.61%	3.98%	4.23%	7.45%	3.43%	2.93%	4.61%
9		May-18	-1.59%	-4.05%	-2.14%	1.62%	-0.64%	0.28%	6.86%	-8.50%	0.18%	3.16%	-6.10%
10		Jun-18	-0.54%	-2.20%	2.95%	-1.05%	-1.01%	-2.38%	-3.58%	-0.72%	0.08%	0.38%	2.56%
11		Jul-18	1.26%	4.73%	1.95%	0.95%	3.41%	4.08%	-0.30%	2.95%	3.88%	4.74%	3.57%
12		Aug-18	-1.72%	-6.83%	-4.20%	3.15%	-1.43%	-2.48%	-1.27%	-8.98%	-1.38%	-1.59%	-5.38%
13		Sep-18	-0.30%	3.26%	-2.43%	-0.41%	1.40%	-1.53%	-3.41%	2.50%	-3.27%	-3.01%	0.14%
14		Oct-18	-8.56%	-3.95%	-9.53%	-7.23%	-7.10%	-6.15%	-5.83%	-7.10%	-5.45%	-7.53%	-4.46%
15		Nov-18	-2.85%	-4.18%	1.95%	-4.94%	-1.70%	-1.97%	-5.65%	0.83%	2.89%	-3.07%	1.93%
16		Dec-18	-6.22%	-12.48%	-10.16%	-1.66%	-5.48%	-6.66%	-5.91%	-4.36%	-7.02%	-2.85%	-4.71%
17		Jan-19	6.05%	8.50%	9.43%	7.86%	5.45%	6.31%	4.79%	8.14%	6.85%	4.60%	6.15%
18		Feb-19	4.46%	3.55%	3.35%	2.04%	4.94%	2.52%	6.63%	4.22%	5.56%	3.89%	2.85%
19		Mar-19	3.35%	-1.74%	4.62%	-0.12%	1.85%	-0.05%	1.46%	3.48%	2.56%	3.27%	-0.23%
20		Apr-19	4.45%	6.74%	3.62%	-0.42%	4.81%	7.17%	6.59%	2.28%	6.13%	1.89%	4.42%
21		May-19	-5.18%	-10.33%	-7.56%	-4.03%	-5.59%	-5.56%	-5.03%	-7.67%	-5.48%	-3.62%	-5.52%
22		Jun-19	4.95%	3.86%	4.17%	3.38%	6.11%	5.05%	2.02%	7.45%	4.26%	2.00%	2.57%
23		Jul-19	-0.49%	1.36%	8.48%	1.35%	-0.18%	-1.28%	0.71%	0.85%	3.18%	1.79%	-2.04%
24		Aug-19	-1.41%	-4.01%	-4.04%	-2.29%	-0.62%	-1.88%	-1.15%	-0.31%	0.12%	-0.58%	-2.41%
25		Sep-19	2.31%	4.03%	3.80%	3.60%	3.48%	3.49%	4.35%	3.84%	3.65%	4.92%	5.13%
26		Oct-19	1.04%	4.31%	-7.14%	-2.41%	1.25%	3.50%	4.76%	2.19%	-1.12%	2.24%	-0.12%
27		Nov-19	3.92%	-0.70%	2.45%	-0.65%	2.92%	3.01%	7.07%	1.78%	3.55%	0.29%	0.75%
28		Dec-19	0.75%	1.31%	0.95%	3.22%	1.13%	0.11%	2.61%	0.91%	1.90%	2.94%	2.31%
29		Jan-20	-0.23%	-5.01%	-2.29%	2.45%	-2.15%	-1.82%	-3.20%	-0.91%	-1.72%	6.08%	-1.48%
30		Feb-20	-7.16%	-10.06%	-15.97%	-6.81%	-8.06%	-8.37%	-6.38%	-5.88%	-7.10%	-5.93%	-6.50%
31		Mar-20	-14.93%	-31.60%	-15.97%	-13.29%	-17.61%	-16.93%	-16.14%	-22.45%	-11.07%	-10.92%	-21.99%
32		Apr-20	6.88%	15.33%	6.77%	9.60%	4.58%	10.01%	8.73%	2.02%	8.86%	1.57%	1.60%
33		May-20	5.41%	0.50%	-0.79%	7.41%	3.33%	7.24%	5.60%	4.11%	5.67%	8.34%	2.69%
34		Jun-20	5.65%	1.77%	4.10%	0.21%	5.00%	5.29%	1.82%	6.76%	6.07%	-0.45%	3.27%
35		Jul-20	-0.18%	-5.97%	-0.11%	1.82%	-2.40%	-0.22%	3.27%	-1.51%	-0.91%	-3.14%	-4.79%
36		Aug-20	4.74%	4.28%	0.39%	7.17%	3.60%	5.25%	5.17%	2.43%	1.65%	-1.10%	0.50%
37		Sep-20	-0.82%	-7.02%	-2.56%	-1.78%	-2.66%	-1.13%	1.01%	-3.79%	0.76%	-3.42%	-3.68%
38		Oct-20	-5.33%	-0.17%	-5.36%	-7.05%	-4.15%	-9.63%	-0.12%	-6.27%	-1.73%	-2.07%	-3.39%
39		Nov-20	15.22%	28.60%	19.78%	13.26%	19.70%	14.09%	7.07%	23.51%	11.10%	10.09%	26.13%
40		Dec-20	1.90%	9.99%	0.04%	0.26%	0.54%	3.67%	1.25%	1.35%	3.98%	8.06%	0.46%
41		Jan-21	-2.77%	2.99%	-1.69%	2.72%	-2.40%	-1.13%	-2.98%	-3.30%	3.18%	-1.25%	-3.87%
42		Feb-21	4.00%	4.21%	-1.00%	-1.14%	5.08%	2.17%	0.74%	5.80%	3.92%	-3.83%	5.32%
43		Mar-21	5.22%	6.00%	4.24%	1.11%	6.00%	7.50%	12.18%	8.15%	7.97%	4.94%	3.80%
44		Apr-21	3.86%	1.51%	4.37%	4.18%	3.65%	1.12%	-1.52%	-1.89%	2.77%	-1.12%	2.94%
45		May-21	3.09%	11.08%	2.66%	2.21%	3.10%	1.73%	0.86%	4.78%	0.79%	3.88%	4.01%
46		Jun-21	1.36%	-1.99%	1.04%	3.15%	1.15%	1.31%	2.19%	-0.09%	2.76%	-4.36%	-2.52%
47		Jul-21	3.42%	2.33%	-1.97%	5.94%	1.70%	0.13%	1.77%	0.72%	4.58%	-0.74%	-1.80%
48		Aug-21	1.92%	6.13%	1.25%	0.84%	1.10%	1.79%	6.39%	1.77%	7.37%	7.48%	2.07%
49		Sep-21	-3.64%	4.02%	-2.73%	-7.21%	-2.51%	-3.85%	-5.51%	-1.25%	-5.76%	-0.61%	-1.24%
50		Oct-21	5.32%	1.30%	3.29%	1.24%	4.61%	2.31%	1.56%	5.33%	6.92%	6.92%	4.01%
51		Nov-21	-0.04%	-0.15%	-4.56%	-2.32%	-1.54%	-3.90%	-6.72%	-3.85%	-3.17%	-3.45%	-8.23%
52		Dec-21	3.46%	5.99%	5.17%	6.09%	5.98%	4.50%	8.23%	6.23%	1.93%	0.56%	5.26%
53		Jan-22	-10.15%	-0.73%	-3.65%	-4.97%	-2.42%	-2.30%	-5.83%	-1.17%	-10.35%	-2.60%	-0.15%
54		Feb-22	-5.58%	-15.20%	-0.94%	-7.02%	-4.65%	-7.02%	-5.14%	-5.03%	-5.12%	-0.35%	-1.75%
55		Mar-22	2.81%	-2.85%	1.79%	1.91%	0.30%	-1.92%	-8.61%	-2.05%	-0.71%	7.74%	-0.13%
56		Apr-22	-4.64%	-1.37%	-0.30%	0.98%	-1.29%	-3.08%	1.12%	-1.88%	-5.14%	-1.08%	2.70%
57		May-22	-1.96%	2.88%	-4.30%	-0.75%	0.09%	2.16%	-0.88%	1.81%	-0.72%	3.58%	3.25%
58		Jun-22	-9.78%	-13.18%	-3.42%	-5.38%	-8.18%	-11.34%	-14.58%	-12.27%	-8.31%	-2.47%	-8.09%
59		Jul-22	12.66%	3.97%	2.49%	6.14%	9.11%	4.94%	9.84%	5.10%	13.24%	7.36%	0.95%
60		Aug-22	-7.25%	-8.04%	-6.10%	-0.70%	-5.03%	-5.06%	3.30%	-3.31%	-8.84%	-2.17%	-2.87%
61		Sep-22	-6.25%	-5.40%	-3.81%	-6.06%	-6.15%	-6.37%	-11.12%	-3.96%	-7.64%	-9.60%	-6.47%
62		Oct-22	5.68%	9.98%	7.15%	3.28%	8.51%	8.39%	10.13%	10.33%	4.92%	2.82%	8.62%
63		Nov-22	8.22%	12.99%	5.18%	6.75%	7.32%	9.40%	4.04%	9.00%	12.65%	4.38%	5.14%
64		Dec-22	-3.90%	-3.18%	-0.18%	-3.17%	-3.69%	-3.47%	-2.69%	-3.55%	-6.00%	0.26%	-1.21%

Fig. 9.1 The inputs of the RBSA

fund (see cells O3:X3) can explain 86.53% of the actual fund returns variance. At this point, in order to move from an invented style benchmark to one obtained by using the RBSA methodology properly, we need to find out values for the regression

	N	O	P	Q	R	S	T	U	V	W	X	Y
1		STYLE ANALYSIS IMPLEMENTATION										
2		MSCI AUSTRIA	MSCI BELGIUM	MSCI FINLAND	MSCI FRANCE	MSCI GERMANY	MSCI IRELAND	MSCI ITALY	MSCI NETHER-LANDS	MSCI PORTUGAL	MSCI SPAIN	SUM
3	INITIAL STYLE WEIGHTS	10.00%	10.00%	10.00%	10.00%	10.00%	10.00%	10.00%	10.00%	10.00%	10.00%	100.00%
4		STYLE RETURNS	RESIDUAL RETURNS	RESIDUAL RETURNS VARIANCE		R-SQUARED						
5	Jan-18	3.15%	0.70%	0.00041		86.53%						
6	Feb-18	-2.93%	-1.00%	ST.DEV.								
7	Mar-18	-1.54%	-1.01%	2.0134%								
8	Apr-18	3.80%	-0.64%									
9	May-18	-0.93%	-0.66%									
10	Jun-18	-0.50%	-0.04%									
11	Jul-18	3.00%	-1.73%									
12	Aug-18	-3.04%	1.31%									
13	Sep-18	-0.68%	0.37%									
14	Oct-18	-6.43%	-2.12%									
15	Nov-18	-1.39%	-1.46%									
16	Dec-18	-6.13%	-0.09%									
17	Jan-19	6.81%	-0.76%									
18	Feb-19	3.96%	0.51%									
19	Mar-19	1.51%	1.84%									
20	Apr-19	4.32%	0.13%									
21	May-19	-6.04%	0.86%									
22	Jun-19	4.09%	0.86%									
23	Jul-19	1.42%	-1.91%									
24	Aug-19	-1.72%	0.30%									
25	Sep-19	4.03%	-1.72%									
26	Oct-19	0.75%	0.29%									
27	Nov-19	2.05%	1.88%									
28	Dec-19	1.74%	-0.98%									
29	Jan-20	-1.01%	0.78%									
30	Feb-20	-8.10%	0.94%									
31	Mar-20	-17.80%	2.86%									
32	Apr-20	6.91%	-0.03%									
33	May-20	4.41%	1.00%									
34	Jun-20	3.39%	2.26%									
35	Jul-20	-1.40%	1.21%									
36	Aug-20	2.93%	1.81%									
37	Sep-20	-2.42%	1.60%									
38	Oct-20	-3.99%	-1.34%									
39	Nov-20	17.33%	-2.12%									
40	Dec-20	2.96%	-1.06%									
41	Jan-21	-0.77%	-2.00%									
42	Feb-21	2.13%	1.88%									
43	Mar-21	6.19%	-0.97%									
44	Apr-21	1.60%	2.26%									
45	May-21	3.51%	-0.42%									
46	Jun-21	0.26%	1.10%									
47	Jul-21	1.27%	2.15%									
48	Aug-21	3.62%	-1.70%									
49	Sep-21	-2.66%	-0.97%									
50	Oct-21	3.75%	1.57%									
51	Nov-21	-3.79%	3.75%									
52	Dec-21	4.99%	-1.54%									
53	Jan-22	-3.42%	-6.73%									
54	Feb-22	-5.22%	-0.35%									
55	Mar-22	-0.45%	3.26%									
56	Apr-22	-0.93%	-3.71%									
57	May-22	0.71%	-2.67%									
58	Jun-22	-8.72%	-1.06%									
59	Jul-22	6.31%	6.34%									
60	Aug-22	-3.88%	-3.37%									
61	Sep-22	-6.66%	0.41%									
62	Oct-22	7.41%	-1.74%									
63	Nov-22	7.68%	0.54%									
64	Dec-22	-2.69%	-1.21%									

Fig. 9.2 The start of the RBSA in Excel

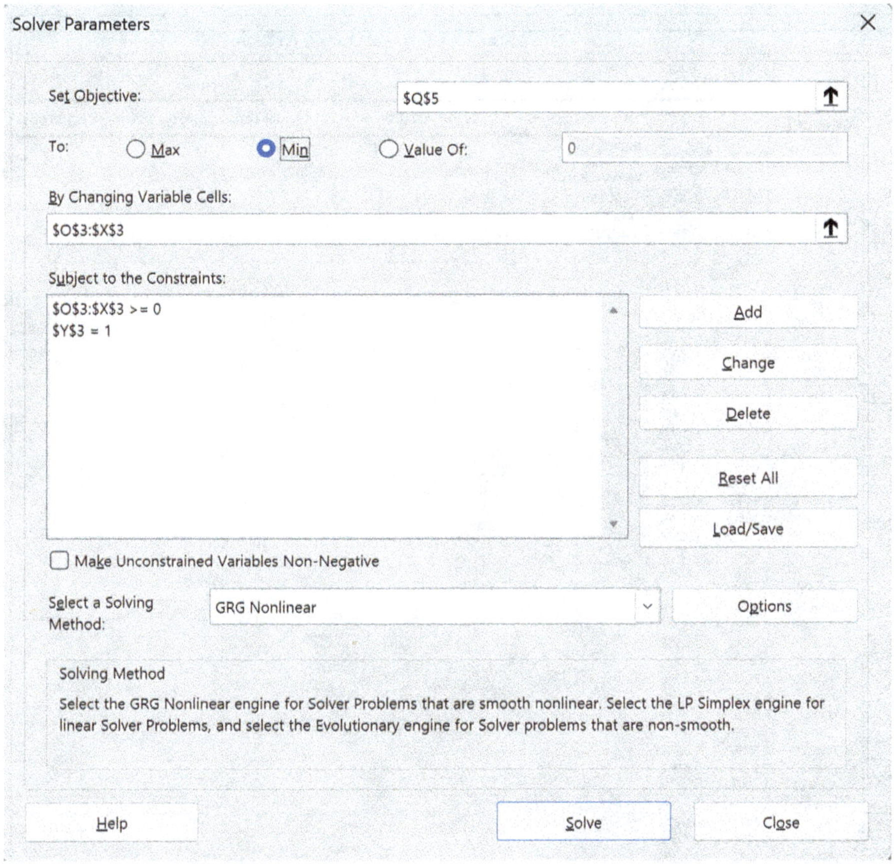

Fig. 9.3 The Excel Solver window

coefficients such that the variance of ε_t is as far below 0.000405382 as possible. For this, the Solver function in Excel is extremely useful.[15] Specifically, and as illustrated in Fig. 9.3, we need to enter the reference to the target cell (Q5) into the Solver window and choose "Min" to obtain the minimisation. By entering O3:X3 into the "by changing cells" box, we specify that this must occur by modifying the invented style weights. With the "Add" button, we also set the constraints to apply to the style weights: *O3*:*X3* ≥ 0 and *Y3* = 1. Finally, we click the "Solve" button to initialise the search for the optimal solutions (i.e. those that minimise the target cell) describing the fund's implicit asset allocation. The output is shown in Fig. 9.4, and the resulting estimated style regression model is given in (9.14):

[15] A helpful guide to using the Excel "Solver" for RBSA can also be found in Atkinson and Choi (2001).

	N	O	P	Q	R	S	T	U	V	W	X	Y
1		STYLE ANALYSIS IMPLEMENTATION										
2		MSCI AUSTRIA	MSCI BELGIUM	MSCI FINLAND	MSCI FRANCE	MSCI GERMANY	MSCI IRELAND	MSCI ITALY	MSCI NETHERLANDS	MSCI PORTUGAL	MSCI SPAIN	SUM
3	STYLE WEIGHTS	0.00%	0.00%	8.36%	51.90%	0.00%	0.00%	0.00%	39.74%	0.00%	0.00%	100.00%
4		STYLE RETURNS	RESIDUAL RETURNS	RESIDUAL RETURNS VARIANCE		R-SQUARED						
5	Jan-18	2.88%	0.96%	0.000186		93.84%						
6	Feb-18	-2.26%	-1.67%	ST.DEV.								
7	Mar-18	-1.69%	-0.86%	1.3623%								
8	Apr-18	5.31%	-2.15%									
9	May-18	-0.12%	-1.47%									
10	Jun-18	-0.58%	0.04%									
11	Jul-18	3.39%	-2.13%									
12	Aug-18	-1.03%	-0.70%									
13	Sep-18	-0.61%	0.31%									
14	Oct-18	-6.46%	-2.10%									
15	Nov-18	-0.15%	-2.70%									
16	Dec-18	-5.77%	-0.45%									
17	Jan-19	6.21%	-0.16%									
18	Feb-19	4.95%	-0.48%									
19	Mar-19	1.97%	1.39%									
20	Apr-19	4.90%	-0.44%									
21	May-19	-5.42%	0.24%									
22	Jun-19	5.14%	-0.20%									
23	Jul-19	1.28%	-1.77%									
24	Aug-19	-0.47%	-0.95%									
25	Sep-19	3.56%	-1.25%									
26	Oct-19	0.00%	1.03%									
27	Nov-19	2.87%	1.05%									
28	Dec-19	1.61%	-0.86%									
29	Jan-20	-1.60%	1.37%									
30	Feb-20	-7.58%	0.41%									
31	Mar-20	-14.65%	-0.28%									
32	Apr-20	6.70%	0.17%									
33	May-20	4.60%	0.81%									
34	Jun-20	5.03%	0.62%									
35	Jul-20	-1.46%	1.27%									
36	Aug-20	3.12%	1.62%									
37	Sep-20	-1.23%	0.40%									
38	Oct-20	-3.43%	-1.90%									
39	Nov-20	15.75%	-0.53%									
40	Dec-20	1.88%	0.02%									
41	Jan-21	0.25%	-3.02%									
42	Feb-21	4.10%	-0.09%									
43	Mar-21	6.37%	-1.15%									
44	Apr-21	3.34%	0.52%									
45	May-21	2.11%	0.98%									
46	Jun-21	1.96%	-0.60%									
47	Jul-21	3.20%	0.22%									
48	Aug-21	3.57%	-1.65%									
49	Sep-21	-4.19%	0.56%									
50	Oct-21	5.25%	0.07%									
51	Nov-21	-2.25%	2.21%									
52	Dec-21	4.38%	-0.92%									
53	Jan-22	-5.78%	-4.36%									
54	Feb-22	-5.03%	-0.54%									
55	Mar-22	0.03%	2.77%									
56	Apr-22	-2.63%	-2.02%									
57	May-22	-0.30%	-1.66%									
58	Jun-22	-8.00%	-1.78%									
59	Jul-22	10.50%	2.15%									
60	Aug-22	-6.18%	-1.07%									
61	Sep-22	-6.74%	0.49%									
62	Oct-22	6.65%	-0.97%									
63	Nov-22	9.39%	-1.17%									
64	Dec-22	-4.56%	0.66%									

Fig. 9.4 The output of the RBSA

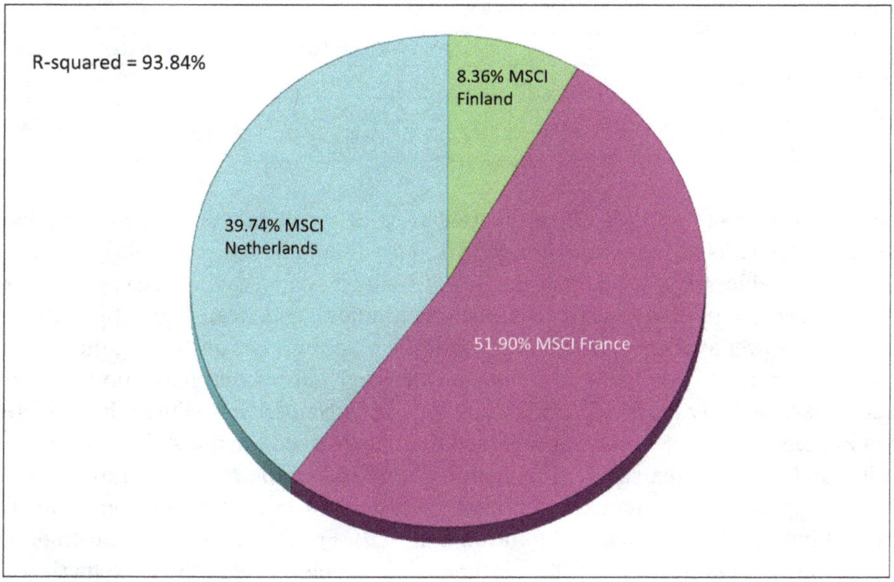

Fig. 9.5 The style benchmark

$$R_{\text{Euro area large cap equity fund}} = 0.083552 \cdot R_{\text{MSCI Finland}} + 0.519017 \cdot R_{\text{MSCI France}} + \\ + 0.397429 \cdot R_{\text{MSCI Netherlands}} \quad (9.14)$$

It is well-known that a pie chart is the typical representation of the style benchmark, as shown in Fig. 9.5.

Over the period analysed, the average composition of the mimicking portfolio suggested by the estimated model implies a variance of the residual (selection) returns of 0.000185593, corresponding to a volatility of 1.3623%. Obviously, these are lower than the previous values, while the goodness of fit level is higher, with R^2 equal to 93.84%. As stated in Sect. 9.2.1, since we use RBSA to identify the real underlying relationship between the returns on the fund and the returns on the asset class benchmarks chosen as explanatory variables, it is easy to understand why discovering the volatility associated with the estimate of each style weight is of such interest. It goes without saying that the procedure described in the previous section is invaluable in checking the accuracy of the estimates. Therefore, for each country benchmark constituting the MSCI EMU with non-zero weight, a style index was identified which potentially includes the nine remaining benchmarks. After that, the monthly values of the unexplained Sharpe style index volatility can be determined for each index, and the results are given in Table 9.2.

Table 9.2 reveals that MSCI Finland is the index for which the attempt to explain its performance as a linear combination of exposures to the remaining indices is the least successful. In fact, it is here that we find the highest value for the standard deviation in the residual returns derived from the monthly difference between the

Table 9.2 The unexplained Sharpe style index volatility

	Unexplained Sharpe style index volatility
MSCI FINLAND	2.7004%
MSCI FRANCE	1.2087%
MSCI NETHERLANDS	2.3709%

performance of the index itself and that of its style benchmark calculated using the other benchmarks as regressors in a RBSA. This should not be particularly surprising, since Table 9.1 suggests that the MSCI Finland is among the indexes with the lowest average correlation value with the other indices, and, even more importantly, it shows a lower average correlation towards the other indices constituting the MSCI EMU compared to what can be found for the other indices that make up the style benchmark, namely the MSCI France and the MSCI Netherlands, although these are not extreme differences. Having obtained the unexplained Sharpe style index volatility, and knowing the standard deviation (1.3623%) of the residual returns of the fund compared to its mimicking portfolio, we are now in possession of the fundamental inputs needed to apply the formula in (9.9). For the purposes of illustration, the level of uncertainty of the estimated coefficient for MSCI Netherlands (0.397429) in the RBSA for the fund can be expressed as (9.15):

$$\sigma_{w_{\text{MSCI Netherlands}}} = \frac{1.3623\%}{2.3709\%\sqrt{60-9-1}} = 8.1260\% \tag{9.15}$$

The 90% confidence interval for the style weight associated with MSCI Netherlands is calculated as in (9.16):

$$\begin{aligned}&(39.7429\% - 1.645 \times 8.1260\%, 39.7429\% + 1.645 \times 8.1260\%)\\ &\quad = (26.3769\%, 53.1090\%)\end{aligned} \tag{9.16}$$

The regression coefficient for MSCI Netherlands can thus be taken as being significantly different from zero.[16] The results of (9.15) and (9.16) for each regressor in the fund's RBSA are summarised in Table 9.3. These figures enable us to

[16] A t-test leads to the same conclusion. This requires, first, the calculation of the t-ratio, given (in the case of MSCI Netherland) by:

$$t\text{-ratio} = \frac{39.7429\%}{8.1260\%} = 4.890854$$

Second, the critical value must be calculated. This can be done by using the inverse of Student's t function. In the present case, given a confidence level of 90% and again using Excel, we need to calculate T.INV(10%;50), which gives 1.6759. Clearly, the t-ratio is greater than the critical value, so the null hypothesis of a style weight statistically indistinguishable from a null value can be rejected.

Table 9.3 Standard deviation and confidence interval for the style weights

	Standard deviation of the *i*-th style weight (%)	90% confidence interval (%)
MSCI FINLAND	7.1345%	(−3.3800%; 20.0905%)
MSCI FRANCE	15.7827%	(25.9416%; 77.8620%)
MSCI NETHERLANDS	8.1260%	(26.3769%; 53.1090%)

conclude that only two of the three market exposures identified by the RBSA are statistically significant, namely MSCI France and MSCI Netherlands.[17]

9.4 Rolling Returns-Based Style Analysis

The detailed discussion of RBSA in Sect. 9.2 highlights that this methodology seeks to identify the fixed asset allocation that best explains a fund's returns over the specified period of time. For this reason, this RBSA is also called average style analysis. However, the fact that there are scientific studies that document the presence of style change or style drift in the behaviour of funds along with the practitioners' awareness of a more or less assiduous use by fund managers of dynamic portfolio management strategies, explains why there is an interest in adapting the application of RBSA so that it takes into account the time-varying behaviour of a fund's exposures to the indices representing the asset classes.[18] In this case, the solution proposed is called rolling style analysis, the most important features of which are the following:

- Implementation of a series of RBSAs in the classical format.
- Use, in each of these, of a fixed number of observations for the historical returns of both the fund and the asset class benchmarks.
- Estimation of the regression coefficients and, thus, of the style exposures with reference to rolling time periods, i.e. fixed length estimation windows that move

[17]It is interesting to note that if the RBSA had been performed using only the performance of the two statistically significant asset class benchmarks as explicative variables, the overall quality of the model would not have been particularly affected. In fact, R^2 would have fallen only slightly (from 93.84% to 93.61%). A second situation which, in that case, would have changed is the selection ratio since the average value of the differential returns not explained by the style benchmark would have worsened slightly, going from −0.3621% to −0.3703%, compared to an increase in their volatility which changed from 1.3623% to 1.3875% with the net effect of a slight worsening of the negative selection ratio.

[18]See, for example: Chan et al. (2002) and Swinkels and van der Sluis (2006). There can be different motives for style changes. They may be attributable to economic reasons, such as the pursuit of market timing or volatility timing policies, or to changes in the management team of the investment vehicle.

over time. Compared to the previous period, each estimation period discards the oldest data and includes the newest figures.

On this basis, and having monthly data for a 10-year period (therefore for 120 months), we can make a dynamic examination of a fund's investment style compared to a pre-defined set of indices applying style analysis based on rolling estimation windows, each including 60 observations. In the present case, the first regression identifies the mix of benchmark exposures that most closely resembles the investment style of the fund for the first 60 months available. Once this style benchmark has been determined, a second style regression is estimated for the period that goes from the 2nd month to the 61st month. This procedure is then repeated until only 60 data remain (evidently those representing the last 5 calendar years), given the data sample available.

Obviously, the output of all these style regressions and, therefore, the output of rolling style analysis cannot be illustrated in a pie chart. In this case, we need to use an area graph, the so-called exposure distribution area graph. This type of chart shows the percentages of the weights vertically and the reference dates horizontally, so that by monitoring the height of each colour band, we can follow the development over time of the fund's estimated exposure to the asset class benchmarks represented by the different colours. In order to make the connection between this area graph and the classical pie chart used for standard RBSA, it should be made clear that for any date on the horizontal axis of the area graph, the height of the colour band corresponds to the slice of the same colour in the pie chart generated by the estimates for the same time period.[19] In other words, the exposure distribution area graph shows in the vertical plane a view of the composition of a series of pie charts.

Regarding the information provided by rolling style analysis, it must be said that the visual representation of the exposure distribution area graph offers an immediate perception of the stability or instability of the fund's investment style over time. In the former case, i.e. in the absence of any particular style drifts, the surface of the area graph appears relatively even with little fluctuation in the colour bands; in this instance, there is a certain coherence between the benchmark exposures estimated using rolling windows and those expressed in a pie chart of a standard RBSA performed using the overall sample period. In the latter case, this is obviously no longer true, and the distribution area graph illustrates extremely volatile benchmark exposures.

From the point of view of an institutional investor (e.g. an insurance company offering unit-linked products or a pension fund with assets to be managed by assigning different mandates), it is clear that it is important to have the opportunity (or not) to rely on the fact that the fund/the asset manager does or does not adhere to a certain investment style over time.

[19] For a practical application, see Sect. 9.5.

If the distribution area graph shows instability, it is best not merely to note the variability in style exposures, but also to look for the potential sources.[20] These may be identified in:

- The ways in which the fund is managed. In particular, the reason could be the implementation of a heightened level of tactical asset allocation, which, since it is achieved by switching into and out of certain asset classes, can mean that the best mimicking portfolio changes over time. Evidence of a high turnover ratio[21] can help to identify this as the cause of the inconsistency in the area graph.
- Changes in the main features or nature of the securities in which the fund invests. This could be the reason if the fund's portfolio is highly concentrated in a small number of holdings, and there is a very low turnover ratio.
- The use of noisy data, i.e. the presence of errors in the data on which the estimation of the different style regressions is based.
- A poor selection of the asset class benchmarks used as explanatory variables. In effect, the style drifts can be the result of the use of non-mutually exclusive indices. This means that in trying to deduce style exposures coherent with the fund's performance, the RBSA estimated using rolling windows will oscillate from one period to the next between redundant indices when assigning the style weights. On the other hand, if the selected indices are not exhaustive, the quadratic programming algorithm at the heart of the RBSA is "obliged" to allocate the performance connected to the category (or categories) not covered by the independent variables to that benchmark which at that time provides the best fit.

The rolling style analysis discussed here has established itself as the way to visualise the time-varying properties of the style coefficients. We must admit that there are no particular theoretical arguments underlying this method, since, paradoxically, it also possesses a kind of internal contradiction: it aims to provide time-varying benchmark exposures as an output, but assumes that the sensitivities towards the indices be constant throughout the rolling window, the length of which is arbitrarily chosen.[22] Therefore, the fact that rolling style analysis has become a sort of common practice is hardly attributable to its theoretical robustness, but rather

[20] On this point, see Lucas and Riepe (1996).

[21] The turnover ratio can be seen as a measure of the intensity of trading, i.e. the portfolio rotation undertaken by a fund. This is generally determined by dividing the smallest of acquisitions and sales in a given period by the mean value of the fund's assets in the same period.

[22] This idea is shared by Annaert and Van Campenhout, who write: "To visualize the time-varying property of style coefficients, it is common practice to estimate Sharpe's model over rolling windows [. . .] It is then possible to graph the style exposures through time. This approach might help to establish a first impression of the time variation of style coefficients. Nevertheless, one still assumes that the sensitivities are constant over the arbitrarily chosen length of the rolling window. Besides, the approach of overlapping rolling windows and the fact that observation is equally weighted cause a delay in the recognition of a style change" Annaert and Van Campenhout (2007), p. 637.

to its extreme simplicity of use compared to alternative solutions proposed by various authors, where complexity of understanding and implementation constitutes a deterrent to wider adoption.[23]

9.5 An Empirical Application of Rolling Returns-Based Style Analysis

This section presents an empirical application of a rolling style analysis with reference to the Euro Area Large Cap Equity fund discussed in Sect. 9.3. For this fund, monthly performance data from January 2013 to December 2022 are available, as are returns for the ten previously selected benchmarks (MSCI Austria, MSCI Belgium, MSCI Finland, MSCI France, MSCI Germany, MSCI Ireland, MSCI Italy, MSCI Netherlands, MSCI Portugal, and MSCI Spain.). As a rolling style analysis is based by definition on a rolling sample approach, starting from data series of length T (with $T = 120$), we need to choose an estimation window of length M, where M is, obviously, less than T. In the example $M = 60$, so each style regression is estimated over a period of 5 years. The monthly updating of the 5-year estimation window with the most recent monthly data for the fund and the indices, together with the discarding of the oldest data, allows us to estimate a total of $T - M + 1$ style regressions.[24] In the present case, this amounts to 61 style regressions. As we know, the exposure distribution area graph can provide a combined view of this type of output. The graph is shown in Fig. 9.6.

The result of the rolling style analysis presented in the area graph shows an R^2 of 93.80%. This is obtained as the mean value of the R^2 of each of the 61 style regressions shown in Fig. 9.6 as an output on the vertical axis, with reference to each month from the 60th to the 120th of the available decade (i.e. from December 2017 to December 2022).[25] Looking carefully at the exposure distribution area graph, the fund's investment style appears rather unstable. In fact, its performances appear to be attributable to implicit exposures towards rather different explanatory benchmarks if one observes the first part of the graph compared to the second part

[23] The alternative methodologies proposed include the use of the Kalman Filter, which seeks to explicitly model the time-varying expositions. For further details, see Swinkels and van der Sluis (2006).

[24] Rolling style analysis, as traditional RBSA, can be implemented using Excel spreadsheets. However, the number of style regressions that have to be estimated makes it more advisable to use a programming code written, for example, in MATLAB.

[25] Although the present section is not concerned with the discussion and implementation of this methodology, it should be noted that, as for traditional RBSA, the issue of the assessment of the accuracy and reliability of the single estimated coefficients is also relevant for rolling style analysis. Obviously, in this case, the verification of whether or not the estimated parameters are statistically significant and/or the identification of the respective confidence intervals must be repeated for each of the $T - M + 1$ style regressions.

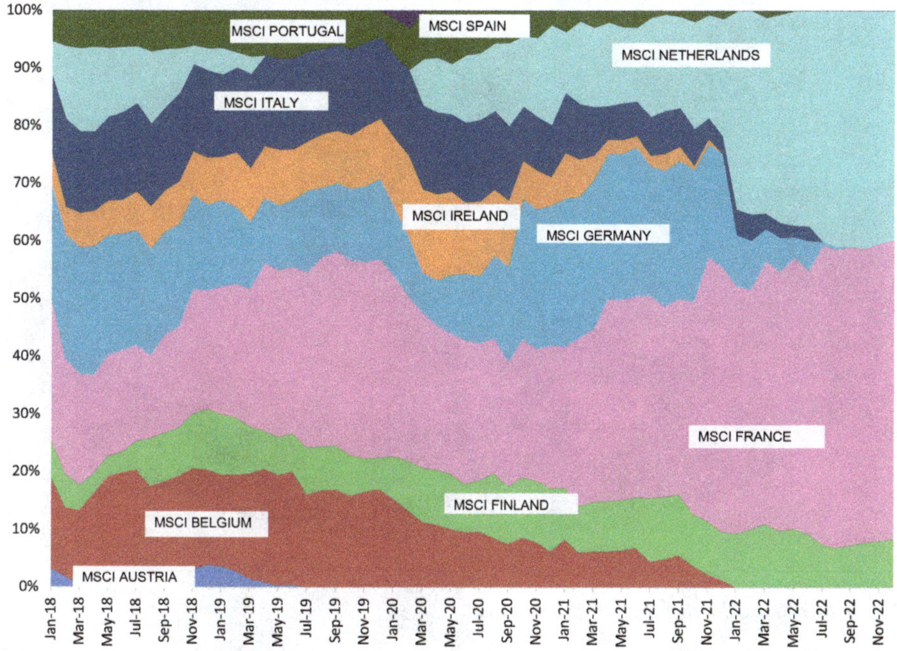

Fig. 9.6 The output of the rolling style analysis

and, in particular, compared to the final section when the implicit asset allocation of the fund clearly recognises a dominant role for MSCI France and MSCI Netherlands. The implicit exposure to the first index has strengthened over time; on average out of the 61 style regressions carried out it was worth 31.08%, but it reached a maximum value of 51.88% in correspondence with the estimation time interval which ended in December 2022, and a minimum value of 14.44% corresponding to the time window ending in August 2018. For the MSCI Netherlands, an average implicit exposure of 14.74% was found, but while it reached the value of 41.18% in the estimation window ending in October 2022, its value was zero approximately when using estimation windows that ended in the months of 2019. Likewise, the exposures to the MSCI Germany, MSCI Italy, and MSCI Belgium indexes are extremely volatile. There are, in fact, style regressions in which the implicit asset allocation of the fund categorically excludes these asset classes (there is, in fact, no trace of their presence in the final part of the exposure distribution area graph) and others in which their style weight is not entirely negligible, reaching, respectively, a maximum value of 25.87%, 16.33%, and 19.65%.

If we perform a style regression for the entire period for which data are available, i.e., 120 months, a long-term style benchmark can be identified. For the fund under consideration, this is shown in Fig. 9.7. The fund's deviations from the benchmark

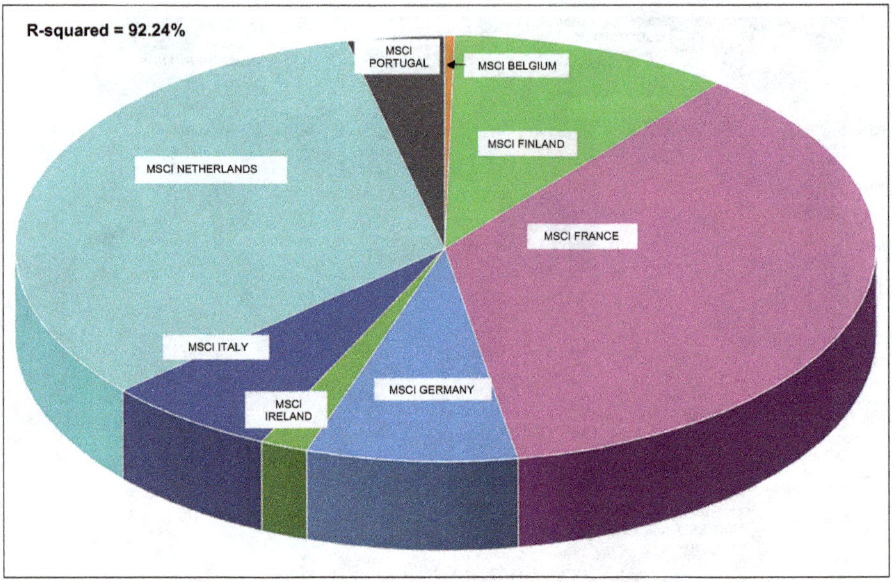

Fig. 9.7 The fund's long-term style benchmark

exposures derived from the estimates based on shorter-term rolling windows can be taken as proof of market timing activity undertaken during the period.[26]

As well as helping to monitor the evolution over time of a fund's investment style, rolling style analysis can be a useful tool to establish whether funds in the same category and/or declaring the same standardised benchmark have the same or a different implicit asset allocation over time. In this respect, it is interesting to consider the output of the rolling style analysis of a "competitor" to the fund assessed above, carried out over the same overall timeframe and with rolling windows of comparable size. In this case, the overall goodness of fit of rolling style analysis is more satisfactory, with an R^2 of 94.03% The relative exposure distribution area graph is shown in Fig. 9.8. On the basis of the graph, we can reasonably claim that the performance of the second fund, compared to the previous one, shows a lower erratic nature of implicit exposures towards the explanatory benchmarks and, in general, their greater involvement across style regressions. The most noticeable element of diversity that deserves to be stressed, however, is the primary and persistent role assumed by MSCI Germany and MSCI Ireland in explaining the periodic returns of the fund. Therefore, in the situation examined, the assumption of being faced with two apparently not dissimilar collective investment vehicles is confirmed since the same standardised benchmark (the MSCI EMU) is adopted, but

[26]It is also possible to quantify a market timing return for each month *t*, as the difference between the justified return of the short-term style benchmark applicable to that month and the return that the long-term style benchmark would have delivered.

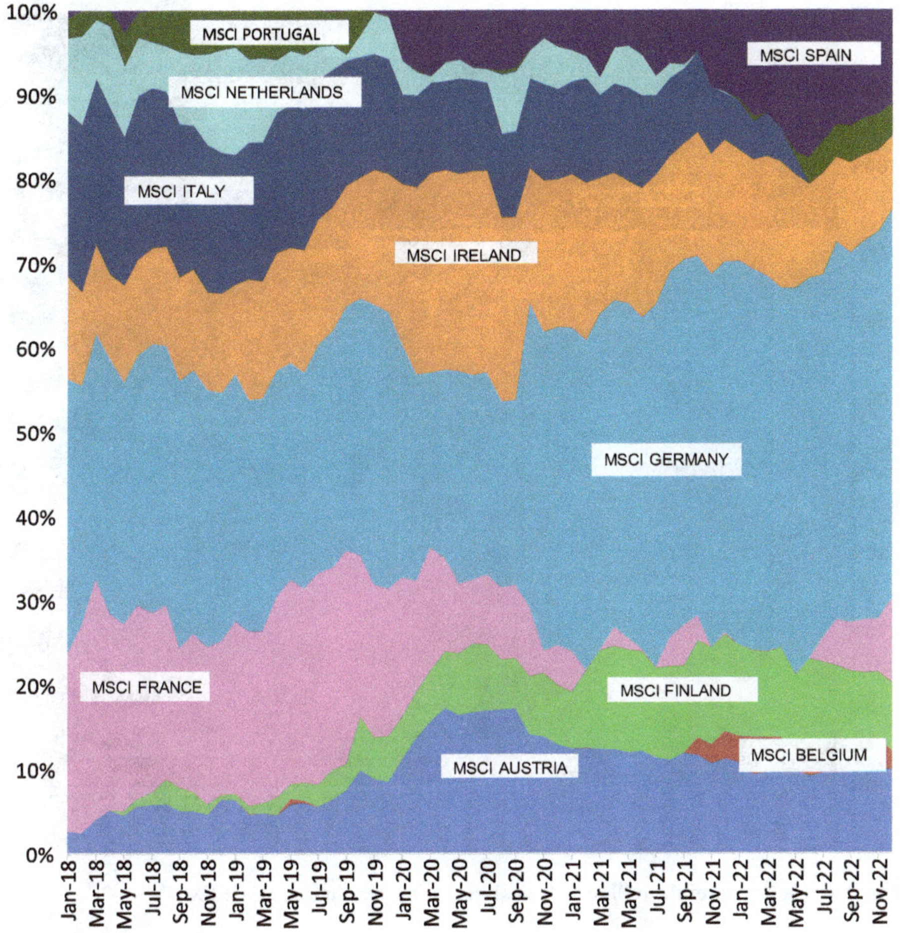

Fig. 9.8 The output of the rolling style analysis for a competitor fund

their observed results find a credible explanation in the adoption of an implicit asset allocation far from homogeneous.

To conclude this empirical section, it is interesting to point out that the methodology used so far, i.e. the rolling style analysis, can also be used for the purpose of investigating the implicit asset allocation in the performance of the standardised reference benchmark. This time we have then made use of the same explanatory benchmarks for the countries in the Eurozone already used as regressors within a rolling style analysis where the role of dependent variable is assigned to the time series of returns for the MSCI EMU. Figure 9.9 shows the results obtained.

At this point, comparing Figs. 9.6 and 9.8 with Fig. 9.9, we immediately deduce that the two collective investment vehicles considered above represent actively managed products and that, as already mentioned, the active decisions made by one are likely to be different from those made by the other .

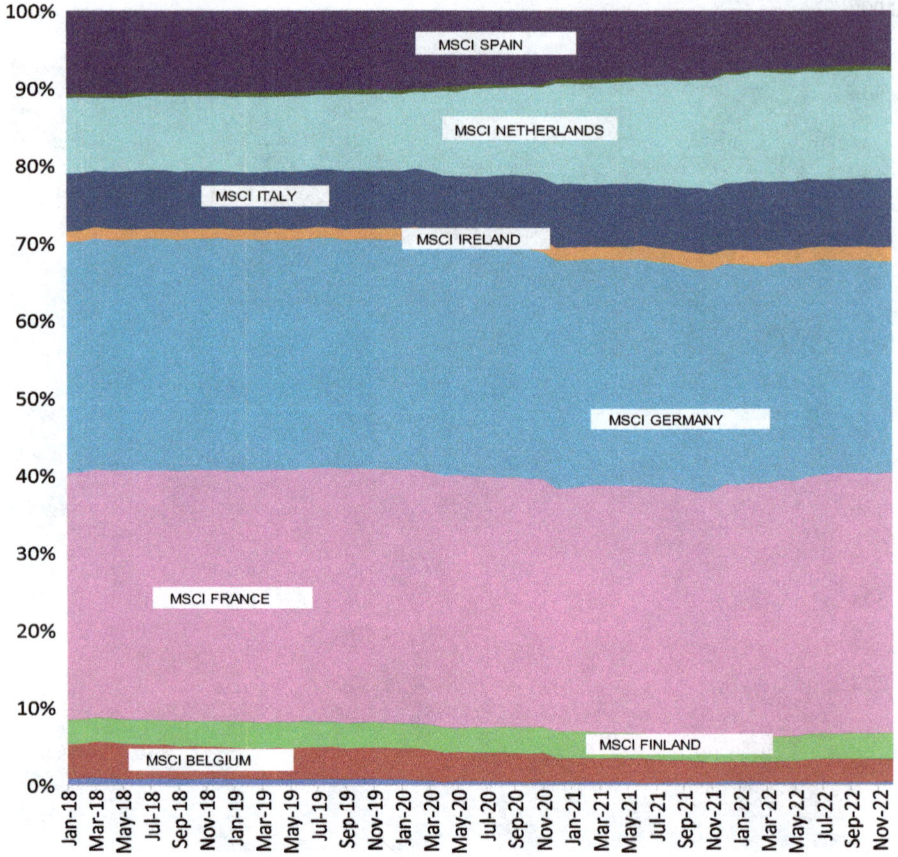

Fig. 9.9 The output of the rolling style analysis for the benchmark MSCI EMU

References

Agarwal V, Naik NY (2000) Generalized style analysis of hedge funds. J Asset Manag 1(1):93–109

Annaert J, Van Campenhout G (2007) Time variation in mutual fund style exposures. Rev Financ 11(4):633–661

Atkinson SM, Choi YK (2001) Investment style of portfolio management: excel applications. J Appl Financ 11:61–69

Chan LK, Chen HL, Lakonishok JK (2002) On mutual fund investment styles. Rev Financ Stud 15(5):1407–1437

DeRoon FA, Nijman TE, ter Horst JR (2004) Evaluating style analysis. J Empir Financ 11(1):29–53

DiBartolomeo D, Witkowski E (1997) Mutual funds misclassification: evidence based on style analysis. Financ Anal J 53(5):32–43

Domian DL, Reichenstein W (2008) Returns based style analysis of high-yield bonds. J Fixed Income 17(4):72–87

Kim TH, White H, Stone D (2005) Asymptotic and Bayesian confidence intervals for Sharpe-style weights. J Financ Econ 3(3):315–343

Lobosco A, DiBartolomeo D (1997) Approximating the confidence intervals for Sharpe style weights. Financ Anal J 53(4):80–85

Lucas L, Riepe MW (1996) The role of returns-based style analysis: understanding, implementing and interpreting the technique. Working paper, Ibbotson Associates.

Sharpe WF (1988) Determining a fund's effective asset mix. Invest Manag Rev 2(6):59–69

Sharpe WF (1992) Asset allocation: management style and performance measurement. J Portfolio Manag 18(2):7–19

Swinkels L, van der Sluis PJ (2006) Returns-based style analysis with time-varying exposures. Eur J Financ 12(6–7):529–552

Chapter 10
Performance Attribution

Maria Debora Braga

10.1 Performance Attribution: Definition, Objectives, and Assumptions

Over the years, performance attribution has become widely known and used among institutional investors and financial advisors, who see this tool as a means of providing important feedback by monitoring and assessing an investment process realised or proposed (at least in principle) in the interest of private end-investors. This approach can be better understood by investigating the positioning and the objective of performance attribution.

With regard to the former, it must be said that in logical and operational terms, performance attribution occurs after the measurement, over a given time period, of the relative performance of the actual portfolio and the benchmark portfolio, where the benchmark portfolio represents a specific strategic asset allocation (or policy portfolio). Taking this result as its starting point, its objective can be described in brief as the decomposition of relative performance. More specifically, it can be said that the aim of performance attribution is to identify the causes of any incongruence in direction and/or size between the overall result of the actual portfolio and that of its reference configuration. Finally, in technical terms, the objective of performance attribution analysis is to highlight and distinguish the management choices and investment decisions that have caused the dissimilar result and assign a value, called attribution effect, to each determining factor, expressing the respective contribution

M. D. Braga (✉)
Department of Economics and Political Science, University of "Valle d'Aosta", Aosta, Italy

SDA Bocconi School of Management, Bocconi University, Milan, Italy
e-mail: m.braga@univda.it; mariadebora.braga@sdabocconi.it

I. Basile, P. Ferrari (eds.), *Asset Management and Institutional Investors*, Contributions to Finance and Accounting,
https://doi.org/10.1007/978-3-031-59819-7_10

to relative performance. In the light of this, it goes without saying that the aggregation of the attribution effects leads then to relative performance.

As regards the usefulness of performance attribution, we should bear in mind that it can provide answers to questions such as: 'Overall, did the tactical views that were formulated (e.g. by the investment committee of an institutional investor or by a financial advisor) regarding recommended short-term overweight/underweight prove to be effective or not? In which specific investment areas did they generate value added or instead, result in a loss'? Or again: 'Did the asset manager exhibit significant skill in selecting securities or collective investment vehicles to get exposure to the markets? Was this managed in all investment categories or only in some'?

The objective pursued by performance attribution means that, although it is a retrospective analysis (as seen in the previous examples of key questions to which it can provide an answer), it is not an end in itself, but can act as a stimulus to strengthen the prior investment process by identifying both its strong and weak points.

Once we have understood why performance attribution should be used, it is obviously necessary to look more closely at its implementation. As a first step, this involves addressing three fundamental issues, which can be listed as follows:

- Which notion of relative performance measurement to adopt.
- What type of performance to consider for the actual portfolio (and, consequently, for the strategic portfolio or policy portfolio).
- Which time-frame to use in performance attribution (single-period attribution versus multi-period attribution).[1]

Each of these issues will be discussed in detail, respecting the order followed above. Firstly, with respect to the definition of relative performance, there are two possible formulations: arithmetic relative performance and geometric relative performance. Their measurement derives from the algebraic expressions (10.1) and (10.2), respectively:

$$\mathrm{PR}_t^{\mathrm{ARIT}} = R_{a,t} - R_{b,t} \tag{10.1}$$

$$\mathrm{PR}_t^{\mathrm{GEO}} = \frac{1 + R_{a,t}}{1 + R_{b,t}} - 1 \tag{10.2}$$

where:

$\mathrm{PR}_t^{\mathrm{ARIT}}$	= Arithmetic relative performance for the period t
$\mathrm{PR}_t^{\mathrm{GEO}}$	= Geometric relative performance for the period t
$R_{a,t}$	= Active return, i.e. actual portfolio performance for the period t
$R_{b,t}$	= Passive return, i.e. performance of the benchmark or policy portfolio for the period t; this can also be called policy return

[1] Fischer and Wermers (2013).

The two metrics thus differ in their mathematical form, since arithmetic relative performance is a difference, while geometric relative performance is a ratio. However, more than the analytical difference, it is the conceptual difference which is crucial, and the following examples serve to illustrate this distinction.

For the first case, let us imagine that an asset manager obtains a performance of +10% in a period t, compared to a strategic asset allocation performance of −5%. In the arithmetic version, relative performance equals +15%, while geometric relative performance is +15.7895%. In the second case, consider an asset manager A who obtains a 20% return versus a 12% policy return. Furthermore, let us suppose that an asset manager B obtains a performance of 10%, while the reference asset allocation produces a 2.5% return. According to (10.1), the relative performance of asset manager A is better than that of asset manager B (8% compared to 7.5%), while (10.2) leads to the opposite conclusion (7.1429% compared to 7.3171%).

These examples illustrate clearly that the arithmetic approach reflects the comparison between the added or lost value in the period t compared with what would have been added or lost in the same period by the benchmark portfolio, assuming an equivalent starting value. The geometric approach, on the other hand, evaluates the same positive or negative contribution resulting during the period t against the final value generated by an investment policy fully replicating strategic asset allocation weights. Moreover, the examples show that the two criteria can lead to contradictory results. Nevertheless, there is a link between the arithmetic and the geometric approaches, which can be highlighted by writing (10.2) as shown in (10.3):

$$\mathrm{PR}_t^{\mathrm{GEO}} = \frac{R_{a,\,t} - R_{b,\,t}}{1 + R_{b,\,t}} \tag{10.3}$$

The Eq. (10.3) expresses geometric relative performance as the 'discounted value' of arithmetic performance.

Once we have established that the two metrics are based on a different conception of value added, it becomes clear that it would be inappropriate to speak of one or the other as 'right' or 'wrong'. In operational terms, it is necessary to choose one of the two valid options and ensure that it is applied constantly and coherently. Opinions among practitioners regarding the greater intuitiveness and proximity to the classical idea of value added or excess return found in arithmetic performance attribution often (as also in the case of the present chapter) lead to a preference for this approach. However, as is clear from the observations above, there are also elements in favour of geometric performance attribution, which are generally summarised under the following three terms: proportionality,[2] convertibility,[3] and compoundability.[4]

The second issue that needs to be addressed concerns the notion of return to consider for the actual portfolio within the chosen performance attribution methodology, in order to make a reasonable comparison with the return of the reference portfolio representing a given strategic asset allocation. The most appropriate decision can be indicated clearly and without any ambiguity. As may be recalled, Chap. 8 analysed and discussed in detail the procedure for calculating the money-weighted rate of return and the time-weighted rate of return. However, considering the objectives of performance attribution analysis, it is easy to conclude that the performance metric to compare with the policy return must be immune from the influence of external cash flows (incoming and/or outgoing) that escape the portfolio or asset manager decisions, as these are largely dependent on investor decisions. It is reasonable, then, that performance attribution should seek to reveal what has made the time-weighted rate of return of the actual portfolio different from that of the strategic (policy) portfolio.[5]

The third and last issue to consider is which time-frame to use for performance attribution. Its actual implementation normally involves using two perspectives (a single-period and a multi-period time-frame) rather than choosing between the two. In the real world, an institutional investor communicates, analyses, and judges the results obtained in a conventional reporting period, or in one which is considered reasonable, or in one dictated by normative and regulatory requirements (e.g. a

---

[2]Proportionality highlights the capacity of geometric relative performance to provide a version of value added that is scaled to the return of the passive portfolio. More pragmatically, we can say that with geometric performance attribution, a situation in which the actual portfolio and the passive portfolio register a performance of +11% and +10%, respectively, and one in which they realise −10% and −11%, respectively, are different. In effect, in the first case, geometric relative performance is +0.91%, while in the second it is +1.12%. It should be remembered that with arithmetic relative performance, the two cases would give the same result: a + 1% value added on the part of the managed portfolio. For further details, see Bacon (2022).

[3]Attribution of convertibility to geometric relative performance means it is independent and indifferent to the choice of a specific currency to express the results returned.

[4]Compoundability in geometric relative performance refers to the question of multi-period performance attribution addressed later in this chapter. Specifically, it describes the possibility to multiply this metric (available for consecutive periods t) across time to obtain a multi-period, geometric relative performance, $\mathrm{PR}_T^{\mathrm{GEO}}$. This linking over multiple periods of geometric relative performance can be expressed as follows:

$$\mathrm{PR}_T^{\mathrm{GEO}} = \left(\frac{1+R_{a,1}}{1+R_{b,1}}\right)\cdot\left(\frac{1+R_{a,2}}{1+R_{b,2}}\right)\cdot\left(\frac{1+R_{a,3}}{1+R_{b,3}}\right)\cdot\ \ldots\ \cdot\left(\frac{1+R_{a,t}}{1+R_{b,t}}\right)\cdot\ \ldots\ \cdot\left(\frac{1+R_{a,T}}{1+R_{b,T}}\right)-1$$

[5]The performance of the benchmark portfolio is ‘spontaneously’ time-weighted, given that it describes a static portfolio asset allocation that serves as a medium/long-term reference and is by its very nature immune to the effects of additional contributions and/or withdrawals made by the client.

quarter or half-year). However, in order to measure attribution effects, it can be hazardous (or at least questionable) to consider this reporting period as a single, indivisible length of time, since during the course of the reporting period, changes may occur in the composition of the portfolio which are not solely attributable to the performance of the different asset classes, but may be due to external circumstances, i.e. the result of deliberate actions of a varying nature that inevitably influence the composition of the portfolio.[6] Reference to an indivisible reporting period ignores these circumstances and incorrectly indicates financial dynamics as the sole origin of the change in portfolio allocation from the beginning to the end of the period. To avoid this distortion, it is best to calculate first the single-period performance attribution and attribution effects, taking as the single-period time-frame the interval within which there is usually no portfolio rebalancing (e.g. a day or a month).[7] Once the single-period performances have been measured and the components of those single-period performances identified, we need to combine these results over time, in order to identify and explain the positive or negative 'extra' return for the entire reporting period. This is the same as saying that it is wise to envisage the 'co-existence' of single-period and multi-period performance attribution analyses.[8]

Having reviewed the pre-requisites for actual performance attribution, we must now look in detail at the methodology that can be considered the standard for asset management industry using a single-period, arithmetic approach, i.e. the Brinson model.

10.2 Performance Attribution Using the Brinson Model

The basis of the Brinson et al. model can be traced back to two papers published a few years apart in the *Financial Analysts Journal* under the titles 'Determinants of Portfolio Performance'[9] and 'Determinants of Portfolio Performance II: An Update'.[10]

Obviously, a model which aims to break down portfolio performance into additive parts starts by identifying those activities constituting the investment

---

[6] As an example of decisions able to modify the vector of portfolio weights, we can mention the tactical choices formulated either independently by the institutional investor or with the support of a financial advisor. On the other hand, if we think of a delegated portfolio management service, the change in the weights vector could result from further income or costs selectively attributable to certain financial assets or collective investment vehicles. These latter cases are more plausible: the greater the level of customisation of the management service, the greater the investor's freedom to issue 'instructions' to the manager.

[7] Single-period analysis may also refer to periods defined by decisions that imply a change in the vector of portfolio weights. In this case, the length of the periods would be irregular.

[8] Singer and Karnosky (1995).

[9] Brinson et al. (1986).

[10] Brinson et al. (1991).

Fig. 10.1 The quadrant framework proposed by Brinson et al. (1986)

(IV) Actual (active) portfolio return	**(II)** Policy and timing return
(III) Policy and security selection return	**(I)** Policy return (Passive portfolio benchmark return)

process or, similarly, by establishing the different types of decision that can significantly influence overall portfolio performance and variability.

According to the model and the original terminology of Brinson et al., the activities that can contribute to performance are the following:

- Investment policy, i.e. strategic asset allocation. With reference to Chaps. 4 and 5, it is clear that this refers to actions that define a portfolio's long-term composition, i.e. its structure in terms of normal and passive weights.
- Market timing, also called active asset allocation or, more simply, tactical asset allocation. This occurs following the deliberate and conscious assumption of temporary deviations from the normal portfolio weights realised either through overweights or underweights aimed at seeking to gain benefit from short-term forecasts of asset class performance.
- Security selection, understood as the selection of financial instruments within the asset classes. This results in changes in the composition of one or more asset classes compared to the composition of the respective reference index, in order to profit from mispricing, which scientific research and micro-analysis suggest is possible within a given investment category.

The solution suggested by Brinson et al. of measuring the partial results attributable to the above decision-making factors is known as the 'quadrant system or framework'. The name derives from the fact that the introductory step of the model involves comparing different return configurations shown in a 2×2 matrix (i.e. in four quadrants) so as to reveal the attribution effects. In two cases, the content of the four quadrants corresponds to real return configurations, while in the other two, 'notional' return configurations are shown. Figure 10.1 describes the contents of the 'quadrant system', maintaining the original numbering used by the authors.

In order to calculate these return configurations, we need the passive and actual weights[11] and the passive and actual returns for each of the N asset classes. If we take:

[11] If, in order to implement a portfolio, in addition to individual financial instruments, we also use other investment vehicles that can assume exposures to a number of asset classes, then the use of returns-based style analysis may be indispensable to an understanding of the real structure of the vector of actual weights. On this aspect, see Chap. 9.

Fig. 10.2 Calculation of the values in the Brinson et al. (1986) quadrants

(IV) Actual (active) portfolio return $\sum_{i=1}^{N} w_{a,t}^{i} \cdot r_{a,t}^{i}$	**(II)** Policy and timing return $\sum_{i=1}^{N} w_{a,t}^{i} \cdot r_{b,t}^{i}$
(III) Policy and security selection return $\sum_{i=1}^{N} w_{b,t}^{i} \cdot r_{a,t}^{i}$	**(I)** Policy return (Passive portfolio benchmark return) $\sum_{i=1}^{N} w_{b,t}^{i} \cdot r_{b,t}^{i}$

$r_{b,t}^{i}$	= Passive return of asset class *i* in the period *t*
$r_{a,t}^{i}$	= Actual return of asset class *i* in the period *t*
$w_{b,t}^{i}$	= Passive weight of asset class *i* in the period *t*
$w_{a,t}^{i}$	= Actual weight of asset class *i* in the period *t*

We can specify an analytical formulation for the content of each quadrant.

Starting with the configurations of actual returns, i.e. quadrants (I) and (IV), the algebraic expressions are as in (10.4) and (10.5), respectively:

$$R_{b,\ t} = \sum_{i=1}^{N} w_{b,\ t}^{i} \cdot r_{b,\ t}^{i} \tag{10.4}$$

$$R_{a,\ t} = \sum_{i=1}^{N} w_{a,\ t}^{i} \cdot r_{a,\ t}^{i} \tag{10.5}$$

For the configurations of the 'notional' returns—quadrants (II) and (III)—the calculation is given by the expressions (10.6) and (10.7), respectively:

$$R_{\text{policy+timing},\ t} = \sum_{i=1}^{N} w_{a,\ t}^{i} \cdot r_{b,\ t}^{i} \tag{10.6}$$

$$R_{\text{policy+security selection},\ t} = \sum_{i=1}^{N} w_{b,\ t}^{i} \cdot r_{a,\ t}^{i} \tag{10.7}$$

The availability of these formulas allows us, first of all, to present the 'quadrant framework' as in Fig. 10.2. Furthermore, the presentation of the calculation methods makes it easy to understand why the returns in quadrants (II) and (III) are not authentic: in both cases, the deviation from the strategic policy portfolio is permitted only in one direction. Quadrant (II) reflects 'bets' regarding allocations, but imagines

Table 10.1 Measurement of performance attribution effects starting from the 'quadrant system'

Performance attribution effects and relative performance	Calculation using quadrants
Market timing effect (or tactical asset allocation)	II − I
Security selection effect (or stock picking)	III − I
Interaction effect	IV − III − II + I
Total relative performance	IV − I → (II − I) + (III − I) + (IV − III − II + I)

returns in line with the respective benchmark for each area of investment, while quadrant (III) acknowledges the actual returns in the investment categories, but assumes allocations to these categories in line with those of the policy portfolio. Finally, the algebraic expressions unequivocally suggest that the objective of performance allocation—'to delineate investment responsibility and measure performance contribution'[12]—can be correctly translated into the search for the reasons for the lack of equality between quadrants (IV) and (I), given that the difference between the two corresponds to relative performance, PR_t^{ARIT}.

Having illustrated the quadrants, the next step is to highlight the individual attribution effects into which relative performance can be divided. This is achieved by simple subtractions/additions of the return configurations in the different quadrants shown in Fig. 10.2. Table 10.1 summarises these algebraic steps:

The Brinson et al. performance attribution model envisages three possible attribution effects, i.e. three potential sources of relative performance. Before expressing any considerations regarding the effects, it is useful to look beyond their measurement using algebraic operations on the quadrants and describe a direct calculation of each effect that is still based on the information in Table 10.1. Proceeding as indicated, the tactical asset allocation (TAA_t) contribution, or market timing effect, is obtained by (10.8):

$$TAA_t = \sum_{i=1}^{N} w_{a,t}^i \cdot r_{b,t}^i - \sum_{i=1}^{N} w_{b,t}^i \cdot r_{b,t}^i = \sum_{i=1}^{N} \left(w_{a,t}^i - w_{b,t}^i\right) \cdot r_{b,t}^i \tag{10.8}$$

The security selection effect (SE_t) is defined by the procedure indicated in (10.9):

$$SE_t = \sum_{i=1}^{N} w_{b,t}^i \cdot r_{a,t}^i - \sum_{i=1}^{N} w_{b,t}^i \cdot r_{b,t}^i = \sum_{i=1}^{N} \left(r_{a,t}^i - r_{b,t}^i\right) \cdot w_{b,t}^i \tag{10.9}$$

Finally, the interaction effect (I_t) is derived as described in (10.10):

[12] Brinson et al. (1986), p. 39.

$$I_t = \sum_{i=1}^{N} w_{a,\ t}^{i} \cdot r_{a,\ t}^{i} - \sum_{i=1}^{N} w_{b,\ t}^{i} \cdot r_{a,\ t}^{i} - \sum_{i=1}^{N} w_{a,\ t}^{i} \cdot r_{b,\ t}^{i} + \sum_{i=1}^{N} w_{b,\ t}^{i} \cdot r_{b,\ t}^{i} =$$
$$= \sum_{i=1}^{N} \left[\left(w_{a,\ t}^{i} - w_{b,\ t}^{i} \right) \cdot \left(r_{a,\ t}^{i} - r_{b,\ t}^{i} \right) \right] \tag{10.10}$$

Obviously, relative performance can now be expressed as the sum of the attribution effects, as indicated in (10.11):

$$\begin{aligned} \mathrm{PR}_t^{\mathrm{ARIT}} = R_{a,t} - R_{b,t} &= \sum_{i=1}^{N} w_{a,\ t}^{i} \cdot r_{a,\ t}^{i} - \sum_{i=1}^{N} w_{b,\ t}^{i} \cdot r_{b,\ t}^{i} \\ &= \mathrm{TAA}_t + \mathrm{SE}_t + I_t \end{aligned} \tag{10.11}$$

At this point, if we look at the formulas (10.8), (10.9), and (10.10), an important difference can be noted between the tactical asset allocation and security selection effects, on the one hand, and the interaction effect, on the other. In the former cases, the attribution effect is unambiguously and exclusively determined by an active decision not to track the policy portfolio in terms both of exposure to asset classes [as proven by the weight differential in (10.8)] and of returns from the various areas of investment [as confirmed by the return differential in (10.9)]. Consequently, for any individual asset class, it is easy to state that market timing will provide a positive contribution if an over-exposure for an investment category whose reference index has performed positively has been recommended; similarly, the security selection effect will lead to an added value, if, for example, financial products are selected that turn out to be undervalued, and so generate an active return greater than the passive return of that asset class. In contrast to these attribution effects, in the investment process there is no single specific 'responsibility' or active decision at the root of the interaction effect, as shown by its mathematical formulation as the cross-product between a weight differential and a return differential.

This description of the method to calculate the performance attribution effects reflects the best known and most 'authentic' version of the Brinson et al. model. However, we must also pay attention to the 'variants' of the model which have been applied over time. These variants regard tactical asset allocation (market timing) and interaction effects and are discussed in detail in Sects. 10.2.1 and 10.2.2, below.

10.2.1 The 'Variant' for the Market Timing Effect

Calculation of the tactical asset allocation effect as in (10.8) leads, respectively, to the identification of a positive contribution to relative performance for all overweighted positions in asset classes with a positive passive return, and a negative contribution for all overweighted positions for which the return of the reference benchmark is negative. Obviously, a decision in favour of underweighting these positions results in the opposite contributions to relative performance compared to the two hypotheses considered above.

Reflecting on these examples, it becomes clear that the traditional analytical formulation of TAA_t neglects the obvious and undeniable impact that an active decision on the size of an investment in an asset class has on the relative weight of one or more of the other asset classes in the portfolio. This can be taken as a consequence of the fact that the Brinson et al. model assesses the allocation choices from an absolute perspective, in the sense that it judges *ex post* the tactical shifts depending on whether the corresponding passive return is above or below 0%. However, there are reasons to claim the legitimacy of an alternative perspective, too, which we might call relative. In concrete terms, this means that—assuming we are aiming for a positive relative performance—the assessment of the allocation choices should reflect the decision-maker's capacity to overweight the investment categories which outperform the policy return and underweight those categories that underperform it.

It goes without saying that in this perspective, we need an alternative expression for (10.8) to calculate the market timing effect. Strangely enough, the best solution is to be found in an article published by Brinson and Fachler in the *Journal of Portfolio Management*[13] prior to the publication of the 'canonical' Brinson et al. model. This different analytical formulation of the market timing effect (TAA_t^{BF}) is given in (10.12):

$$\mathrm{TAA}_t^{\mathrm{BF}} = \sum_{i=1}^{N} \left[\left(w_{a,\,t}^i - w_{b,\,t}^i \right) \cdot \left(r_{b,\,t}^i - R_{b,\,t}^i \right) \right] \tag{10.12}$$

Given the discussion above, it is no surprise to see that the distinctive element of (10.12) compared to (10.8) is the insertion of the difference between the passive return of the individual asset class $r_{b,\,t}^{\,i}$ and the passive return of the strategic portfolio as a whole $R_{b,\,t}^{\,i}$. It should be noted that the use of (10.12) does not result in a different overall contribution of tactical asset allocation, but it does modify the contribution of the individual investment category.

The conditioning caused by the presence of the policy return in the calculation of the market timing effect and the fact that the important point is not whether it is positive or negative, but rather whether it is better or worse than the performance of the respective individual asset class, can be more easily understood with the example in Table 10.2. The table shows four combinations of positive or negative values for the return of an asset class *i* and for the policy return, comparing the results generated by the traditional Brinson et al. model (labelled BHB in the table) and the Brinson

[13] Brinson and Fachler (1985).

Table 10.2 Assessment of the tactical asset allocation (market timing) effect using BHB and BF

	$r_i = 2\%, R_b = 4\%$		$r_i = 2\%, R_b = 1\%$	
	BHB	BF	BHB	BF
Overweight	Positive	Negative	Positive	Positive
Underweight	Negative	Positive	Negative	Negative
	$r_i = -2\%, R_b = -4\%$		$r_i = -2\%, R_b = -1\%$	
	BHB	BF	BHB	BF
Overweight	Negative	Positive	Negative	Negative
Underweight	Positive	Negative	Positive	Positive

and Fachler modified version (BF in the table) in cases of overweighting and underweighting.

10.2.2 The 'Variants' for the Interaction Effect

In the breakdown of relative performance suggested by the Brinson et al. method, indication of the so-called interaction effect is often seen as problematic by institutional investors as well as by their financial advisors, because it is rather vague and ambiguous (in conceptual terms, rather than algebraic ones). Mathematically, it is clear that the interaction effect results from the sum of the products of the weight differential and the return differential for each asset class.[14] Logically, however, it is clearly not possible to identify one particular component within the investment process that can be considered responsible for, and the origin of such a performance attribution effect, given that a correct and disciplined planning of the investment process, as summarised at the beginning of Sect. 10.2, does not denote interaction as a deliberate and independent action or as a specific type of active decision.[15]

The worries induced by the difficulty involved in providing an economic and financial interpretation of the interaction effect and even more so, in explaining the effect in an easily understandable way, have often led to proposals to combine interaction in one of the other attribution effects. Before continuing in that direction, however, it is worth pointing out that in practice this component of relative performance can sometimes objectively show a greater influence in performance attribution analysis than that of the well-defined market timing and security selection components.

In this regard, we can consider an example from Laker.[16] The author presents the case of an equity portfolio comprising stocks listed on the Australian Stock

[14] See equation (10.10) in Sect. 10.2.

[15] In this regard, Bacon (2023) has stated: 'Interaction is not part of the investment decision process; you are unlikely to identify any individual portfolio manager responsible for adding value through interaction', p. 258.

[16] Laker (2000).

Table 10.3 Input for Laker's performance attribution example

	Policy (passive) weights (%)	Actual (active) weights (%)	Passive returns (%)	Active returns (%)
Total	100	100	**16.18**	**7.26**
ASX 100	91.26	44.44	15.35	20.68
ASX Small Ordinaries	8.74	55.56	24.82	−3.48

Table 10.4 Calculation of the performance attribution effects for Laker's example

	Market timing effect (%)	Security selection effect (%)	Interaction effect (%)	Total relative performance (%)
Total	**4.43**	**2.39**	**−15.74**	**−8.92**
ASX 100	**0.39** (44.44 – 91.26) × (15.35 – 16.18)	**4.86** (20.68 – 15.35) × 91.26	**−2.49** (44.44 – 91.26) × (20.68 – 15.35)	**2.76**
ASX Small Ordinaries	**4.04** (55.56 – 8.74) × (24.82 – 16.18)	**−2.47** (−3.48 – 24.82) × 8.74	**−13.25** (55.56 – 8.74) × (−3.48 – 24.82)	**−11.68**

Exchange (ASX) for which the reference benchmark is the ASX All Ordinaries Index. The variable used to classify the investment categories is the level of capitalisation. Consequently, both the composition of the managed portfolio and of the benchmark representing its strategic configuration are described by indicating the proportion allocated to the large cap and the small cap categories, identified by the ASX100 and the ASX Small Ordinaries sub-indices, respectively. Table 10.3 shows the values of the inputs necessary to apply the Brinson et al. method, while Table 10.4 gives the attribution effects for the individual investment categories and the portfolio as a whole.

The figures in Table 10.3 show a particularly active positioning of the managed portfolio compared to the benchmark. While the latter is composed almost exclusively of large cap stocks (91.26%), the portfolio has an allocation of over 50% in small cap stocks. The difference in the return of the portfolio (+7.26%) and that of the reference benchmark (+16.18%) is equally evident. This means that in the period analysed above, the decision not to follow the configuration of the ASX All Ordinaries Index has a cost for the portfolio of 8.92% and results in an equal relative negative performance.

The most interesting and marked findings are however in the information given in Table 10.4. This reveals that in the assessment of the tactical asset allocation choices, the overweighting in favour of the small cap stock was rewarding. The market timing effect for this asset class was +4.04%, a value added clearly attributable to the active decision to overweight an asset class whose passive return was significantly better than that of the strategic portfolio.[17]

[17] Laker's example uses the formulation in (10.12) to determine the tactical asset allocation effect.

This last circumstance is completely inverted for the ASX 100 and, as a result, the decision to underweight the large cap stocks produces a positive, albeit less notable (+0.39%), market timing effect. Perhaps a little surprisingly, Table 10.4 also shows a positive outcome for the selection effect. The decisive contribution to this component of relative performance comes from the large cap category and is explained by an actual return substantially greater than the passive return attributable to an investment category in which normal weight is prevalent, namely 91.26%. Moreover, in calculating the selection effect, it should not be forgotten that the very small allocation to small cap stocks in the benchmark justifies a reduced negative effect, even though it refers to an asset class whose actual return was considerably below that of the respective benchmark (−3.48% compared to +24.28%). In contrast to the positive values calculated for the market timing and selection effects, Table 10.4 shows a ruinous result of −15.74% for the interaction effect. Given these figures, we can appreciate the penalty to portfolio performance caused by a combination of a rather reckless choice in the tactical asset allocation for the small cap category and a drastic underperformance of the respective benchmark. It is important to note that due to the very small weight of the small cap category in the general reference benchmark (ASX All Ordinaries Index), this is a loss of value that cannot be understood on the basis of the selection effect alone.

The example from Laker is a case that proves how the interaction effect may sometimes result in a more acute interpretation of the relative performance achieved and stimulate a review and re-thinking of the management choices made in the investment process. In this case, assuming that the bias towards small cap stocks was intentional and not fortuitous, the analysis of performance attribution would suggest a reconsideration of the decision to use an active approach in favour of a composition of that part of the portfolio more in line with the benchmark. To conclude, we can say that the interaction effect has the merit of revealing any possible 'collision' (as in Laker's example) or 'agreement' that may be unintentionally created between different types of investment decisions, as well as clearly showing their consequences.

As already stated above, although the interaction effect can provide useful insights, it must be admitted that there is a widespread preference among practitioners not to explicitly recognise this element. Returning to our observations at the beginning of this section, it is clear that avoiding disclosure of the interaction effect implies its formal inclusion in one of the other attribution effects.[18] The algebraic procedure to follow if we decide to include the interaction effect in the market timing effect so as to create a new tactical asset allocation effect (indicated as TAA_t^*) is described in (10.13):

[18] Solutions other than an analytically defined inclusion in the remaining attribution effects, such as a random or equal allocation between the market timing and selection effects, must be rejected.

$$
\begin{aligned}
\mathrm{TAA}_t^* &= \sum_{i=1}^{N}\left[\left(w_{a,t}^i - w_{b,t}^i\right)\cdot\left(r_{b,t}^i - R_{b,t}^i\right)\right] + \sum_{i=1}^{N}\left[\left(w_{a,t}^i - w_{b,t}^i\right)\cdot\left(r_{a,t}^i - r_{b,t}^i\right)\right] = \\
&= \sum_{i=1}^{N}\left[\left(w_{a,t}^i - w_{b,t}^i\right)\cdot\left(r_{a,t}^i - R_{b,t}^i\right)\right]
\end{aligned}
\tag{10.13}
$$

Essentially, compared to the original formulation of the market timing effect, the passive return for each asset class is replaced by the actual return. If, on the other hand, the interaction effect is to be included in the selection effect, we need, as before, to take the respective calculations and sum them algebraically as indicated in (10.14). This results in a modified definition of the security selection effect, indicated as (SE_t*):

$$
\begin{aligned}
\mathrm{SE}_t^* &= \sum_{i=1}^{N}\left(r_{a,t}^i - r_{b,t}^i\right)\cdot w_{b,t}^i + \sum_{i=1}^{N}\left[\left(w_{a,t}^i - w_{b,t}^i\right)\cdot\left(r_{a,t}^i - r_{b,t}^i\right)\right] \\
&= \sum_{i=1}^{N}\left(r_{a,t}^i - r_{b,t}^i\right)\cdot w_{a,t}^i
\end{aligned}
\tag{10.14}
$$

Comparing the result obtained with the original algebraic expression, it is easy to see that in practice, this transformation involves replacing the normal weight for each asset class with the actual weight.

For the sake of completeness, Tables 10.5 and 10.6 show the results of performance attribution obtained when the interaction effect is included in either the timing

Table 10.5 Calculation of the performance attribution effects if the interaction effect is included in the market timing effect

	Market timing effect (%)	Security selection effect (%)	Total relative performance (%)
Total	**−11.31**	**2.39**	**−8.92**
ASX 100	**−2.11** (44.44 – 91.26) × (20.68 – 16.18)	**4.86** (20.68 – 15.35) × 91.26	**2.75**
ASX Small Ordinaries	**−9.20** (55.56 – 8.74) × (−3.48 – 16.18)	**−2.47** (−3.48 – 24.82) × 8.74	**−11.67**

Table 10.6 Calculation of the performance attribution effects if the interaction effect is included in the selection effect

	Market timing effect (%)	Security selection effect (%)	Total relative performance (%)
Total	**4.43**	**−13.35**	**−8.92**
ASX 100	**0.39** (44.44 – 91.26) × (15.35 – 16.18)	**2.37** (20.68 – 15.35) × 44.44	**2.76**
ASX Small Ordinaries	**4.04** (55.56 – 8.74) × (24.82 – 16.18)	**−15.72** (−3.48 – 24.82) × 55.56	**−11.68**

effect or the selection effect. In the first case, the market timing effect would change its sign from positive to negative for both investment categories; overall, the effect would become responsible for a loss of value of 11.31%. There is nothing surprising in this. In this arrangement, the market timing effect would absorb the losses derived from having, on the one hand, drastically decreased exposure to an asset class which, in terms of performance, dominates the policy return, and, on the other, from conceding too much space (compared to that implied in the benchmark) to an asset class whose actual (active) result is significantly below the overall passive return. In the second case, the selection effect would change its sign, with a value of −13.35% compared to +2.39% obtained with the classical Brinson et al. model. This should be attributed to the extremely disappointing result of the small cap stock (notwithstanding a noteworthy performance by the relative benchmark) that affects a dominant portion of the portfolio.

It must be said that there is no generalised consensus that allows us to consider one possible 'absorption' of the interaction effect as being superior to another. Nevertheless, given an investment process that, among institutional investors at least, is typically organised following a top-down approach,[19] there is no doubt that the subject responsible for stock/bond/fund picking choices makes these decisions only after the amounts of money assigned to the investment categories have been defined. For that reason, in the specific situation described here, it is recommended that the interaction effect be incorporated in the selection effect.

10.3 Integration of the Currency Effect into Performance Attribution

This discussion of the Brinson et al. model and its variants is based on the assumption that all returns were in the investor's base currency, i.e. his/her home currency. However, this is an unjustified assumption, if we remember that the recommendation in Sect. 4.4 to maintain investor access to international diversification when undertaking strategic asset allocation leads to situations of currency exposure. In such cases, the exclusive use of returns expressed in the base currency obviously produces misleading results, as it fails to distinguish the contribution to relative performance generated by choices of tactical asset allocation and selection of financial products from the positive or negative impact of variations in exchange rates between foreign currencies and the base currency.

Given this premise, the present section aims to provide an analytical framework for performance attribution that can measure the currency effect separately and ensure that its omission does not distort the other attribution effects. To this end, consider the example of an institutional investor domiciled in the Eurozone whose strategic portfolio includes exposures not only to the home currency (through

[19]For a description of the top-down approach, see Chap. 3.

Table 10.7 Passive and actual portfolios of a hypothetical institutional investor

	Actual (active) weights (%)	Passive weights (%)
Eurozone equity	13.00	10.00
USA equity	40.00	46.00
UK equity	47.00	44.00

Table 10.8 Variation in spot exchange rates

	Initial spot exchange rate	Final spot exchange rate
EUR/USD	1.30	1.24
EUR/GBP	0.85	0.87

investments in the Eurozone Equity asset class), but also to the US dollar (through investments in the USA Equity asset class) and to the British pound sterling (through investments in the UK Equity asset class). Let us further imagine that this institutional investor makes a tactical decision to move away from the normal weights of the investment categories, thereby modifying his/her currency exposure compared to that implicit in the benchmark portfolio. The configurations of the passive and the tactical portfolios are summarised in Table 10.7.

As a result of currency exposure, it is evident that for a Eurozone institutional investor, the investment in USA Equity and UK Equity determines a result that does not correspond to that registered in the respective home markets by the financial instruments that, on the basis of an indexing choice or of an active selection, in practice represent the investment in those asset classes. To use more technical language, there is a difference between the local return of those positions and the base currency return[20] realised by the institutional investor. The reason is that spot exchange rates have variations in the single period examined.

To continue our example, we thus need to consider the movement in exchange rates during the period under consideration. Table 10.8 gives this information using the 'direct quotation' convention for the euro. In other words, the values indicate the amount of foreign currency (dollars or pounds sterling) equal to one euro, which serves as the fixed term of reference.

Using this convention to express the exchange rates, the calculation of the currency return expresses the appreciation or depreciation in percentage terms of the home currency compared to the foreign currency.

Following this argument, (10.15) and (10.16) determine the currency return, generically indicated as s_t^{EURvsX}.

$$s_t^{\mathrm{EUR\ vs\ USD}} = \frac{1.24}{1.30} - 1 = -4.62\% \tag{10.15}$$

[20] For a closer examination of this issue, see Chap. 18.

$$s_t^{\text{EUR vs GBP}} = \frac{0.87}{0.85} - 1 = +2.35\% \quad (10.16)$$

Given these results, it is clear that in the single-period timeframe of this example, the institutional investor's base currency has depreciated compared to the dollar, while it has appreciated compared to the pound sterling.[21]

At this point, we need a well-ordered view of all the information necessary to conduct a performance attribution analysis in the presence of currency exposure. In addition to the actual and passive portfolio weights and the currency returns already discussed above, this includes the actual returns in the local currency of the different asset classes and those of the respective benchmarks, as well as the actual and passive returns expressed in the investor's base currency. It should be noted that when using the 'direct quotation' convention, the return in the base currency ($r_t^{\text{base currency}}$) is generally calculated according to (10.17)[22]:

$$r_t^{\text{base currency}} = \frac{1 + r_t^{\text{local}}}{1 + s_t^{\text{EUR vs X}}} - 1 \quad (10.17)$$

Table 10.9 summarises all the relevant information and provides the first signs of the favourable or unfavourable impact of currency exposure. Without claiming to be exhaustive, let us consider the investment in USA Equity: in this area, the managed portfolio bears a loss in the local currency of over 4%. It is only the depreciation of

[21] As already said, the method used to express exchange rates in the example is that which is most similar to operational practice in the Eurozone. However, we are aware that the literature has shown a greater preference for the 'indirect quotation' convention (amount of home currency needed for a unit of foreign currency). Using this method, the figures in the first line of Table 10.8 would be replaced by 0.7692 (1/1.30) and 0.8065 (1/1.24), while the figures in the second line would be replaced by 1.1765 (1/0.85) and 1.1494 (1/0.87). The currency return would also be different, giving +4.84% (0.8065/0.7692–1) for the US dollar and −2.30% (1.1494/1.1765–1) for the pound sterling. Confirmation of the use of this approach in the academic community can be found in Chap. 18. It is, however, of fundamental importance to clarify that in this case, the calculated currency returns must not be read as a percentage measurement of the depreciation or appreciation of the euro compared to the foreign currency, but rather as a percentage measurement of the appreciation or depreciation of the foreign currency compared to the home currency in the single period under examination. Together with a wish to keep to the same method used by practitioners, our preference in this chapter for the 'direct quotation' is motivated by the fact that relative performance (as will become apparent) is derived from the comparison between the return of the actual portfolio in the investor's base currency and the return of the benchmark portfolio (again in the investor's base currency). Since this relative performance will also include a currency effect, it seems reasonable to measure the latter from the point of view of the base currency, so as to be able to say that 'the base currency has appreciated or has depreciated by x %'. It is obviously necessary, then, to be able to determine whether this is a positive or negative thing for the investor who has chosen exposures to positions not denominated in his/her base currency.

[22] If we had used the 'indirect quotation' convention, the return in the base currency would have been calculated as follows: $r_t^{\text{base currency}} = r_t^{\text{local}} + s_t^{\text{EUR vs X}} + r_t^{\text{local}} \cdot s_t^{\text{EUR vs X}}$, obtaining, obviously, the same results. On this issue, see Chap. 18.

Table 10.9 The inputs for the multi-currency performance attribution analysis

	Actual (active) weights (%)	Passive weights (%)	Actual returns in local currency (%)	Passive returns in local currency (%)	Base currency returns (%)	Actual returns in base currency (%)	Passive returns in base currency (%)
Eurozone equity	13.00	10.00	7.00	6.00	0.00	7.00	6.00
USA equity	40.00	46.00	−4.50	−3.00	−4.62	0.12	1.69
UK equity	47.00	44.00	3.30	2.00	2.35	0.93	−0.34

the euro against the dollar that in the end makes it possible for the institutional investor to realise a slightly positive result of +0.12%; this would have been impossible for a US competitor who had made the same choices in terms of financial instruments for that asset class.

Once the inputs necessary for performance attribution have been made available, the first step—as well known—is to determine relative performance. This value, indicated as $PR_t^{\text{ARIT., base currency}}$, is expressed in the institutional investor's home currency and is derived from the difference between the return of the actual portfolio in the base currency $\left(R_{a,t}^{\text{base currency}}\right)$ and that of the benchmark portfolio in the same base currency $\left(R_{b,t}^{\text{base currency}}\right)$. Expressions (10.18), (10.19), and (10.20) show the method used to calculate these values and the relative application to the example just described:

$$R_{a,t}^{\text{base currency}} = \sum_{i=1}^{N} w_{a,t}^{i} \cdot r_{a,t}^{\text{base currency},i} = \\ = 13\% \cdot 7\% + 40\% \cdot 0.12\% + 47\% \cdot 0.93\% = 1.393\% \tag{10.18}$$

$$R_{b,t}^{\text{base currency}} = \sum_{i=1}^{N} w_{b,t}^{i} \cdot r_{b,t}^{\text{base currency},i} = \\ = 10\% \cdot 6\% + 46\% \cdot 1.69\% + 44\% \cdot (-0.34\%) = 1.277\% \tag{10.19}$$

$$PR_t^{\text{ARIT., base currency}} = R_{a,t}^{\text{base currency}} - R_{b,t}^{\text{base currency}} = 1.393\% - 1.277\% \\ = 0.166\% \tag{10.20}$$

The result of (10.20) demonstrates a value added as a result of the discretional decisions made by the institutional investor. As we have to admit that currency changes may have played a positive or negative role in this result, we need to quantify 'how much' currency return is behind the performance of the portfolio and the benchmark.

To this end, the calculation of both these results in the local currency is a preparatory step. The expression (10.21) gives the performance in the local currency of the actual portfolio $\left(R_{a,t}^{\text{local}}\right)$:

$$R_{a,t}^{\text{local}} = \sum_{i=1}^{N} w_{a,t}^{i} \cdot r_{a,t}^{\text{local},i}$$
$$= 13\% \cdot 7\% + 40\% \cdot (-4.50\%) + 47\% \cdot 3.30\% = 0.661\% \tag{10.21}$$

while (10.22) provides the equivalent calculation for the return of the reference strategic asset allocation:

$$R_{b,t}^{\text{local}} = \sum_{i=1}^{N} w_{b,t}^{i} \cdot r_{b,t}^{\text{local},i}$$
$$= 10\% \cdot 6\% + 46\% \cdot (-3\%) + 44\% \cdot 2\% = 0.100\% \tag{10.22}$$

At this point, an approximate estimate of the currency contribution (CC_t) can be obtained using the following reasoning. If in the base currency the actual portfolio beats the reference strategic composition by 0.166%, and this advantage would be equal to 0.561% = (0.661% − 0.100%) in terms of the local returns, we can reasonably assume that currency exposure is responsible for a loss of value equal to −0.395% = 0.166% − 0.561%. In order to generalise the logic of this procedure, we can use the analytical formulation given in (10.23):

$$\text{CC}_t = \text{PR}_t^{\text{ARIT, base currency}} - \left(R_{a,t}^{\text{local}} - R_{b,t}^{\text{local}}\right) =$$
$$= 0.166\% - (0.661\% - 0.100\%) = -0.395\% \tag{10.23}$$

In fact, (10.23) could also be re-written as in (10.24):

$$CC_t = \sum_{i=1}^{N} w_{a,t}^{i} \cdot r_{a,t}^{\text{base currency, } i} - \sum_{i=1}^{N} w_{b,t}^{i} \cdot r_{b,t}^{\text{base currency, } i}$$
$$- \left(\sum_{i=1}^{N} w_{a,t}^{i} \cdot r_{a,t}^{\text{local, } i} - \sum_{i=1}^{N} w_{b,t}^{i} \cdot r_{b,t}^{\text{local, } i}\right) = \sum_{i=1}^{N} w_{a,t}^{i} \cdot \left(r_{a,t}^{\text{base currency, } i} - r_{a,t}^{\text{local, } i}\right)$$
$$- \sum_{i=1}^{N} w_{b,t}^{i} \cdot \left(r_{b,t}^{\text{base currency, } i} - r_{b,t}^{\text{local, } i}\right) \tag{10.24}$$

Various arguments can be provided to explain this penalising effect of currency exposure. First, the portfolio is underweighted in the investment category that benefits from the depreciation of the euro compared to the foreign currency (USA Equity). In this category, moreover, the difference between the return in the base currency and that in the local currency is lower than the result that would have been obtained in the indexed version of the investment in that asset class. Second, the portfolio suffers from the strengthening of the euro and the concomitant weakening of the pound sterling as a result of the over-exposition to the UK Equity asset class. However, if we relinquish for a moment our focus on the question of currency exposure, it should also be said that this is the asset class for which active selection of financial instruments (as opposed to adherence to the composition of the corresponding benchmark) is more rewarding in terms of local return.

Once the currency effect has been isolated (CC_t), the remaining attribution effects traditionally envisaged by performance attribution following the Brinson et al. model or the relative variants can be determined with the analytical expressions introduced in the preceding sections. We must simply be careful to use the returns expressed in the local currency. In detail, then, the market timing effect for an individual asset class is calculated by adapting expression (10.12) as shown in (10.25):

$$\text{TAA}_t^{\text{BF},i} = \left[\left(w_{a,t}^i - w_{b,t}^i\right) \cdot \left(r_{b,t}^{\text{local},\ i} - R_{b,t}^{\text{local}}\right)\right] \tag{10.25}$$

On the other hand, the selection effect for the individual investment category is measured in such a way as to include the interaction effect by means of a slight modification to expression (10.14), as given in (10.26):

$$\text{SE}_t^{*,\ i} = \left(r_{a,t}^{\text{local},\ i} - r_{b,t}^{\text{local},\ i}\right) \cdot w_{a,t}^i \tag{10.26}$$

The performance attribution effects in this example of an institutional investor in the Eurozone facing currency exposure and measured consistently with the expressions given above are summarised in Table 10.10.

For the sake of completeness, even though it may go beyond the objectives of the present chapter, it should be pointed out that the question of performance attribution in the presence of currency exposure can be further extended or analysed in greater detail. Of the various additional options, we briefly recall:

- The possibility, suggested by Ankrim and Hensel, of conducting a sharper, more scrupulous analysis of currency returns (and, consequently, also of the currency effects) by disaggregating them into an unpredictable component, called currency surprise, and a component termed forward premium or discount that is known and pre-determined at the start of the single period, since it is linked to the differential between the interest rates of the two different currency areas.[23]
- The opportunity to highlight the impact on relative performance of various choices regarding hedging that may be adopted to manage currency exposure (full hedge compared to partial hedge).

[23] Ankrim and Hensel (1994) and Menchero and Davis (2009). See also Chap. 18.

Table 10.10 Performance attribution effects in the presence of currency exposure

Performance attribution effects and relative performance	%
Currency effect	**−0.395**
USA equity	−0.311
UK equity	−0.084
Market timing effect	**0.420**
Eurozone equity	0.177
USA equity	0.186
UK equity	0.057
Security selection effect	**0.141**
Eurozone equity	0.130
USA equity	−0.600
UK equity	0.611
Relative performance	**+0.166**

10.4 Multi-Period Arithmetic Performance Attribution

The preceding sections have provided an exhaustive exploration of single-period, arithmetic performance attribution analysis. We must, however, recall a point made in Sect. 10.1, namely that institutional investors and financial advisors generally have to explain and illustrate their performance over a longer reporting period comprising a number of single-period time-frames. For this reason, it is not possible to avoid completely the issue of multi-period, arithmetic performance attribution.[24]

Just touching upon the main points, a first objective is to measure multi-period relative performance, i.e. for a number T of periods. For this, multi-period performance of the actual portfolio ($R_{a,T}$) is defined as in (10.27):

$$R_{a,\ T} = \prod_{t=1}^{T} (1 + R_{a,\ t}) - 1 \tag{10.27}$$

while multi-period performance of the policy or passive portfolio ($R_{b,T}$) is given by (10.28):

$$R_{b,\ T} = \prod_{t=1}^{T} (1 + R_{b,\ t}) - 1 \tag{10.28}$$

Starting from (10.27) and (10.28), multi-period, arithmetic relative performance ($\mathrm{PR_T}^{\mathrm{ARIT}}$) is expressed mathematically as the difference between the so-called linked or compounded returns, as indicated in (10.29):

[24]Davies and Laker (2001) and Menchero (2004).

$$\mathrm{PR}_T^{\mathrm{ARIT}} = R_{a,\ T} - R_{b,\ T} \tag{10.29}$$

As in the case of single-period analysis, performance attribution is concerned with breaking down $\mathrm{PR_T}^{\mathrm{ARIT}}$. Use of geometric compounding of the single-period performances to obtain a comparable multi-period result for both the actual and passive portfolios leads to an excess return (positive or negative) for the entire period *T* that cannot be equalled by the algebraic sum of the relative performances of the single periods *t*. In other words, the inequality shown in (10.30) is verified:

$$R_{a,\ T} - R_{b,\ T} \neq \sum_{t=1}^{T} (R_{a,\ t} - R_{b,\ t}) \tag{10.30}$$

Obviously, (10.30) also implies that the single attribution effects cannot simply be added together over multiple periods without introducing unexplained residuals or improper components into the attribution effects. Essentially, (10.30) can be re-expressed as shown in (10.31):

$$R_{a,\ T} - R_{b,\ T} \neq \sum_{t=1}^{T} (\mathrm{TAA}_t + \mathrm{SE}_t + I_t) \tag{10.31}$$

Having acknowledged the difficulties or impracticability of any simple, automatic 'conversion' of an arithmetic, single-period approach into a multi-period approach, it is correct to say that the challenge faced by anyone wishing to implement multi-period, arithmetic performance allocation consists of looking for or selecting methodologies that allow attribution effects to be linked over time in such a way as not to lose the additive force of the multi-period attribution effect. The methodologies referred to here are termed linking algorithms and smoothing algorithms, and their application usually verifies the Eq. (10.32):

$$R_{a,\ T} - R_{b,\ T} = \sum_{t=1}^{T} \text{linking coefficient}_t \cdot (R_{a,t} - R_{b,t}) \tag{10.32}$$

as well as the Eq. (10.33):

$$\begin{aligned} R_{a,\ T} - R_{b,\ T} = {} & \sum_{t=1}^{T} \text{linking coefficient}_t \cdot \mathrm{TAA}_t + \\ & + \sum_{t=1}^{T} \text{linking coefficient}_t \cdot \mathrm{SE}_t + \sum_{t=1}^{T} \text{linking coefficient}_t \cdot I_t \end{aligned} \tag{10.33}$$

Although examining the various linking algorithms goes beyond the scope of the present chapter,[25] it is still interesting to illustrate the properties or requisites that such methodologies should hopefully satisfy in order to deliver a reasonable multi-period performance attribution. These properties can be summarised as follows:

- Being residual-free and fully linkable; this means that, once 'rectified' by the linking coefficient, addition of the single-period attribution effects must return a multi-period relative performance without generating unexplained residuals.
- Being commutative, in other words, the smoothing algorithm chosen must ensure that the multi-period relative performance and the attribution effects are unaffected by variations in the order of the single periods t.
- Metric preservation; this feature guarantees that, if two investment decisions made in different periods t generate the same relative performance, then they should, after rectification by the linking coefficient, contribute in a similar measure to multi-period relative performance.

References

Ankrim E, Hensel C (1994) Multicurrency performance attribution. Financ Anal J 50(2):29–35
Bacon C (2022) Practical risk-adjusted performance measurement, 2nd edn. Wiley, Chichester
Bacon C (2023) Practical portfolio performance measurement and attribution, 3rd edn. Wiley, Chichester
Brinson GP, Fachler N (1985) Measuring non U.S. equity portfolio performance. J Portfolio Manag 11(3):73–76
Brinson GP, Hood LR, Beebower GL (1986) Determinants of portfolio performance. Financ Anal J 42(4):39–44
Brinson GP, Singer BD, Beebower GL (1991) Determinants of portfolio performance II: an update. Financ Anal J 47(3):40–48
Davies O, Laker D (2001) Multiple-period performance attribution using the Brinson model. J Perform Meas 6(1):12–22
Fischer B, Wermers R (2013) Performance evaluation and attribution of security portfolios. Academic, Oxford
Laker D (2000) What is this thing called interaction? J Perform Meas 5(1):43–57
Menchero J (2004) Multiperiod arithmetic attribution. Financ Anal J 60(4):76–91
Menchero J, Davis B (2009) Multi-currency performance attribution. J Perform Meas 14(1):45–55
Singer BD, Karnosky DS (1995) The general framework for global investment management and performance attribution. J Portfolio Manag 21(2):84–92

[25] For a review of different methodologies for linking attribution effects over time, see Menchero (2004).

Part IV
Portfolio Diversification Towards Alternative Asset Classes

Chapter 11
Portfolio Diversification Policies: Alternative Asset Classes

Ignazio Basile

11.1 Alternative Investment Features

Alternative investments have been subject for debate for a long time, both in scientific literature and in practice, from two points of view: on the one hand, from the defining perspective, which is aimed to determine what is understood as an alternative investment; on the other hand, from the management perspective, based on defining rationally which weight should be attributed to these asset classes within the context of investment policies.[1]

The defining issue has tackled and solved in various ways, namely by considering the following as alternative investments:

1. All that is not identified with traditional asset classes (money market, bonds, and equity).[2]
2. Innovative management strategies of traditional asset classes.[3]
3. Instruments which are negotiated outside the regulated markets.[4]

[1] Abbink (2010), Anson et al. (2011), and Chambers et al. (2021). In particular, as highlighted by Abbink, an efficient classification criterion should not be based solely on the reference asset classes but should also intersect further dimensions of the vehicle used: investment strategy, localisation, and technical-judicial characteristics.

[2] In this case, hedge funds investing in shares and bonds and private equity would be excluded.

[3] Such a definition would exclude indirect investment in the real estate asset class.

[4] Reference to regulated markets would, for example, lead to the exclusion of listed private equity funds and real estate funds, Exchange Traded Funds (ETFs) specialised in alternative investments (e.g. commodities and hedge fund strategies) and Exchange Traded Commodities (ETCs).

I. Basile (✉)
Department of Economics and Management, University of Brescia, Brescia, Italy
e-mail: ignazio.basile@unibs.it

I. Basile, P. Ferrari (eds.), *Asset Management and Institutional Investors*, Contributions to Finance and Accounting,
https://doi.org/10.1007/978-3-031-59819-7_11

Nevertheless, they do not match with the actual investment policies of institutional investors, for all the above definitions exclude investments that have a significant weight in their asset allocation.[5]

Differently, from the specific perspective of a domestic institutional investor, the principle underlying the law that governs EU management companies would sound to be more effective. This principle, which takes the harmonisation criterion as a discriminant, considers as alternative investments all the collective investment vehicles (CIVs) that are not regulated by the UCITS Directive.[6] Therefore, the following investments should be considered alternative:

- Hedge funds.
- Venture capital and private equity funds.
- Private debt funds.
- Funds and other investment vehicles specialised in real estate.
- Infrastructural funds.
- Commodity vehicles.
- Investment trusts.
- Other types of institutional funds.[7]

Given the complexity and heterogeneity of the phenomenon investigated, official reliable statistics are not available at the moment all over the world.

Recently, however, the International Organization of Securities Commissions (IOSCO) has provided some insights into the relative size of the alternative fund industry compared to that of open-end funds, resulting from a data collection aggregated from the main jurisdictions worldwide.[8] The market share of alternative investment funds, at the time of the Report, amounted to 13.5%, compared to a weight of traditional funds of 86.5%.

In the following sections, attention will be focused on the main types of alternative investments (Fig. 11.1).

11.2 Alternative Investments from a Portfolio Perspective

The inclusion of alternative investments in the asset allocation of an institutional investor satisfies the primary need to achieve the optimal risk-return trade-off of a portfolio. Historically, alternative asset classes have proven to be more successful than the traditional instruments in rewarding risk when measured exclusively by

[5] As Abbink (2010) points out, the boundary lines between traditional and alternative investments are becoming increasingly less defined since the two worlds are extremely permeable.

[6] For a detailed description of the UCITS instruments, see Chap. 2

[7] With respect to the definitions used in literature, only credit derivatives and currency management would be excluded.

[8] IOSCO (2023).

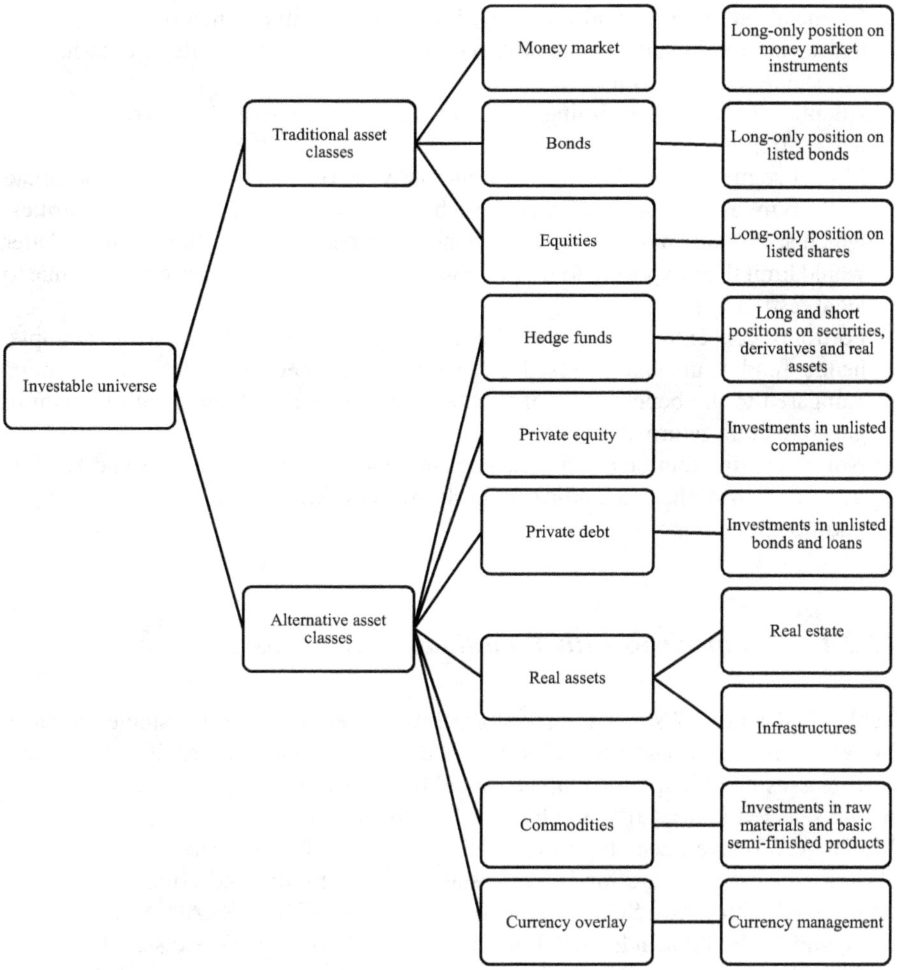

Fig. 11.1 Main traditional and alternative asset classes

variance. In addition, alternative investments have demonstrated to be sufficiently uncorrelated to equity and fixed income asset classes.[9] Benefits can therefore be gained both at the strategic asset allocation level via reduction of the market risk exposure, and at the tactical asset allocation level by generating extra returns thanks to dynamic techniques of active management.[10]

The main drivers of the managers' efficiency in making alternative investments to over-perform the reference markets are briefly identified in literature as follows:[11]

[9] Jackwerth & Slavutskaya (2016), Jobman (2002), KPMG (2015).
[10] Dorsey (2007), Jackwerth & Slavutskaya (2016).
[11] Jobman (2002), Anson (2006).

- Financialisation of the real economy: It makes it possible to incorporate rights on real assets (real estate, commodities) into financial instruments, increasing the investable universe as a result.
- Adoption of strategies aiming to obtain stable returns over time, irrespective of market trends.
- Market segmentation: Managers of alternative investments have the appropriate skills to invest in market segments, such as commodities or distressed securities, which are usually avoided by traditional investment vehicles. In turn, these latter would limit their exposure to illiquid assets or non-regulated markets partly due to legal restrictions.
- Portfolio concentration: Some alternative investment vehicles (for example, hedge funds) are characterised by a higher concentration of their portfolios compared to the benchmarks of traditional investment vehicles, with the aim to generate extra returns.
- Non-linear distribution of returns: The presence of marked skewness and kurtosis may lead to higher risk-adjusted performances, when measured according to traditional metrics.

11.2.1 Comparison with Traditional Asset Classes

Tables 11.1 and 11.2 show some historical data referred to the investable universe, by separating traditional asset classes and alternative asset classes. The traditional asset classes include money market, global bond market, and global equity market, whose selected indices JPM Cash EU 6 Month, Bloomberg Global Aggregate, and MSCI ACWI have been chosen respectively as a benchmark. The alternative asset classes— namely, hedge funds, real estate, private equity, and commodities—are represented with Credit Suisse Hedge Fund, LPX 50, FTSE EPRA/NAREIT Developed, and S&P GSCI indices.[12] Returns, historical volatility, basic statistics of the traditional and alternative asset classes, and correlation matrix are provided for each index. As from the matrix, we can see how major benefits deriving from diversification can be obtained for portfolios that contain money market or bond investments.[13] This hypothesis is confirmed by observing the efficient frontier provided in Fig. 11.2, built on monthly returns of money market, bonds, equities, hedge funds, private equity, real estate, and commodities over a period of time ranging from 01/1994 to 04/2023.[14]

[12] All indices were considered here as total return and in Euros.

[13] KPMG (2015), Baker and Filbeck (2013).

[14] The analysis was made for the period 1/1994–4/2023 in order to cover the past as long as possible, depending on the availability of CS Hedge Fund Indexes, that are calculated as from December 31, 1993.

Table 11.1 Risk-return profiles of traditional and alternative asset classes (1994–2023)

	Money market	Bonds	Equity	Hedge funds	Private equity	Real estate	Commodities
Yearly data	JPM Cash EU 6 Month	Bloomberg Global Aggregate	MSCI ACWI	Credit Suisse Hedge Fund	LPX 50	FTSE EPRA/NAREIT Developed	S&P GSCI
Geometric mean	2.41%	3.87%	7.35%	5.88%	8.42%	6.62%	–0.69%
Arithmetic mean	2.41%	4.10%	8.48%	6.10%	11.05%	8.05%	1.59%
Standard deviation	0.67%	6.82%	14.96%	6.66%	22.83%	16.77%	21.15%
Skewness	0.59	0.58	–0.53	–0.26	–0.17	0.71	–0.42
Kurtosis	2.83	3.51	3.66	6.94	7.66	6.26	4.76

Source: Author's calculation using data from Morningstar

Table 11.2 Correlation matrix of traditional and alternative asset classes (1994–2023)

	Money market	Bonds	Equity	Hedge funds	Private equity	Real estate	Commodities
	JPM Cash EU 6 Month	Bloomberg Global Aggregate	MSCI ACWI	Credit Suisse Hedge Fund	LPX 50	FTSE EPRA/ NAREIT Developed	S&P GSCI
JPM Cash EU 6 Month	1						
Barclays Global Aggregate	0.1570	1					
MSCI ACWI	−0.1066	0.2809	1				
Credit Suisse Hedge Fund	0.0834	−0.0098	0.5857	1			
LPX 50	−0.1432	0.0784	0.8203	0.5879	1		
FTSE EPRA/ NAREIT Developed	−0.0933	0.2874	0.7691	0.4895	0.7274	1	
S&P GSCI	−0.0753	0.0414	0.3393	0.3888	0.3449	0.2795	1

Source: Author's calculation using data from Morningstar

While still limiting the analysis to the theoretical framework originally defined by Markowitz, for each risk profile, a diversified portfolio with an allocation in alternative instruments is dominant over single traditional asset classes, particularly global equity, as shown by the absence of the MSCI ACWI index in the portfolios on the efficient frontier represented in Fig. 11.2. Nevertheless, efficient portfolios thus obtained overlook three key elements:

– The fact that the construction of the efficient frontier generally requires the use of weight constraints to fix the minimum and/or maximum limit of the different asset classes. In this context, a maximum threshold is always set for alternative investments.
– The circumstance that the frontier obtained in this way operates in terms of mean-variance optimisation, overlooking the two moments of higher order (skewness and kurtosis) of return distribution.

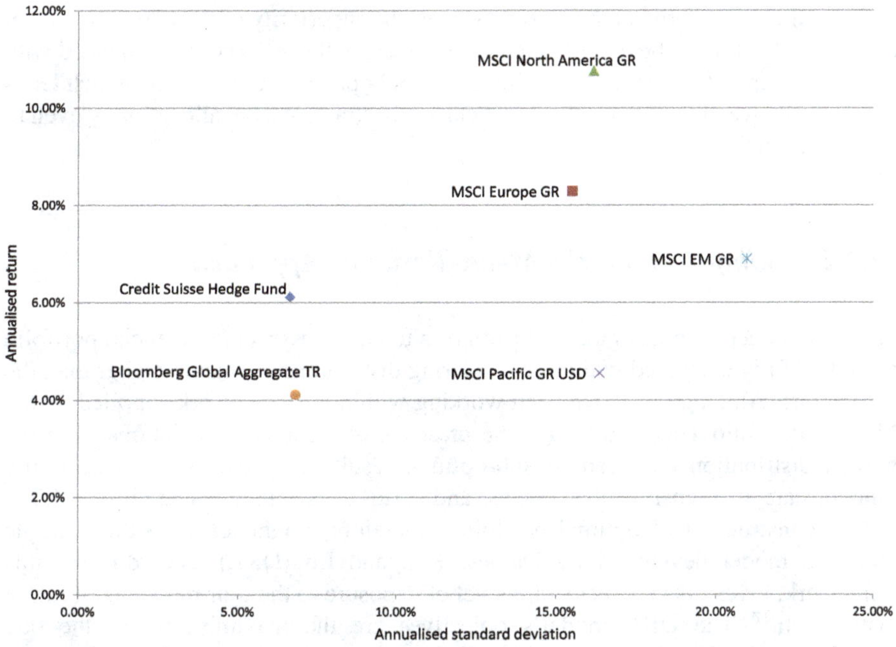

Efficient portfolios		Weights on the efficient frontier						
		Money market	Bonds	Equity	Hedge Fund	Private Equity	Real Estate	Commodities
Portfolio Return	Portfolio Risk	JPM Cash EU 6 Month TR	Bloomberg Global Aggregate TR	MSCI ACWI GR	Credit Suisse Hedge Fund	LPX 50 TR	FTSE EPRA/NAREIT Developed TR	S&P GSCI ER
2.45%	0.65%	99.35%	0.00%	0.07%	0.00%	0.39%	0.02%	0.16%
3.39%	1.73%	69.68%	9.21%	0.00%	20.07%	1.03%	0.00%	0.00%
4.35%	3.22%	38.82%	18.97%	0.00%	40.92%	1.28%	0.00%	0.00%
5.31%	4.75%	7.96%	28.73%	0.00%	61.78%	1.53%	0.00%	0.00%
6.26%	6.60%	0.00%	16.41%	4.96%	71.08%	7.54%	0.00%	0.00%
7.23%	9.13%	0.00%	0.00%	13.72%	70.11%	16.17%	0.00%	0.00%
8.18%	12.18%	0.00%	0.00%	16.20%	49.51%	34.29%	0.00%	0.00%
9.14%	15.56%	0.00%	0.00%	18.68%	28.91%	52.41%	0.00%	0.00%
10.09%	19.09%	0.00%	0.00%	21.16%	8.31%	70.52%	0.00%	0.00%
11.05%	22.79%	0.00%	0.00%	0.00%	0.00%	100.00%	0.00%	0.00%

Fig. 11.2 Efficient frontier (traditional and alternative asset classes, 1994–2023). *Source*: Author's calculation using data from Morningstar

– The trend of all traditional and alternative asset classes to significantly increase their degree of positive correlation both in crisis and bull market phases.[15]

[15] This trend does not undermine the basic assumptions of modern portfolio theory. Instead, it underlines the need to pay greater attention to the portfolio construction phase and to the reference time horizon. On this topic, see Geczy (2014).

From the risk-return perspective, however, the superiority of the alternative asset classes (hedge funds above all) over traditional asset classes is counterbalanced with a negative gap at the transparency, liquidity, and operating risks levels, which leads to reduce the weight of alternative asset classes in strategic asset allocation or even to discourage their use.[16]

11.2.2 Going Beyond the Mean-Variance Approach

The benefits deriving from the inclusion of alternative assets in a financial portfolio cannot be fully evaluated without considering distribution moments higher than the second one. Although we have been working within the framework supplied by the Modern Portfolio Theory till now, the presence of significant deviations from the normal distribution can lead to sub-optimal results for investors, whose utility functions are also sensitive to skewness and kurtosis, as demonstrated in Sect. 4.2.[17]

The construction of optimal portfolios, therefore, might resort to the multiple objective model developed by Davies, Kat, and Lu (DKL): according to this framework, investors can decide the level of exposure to the four moments of return distribution.[18] The DKL model's objectives are the maximisation of the odd moments (mean and skewness) and at the same time the minimisation of the even moments (variance and kurtosis).[19] A two-stage process is needed in order to achieve the result.

The first stage determines the optimal portfolios as a function of three distinct objectives, obtaining efficient solutions for each one, given a level of variance and the usual weight constraints of the components (non-negativity and budget constraint):

- Mean-variance-efficient: Maximised return is indicated by Z_1*.
- Skewness-variance-efficient: Maximised skewness is indicated by Z_3*.
- Kurtosis-variance-efficient: Minimised kurtosis is indicated by Z_4*.

In the second stage, the estimate of the Mean-Variance-Skewness-Kurtosis-efficient (MVSK-efficient) portfolio is achieved: it is constructed on the basis of the

---

16 On the need to use risk measures that take into account all the risk factors to which all asset classes, traditional and alternative, are exposed, refer to Pedersen et al. (2014).

17 It should be remembered that Markowitz (1952) himself had contemplated the convenience of whether to include or not such moments in the context of optimisation, reaching a negative conclusion due to the potential irrationality of some of the attainable results. Nevertheless, if the skewness and kurtosis aversion coefficients are bound within some given limits, surpassing the mean-variance model may give results that are consistent with the hypothesis of investors' rationality.

18 Davies et al. (2009).

19 It should be remembered that positive skewness implies greater frequency of positive extreme returns with respect to Gaussian distribution.

three optimal values of mean, skewness, and kurtosis calculated in the first stage and on the basis of the priority assigned to each of these moments by the investor.

If the return of the optimal MVSK portfolio is indicated as Z_1, its skewness as Z_3, and its kurtosis as Z_4, the distance d_i of each of these moments from those calculated in the first stage should be measured, as follows:

$$d_1 = Z_1^* - Z_1$$

$$d_3 = Z_3^* - Z_3$$

$$d_4 = Z_4 - Z_4^*$$

As a result, the optimal MVSK portfolio for a certain level of variance is produced by the minimisation of the Z function, obtained by acting on the vector $\mathbf{w}$ of the portfolio weights, since distances d_i and the relative preference parameters (k_i) are known:

$$\min_{\mathbf{w}} Z = (1 + d_1)^{k_1} + (1 + d_3)^{k_3} + (1 + d_4)^{k_4}$$

given the constraints:

$$d_1, d_3, d_4 \geq 0$$

Parameters k_i are indicative of the investor's degree of preference towards a specific form of efficiency (e.g. k_1 = mean-variance). The higher a specific k_i compared with the other two, the greater the relative importance attributed to this distribution moment by the investor. For example, according to Markowitz's original model, rational subjects will have k_3 and k_4 as equal to zero since they are not interested in skewness and kurtosis. On the other hand, investors who are focused on both mean-variance and kurtosis will set only k_3 as equal to zero.

Given the statistical characteristics of the alternative asset classes shown in Table 11.1, the DKL model provides for the achievement of a relative optimum which is different compared to that *à la* Markowitz, since the weight attributed to alternative investments may not coincide with the one which is defined by the Modern Portfolio Theory.

11.3 The New European Union Regulations

Following the global crisis experienced by financial markets since 2007, which was also attributed to the hedge fund industry in terms of poor transparency, excessive fund managers' risk appetite, and lack of a harmonised regulatory framework, in 2011, the European Union authorities issued the "Alternative Investment Fund

Manager Directive" (AIFMD), aiming at restoring credibility and stability to the entire alternative investments industry, and so reducing the systemic risk connected thereto.[20] The range of application is not limited to hedge funds but extends to the management of all funds beyond the perimeter of application of the UCITS IV Directive, since these are all considered equally able of amplifying the risks and instability of the overall financial system in the event of similar financial turbulence.[21]

The AIFMD defines a harmonised regulatory framework regarding the authorisation, operation, and transparency of all alternative investment funds managers. Moreover, it introduces a passport for marketing of units or shares of EU AIFs in all Member States and strict rules for cross-border management.

The objective of the new EU Directive is to create for investment funds managers an internal market other than the UCITS market and to establish for them a harmonised regulatory and supervisory framework.

The AIFMD focuses on alternative investment funds managers without directly providing regulation for alternative investment funds (AIFs), which continue to be regulated and supervised at a national level. In fact, with reference to the different categories of CIVs included in the definition of alternative investment funds, it would have been impossible to regulate the structure and composition of their portfolio and to identify rules which are common to alternative investment funds that are so different. For this reason, at the national level each Member State can issue regulations or continue to regulate autonomously the domestic alternative investment funds, in compliance with the common principles established by the Directive as summarised in Table 11.3. Given the complexity of the issue, the AIFMD foresees a long transitory period.

The case history is very complex, in relation to the country of origin of the alternative fund manager (the so-called AIFM) and to the country where the AIFs were established.

In general, for the purposes of distribution to professional investors, the country (EU or non-EU) of the AIF does not appear to be relevant if the AIFM has been authorised by the authorities of an EU country. For non-EU AIFMs, the authorisation regime depends on whether the AIF is EU or non-EU and is conditioned by the willingness to collaborate on the basis of reciprocity of the competent authorities of the country of origin of the non-EU AIFM.

Instead, the authorisation to market AIFs to retail customers is subject to prior approval by the authorities of the country where the AIFM aims to distribute its products and is valid only at a national level, without therefore extending automatically to other member countries.

[20] Directive 2011/61/CE. For a detailed description of the different regulatory regimes and relative economic-financial implications, see Athanassiou (2012) and Cumming et al. (2013).

[21] As well as excluding harmonised CIVs, the definition of alternative investment fund also excludes holding companies, joint ventures, special purpose vehicles, pension funds, and family offices.

Table 11.3 Summary of the principles of the European Union AIFM Directive

Authorisation framework
– Managers of alternative investment funds (AIFMs) who exceed the threshold of 100 million euros of cumulative AIFs asset under management shall apply for specific authorisation from the competent authorities of their home Member States. The threshold increases to 500 million euros, if AIFs are unleveraged and do not grant investors redemption rights during a period of 5 years.[a] – A less strict regulatory and supervisory framework is provided for AIFMs whose cumulative AIFs asset under management fall below the threshold of 100 million. Lower organisational requirements and transparency provisions are needed, with the possibility of voluntarily complying with the more stringent regulations provided by the Directive for managers of larger funds. – AIFMs shall have an initial minimum capital, according to whether they act as external managers (125,000 euros) or internal managers (300,000 euros). An additional amount, which is proportional to the assets under management, shall be added to cover professional liability risks. – Authorisation is different from that of non-alternative fund managers (UCITS IV Directive). – Legal domicile of the fund and type of fund (open-end or closed-end) are of no significance.
Marketing
– Authorised AIFMs may market their alternative funds both in their own countries and in other EU Member States. The passport is valid only for professional investors, while the possibility of extending it to the retail market is discretionary for each country.
Organisation
– Rules of conduct similar to those established for the UCITS IV Directive. – Requirement to adopt policies and organisation that are adequate for the management of typical risks of asset management. – Requirement to appoint a depositary bank and a prime broker which must comply with minimum regulatory standards and obligations. – Requirement to adopt the necessary measures to prevent conflicts of interest. – Compliance with the specific requirements regarding leverage (maximum level indicated in the prospectus) and liquidity.
Transparency
– Compulsory periodic report provided to investors and supervisory authorities, varying according to the risk involved. – Assets shall be evaluated properly by an independent subject and in accordance with the domestic laws.

[a] The lock-up clauses, widespread in the hedge fund industry, require that the participant does not apply for reimbursement before a set expiry date, usually longer than 3 years. This constraint provides for a more efficient portfolio management and proves to be particularly useful in limiting liquidity risk

Based on the ESMA findings in the 30 countries of the European Economic Area, at the end of 2020, 30,035 AIFs were distributed to professional investors, of which 80% had a European passport and only 13% were also distributed at retail (Table 11.4).

As many as 62% of the AIFs overall, due to classification difficulties, were not qualified as a fund of funds, hedge fund, real estate, or private equity fund, but it was included in the residual category "others".

On average, AIFs, with the exception of hedge funds (548% gross leverage compared to NAV), did not use any financial leverage significantly (151% gross leverage compared to NAV), limiting the usage of credit lines in particular (on average 21% of NAV with peaks of 250% for hedge funds).

Table 11.4 Alternative investment funds

	Funds of funds	Real estate	Hedge funds	Private equity	Others	Total
Size						
Number	5362	3978	930	4992	14,324	30,035
Net asset value (bln EUR)	873	766	89	363	3652	5899
Market share	15%	13%	2%	6%	62%	100%
Leverage (% of net asset value)						
Gross leverage	120%	141%	548%	117%	154%	151%
Borrowing	9%	12%	250%	5%	25%	21%
Distribution						
EU passport	90%	84%	76%	51%	86%	80%
Retail market	19%	23%	10%	6%	10%	13%

Source: ESMA (2022)

The organisational impact on Member States' management companies should be basically limited, even if in some cases national laws deviate markedly from European legislation. Instead, at the competition level, the consequences could be more significant since many countries have always been net importers of all forms of CIVs and the assignment of the European passport to alternative funds may knock down the fragile protective barrier currently shielding the internal markets of hedge funds and other non-harmonised instruments. It should also be pointed out that the validity of the EU-wide passport is unlimited only for professional and qualifying investors, while it is subject to the authorisation of the national authorities with regard to retail investors.

On March 2024, the AIFMD was amended by the so-called AIFMD II on the following matters:

- Delegation: additional requirements in terms of supervision and administration for functions to be delegated.
- Authorisation: AIFM will be required to provide its National Competent Authority (NCA) with more detailed information.
- Reporting to NCAs and Disclosure to investors: documentation and information requirements were expanded.
- Loan origination regime: new provisions to all AIFs that originate loans on a significant basis.
- Liquidity management tools: new obligations, powers and specific tools (suspension of subscriptions and redemptions; redemption gate; notice period, redemption fees; swing and dual pricing; anti-dilution leverage; redemptions in kind; side pockets) for managers of open-end AIFs.
- Depositaries: NCAs may allow AIFMs to appoint a depositary established in a member State different from domestic one.
- Non-EU AIFM Marketing in EU: increased focus from EU and national supervisory authorities.

References

Abbink JB (2010) Alternative assets and strategic allocation. Rethinking the institutional approach. Wiley and Bloomberg Press, Hoboken

Anson M (2006) Handbook of alternative assets, 2nd edn. Wiley, Hoboken

Anson MJP, Fabozzi FI, Jones FJ (2011) The handbook of traditional and alternative investment vehicles: investment characteristics and strategies. Wiley, Hoboken

Athanassiou P (ed) (2012) Research handbook on hedge funds, private equity and alternative investments. Edward Elgar, Cheltenham

Baker HK, Filbeck G (eds) (2013) Alternative investments: instruments, performance, benchmarks, and strategies. Wiley, Hoboken

Chambers DR, Kazemi HB, Black KH (2021) Alternative investments. An Allocator's approach, 4th edn. Wiley, Hoboken

Cumming D, Na D, Johan SA (2013) Hedge fund structure, regulation, and performance around the world. Oxford University Press, Oxford

Davies RJ, Kat HM, Lu S (2009) Fund of hedge funds portfolio selection: a multiple-objective approach. J Deriv Hedge Funds 15(2):91–115

Dorsey A (2007) Active alpha: a portfolio approach to selecting and managing alternative investments. Wiley, Hoboken

ESMA (2022) EU alternative investment funds. ESMA annual statistical report, Paris

Geczy C (2014) The new diversification: open your eyes to alternatives. J Portfolio Manag 40(5): 146–155

IOSCO. (2023). IOSCO investment funds statistics report, Madrid.

Jackwerth JC, Slavutskaya A (2016) The Total benefit of alternative assets to pension fund portfolios. J Financ Mark 31(C):25–42

Jobman D (2002) The handbook of alternative investments. Wiley, Hoboken

KPMG (2015) AIFMD transposition state of play across EU member states & EEA countries, Luxembourg

Markowitz HM (1952) Portfolio selection. J Financ 7(1):77–91

Pedersen N, Page S, He F (2014) Asset allocation: risk models for alternative investments. Financ Anal J 70(3):34–45

Chapter 12
Hedge Funds

Ignazio Basile

12.1 Hedge Fund Features

Hedge fund refers to all collective investment vehicles (CIVs) which aim to achieve positive risk-adjusted returns, irrespective of the trends of the target markets, through active investment techniques and often with high financial leverage. This result is made possible by the absence (or lower impact) of legislative restrictions on investment policies, which in turn gives hedge fund managers a wide margin of discretion in the adoption of hedging or investment strategies that are deemed to be suitable to the objectives pursued. In other words, the achievement of risk-return combinations which result in being more efficient when compared to those typically pursued by managers of traditional investment funds depends, according to the Modern Portfolio Theory, on the availability of a generally unlimited investable universe, on the possibility of using financial leverage, and on the opportunity of using derivative instruments and short selling in order to take short positions (Table 12.1).

In addition to the traditional technical/legal solutions that are typical of CIVs, hedge funds often take the form of a limited partnership, or trust, according to their domicile and relevant legislation.[1]

Hedge funds have overcome the legal restrictions regarding financial leverage and short selling imposed by the 1940 Investment Company Act for traditional investment funds due to the atypical nature of their legal structure within the US law since their origins in the 1940s. Moreover, they are able to implement innovative

[1] Fagetan (2021).

I. Basile (✉)
Department of Economics and Management, University of Brescia, Brescia, Italy
e-mail: ignazio.basile@unibs.it

I. Basile, P. Ferrari (eds.), *Asset Management and Institutional Investors*, Contributions to Finance and Accounting,
https://doi.org/10.1007/978-3-031-59819-7_12

Table 12.1 Hedge funds: long and short positions

	Long (% NAV)	Short (% NAV)	Net (% NAV)
Cash	24.50%	13.53%	10.97%
Equities	63.44%	19.92%	43.52%
Corporate bonds	6.61%	1.89%	4.72%
Sovereign bonds	49.95%	40.39%	9.56%
Municipal bonds	0.81%	0.03%	0.78%
Convertible bonds	3.97%	0.09%	3.88%
Securitised bonds	5.27%	1.98%	3.29%
Loans	9.21%	0.41%	8.80%
Real estate	3.43%	0.00%	3.43%
Commodities	0.19%	0.00%	0.19%
Collective investment schemes	6.81%	0.04%	6.77%
Physical investments	0.35%	0.11%	0.24%
Equity derivatives	27.99%	24.96%	3.03%
Credit derivatives	8.39%	10.10%	−1.71%
Commodity derivatives	7.17%	3.82%	3.35%
Other derivatives	2.54%	1.43%	1.11%
Others	2.22%	0.12%	2.10%

Source: IOSCO (2023)

styles of management, which are uncorrelated from the market trends and useful to exploit market anomalies and imperfections.[2]

Although hedge fund managers have no restrictions—potentially and theoretically at least—they are anyway specialised in adopting strategies which cannot always be referred to specific asset classes, unlike traditional CIVs, but to management styles that, in practice, have become the most widely used classification criteria. Together with the pursuit of absolute returns without reference to a particular benchmark, there are other specific aspects at the organisational/managerial level that are common to hedge fund managers, such as:[3]

- The historical absence or limited weight of regulatory constraints.[4]
- The offshore location of the management activities.
- The lack of transparency according to the management philosophy adopted and to the composition of investments, in order not to transfer know-how to competitors.[5]
- The extensive use of outsourcing for all the operations that are not strictly related to investment management.

[2] The first person to describe the typical philosophy of hedge fund managers is thought to have been Jones (1949). His very simple logical scheme contained all the basic elements that still characterise the industry today: short selling, hedging strategies, leverage, incentive fees, co-investment by the hedge fund manager. See Zask (2013) for a historical excursus.

[3] Lhabitant (2006), Schneeweis et al. (2011), and Baker-Filbeck (2017).

[4] See Sect. 11.3.

[5] Hedges (2005).

- The close relationship with one or maximum two prime brokers that guarantee exclusive access to the markets and to financial leverage.[6]
- The adoption of closed-end or, as an alternative, open-end vehicles with different liquidity terms (monthly, quarterly, and semi-annually, or even longer according to the strategy pursued, notice period, lock-up clauses, provision of gates to slow exit for investors, creation of side pockets to segregate the most illiquid assets, assigned to all investors, comprehensive of those who ask for redemption).[7]
- The managers' acceptance of a high reputational risk.
- The lower average size if compared with traditional investment funds, especially for niche strategies.

Another significant and distinctive peculiarity of hedge funds is the adoption of managers' remuneration schemes based on their participation to performances that are defined on the basis of the high-water mark mechanism.[8] When the contract foresees this clause, the manager receives the performance fee provided that the NAV of the hedge fund exceeds the maximum value reached before the last calculation of the performance fee. This value is precisely known as the high-water mark. Therefore, hedge fund managers have a call option on the fund NAV implicitly, whose strike price is the high-water mark. With this option-like provision, a potential conflict of interest between managers and investors may arise because the value of an option increases at the rising of any underlying asset volatility. In contrast to managers, to participants this represents an increase in the investment's level of riskiness.[9]

Nevertheless, this issue may be mitigated both by the managers' strong aversion to any reputational risk, encouraging them to limit excessively opportunistic behaviour in the NAV manipulation, and by the fact that the high-water mark clause involves the assignment of a potentially unlimited number of call options to the managers, one for each new absolute maximum.

If the analysis is extended to an inter-temporal perspective, it may reveal that excessive high-water mark causes the manager's right to receive a commission, on

[6]The prime broker is a subject who is specialised in a combination of integrated services that range from orders handling to a series of additional services such as financing, securities lending, custody services, research, information technology support, risk monitoring, clearing operations, etc.

[7]In order to improve the conditions of liquidity in hedge funds investments, institutional investors make increasingly greater use of the so-called managed accounts, set within highly sophisticated management platforms. Managed accounts are segregated accounts managed by hedge fund managers that try to replicate the management policies of the master fund for one or more feeder funds. They guarantee greater transparency in the positions hold, the possibility to apply specific restrictions, better liquidity terms, and more competitive fee structure. However, the organisational costs of the platform set a minimum size threshold, above which the managed account may be profitable for the platform manager.

[8]Agarwal et al. (2009).

[9]Goetzmann et al. (2003). The authors highlighted that increase in volatility brings maximum benefits to the manager when the NAV is close to the high-water mark, in line with what occurs with "at the money" options.

Table 12.2 Correlation matrix hedge funds vs traditional asset classes (1994–2023)

	Credit Suisse HF Index	MSCI AC World GR	Bloomberg Global Aggregate TR
Credit Suisse Hedge Fund Index	1		
MSCI AC World GR	0.5857	1	
Bloomberg Global Aggregate TR	0.2809	−0.0098	1

Source: Author's calculation using data from Morningstar

the one hand, and raises the strike price of the next implicit option on the other. As a result, if the manager has created no real value but only volatility, he/she runs the risk of no longer being able to exceed the new maximum. In addition, the increased volatility exposes the fund to a greater probability of incurring losses in its NAV, with consequent drawdown from the high-water mark.[10]

A further critical and often neglected point, in the way the high-water mark works, is the determination of an inequitable allocation of performance fees to the subscribers of the hedge fund. This issue happens whenever the NAV that is identified as high-water mark is the same for all the investors and is applied in a standardised manner to the fund, although to the subscribers should be given different high-water marks according to the moment when they invested in the fund.[11]

12.2 Management Strategies

Despite these issues, hedge funds have been gaining wide diffusion in recent years, especially with institutional investors, who have been attracted by the managers' ability both to control the risk and to be uncorrelated to the market trends, significantly improving the risk-return profile of the overall asset allocation.[12] This theory is confirmed by the long-term correlation coefficients between the Credit Suisse Hedge Fund Index, which is representative of the general hedge fund market trend, and the typical benchmark indices of the equity and bonds market for the period January 1994–April 2023 provided in Table 12.2.[13]

As shown in Fig. 12.1, hedge funds managers have not disappointed investors' expectations in the long term, ensuring an average positive performance, while lowering the portfolio volatility. As a result, the superiority of the risk-return

[10] Panageas and Westerfield (2009).

[11] Concretely, some hedge funds adopt an equalisation mechanism that determines the debit or credit amounts of each investor in order to limit the different treatment among investors.

[12] Schneeweis et al. (2002).

[13] The correlations are based on historical series of indices in US dollars.

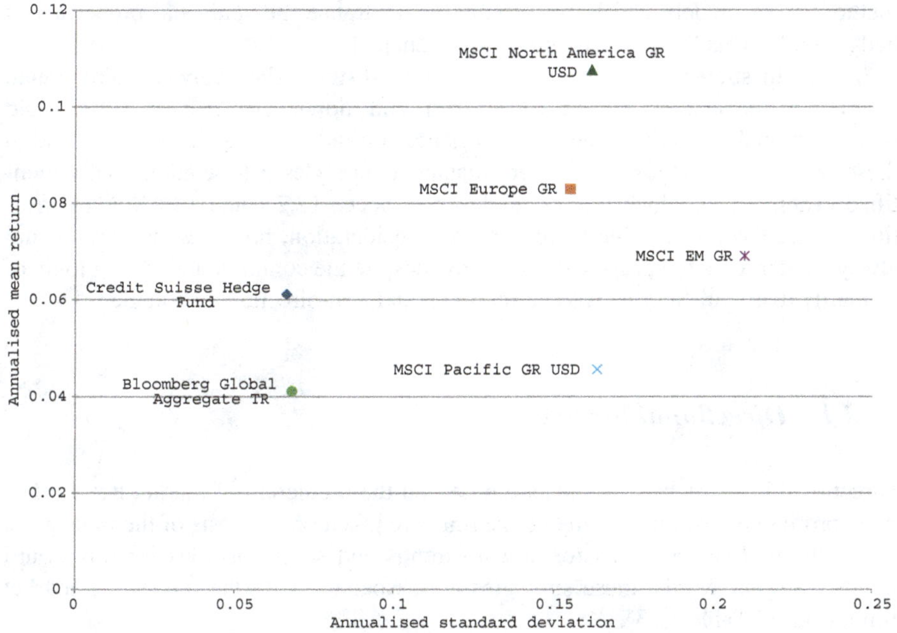

Fig. 12.1 Performance and volatility: hedge funds and traditional asset classes (1994–2023). Source: Author's calculation using data from Morningstar

combination of the hedge fund asset classes compared to traditional asset classes has prevailed, which was by no means obvious. With the exclusion of very short periods of observation, lower volatility seems not to have produced negative effects on returns.

With regard to performance, as a natural result of the complexity of the management choices involved, hedge funds usually show:

- Non-normal distribution of returns.[14]
- Returns which are more stable if compared to traditional asset classes.
- Greater tail risk exposure.
- Average lower drawdowns and historically shorter relative recovery periods.[15]

The above-mentioned peculiarities have significant and specific implications for the assessment of the performance profile, as they seem to require not only a critical review of the traditional measures of risk and return but also the integration of these

[14]Lhabitant (2004), Gregoriou et al. (2005) and Lo (2010).

[15]Drawdown indicates the accumulated losses incurred by an investment, whereas the recovery period is the time required to recover completely from the drawdown. The maximum drawdown is the maximum loss during the lifetime of an investment. Sancetta and Satchell (2004).

methods with models which are capable to neutralise the statistical properties of hedge funds. This topic will be the core of Chap. 13.

The main strategies of hedge funds can be distinguished between directional, attempting to take advantage of upward and downward market trends, and non-directional, founding their performance on arbitrage techniques.[16] Within these two macro-categories, different management styles can be additionally identified (Table 12.3), which will be analysed in Sects. 12.2.1 and 12.2.2. Classifications offered within the literature deserve consideration, however, as they do not always match with the proposed taxonomy, despite the common underlying logic of primarily distinguishing between directional and non-directional strategies.[17]

12.2.1 Directional Strategies

Directional investment strategies are based on the prediction of market trends. The most profitable positions for the hedge fund are taken on the basis of the manager's expectations: long positions for upward trends and short positions for downward trends. In other words, these strategies are associated with the manager's market timing skills (Table 12.3).

The first directional strategy is Equity Long/Short, where the allocation of capital between long and short positions aims to regulate portfolio exposure to the systematic risks of the stock market. The manager defines a longer or shorter net exposure and a different degree of financial leverage (gross exposure) on the basis of the forecast trends.[18]

The Managed Futures style focuses on derivatives instruments, with real assets (commodities) and financial assets (equity, bonds, interest rates, etc.) being used as underlying assets. The manager's discretionary choices may be supported by quantitative models and have a variable reference time horizon.

The Global Macro style is based on an accurate top-down macroeconomic analysis of the trend of markets fundamentals and possible macroeconomic gaps. The asset classes taken into consideration are widely diversified and the investment choices make broad use of derivatives and financial leverage aiming at maximising return.

Short Selling, also known as Short Bias, involves taking a short position in securities that are bound to fall in prices in the opinion of the hedge fund manager. This strategy requires the presence of specific lending facilities, supplied by the prime broker, which guarantee timely access to the lending of those securities that are short sold, for significant amounts and for customised periods, according to the

[16] Aragon and Strahan (2012) and McCray (2005).

[17] Brown and Goetzmann (2003), Amenc and Martellini (2003), Phillips and Surz (2003), and Stefanini (2006).

[18] Very often long/short asset managers operate as *activists*, engaging in the management of companies, achieving on average positive performance. Brav et al. (2015), Becht et al. (2017).

Table 12.3 Hedge funds: management strategies

	Strategy	Management policies	Performance determinants
Directional strategies	Long/short equity	Main focus on equity markets, investing both directly in stocks and indices and in derivatives instruments, building a portfolio of long, short, or less frequently, neutral positions. A certain directionality is maintained in relation to the expected market trends.	Direction of equity markets Financial leverage Stock picking (long and short positions)
	Managed futures	Main focus on futures or forward contracts representative of commodities, bonds, shares, precious metals, currencies, or indices, in an attempt to exploit trends in the very short term, short and medium term, on the basis of both discretional and non-discretional models.	Direction of markets Effectiveness of technical analysis instruments Efficiency of derivatives markets
	Global macro	Main focus on extremely liquid markets with long and short positions on stocks, bonds, currencies, and raw materials, chosen using the top-down approach, in an attempt to exploit any opportunity offered by the financial markets.	Direction of global markets Efficiency of derivatives markets Forecasting ability Volatility
	Short selling	Main focus on equity markets, taking exclusively short positions in order to follow downward trends and/or reduce the overall volatility of the portfolio.	Direction of markets Effectiveness of trading system Lending facilities
	Emerging markets	Geographically specialised funds that mainly take long positions on securities and indices in emerging countries, with no concentration or financial leverage limits.	Direction of markets Access to markets Opportunities of hedging and short selling Volatility
Non-directional strategies	Fixed income arbitrage	Long and short positions on fixed income with similar characteristics such as government bonds, corporate bonds, swaps, guaranteed securities, debts of emerging countries temporarily involved in mispricing and inefficiencies. Given the limited contribution of each arbitrage operation, this strategy makes wide use of financial leverage.	Spread trades Arbitrage on forward rates curve Financial leverage Effectiveness of interest rate risk immunisation policies
	Equity market neutral	This strategy consists of the acquisition of securities considered undervalued and, simultaneously, the short sale of securities considered overvalued or of indices, in order to eliminate systematic risk. In this case too, given the limited contribution of each arbitrage operation, the strategy makes wide use of financial leverage.	Covered arbitrage on securities, derivatives, indices Liquidity Financial leverage Volatility Stock picking

(continued)

Table 12.3 (continued)

	Strategy	Management policies	Performance determinants
	Convertible arbitrage	This strategy is based on arbitrage between shares and convertible bonds of the same issuer, with the short position taken on the former, and the long position on the latter, in order to exploit mispricing.	Coupons on bonds Delta between interest and dividends on short positions Volatility Credit spread
	Event driven	This strategy exploits the opportunities that emerge following extraordinary operations such as mergers, acquisitions, and leveraged buyouts, using arbitrage on the bonds of the companies involved. Typically, long positions are taken for the target company securities, and short positions for the bidder company securities. The event driven strategy has numerous sub-strategies.	Market cycles Deal target success Portfolio concentration Development of private equity market
	Distressed securities	This strategy uses a net long position on shares, bonds, and other securities of companies in temporary or permanent financial distress and which nonetheless show a good probability of overcoming the crisis or have been excessively penalised by the market.	Negative economic situation Downgrade of illiquid securities Development of high yield market Availability of law, tax, and management expertise Credit derivatives market efficiency

predictable downward trend of the selected securities. Unfortunately, for this strategy, which, moreover, has marginal weight in institutional portfolios, there is no reliable benchmark.

Hedge funds specialised in Emerging Markets invest in financial instruments of issuers residing in emerging and "frontier" countries.[19] Long positions are usually taken, and the manager's ability lies not only in the selection of each single security belonging to traditional equity and bonds asset classes, but also in the choice of the countries which are deemed to over-perform global economy. Given the difficulty in finding derivatives and lending services of securities negotiated in emerging and "frontier" markets, the choice of the prime broker is of greatest importance for the hedge fund.

[19] The so-called frontier markets include those markets that cannot still be considered emerging markets because of the degree of inefficiency or the lack of liquidity of the market.

12.2.2 Non-Directional Strategies

The very name "hedge fund" evokes the concept of hedging and consequently the implementation of investment policies that are not linked to financial market trends. Non-directional investment techniques involve the full hedging of market risks, and at the same time, the exploitation of inefficiencies in specific financial activities.

As shown in Table 12.3, five main non-directional strategies can be identified.

First of all, there is the Fixed Income Arbitrage, focused on exploiting mispricing and inefficiencies in the fixed income market (bonds, government bonds, swaps, etc.), via arbitrage operations. Since the markets where these financial assets are traded show limited deviation from full efficiency, the fund needs high financial leverage to achieve significant performance. In order to immunise the portfolio from market risks, the asset manager must pay particular attention to the interest rate risk management.

The purpose of the Equity Market Neutral strategies is to remove the systematic risk (beta = 0) and maximise alpha. In other words, they seek to minimise the market risk and maximise the return component thanks to the managers' stock picking activity. In order to achieve this goal, the fund purchases securities that are considered undervalued and at the same time short sells the same amount of securities considered overvalued. Exposure to the market is therefore close to zero. Financial leverage is necessarily higher given the limited return of each operation.

Hedge funds belonging to the Convertible Arbitrage category base their investment strategy on realising arbitrages between shares and convertible bonds of the same issuer in order to exploit mispricing via the acquisition of the undervalued security and the simultaneous short sale of the overvalued security.

The Event Driven strategy is used by managers who focus on companies affected by special events regarding finance operations such as mergers, acquisitions, and leveraged buyouts. The implementation of this strategy requires the use of arbitrage techniques on the shares of the companies involved, taking advantage of any market mispricing. Market experience recommends the adoption of a long position in the target company and a short position in the bidder company's shares, in the case of mergers and acquisitions. Further subcategories identify Event Driven investment strategies for other operations, for example the Initial Public Offering on the market of US small cap (Regulation D).

Distressed Securities funds invest in bonds or shares of companies experiencing financial difficulties, which may be temporary, or lasting in time, even reaching conditions close to liquidation or bankruptcy. The manager does not take any directional position. On the contrary, he/she will buy securities at a price that will guarantee a high return. If the issuing company is able to get through the crisis, he/she will benefit from the sharp increase in share prices. Likewise, if the issuer goes bankrupt, the securities guarantee a positive performance because they were purchased, in the wake of excessive penalisation by the market, at a lower price than the company's liquidation value.

12.2.3 Management Strategies Performance

The fact that each hedge fund strategy refers to one or more asset classes and that the techniques exploited by managers to generate performance and control risk vary according to the management style adopted gives the reasons for the benefits of diversification highlighted at portfolio level if they include hedge funds. The combination of traditional and alternative asset classes is further amplified if the manager avoids the concentration of investments in hedge funds belonging to a single strategy and instead diversifies them on the basis of the target return and target volatility pursued.[20]

Table 12.4 shows the correlation matrix between hedge funds strategies, as defined by Credit Suisse, with the exception of short selling, where it is clear how the benefits obtained from diversification can be achieved by combining directional strategies with non-directional strategies, operating within the framework of the two subclasses, or choosing managers that are reciprocally uncorrelated, within the framework of a single strategy.[21]

As with the correlations among the different management strategies, it may be useful to analyse the trend of these correlations over the time with the Global Equity asset class, represented by the MSCI World Gross Total Return in US dollars (Fig. 12.2).

Correlations calculated on rolling periods of 36 months have incremented over the years. On the one hand, markets are increasingly efficient, reducing arbitrage opportunities and the possibility for managers to differentiate their strategies and to realise extra-performance based on their own expectations; on the other, every time the market is widely characterised by unequivocal upward or downward trends, all hedge fund managers tend rationally to exploit these market signals by increasing their net positions (net long or net short, respectively).

It can also be easily understood how the cyclical and unpredictable market trends can be favourable to each strategy according to the moment and that no strategy is structurally dominant within the hedge fund universe in terms of performance. A dynamic reallocation process among different strategies may therefore make it possible to achieve significant extra-performance over the time. This phenomenon is described in Fig. 12.3, which provides the cumulative performance of the different strategies represented in the Credit Suisse indices, shown in US dollars and in increasing periods from 1 to 10 years.

At least in the long term, all hedge fund strategies have been able to reward the investors' target risk in a satisfactory way (Table 12.5).

[20] Lederman and Klein (1995) and Anderson et al. (2010).

[21] It is obvious that this conclusion is conditioned by the adoption of a long period reference horizon. Correlation coefficients on shorter horizons, especially during particularly negative market phases, tend to increase, even significantly, especially for *net long* directional strategies (Baesel et al. 2013).

Table 12.4 Correlation matrix between the hedge fund strategies (1994–2023). Source: Credit Suisse Liquid Alternative Beta and Hedge Fund Indices (1994–2023)

	Credit Suisse Hedge Fund Index	Credit Suisse Convertible Arbitrage	Credit Suisse Event Driven	Credit Suisse ED Distressed	Credit Suisse ED Multi-Strategy	Credit Suisse ED Risk Arbitrage	Credit Suisse Emerging Markets	Credit Suisse Equity Market Neutral	Credit Suisse Fixed Income Arbitrage	Credit Suisse Global Macro	Credit Suisse Long/Short Equity TR	Credit Suisse Managed Futures	Credit Suisse Multi-Strategy
Credit Suisse Hedge Fund Index	1.000												
Credit Suisse Convertible Arbitrage	0.569	1.000											
Credit Suisse Event Driven	0.708	0.623	1.000										
Credit Suisse ED Distressed	0.763	0.638	0.826	1.000									
Credit Suisse ED Multi-Strategy	0.525	0.502	0.600	0.685	1.000								
Credit Suisse ED Risk Arbitrage	0.705	0.480	0.622	0.644	0.502	1.000							
Credit Suisse Emerging Markets	0.319	0.231	0.334	0.266	0.204	0.192	1.000						
Credit Suisse Equity Market Neutral	0.765	0.657	0.925	0.974	0.702	0.657	0.303	1.000					

(continued)

Table 12.4 (continued)

	Credit Suisse Hedge Fund Index	Credit Suisse Convertible Arbitrage	Credit Suisse Event Driven	Credit Suisse ED Distressed	Credit Suisse ED Multi-Strategy	Credit Suisse ED Risk Arbitrage	Credit Suisse Emerging Markets	Credit Suisse Equity Market Neutral	Credit Suisse Fixed Income Arbitrage	Credit Suisse Global Macro	Credit Suisse Long/Short Equity TR	Credit Suisse Managed Futures	Credit Suisse Multi-Strategy
Credit Suisse Fixed Income Arbitrage	0.554	0.762	0.529	0.524	0.352	0.434	0.318	0.545	1.000				
Credit Suisse Global Macro	0.798	0.335	0.386	0.438	0.254	0.423	0.105	0.421	0.382	1.000			
Credit Suisse Long/Short Equity TR	0.838	0.472	0.668	0.732	0.569	0.665	0.251	0.740	0.400	0.446	1.000		
Credit Suisse Managed Futures	0.224	−0.072	−0.013	0.009	−0.024	−0.024	0.018	0.001	−0.060	0.350	0.077	1.000	
Credit Suisse Multi-Strategy	0.578	0.679	0.538	0.605	0.427	0.332	0.371	0.606	0.607	0.342	0.524	0.128	1.000

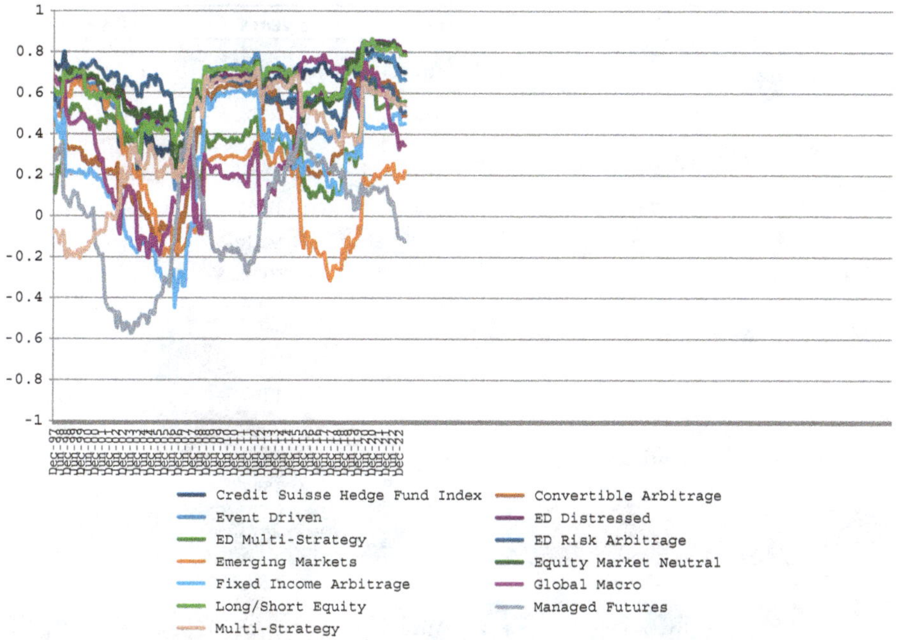

Fig. 12.2 Correlation between hedge fund strategies and "global equity" asset class (1994–2023, estimates made on rolling periods of 36 months). Source: Author's calculation using data from Morningstar

The measurement of the hedge fund strategy performances cannot ignore the inherent limits of the available databases. There are four types of distortions:[22]

- *Selection Bias*: The managers autonomously declare the category their hedge funds belong to, without any analysis by independent parties. In addition, managers may be incentivised to supply information only on funds that achieved the best performance and that have not reached the asset under management target amount yet. The effect on returns can be twofold: a potential effect of over-estimation when data have only been provided on best performing funds; a possible effect of under-estimation, when the best funds, whose subscription of new shares has been closed, do not wish to be publicised and therefore do not make their data known.
- *Backfill Bias* (or *Instant History Bias*): When funds supply their return for the first time, they usually provide their track record as well. The managers may therefore choose to publish only data regarding the time period corresponding to a particularly positive performance, thereby generating an over-estimation of the returns of the hedge fund sector.

[22] Ackermann et al. (1999) and Baker-Filbeck (2017).

1 year	2 years	3 years	5 years	10 years
ED Risk Arbitrage	Managed Futures	ED Distressed	Global Macro	Long/Short Equity
3.23%	12.96%	34.89%	23.83%	39.17%
Emerging Markets	Global Macro	Equity Market Neutral	Managed Futures	Multi-Strategy
2.51%	7.03%	30.47%	19.63%	38.04%
Long/Short Equity	Emerging Markets	Global Macro	ED Multi-Strategy	Global Macro
1.31%	3.43%	30.18%	15.43%	30.59%
Fixed Income Arbitrage	Multi-Strategy	Long/Short Equity	Long/Short Equity	ED Risk Arbitrage
0.12%	1.53%	23.62%	9.47%	25.77%
ED Multi-Strategy	Fixed Income Arbitrage	ED Multi-Strategy	Multi-Strategy	ED Multi-Strategy
-1.82%	-0.40%	23.55%	9.35%	24.76%
Convertible Arbitrage	Long/Short Equity	ED Risk Arbitrage	ED Distressed	Fixed Income Arbitrage
-3.64%	-2.19%	22.11%	9.22%	19.77%
Equity Market Neutral	Event Driven	Event Driven	Equity Market Neutral	Convertible Arbitrage
-4.32%	-2.38%	21.29%	7.64%	17.39%
Multi-Strategy	Convertible Arbitrage	Managed Futures	Convertible Arbitrage	Event Driven
-4.35%	-3.60%	19.25%	7.23%	16.76%
ED Distressed	ED Multi-Strategy	Multi-Strategy	Fixed Income Arbitrage	Equity Market Neutral
-4.37%	-3.73%	18.15%	4.69%	15.56%
Event Driven	Equity Market Neutral	Convertible Arbitrage	ED Risk Arbitrage	ED Distressed
-5.45%	-3.98%	14.55%	4.39%	14.07%
Managed Futures	ED Distressed	Emerging Markets	Event Driven	Managed Futures
-5.78%	-4.52%	13.25%	1.53%	12.84%
Global Macro	ED Risk Arbitrage	Fixed Income Arbitrage	Emerging Markets	Emerging Markets
-12.44%	-6.55%	11.91%	-3.54%	2.91%

Fig. 12.3 Time horizon and performance: the ranking of hedge fund strategies. Source: Credit Suisse Liquid Alternative Beta and Hedge Fund Indices (2023)

– *Survivorship Bias*: This occurs when funds in the databases cease to supply data, usually following significant losses. This phenomenon generates an over-estimation of the returns deriving from the different strategies, since the worst funds are systematically excluded from the sample. If the fund no longer provides

Table 12.5 Performance and volatility of hedge fund strategies (1994–2023)

	Mean	Standard deviation
Credit Suisse Hedge Fund Index	6.10%	6.66%
Credit Suisse Convertible Arbitrage	5.09%	6.12%
Credit Suisse Event Driven	6.89%	6.26%
Credit Suisse ED Distressed	5.93%	7.29%
Credit Suisse ED Multi-Strategy	4.54%	4.39%
Credit Suisse ED Risk Arbitrage	6.04%	12.77%
Credit Suisse Emerging Markets	3.23%	8.60%
Credit Suisse Equity Market Neutral	6.24%	6.58%
Credit Suisse Fixed Income Arbitrage	3.88%	5.04%
Credit Suisse Global Macro	8.27%	8.81%
Credit Suisse Long/Short Equity	6.88%	8.85%
Credit Suisse Managed Futures	4.22%	11.10%
Credit Suisse Multi-Strategy	5.81%	4.95%

Source: Credit Suisse Liquid Alternative Beta and Hedge Fund Indices (2023)

its data but is still active, it is defined as a defunct fund. If it has ceased activity instead, it is defined as a dead fund.[23]

– *Liquidation Bias*: This is generated by the fund managers' common practice to stop supplying information on funds just before their liquidation, once a decision has been made to cease their activity. This practice makes it impossible to know the NAV of the funds under liquidation.[24]

Together with these limits—inherent to the hedge funds' databases—a further distortion should be considered, which is intrinsically connected to this asset class, that is to say, the illiquidity bias or the stale pricing bias.[25] The phenomenon derives from the presence of illiquid assets or unlisted assets held in hedge fund portfolios.[26] The market price of these assets is not known for a continuous period if they are illiquid, or simply it is not available if they are unlisted. In both cases, the frequency of publication of the hedge fund NAV may be higher than that of the asset evaluation process. The direct consequence of this time discrepancy is the maintenance of assets with a fixed and constant price in the fund NAV, generating a reduction of the effective volatility of the fund and the onset of autocorrelation phenomena (or serial correlation).[27] This distortion becomes particularly significant in hedge funds that focus on very small cap shares, in bonds from issuers in emerging countries, in OTC bonds, and in distressed securities.

[23] Baker and Filbeck (2013).

[24] Teo (2011).

[25] Gregoriou et al. (2005).

[26] Schaub and Schmid (2013).

[27] The impact of this bias on the evaluation of hedge fund performances is analysed in Chap. 13.

References

Ackermann C, McEnally R, Ravenscraft D (1999) The performance of hedge funds: risk, return, and incentives. J Financ 54(3):833–874

Agarwal V, Naveen D, Naik N (2009) Role of managerial incentives and discretion in hedge fund performance. J Financ 64(5):2221–2256

Amenc N, Martellini L (2003) The brave new world of hedge fund indexes. EDHEC-Risk Institute, Lille-Nice

Anderson SC, Born JA, Schnusenberg O (2010) Closed-end funds, exchange-traded funds and hedge funds. Origins, functions, and literature. Springer, Heidelberg

Aragon GO, Strahan PE (2012) Hedge funds as liquidity providers: evidence from the Lehman bankruptcy. J Financ Econ 103(3):570–587

Baesel JB, González-Heres JF, Chen P, Shin SS (2013) Hedge fund benchmarking: equity correlation regimes and alpha. J Altern Invest 15(4):24–47

Baker HK, Filbeck G (eds) (2013) Alternative investments. Instruments, performance, benchmarks, and strategies. Wiley, Hoboken

Baker HK, Filbeck G (eds) (2017) Structure, strategies and performance. Oxford University Press, New York

Becht M, Franks J, Grant J, Wagner FH (2017) Returns to Hedge funds activism: an international study. Rev Financ Stud 30(9)

Brav A, Jiang W, Kim H (2015) Recent advances in research on hedge fund activism: value creation and identification. Ann Rev Financ Econ (7):479–595

Brown SJ, Goetzmann WN (2003) Hedge funds with style. J Portfolio Manag 29(2):101–112

Credit Suisse Liquid Alternative Beta and Hedge Fund Indices (2023)

Fagetan AM (2021) The regulation of hedge funds. A global perspective. Palgrave Macmillan, Cham

Goetzmann WN, Ingersoll JA, Ross SA (2003) High-water marks and hedge fund management contracts. J Financ 58(4):1685–1718

Gregoriou G, Hübner G, Papageorgiou N, Rouah F (2005) Hedge funds. Insights in performance measurement, risk analysis, and portfolio allocation. Wiley, Hoboken

Hedges JR IV (2005) Hedge fund transparency. Eur J Financ 11(5):411–417

IOSCO (2023) IOSCO investment funds securities survey

Jones AW (1949) Fashions in forecasting. Fortune 88–91:180–186

Lederman J, Klein RA (1995) Hedge funds. Investment and portfolio strategies for the institutional investor. McGraw-Hill, New York

Lhabitant FS (2004) Hedge funds. Quantitative insights. Wiley, Hoboken

Lhabitant FS (2006) Handbook of hedge funds. Wiley, Hoboken

Lo AW (2010) Hedge funds. An analytic perspective, 2nd edn. Princeton University Press, Princeton

McCray SA (2005) Hedge fund course. Wiley, Hoboken

Panageas S, Westerfield MM (2009) High-water marks: high risk appetites? Convex compensation, long horizons, and portfolio choice. J Financ 64(1):1–36

Phillips KS, Surz RJ (eds) (2003) Hedge funds. Definitive strategies and techniques. Wiley, Hoboken

Sancetta A, Satchell SE (2004) Calculating hedge fund risk: the draw down and the maximum drawdown. Appl Math Financ 11(3):259–285

Schaub N, Schmid M (2013) Hedge fund liquidity and performance: evidence from the financial crisis. J Bank Financ 37(3):671–692

Schneeweis T, Karavas V, Georgiev G (2002) Alternative investments in the institutional portfolio. Alternative Investment Management Association, London

Schneeweis T, Kazemi H, Szado E (2011) Hedge fund database "deconstruction": are hedge fund databases half full or half empty? J Altern Invest 14(2):65–88
Stefanini F (2006) Investment strategies of hedge funds. Wiley, Hoboken
Teo M (2011) The liquidity risk of liquid hedge funds. J Financ Econ 100(1):24–44
Zask E (2013) All about hedge funds, 2nd edn. McGraw-Hill, New York

Chapter 13
Hedge Fund Performance

Roberto Savona

13.1 The Complexity of Hedge Fund Performance Measurement

Assessing the risk-adjusted performance of hedge funds is a complex issue in finance due to the particular nature of these financial vehicles, which translates into non-Gaussian return distributions. Indeed, hedge funds usually exhibit non-linear option-like exposures to standard asset classes, because of the use of derivatives and because they implement highly dynamic trading strategies.[1] As a result, traditional risk-adjusted performance, such as the Sharpe ratio and/or Jensen's alpha, cannot be applied *tout court* to the hedge funds, since traditional performance measures are conceived and derived under the normality assumption of the return distribution.

Empirical evidence on hedge fund performance is mixed and seems depending on the methodology and the time period inspected in the empirical analysis. Dichev and Yu, for example, highlight an overall performance of the hedge fund sector at the end of 2008 lower than the Standard & Poor's 500 index. Differently, Kosowski et al. and Jagannathan et al., among others, provide evidence that hedge fund industry overperform many market benchmarks. More recently, Bollen et al. document a decline in aggregate hedge fund performance over the period 2008–2016.

Hedge fund returns exhibit the following features, which lead to possible deceptive results when using the classical performance measures:

[1] Fung and Hsieh (1997, 2001), Dichev and Yu (2011), Kosowski et al. (2007), Jagannathan et al. (2010), and Bollen et al. (2021).

R. Savona (✉)
Department of Economics and Management, University of Brescia, Brescia, Italy
e-mail: roberto.savona@unibs.it

I. Basile, P. Ferrari (eds.), *Asset Management and Institutional Investors*, Contributions to Finance and Accounting,
https://doi.org/10.1007/978-3-031-59819-7_13

- Fat tails, namely a significant concentration of return observations on the extremes of the distribution, thus resulting in higher probability of getting large returns relative to Gaussian distributions for which the tails are instead associated to very low probabilities.
- Skewness, which measures the asymmetry of the probability distribution around its mean and potentially showing higher chance to get positive returns (positive skewness) or negative returns (negative skewness).
- Serial correlation, namely smooth return dynamics which translate into biased sample moments (in particular the standard deviation), thereby causing misleading results when using classical risk-adjusted performance ratios.

These "anomalies" led financial economists to explore more advanced methods to estimate both the total risk (standard deviation) and its systematic component (beta coefficient).

Relative to the total risk, many studies used Value at Risk (VaR)-based measures, often expressed in their more robust form when dealing with non-Gaussian returns such as the Conditional VaR (CVaR), defined as the average (expectation) of high losses located on a pre-selected tail of the loss distribution.

Relative to the beta estimation, possible solutions include simple approaches, such as the one suggested in Asness, Krail, and Liew,[2] in which the Capital Asset Pricing Model (CAPM) includes also lagged betas in order to measure the total exposure towards the market factor. More sophisticated methodologies introduce mean-reverting processes and stochastic components to describe the hedge fund time-varying beta dynamics.[3]

Multifactor models Arbitrage Pricing Theory-based or (APT-based) have been used to detect the different risk sources of return dynamics. The point is not trivial, as the hedge funds are absolute return investment vehicles, and as such they are free from any traditional market benchmark. As a consequence, it is complex to regress hedge fund returns onto a suitable number of market indices able to explain the risk/return profile. It is for this reason that some of the methodological procedures we will analyse later in this chapter are both focused on beta estimation and on risk proxies representing the hedge fund's investable universe. Both topics (beta estimation and risk factors) are particularly challenging in light of what we learned from the sub-prime crisis started in August 2007 achieving a systemic dimension on September 2008 with the Lehman crash, when many funds experienced huge losses though their little prior exposure to systematic risk, therewith showing clear evidence about the incompleteness of traditional asset pricing models.

In this chapter, we present and discuss the performance measurement of hedge funds focusing on both traditional (classical risk-adjusted indicators and classical CAPM and APT models) and advanced approaches, which deal with investment style and risk/return characterizations in a dynamical (time-varying) framework.

---

[2] Asness et al. (2001).

[3] Savona (2014).

13.2 Classical Performance Measures

Using classical performance metrics when measuring the risk/return profile of hedge funds requires special adaptations in order to minimize the potential misleading results one can obtain when plainly implementing traditional performance measurement techniques. In the case of risk-adjusted indicators, such as the Sharpe ratio and the Omega ratio, it is essential to introduce some additional assumptions to control for non-Gaussian distributions. In the case of performance measures based on equilibrium models as the CAPM and APT, a special attention must be paid to the explanatory variables used in the regression estimation. In particular, the CAPM should include lagged market returns so as to control for autocorrelation in returns; the APT should include suitable risk factors representing the actual macroeconomic and financial variables we assume as the main drivers of the investment strategies followed by the managers.

13.2.1 Sharpe Ratio

The Sharpe ratio is the most popular risk-adjusted indicator based on two building blocks, expected returns and volatilities. Excess returns, relative to a risk-free rate, are scaled by the total risk, quantified by the standard deviation. Therefore, the accuracy of the Sharpe ratio "hinges on the statistical properties of returns, and these properties can vary considerably among portfolios, strategies, and over time".[4] As is clear, such an issue strongly affects hedge funds which exhibit higher moments return distribution (skewness, kurtosis) far from the Gaussian case as well as serial correlation.

In fact, skewness, kurtosis, and serial correlation significantly affect the risk estimation of hedge funds, possibly leading to biased risk-adjusted performance measures, since higher moments impact on lower estimations of standard deviation, thereby offering a misleading representation of the actual risk profile of fund managers.[5] Consider, for example, investment strategies with a long position on a market benchmark and a set of short positions on call and put options out-of-the

[4]Lo (2002), p. 36.

[5]Amenc et al. (2002) offer a simple example to make effective the point, comparing an equally weighted mixture of a Gaussian with a mean of 0.5 and a standard deviation of 0.5 and a Gaussian with a mean of −0.5 and a standard deviation 1.32. The distribution exhibits the same mean and variance as a standardised Gaussian (0 and 1, respectively), but a skewness of −0.75 and a kurtosis of 6.06, compared to 0 and 3, respectively, for the standardised Gaussian. The authors note that only by taking into account the third and fourth order moments it is possible to distinguish funds with returns that follow those probability distributions. From a theoretical point of view, this statistical evidence leads to the question whether investors are only interested in the first and second order moments, i.e. their mean and variance (as assumed by the Sharpe ratio), or instead look also at higher moments, as it should be the case when skewness and kurtosis are far from the Gaussian case.

money. Such a combination leads to higher ex post performance with a "masked" standard deviation which appears artificially low, since the implicit downside risk is extremely high as the short call and put imply significant risky positions being the potential losses unlimited.

On this point, Lo proposes amending the traditional Sharpe ratio on an annual basis according to the following formulation:[6]

$$\mathrm{SR}_{\mathrm{annual}} = \eta \cdot \mathrm{SR}_{\mathrm{monthly}}$$

with:

$$\eta = \frac{q}{\sqrt{q + 2\sum_{k=1}^{q-1} (q-k)\rho_k}};$$

- $\rho_k = k$-th order of the serial correlation coefficient.
- $q =$ Number of time intervals, which is equal to 12 when expressing the Sharpe ratio on an annual basis.

When returns are independently and identically distributed, the annual Sharpe ratio is $\sqrt{12} \cdot \mathrm{SR}$. with no serial correlation at each lag, η reduces to $\eta = q/\sqrt{q} = \sqrt{q}$. However, with autocorrelated returns, without the correction for the factor η we would overestimate the true Sharpe ratio when there is a tendency for positively autocorrelated returns (in this case, we have $\eta < \sqrt{q}$), or underestimate the true Sharpe ratio when the autocorrelation tends to be negative (in this case, we have $\eta > \sqrt{q}$).

13.2.2 Omega Ratio

The Lo's version of the Sharpe ratio focuses on the variance underestimation when returns are autocorrelated, which may cause a potential bias, because skewness and kurtosis are neglected when measuring the risk profile (standard deviation). Actually, the hypotheses introduced by Lo also deal with this last point, reducing the contribution of the higher moments of the distribution (skewness and kurtosis) to a marginal role when you control for the serial correlation.[7]

Nevertheless, even the Lo's version of the Sharpe ratio is still part of a "mean-variance world", which actually neglects skewness and kurtosis in the evaluation process of the performance of an investment portfolio. A possible solution to this

[6] Lo (2002).

[7] For further technical details, see Lo (2002).

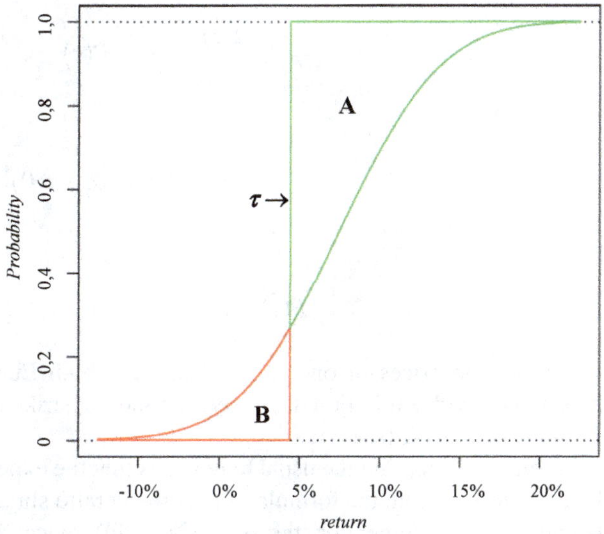

Fig. 13.1 Graphical representation of the Omega ratio

problem is the Omega ratio.[8] Unlike other risk-adjusted performance measures, this indicator takes into account the entire distribution of returns, thus offering the opportunity to deal with distributions exhibiting strong asymmetries and excess kurtosis, as it is the case of hedge funds, which massively use derivative instruments and implement sophisticated option-based strategies.

Computationally, the Omega ratio starts defining a minimum performance threshold τ under which investors negatively evaluate their portfolio performance. In general, such a threshold could be proxied by the risk-free rate or, alternatively, by another value the investor assesses as "non-appropriate" in terms of his/her return targets. Given the entire range of possible return values that the portfolio can assume, the threshold τ allows to split the return distribution between an "appropriate" area, on the right of the threshold showing returns greater than τ (see Fig. 13.1, A area), and a "non-appropriate" area, on the left of the threshold corresponding to portfolio returns lower than τ (see Fig. 13.1, B area).

The Omega ratio provides a relative performance measurement, as classical risk-adjusted measures, being the ratio between the area A and the area B.

The ratio is mathematically expressed through differentials while a more simplified version can be obtained by the option pricing theory. Indeed, the Omega ratio can be also expressed as:[9]

[8] Keating and Shadwick (2002).

[9] Kazemi et al. (2004).

$$\Omega(\tau) = \frac{A}{B} = \frac{C(\tau)}{P(\tau)}$$

with:

$$C(\tau) = e^{-r_f} E[\max(r - \tau, 0)]$$

and

$$P(\tau) = e^{-r_f} E[\max(\tau - r, 0)]$$

which are the prices of one call and one put both European-style options on the investment, with expiration date fixed at 1 month, strike price equals to the threshold τ, and r_f is the risk-free rate.[10]

Therefore, adopting the usual hypothesis that the expected return is approximated by the sample mean, the formula of the omega ratio simplifies and becomes the ratio between, in the numerator, the sum of the differences between the returns greater than the threshold τ and the threshold itself and, in the denominator, the sum of the differences between the threshold τ and the yields lower than it, i.e.:

$$\Omega(\tau) = \frac{\sum_{t=1}^{T} \max(r_t - \tau, 0)}{\sum_{t=1}^{T} \max(\tau - r_t, 0)}$$

A different expression of the Omega ratio having the same results, and similar to the Sharpe ratio, is the so-called Sharpe-Omega ratio, which assumes the following expression:

$$\text{Sharpe } \Omega(\tau) = \frac{(\overline{r} - \tau)}{P(\tau)}$$

where $\overline{r}$ is the expected return over one-period time horizon with continuous compound interest.

In this case, the excess return relative to the threshold—which can be different to the risk-free rate—is scaled by a risk measure representing a portfolio insurance with a minimum return equal to τ.

[10] See Kazemi et al. (2004) for further technical details on call and put pricing approaches to use when the underlying and the strike price are both expressed as returns.

13.2.3 CAPM-Based Measures

As is well known, the CAPM gives a performance measure relative to a benchmark, representing the investment universe where managers take their decisions. Relative to appropriateness of the CAPM when exploring the performance of hedge funds, the major concerns are with the extra return (the Jensen's alpha) and the systematic risk exposure. In fact, both estimates could be seriously biased if the market portfolio does not represent the real investment universe of hedge fund managers. The issue is widely discussed in the literature, especially dealing with the potential bias whenever the model includes a passive market index, when instead the portfolio returns show non-linear payoffs.

To bypass this problem, Asness, Krail, and Liew suggest a CAPM-based expression as follows:[11]

$$(r_t - r_{f,t}) = \alpha + \sum_{k=0}^{K} \beta_k (r_{M,t-k} - r_{f,t-k}) + \varepsilon_t$$

where:

r_t	is the hedge fund return at time t
r_t	is the risk-free rate at time t
α	is the intercept (the Jensen's alpha)
β_k	is k-th sensitivity coefficient relative to the excess return of the market portfolio
ε_t	is the error term at time t
$r_{M,t-k}$	is the return of the market portfolio at time $t-k$
$r_{f,t-k}$	is the risk-free rate at time $t-k$
K	is the number of lags, usually fixed at 3

The empirical evidence is strong enough to consider the model as statistically robust both in terms of explanatory power and estimation of the coefficients (alphas and betas). Using this approach, it is then possible to correctly assess the systematic risk exposure, next providing a better extra performance estimation for the hedge funds (Jensen's alpha).

13.2.4 APT-Based Measures

The main limitation of the CAPM-based modelling is with one single risk factor as proxy of the market portfolio. To overcome possible misrepresentation due to omitted factors and ineffective benchmarks, multifactor modelling is the obvious

[11] Asness et al. (2001).

solution. Many authors explored the multifactor modelling in the hedge fund industry, leading to the identification of a set of major indices which have been proven to be effective in measuring the performance of hedge funds.

Formally speaking, a generic hedge fund return is linearly mapped onto a parsimonious set of risk factors through the following equation:

$$\left(r_t - r_{f,t}\right) = \alpha + \sum_{k=1}^{K} \beta_k \gamma_{k,t} + \varepsilon_t$$

in which $\gamma_{k,t}$ are the k-th risk factors to be selected with the end to better explain the return dynamics shown by hedge funds.

Based on a number of studies that focused specifically on hedge fund risk factors and the abundant literature on equity and bond risk drivers, we can detect the following set of indices:[12]

I. Market factor:

(a) US equity (US equity index return net of risk-free rate).
(b) World equity (World equity index return net of risk-free rate).
(c) Implied volatility (CBOE Volatility Index—VIX—measuring the implied volatilities of S&P 500 call and put options with 30 days maturity).

II. Fundamental factor:

(a) Size (market value of the stocks within the US equity index).
(b) Book-to-market (based on US equity index).
(c) Dividend yield (based on US equity index).
(d) Market volume (changes in the market volume).

III. Macroeconomic factor:

(a) T-bill (3 months US T-bill rate).
(b) Default spread (difference between BAA-rated return and AAA-rated return of long-term bonds).
(c) Term spread (10 years bond yield minus 1 year bond yield).
(d) Oil price (West Texas Intermediate crude oil).
(e) Currency (US dollar exchange rate variation computed against the more traded foreign exchange rates).
(f) Inflation.
(g) Industrial production (logarithmic variation of global industrial production).

Empirical evidence shows good explanatory power when using six risk factors (US equity, world equity, implied volatility, market volume, T-bill, oil price), with adjusted R-squared in some cases quite high, as for "Convertible arbitrage" and "Fixed income arbitrage" hedge fund categories, with values greater than 50%.

[12] Amenc et al. (2003a, 2003b) and Kat and Miffre (2002).

Instead, in other cases the values are low, as for "Equity market neutral" and "Event driven" categories, with values around 15%.

13.3 Advanced Performance Measures

The approaches presented in the previous sections are important examples of how classical models have been modified according to the special features of hedge funds return distributions. However, such approaches are in some way ineffective to describe all possible hedge fund strategies and their non-linear payoffs. A possible solution to this problem is with another class of models, that we can call "advanced", whose main target is to describe the dynamics of hedge funds based on a linear factor-based return generating process. The underlying idea of these models is to "transfer" the non-linearity of hedge funds onto ad hoc explanatory risk factors representing the main strategies assumed by the hedge funds. In doing this, the resulting Jensen's alpha is correctly assessed, being the risk-adjusted performance of the hedge fund manager relative to a set of complex benchmarks covering the space of all major investment strategies available in the industry.

13.3.1 Advanced Measures Based on Style Factors

In general, the hedge fund return generating process assumed by hedge funds can be modelled according to the following equation:

$$r_t = \alpha + \sum_{k=1}^{K} \beta_k \mathrm{SF}_{k,t} + \varepsilon_t$$

where SF_k are the k-th generic style factors towards which the fund is exposed.

Scientific literature has been prolific in this regard, by offering many methodological alternatives based on both single-factor and multi-factor models, thereby contributing for a better performance measurement activity while controlling for the high heterogeneity within the hedge fund industry in terms of investment strategies followed by the managers.

The main methodological approaches used in the literature includes the following style factors:

- Latent factors.
- Peer groups.
- Embedded options.
- Asset-based style factors.

The first alternative to the qualitative style descriptions of the hedge funds is based on returns alone. The rationale is that if two managers use the same trading strategy on the same markets, their returns are correlated to each other, even if they are not correlated to the returns of asset markets. Based on this reasoning, Fung and Hsieh used principal component analysis to group funds based on their correlation with each other, finding that the first five principal components can jointly explain roughly 45% of the cross-sectional variation in hedge fund returns.[13]

The second alternative is quite a standard method to model hedge fund risk, using broad-based indices of hedge funds clustered according to qualitative and quantitative criteria, thereby forming homogeneous peer groups. The empirical literature has provided two main approaches to the construction of such peer groups:

- Using the same information provided by the hedge funds relative to their investment strategy, one can easily compute the equally weighted average of hedge fund returns within the same group or also compute the asset-weighted average, and even calculate their median,[14] then obtaining category benchmarks to be used in exploring the performance of hedge funds. This is what major index providers offer in the market through extensive data collection.
- A second way to compute peer groups is by quantitatively aggregating hedge funds based on their return series with the objective of assembling homogeneous groups in terms of return dynamics. Mathematically, the cluster analysis is the main approach allowing to endogenously identify a natural grouping of funds that has some predictive power in explaining the future cross-sectional dispersion in fund returns. The style factor is simply obtained by computing an equally/asset-weighted average of the fund returns. pertaining to the same cluster.[15]

The approach of embedded options can be traced back to Glosten and Jagannathan, who inspected the performance of highly dynamic managers with non-linear payoffs.[16] The authors suggest an asset pricing model using options, written on the market index, as risk factors, together with the market portfolio. More specifically, they introduce the following equation:

$$r_t = \alpha + \beta_1 R_I + \sum_{i=1}^{N} \delta_i \max(R_I - o_i, 0) + \varepsilon_t$$

in which $\max(R_I - o_i, 0)$ is the payoff at the expiration date of the i-th call option written on the market index I with strike price o_i. Glosten and Jagannathan suggest to use three options in order to better describe the complexity of sophisticated managers with return dynamics difficult to explain using only passive benchmarks. They also

[13] Fung and Hsieh (1997).

[14] The best known index providers are BarclayHedge, Credit Suisse Hedge Fund Indexes, and Hedge Fund Research.

[15] See, for example, Brown and Goetzmann (2003).

[16] Glosten and Jagannathan (1994).

prove that, for investment funds, one embedded option style factor is enough to efficiently describe their returns over time. Note, on this point, that a model with only one synthetic option is quite similar to the Henriksson and Merton model, although the strike price of the embedded option in Glosten and Jagannathan is not the risk-free rate, as in Henriksson and Merton.[17]

Asset-Based Style Factors (ABSF) are benchmarks derived from observed market prices which provide direct descriptions of hedge fund strategies. As discussed, the complex nature of hedge fund strategies reflects on non-linearity in returns, making it difficult to uncover how the returns are generated and where they come from. To reduce this problem, Fung and Hsieh[18] advanced the concept of primitive trading strategies (PTS), namely option-based benchmarks designed to describe the strategy-driven non-linearity and contextualized within Henriksson and Merton's insight, for which the payoff of a perfect market timer should be identical to the payoff from owning a call option on the market. This reasoning allows us to bypass the problem connected to linear models when passive benchmarks are used to describe hedge fund returns: since the intrinsic non-linearity of hedge fund returns is moved onto the explanatory factors, which act as proxies of the complex and often unobservable trading strategies, a simple linear multivariate regression can be used effectively to describe the behaviour of hedge fund returns.

The idea of ABSF comes from Glosten and Jagannathan, who suggest to select an appropriate mathematical formulation of the pricing model after a careful discussion with the manager, so as to best select the risk factors and their functional form. To compute ABSF, Fung and Hsieh used traded options to explicitly model the characteristics of trend-following hedge funds, showing that the returns from these strategies can be replicated by a dynamically managed option-based strategy known as a "lookback straddle". Since the trend follower is a trader that buys an asset at the low and sells it at the high over a given investment horizon, and since this pattern can be described by a payout of a lookback straddle on that asset, then the return of the strategy is isomorphic to the payout of the lookback straddle less the option premium. Using this economic reasoning, Fung and Hsieh relied on lookback options as asset-based style factors for trend-following hedge funds, capturing high degree of explanatory power for hedge funds that adopt this style. In the same way, such an approach is also useful to compute the correct manager's excess return: alpha is indeed estimated by comparing the returns of the hedge fund with ABSFs that describe the expected return of a class of complex hedge fund strategies that cannot be directly observed.

As pointed out by Fung and Hsieh,[19] the research on risk factors can be characterized by a bottom-up or a top-down perspective. With the first approach, the interest is on the understanding of risks in specific hedge fund styles, while the second approach is devoted to inspecting the risk factors of diversified hedge fund

---

[17] Henriksson and Merton (1981).

[18] Fung and Hsieh (2001).

[19] Fung and Hsieh (2004, 2007a, b).

portfolios. A top-down perspective is then conceived with the aim of dealing with all the common risks that transversely affect the hedge fund industry as a whole, allowing the inspection of the inner sources of risk dynamics and their potential impacts on the markets.

Fung and Hsieh address the issue of the top-down approach in modelling the risk factors of hedge funds, by proposing a basic model with eight risk factors that are found to be present in empirical studies on many of the major hedge fund styles. The authors propose the use of four dominant risk classes:

1. Trend-Following Risk Factors: Three out of the five primitive trend-following strategies proxied as pairs of standard straddles and constructed from exchange-traded put and call options as described in Fung and Hsieh,[20] namely (1) Bond Trend-Following Factor (PTFSBD), (2) Currency Trend-Following Factor (PTFSFX), and (3) Commodity Trend-Following Factor (PTFSCOM).
2. Equity-oriented Risk Factors: (4) Equity Market Factor, proxied by the Standard & Poor's 500 index monthly total return (EQUITY_MKT) and (5) Size Spread Factor (SIZE_SPREAD), proxied by Wilshire Small Cap 1750 minus Wilshire Large Cap 750 monthly returns.
3. Bond-Oriented Risk Factors: (6) Bond Market Factor, proxied by the month end-to-month end change in the 10-year treasury constant maturity yield (BOND_MKT) and (7) Credit Spread Factor (CREDIT_SPREAD), proxied by the month end-to-month end change in the Moody's Baa yield less the 10-year treasury constant maturity yield.
4. Emerging Market Risk Factor: (8) the IFC (or MSCI) Emerging Market Index (EM).

Factors (1) to (7) are the original Asset-Based Style Factors (ABSF) advocated by Fung and Hsieh,[21] denominated in this way because all the factors are based on traded securities and their derivatives, while the eighth factor has been recently proposed since the authors plausibly suppose that: "When younger emerging equity markets mature and begin to figure more prominently in the opportunity set of hedge fund managers, it may be important to see if a new P(rimitive)T(rading)S(trategy) in some form of an emerging market stock index is needed".[22]

These eight factors are used to represent hedge fund strategies by running the following regression:

$$R_{i,t} = A_i + \sum_{k=1}^{8} B_{i,k} F_{k,t} + E_{i,t}$$

[20] Fung and Hsieh (2001).

[21] Fung and Hsieh (2004).

[22] Fung and Hsieh (2007a), p. 74.

Here, $R_{i,t}$ is the return of the hedge fund i for time t, A_i is the abnormal performance of hedge fund i, $B_{i,k}$ is the factor loading of hedge fund i on factor k, $F_{k,t}$ is the return of factor k for month t, and $E_{i,t}$ is the error term.

Among all the previous alternatives proposed to inspect hedge fund performance, the asset-based style factor's approach is the more robust in terms of benchmark identification and risk-return assessment. These factors are in fact realized through clear rules, and potentially they can be used to replicate Fung and Hsieh's estimations. Moreover, ABSF have a clear-cut economic framework, then giving a solid scheme of analysis, through which one can first estimate the extra performance of hedge funds and next dig deep into the results by a reasonable performance attribution.[23]

The literature on hedge fund performance is now abundant and massively based on the Fung and Hsieh scheme of the eight risk factors, based on the four asset classes presented before (trend following, equity-oriented, bond-oriented, emerging market).[24]

13.3.2 The SED Measure of Gao-Gao-Song

Gao et al. advance the hypothesis that some hedge funds with managers who have better skills in exploiting rare disaster concerns deliver a superior fund performance while being less exposed to disaster risk.[25] To test such a hypothesis, the authors introduce an ad hoc measure which quantifies the fund's ability to achieve excess returns linked to the disaster concern SED (skill at exploiting rare disaster concerns). They develop a rare disaster concern index that equals the price of a disaster insurance reflecting the ex ante disaster concerns (constructed with out-of-the-money put options on various economic sectorial indices), then measuring the fund's SED based on the covariance between fund excess returns and such an index.

The authors' argument is based on the assumption that where the concern for a disaster is perceived by the market as high, those funds which are able to identify the value of the "fear premium" should obtain higher returns relative to competitors without such a skill. Gao et al. provide empirical evidence on this hypothesis, highlighting a large dispersion of SED values, thereby documenting different abilities of funds in benefitting the fear premium.

Analytically, the disaster risk index proposed by the authors (RIX) to estimate the ex ante market expectation of future catastrophic events is computed through portfolio replication techniques, thus obtaining non-performing portfolios' payoffs using financial instruments traded in the market. More in detail, the RIX is calculated

---

[23] The ABSF can be downloaded at https://faculty.fuqua.duke.edu/~dah7/HFRFData.htm.

[24] Fung and Hsieh (2004, 2007a, b), Fung et al. (2008), and Edelman et al.(2012).

[25] Gao et al. (2018).

as the price difference of two portfolios of synthetic variance swap contracts,[26] replicated by combining call and put options: the first portfolio is characterized by low volatility shocks, while the second takes into account extreme volatility shocks caused by market jumps and associated with the risk of rare events. Mathematically, Gao et al. prove that the RIX can be reduced to the following simplified equation:

$$\mathrm{RIX} = \frac{2e^{r\tau}}{\tau} \int\limits_{K<S_t} \frac{\ln(S_t/K)}{K^2} P(S_t, K, T)dK$$

where r is the risk-free rate, $\tau \equiv T - t$ is the time to maturity, and $P(S_t; K; T)$ is put option with strike K and maturity date T.

Using a dataset of 5864 hedge funds over the period 01/1996–07/2010, Gao et al. document that hedge funds exhibiting high SED values (covariance between hedge fund returns and the RIX index) are moderately exposed to catastrophic risks and exhibit market timing ability, thereby obtaining extra performance in stress scenarios.

13.4 Measuring Hedge Fund Performance: An Empirical Application

In this section, we present an empirical analysis of hedge fund performance based on the Fung-Hsieh (FH) model. Using data from the Credit Suisse AllHedge Index, provided by BarclayHedge, we explore the risk-return profiles of hedge fund industry over the period from January 2010 to May 2023. The Credit Suisse Hedge Fund indices are asset-weighted hedge fund indices computed for ten style categories accounting for at least 85% of the Asset Under Management (AUM) in each group, reported in Table 13.1. We also contrast the results of the FH model with CAPM-based metrics using the CAPM with 3 lags as proposed by Asness, Krail, and Liew.

13.4.1 Data and Preliminary Statistics

Univariate descriptive statistics for the 10 indices (annualized mean, maximum, minimum, annualized standard deviation, skewness, excess kurtosis, and first-order autocorrelation) are reported in Table 13.1. The data shows considerable heterogeneity in the historical risk and return characteristics of the various categories of hedge fund investment styles, in particular denoting high variability in terms of

[26] Variance swaps are OTC contracts having a payoff that is dependent on realized variance.

Table 13.1 Descriptive statistics of hedge fund indices over the period 2010–2023

Index	$Mean_{yr}$	$Median_{yr}$	Max	Min	$StdDev_{yr}$	Skew	Excess Kurt	AR(1)
Hedge fund index	4.37%	4.39%	3.78%	−7.28%	4.77%	−1.1216	5.3472	0.0034
Convertible arbitrage	4.12%	3.53%	3.72%	−6.43%	4.12%	−0.9527	6.1062	0.0030
Emerging markets	4.71%	5.31%	6.34%	−10.27%	8.05%	−0.7722	2.5943	0.0037
Equity market neutral	1.88%	1.24%	4.34%	−4.30%	4.74%	−0.0586	0.8226	0.0017
Event driven	3.87%	5.47%	6.75%	−13.47%	7.03%	−2.0727	13.1420	0.0026
Fixed income arbitrage	4.75%	5.44%	2.67%	−6.51%	3.05%	−3.1210	23.2861	0.0027
Global macro	5.17%	4.33%	8.79%	−6.19%	6.30%	−0.0391	3.8411	0.0035
Long/short equity	5.05%	5.05%	5.09%	−7.41%	7.07%	−0.5318	1.0886	0.0044
Managed futures	3.45%	3.39%	8.49%	−7.50%	10.22%	−0.0181	−0.2995	0.0034
Multi-strategy	5.54%	7.31%	4.19%	−5.98%	4.18%	−1.1933	5.5326	0.0043
Average	4.29%	4.54%	5.42%	−7.53%	5.95%	−0.9881	6.1461	0.0033
S&P 500	12.32%	18.73%	12.82%	−12.35%	14.72%	−0.3684	0.5942	0.0118

Note: The table reports summary statistics for Credit Suisse Hedge Fund indices over the period 01/2010–05/2023. $Mean_{yr}$ is the annualized mean return. Max and Min are the maximum and minimum monthly return, respectively. $StdDev_{yr}$ is the annualized standard deviation. Skew and Excess Kurt are the skewness and the excess kurtosis, while AR(1) is the autocorrelation coefficient computed regressing returns at t against returns at time t–1 minimum and maximum values. Note that the annualized standard deviation is lower than that of the S&P 500 index, as well as for the average returns, with S&P 500 index exhibiting higher values relative to hedge fund indices.

By inspecting higher moments, we note that all indices exhibit negative skewness, and excess kurtosis is positive for all strategies, with very high values for some category (Fixed Income Arbitrage). These figures prove strong departure from Gaussian distributions. Moreover, all the hedge fund styles show positive autocorrelation (AR(1)), thus indicating smooth return dynamics, which can translate into biased sample moments (in particular, the standard deviation). And indeed, the annualized standard deviation is on average 5.95% versus 14.72% we computed for the S&P 500 index over the same time period, thus opening the question about the real risk profiles of hedge funds and their impact on performance. In other terms, we need a more in-depth analysis of the risk-adjusted returns of hedge fund indices, taking into account for both non-Gaussian distribution features and non-linear payoffs. In the next section, we do this by implementing the FH and the 3-lags CAPM models.

13.4.2 Performance Measurement

We run the FH model with the eight risk factors commented in Sect. 13.3 and the CAPM model with 3 lags using the S&P 500 index as proxy for the market return. The results are reported in Table 13.2 Panel A, for the FH model, and Panel B, for the 3-lags CAPM. Let us first inspect the results of the FH model in Panel A. The Adjusted R-squared denotes high variability of the explanatory power of the model, with values ranging from 0.0309 (Equity Market Neutral) to 0.5490 (Emerging Markets) with an average of 0.2957. In terms of overall model's reliability, the F-stat denotes values far from significance only for Equity Market Neutral, which is coherent with their low explanatory power (Adjusted R-squared). The Jensen's alpha, which gives us direct information about extra performance relative to ABSFs, while showing all positive values, is statistically significant only for the aggregate index (Hedge Fund Index) and for the Event Driven category.

The results from the 3-lags CAPM reported in Panel B show a quite different picture about the risk-adjusted performance relative to the FH model. Indeed, the Jensen's alpha is positive and statistically significant only for Global Macro, while Event Driven exhibits a significant negative alpha. The exposure towards the market index in terms of sum of betas denotes positive values for all indices, thereby suggesting a general momentum-based strategy tendency of the industry. Comparing the Adjusted R-squared of the FH model with that of the 3-lags CAPM (the difference between the corresponding diagnostics for each strategy reported in Panel C of the table as "A minus B"), we observe more power of the FH model relative to the 3-lags CAPM. On average, the ABSFs explain about 40 percent more than S&P index with its lags.

Table 13.2 Performance measurement

	Hedge fund index	Convertible arbitrage	Emerging markets	Equity market neutral	Event driven	Fixed income arbitrage	Global macro	Long/short equity	Managed futures	Multi-strategy
Panel A: FH model										
Jensen's alpha	0.0119*	0.0033	0.0141	0.0022	0.0208**	0.0053	0.0133	0.0135	0.0062	0.006
EQUITY_MKT	−0.0413*	0.0108	−0.02	0.0238	−0.0455	−0.0196	−0.1068***	−0.0384	−0.0156	−0.0161
SIZE_SPREAD	0.0284	0.007	−0.0024	−0.0711*	0.0382	0.0202	0.1333**	−0.0074	−0.0049	0.0032
BOND_MKT	−11.431	−12.177	−20.888	−0.5254	−19.394	−0.1162	−33.752	−0.8321	13.238	−0.7697
CREDIT_SPREAD	−31.853	0.746	−35.568	−0.1189	−7.07**	−0.5661	−0.9535	−42.201	−32.622	−0.1447
EM	0.1816***	0.1318***	0.3684***	0.0473	0.2395***	0.0396**	0.1403***	0.2562***	0.1461**	0.1086***
PTFSBD	−0.0005	−0.0019	−0.0127*	−0.0059	−0.0083	−0.0085***	0.0077	−0.0031	0.0317**	−0.0036
PTFSFX	−0.0062	−0.0108**	0.0021	−0.004	−0.0211***	−0.012***	0.0067	−0.0132*	0.0215	−0.0055
PTFSCOM	−0.0068	−0.0081	−0.0124	−0.0055	−0.013	−0.0059	−0.0016	−0.0158*	0.0147	−0.0081
Adjusted R^2 (A)	0.3961	0.3265	0.5490	0.0309	0.4898	0.3260	0.1513	0.4071	0.0695	0.2110
F-statistic	14.1197	10.6966	25.3435	16.380	20.2019	10.6756	45.661	14.7324	24.945	63.488
Prob(F-statistic)	0.0000	0.0000	0.0000	0.1184	0.0000	0.0000	0.0001	0.0000	0.0143	0.0000
Panel B: 3-lags CAPM										
Constant	−0.0004	−0.0002	−0.0015	−0.0013	−0.0033***	0.0009	0.0027*	−0.0013	0.0016	0.0011
Mkt	−0.0003	0.0289	0.0155	0.032	0.0089	−0.0091	−0.0561*	0.009	0.0301	0.0105
Mkt(−1)	0.2636***	0.1739***	0.3616***	0.1524***	0.4025***	0.106***	0.1429***	0.4281***	0.0347	0.1948***
Mkt(−2)	0.0815***	0.0824***	0.1119***	0.0471*	0.1494***	0.092***	0.0119	0.0684***	0.0448	0.092***
Mkt(−3)	0.0268	0.0386**	0.0355	0.0322	0.0678***	0.0541***	0.0019	0.0253	−0.0468	0.0156
Sum of betas	0.3717	0.3241	0.5248	0.2639	0.6287	0.2431	0.1007	0.5309	0.0628	0.3131
Adjusted R^2 (B)	0.6202	0.3622	0.4032	0.1863	0.6787	0.3821	0.1134	0.7501	−0.0130	0.4498
F-statistic	65.0893	23.2870	27.5196	99.858	83.9022	25.2763	60.188	118.8351	0.4951	33.0864
Prob(F-statistic)	0.0000	0.0000	0.0000	0.0000	0.0000	0.0000	0.0002	0.0000	0.7393	0.0000
A minus B	−0.2241	−0.0356	0.1458	−0.1554	−0.1889	−0.0561	0.0380	−0.3430	0.0826	−0.2388

(continued)

Table 13.2 (continued)

	Hedge fund index	Convertible arbitrage	Emerging markets	Equity market neutral	Event driven	Fixed income arbitrage	Global macro	Long/short equity	Managed futures	Multi-strategy
Panel C: Sharpe ratios										
Adjusted Sharpe ratio (Lo's version)	0.7613	0.7002	0.5340	0.3227	0.3611	0.9812	0.6030	0.6657	0.3000	0.8204
Unadjusted Sharpe ratio	0.7796	0.8410	0.5033	0.2586	0.4568	1.3427	0.7161	0.6214	0.2737	0.7676

Note: Panel A reports results from the FH model estimation with 8 risk factors. Panel B reports estimates of CAPM model with 3 lags using the S&P 500 index; Sum of betas is the summation of the 4 coefficients (contemporaneous beta and the three coefficients estimated for the three lags). Panel C reports the difference between the Adjusted R-squared of the FH model and 3-lags CAPM ("A minus B") and the Sharpe ratios both adjusted (using the Lo's version) and unadjusted (using the US T-bill 3-months rate as risk-free rate). ***, **, * denote significance at 0.01, 0.05, and 0.1 level, respectively

Although hedge funds benefitted from a positive market timing during the 2009–2011 sovereign debt crisis in Europe, the sector has partly suffered the effect induced by the COVID-19 pandemic crisis, with substantial negative returns in particular during February 2020.[27]

This is consistent with the findings of Bollen et al., who show that the decline of aggregate hedge fund performance can also be ascribed to an economic environment inhospitable to many hedge fund strategies since the 2008–2009 financial crisis.[28] Moreover, central bank interventions and the impact of heightened regulatory scrutiny (the Dodd-Frank of 21 July 2010) significantly impacted the financial markets. This explains a Jensen's alpha that does not appear statistically different from zero in any hedge fund strategy, except for the aggregate index and the Event Driven category, suggesting that a combination of hedge fund styles in the spirit of fund of funds approach can achieve positive extra performance through diversification of hedge fund strategies.

Finally, Panel C shows the values of the Sharpe ratio, both traditional and Lo's version, using the 3 months U.S. T-bill rate as the risk-free rate and the S&P 500 index as a proxy of the market portfolio. The values highlight the positive Sharpe ratio in the two proposed versions for all hedge fund indices. Interestingly, note that the Lo's version produces corrections both downwards and upwards of the unadjusted Sharpe ratio, depending on the degree of autocorrelation of the returns and the impact on the standard deviation. Indeed, as discussed in the theoretical framework, the adjustment in the Lo's version results in a better estimate of the standard deviation of returns, removing the impact of serial correlation, thereby providing an unbiased Sharpe ratio.

References

Amenc N, Martellini L, Vaissié M (2002) Benefits and risks of alternative investment strategies. EDHEC-Risk and Asset Management Research Centre, Lille-Nice

Amenc N, Curtis S, Martellini L (2003a) The alpha and omega of hedge fund performance. EDHEC-Risk and Asset Management Research Centre, Lille-Nice

Amenc N, El Bied S, Martellini L (2003b) Evidence of predictability in hedge funds returns. Financ Anal J 59(5):32–46

Asness C, Krail R, Lew J (2001) Do hedge funds hedge? J Portfolio Manag 28(1):6–19

Bollen N, Joenväärä J, Kauppila M (2021) Hedge fund performance: end of an era? Financ Anal J 77(3):109–132

[27] On the Euro sovereign debt crisis of 2009–2011, Savona and Ciavolino (2016) provide evidence that hedge funds assumed long positions from December 2008 to March 2011 and short positions from April 2011 to August 2013 relative to sovereign Credit Default Swaps of GIIPS (Greece, Ireland, Italy, Portugal, Spain) as well as of France and Germany. Hedge funds were able to time the sovereign risk, first being capable of and then carrying out a sovereign pair trading strategy by betting on a spread tapering between core countries and GIIPS.

[28] Bollen et al. (2021).

Brown S, Goetzmann W (2003) Hedge funds with style. J Portfolio Manag 29(2):101–112
Dichev I, Yu G (2011) Higher risk, lower returns: what hedge fund investors really earn. J Financ Econ 100(2):248–263
Edelman D, Fung W, Hsieh D, Naik N (2012) Funds of hedge funds: performance, risk and capital formation 2005 to 2010. Financ Markets Portfolio Manag 26(1):87–108
Fung W, Hsieh D (1997) Empirical characteristics of dynamic trading strategies: the case of hedge funds. Rev Financ Stud 10(2):275–302
Fung W, Hsieh D (2001) The risk in hedge fund strategies: theory and evidence from trend followers. Rev Financ Stud 14(2):313–341
Fung W, Hsieh D (2004) Hedge fund benchmarks: a risk based approach. Financ Anal J 60(5): 65–80
Fung W, Hsieh D (2007a) Will hedge funds regress towards index-like products? J Invest Manag 5(2):56–80
Fung W, Hsieh D (2007b) Hedge fund replication strategies: implications for investors and regulators. Banque de France. Financ Stability Rev 10:55–66
Fung W, Hsieh D, Naik N, Ramadorai T (2008) Hedge funds: performance, risk and capital formation. J Financ 63(4):1777–1803
Gao G, Gao P, Song Z (2018) Do hedge funds exploit rare disaster concerns? Rev Financ Stud 31(7):2650–2269
Glosten L, Jagannathan R (1994) A contingent claim approach to performance evaluation. J Empir Financ 1(2):133–160
Henriksson R, Merton R (1981) On market timing and investment performance II. Statistical procedures for evaluating forecasting skills. J Business 54(3):513–533
Jagannathan R, Malakhov A, Novikov D (2010) Do hot hands exist among hedge fund managers? An empirical evaluation. J Financ 65(1):217–255
Kat H, Miffre J (2002) Performance evaluation and conditioning information: the case of hedge funds. Working Paper, ISMA Centre University of Reading
Kazemi H, Schneeweis T, Gupta R (2004) Omega as a performance measure. J Perform Meas 8(3): 16–25
Keating C, Shadwick W (2002) A universal performance measure. J Perform Meas 6(3):59–84
Kosowski R, Naik N, Teo M (2007) Do hedge funds deliver alpha? A Bayesian and bootstrap analysis. J Financ Econ 84(1):229–264
Lo A (2002) The statistics of sharpe ratios. Financ Anal J 59(5):36–52
Savona R (2014) Risk and beta anatomy in the hedge fund industry. Eur J Financ 20(1):1–32
Savona R, Ciavolino E (2016) Sovereign and hedge fund systemic risks. J Altern Invest 18(4): 98–108

Chapter 14
Private Equity

Guido Abate

14.1 Private Equity Features

Equity investment can concern either companies listed on a regulated market or unlisted companies. The former case involves public equity transactions, whilst the latter concerns private equity transactions. If we take a closer look at this initial distinction, private equity can be defined as an institutional investment activity in unlisted target enterprises, implemented by purchasing shares and, potentially, hybrid debt-equity instruments, with the aim of enhancing the value of the target company, so that it can be divested in the medium or long term.

The private equity value chain entails, firstly, a continuing screening activity in order to select potential companies to be enhanced. This objective involves an "active ownership" approach, through the provision not only of financial resources but also of professional, technical, and management expertise, particularly in the following fields:[1]

- Strategic and financial support for investment choices.
- Improvement of management control systems.
- Staff reorganisation.
- Tax planning and optimisation.
- Supply of new managers for the enterprise and new members in the board of directors.

[1] Manigart and Wright (2013), Gompers and Kaplan (2022).

G. Abate (✉)
Department of Economics and Management, University of Brescia, Brescia, Italy
e-mail: guido.abate@unibs.it

I. Basile, P. Ferrari (eds.), *Asset Management and Institutional Investors*, Contributions to Finance and Accounting,
https://doi.org/10.1007/978-3-031-59819-7_14

- Improvement of the reputation with regard to the market and the banking system, thanks to the entry of the financial investor, who reinforces the credibility of the enterprise and its development programmes.
- Entry in networks of relations, through networking activities, including at the international level.

The equity participation methods involve the purchase/subscription of ordinary shares, whether or not accompanied by stock rights, or of debt-equity hybrid instruments. In this second case, private equity transactions make use of the instruments typical of mezzanine finance.

If this financial and human capital is provided in the initial stages of the business life cycle, then the investment activity is known as venture capital. Whilst, according to the US practice, venture capital is a cluster of the broader private equity industry, EU practice clearly separates the two clusters, favouring the special features of investment in start-up businesses.[2] We will use the US terminology in this chapter, as it is increasingly becoming the global standard.

Private equity transactions concerning companies that have gone beyond the initial start-up phase belong to the expansion financing and buyout categories. Expansion financing (or expansion capital) regards companies that are experiencing an initial fast growth and therefore need access to stable sources of financing in order to expand their business. The aim of buyout is instead to impact on the shareholding structure of the target company. Different implementation strategies and different risk and return profiles depend on the stage of the company's life cycle in which the private equity transactions take place. These aspects will be examined in detail in Sect. 14.3, which is dedicated to the different investment management strategies.

Private equity transactions end with the divestment of the participation in the target company. This way out may occur in different modalities, depending on the type of operation, the characteristics of the target, the results achieved and the ownership's objectives:

- Listing of the target on the stock market, resulting in the divestment of the participation by means of an Initial Public Offering (IPO).
- Sale of shareholdings to an enterprise that intends to implement an external growth process (trade sale).
- Buyback of the shareholding by the original owners.
- Sale of the shareholding to other private equity operators specialised in the subsequent life cycle stages of the target enterprise (secondary private equity).
- Total write-off of the shareholding, if the transaction is unsuccessful or the targets have not been met.

[2] Caselli and Negri (2021).

14.2 Investment Vehicles

The "private equity" asset class can be accessed directly, through participation in the equity of one or more companies, or indirectly, using various investment vehicles, each one with specific characteristics. If the instruments are traded on regulated markets, they belong to the listed private equity, otherwise they fall within the unlisted private equity industry.

This section examines the following investment vehicles:

- Private equity funds.
- Special Purpose Acquisition Companies (SPACs).
- Private equity investment companies.
- Other listed private equity vehicles.
- Exchange-Traded Funds (ETFs) of the "private equity" asset class.

In addition to these instruments, many hedge funds and unlisted companies operate in this asset class. However, given the variety of management styles and the different regulatory frameworks governing them in the various jurisdictions, they will not be specifically analysed in this chapter.

14.2.1 Private Equity Funds

Since its origins in the USA in the 1940s, the history of private equity has been characterised by closed-end funds.[3] These US investment vehicles have a defined maturity, generally of 10 years. The working practice typically breaks the life of the fund down into two separate stages:

- Investment period, which is the first 5-year period, when the fund makes the investments and enhances their value.
- Divestment period, which is usually the second 5-year period, when no new investments are made and those already existing are gradually closed as opportunities arise on the market.

[3] Another legal form taken on by investments in this asset class in the English-speaking world is the limited partnership, the general partners of which, with unlimited liability, are managers of the company and therefore manage its portfolio, whereas limited partners can be compared to shareholders of investment funds. The general partner often corresponds to a company specialised in managing limited private equity partnerships, and is therefore known as a private equity firm. Therefore, if we are to look beyond the purely legal aspects, limited partnerships can be considered to be just like undertakings for collective investment, in economic terms, and will be treated as such for the rest of this chapter, whereas general partners (private equity firms) cover a similar role to that of managing companies.

If the manager has not divested all the shares by the contractual deadline of the fund, then the US legislation allows a further extension period, known as grace period, for the sales to be completed.

Fund share subscription, payment, and redemption dates and methods are established by the fund rules or instruments of incorporation and therefore do not depend, unlike open-end funds, on the discretion of the shareholders. Despite these limitations, closed-end funds offer some advantages to investors. These funds can invest in unlisted financial instruments and borrow, in order to increase return for shareholders through financial leverage. This freedom in asset allocation is linked to the fact that the resources available for closed-end funds are less variable than those for open-end funds.

The shares can be listed on a regulated market to enable investors to convert their shareholdings into cash during the life of the closed-end fund. Alongside the aim of ensuring liquidity for investments in closed-end funds, listing may meet financing needs, especially if the transaction takes the form of a primary offering. In this case, the shares of the fund are subscribed in conjunction with the listing process and therefore the raising of funds from investors can address a wider audience than it does in private placements.

In the case of closed-end funds, shares can only be redeemed at maturity, but so-called semi-closed funds can also be set up. These funds have established time windows during which shareholders can be reimbursed, for a total amount not exceeding new subscriptions or divestments of assets carried out by the fund.

Funds can issue shares more than once, thus collecting financial resources in line with the initial investment acquisition process.[4] A private equity fund may not immediately be in a position to allocate the liquidity collected initially to target shareholdings, as the investment opportunities may be identified at different times and, in any case, the financial requirements of the target companies rarely reach their maximum on entry of the private equity fund. The time delay in issuing shares enables therefore an improved matching between cash inflows and cash outflows for newly incorporated funds.

There is no one type of mutual fund specialised in private equity investment in the European Union, and the investment strategies typical of this asset class prevent the funds of this cluster from complying with the requirements of the UCITS directives.[5] In order to overcome market segmentation of funds at the national level, at least partly, the category of "European venture capital funds" (EuVECA) was introduced in the EU system with Regulation (EU) No. 345/2013,[6] the aim of which is to improve access to capital markets by Small and Medium-sized Enterprises (SMEs) demonstrating growth and development potential. SMEs are defined as unlisted

[4]The same result can be reached by staggering outflows over time, i.e. requests to pay amounts owing to the fund by the shareholders.

[5]See Chap. 2 for a more in-depth examination of UCITS compliant funds.

[6]Regulation (EU) No 345/2013 of the European Parliament and of the Council of 17 April 2013 on European venture capital funds.

organisations with fewer than 250 employees and an annual turnover of less than 50 million euros or a balance sheet not exceeding 43 million euros.

The European Regulation allows fund managers to apply on a voluntary basis for the EU-wide EuVECA passport, provided that they comply with the following requirements:

- Their assets under management in total do not exceed the threshold of 500 million euros.
- They are established in the European Union.
- They are subject to registration with the competent authorities of their home Member State.
- They manage portfolios of qualifying venture capital funds.

This passport only applies to those qualifying venture capital funds which, in turn, meet the following restrictions:

- They intend to invest at least 70% of their capital in so-called qualifying assets, i.e. equity or quasi-equity instruments issued by SMEs.[7]
- They never invest more than 30% of their capital in non-qualifying assets.
- They are established within the territory of a Member State.
- The SMEs must be either established in the EU or in a "white list" country.

EuVECA funds of funds can also be established, but, in this case, the underlying funds cannot invest more than 10% of their capital in other EuVECA funds.

Short-term borrowings (bridge financings) that do not exceed the fund's uncalled committed capital are allowed, but systematic recourse to leverage is excluded.

Mainly owing to the risk inherent in equity investments in companies that have not yet reached the stage of maturity, the EuVECAs can only be subscribed by:

- Professional investors or subjects who request to be treated as professional investors, pursuant to the MiFID rules.
- Investors who undertake to subscribe at least 100,000 euros and declare in writing that they are aware of the risks posed by investing in the fund.

There are many benefits for funds with a EuVECA passport, as their legal status is the same in all European Union countries and therefore they are not subject to restrictions imposed by national legislators, allowing a more accessible cross-border placement and greater simplicity when investing in unlisted companies.

Managers of private equity funds are compensated through two types of fees. Management fees are calculated as a percentage either of the Gross Asset Value (GAV) or the Net Asset Value (NAV) and are around 2% of the reference aggregate, with peaks of up to 3.5%. In the private equity industry, performance fees are also known as carried interest, borrowing a term initially used only by limited partnerships to indicate the remuneration of shares held by the general partners. They

[7] Quasi-equity is defined as secured or unsecured loans by the fund to a company, in which the fund already holds qualifying investments. Quasi-equity must not exceed 30% of the capital of the fund.

Table 14.1 An example of carried interest in a private equity fund

A closed-end investment fund, with a 10-year maturity, has been subscribed by investors at a per-unit NAV of 100,000 euros. The hurdle rate established by contract is 5% compounded annually, whilst the carried interest of the manager is 20% of any profit above the minimum target profit. The manager is credited with this compensation as a single payment on the winding up of the fund. The investment of the initial NAV at the hurdle rate leads to a minimum profit at maturity of: $100{,}000\ (1 + 5\%)^{10} = 162{,}889.46$ euros.

For example, let us assume that one of the two following alternatives may arise at maturity: in the first one, NAV is equal to 315,000 euros; in the second one, NAV reaches 140,000 euros.

The two following fee profiles correspond to each one of the scenarios:

- In the first scenario, the NAV at the maturity of the fund, equal to 315,000 euros, is greater than the minimum profit and therefore the carried interest is paid to the manager, according to the following calculation: max [315,000.00 – 162,889.46; 0] · 20% = 30,422.11 euros.
- In the second scenario, the NAV at the maturity of the fund, equal to 140,000 euros, is lower than the minimum profit and therefore no carried interest can be paid out: max[140,000.00 – 162,889.46; 0] 20% = 0 euro.

generally represent 20% of the capital gain obtained by the shareholder, typically within a range of minimum 15% and maximum 40%. This capital gain can be calculated against the original invested capital, or, more often, against a target minimum profit. This profit is equal to a figurative amount generated by the investment of the first-issue NAV at a pre-defined minimum threshold rate, known as the hurdle rate. Table 14.1 shows an example of the calculation of the fees of a private equity fund.

The NAV calculation of closed-end funds shows significant differences compared to that of open-end funds. Whilst the latter allocate their financial resources to listed instruments or instruments whose value can be calculated objectively, private equity funds invest in unlisted companies, whose market price is not available. To prevent distortions owing to the information asymmetry between the fund manager and the investor, the EuVECA Regulation provides that the rules for the valuation of assets shall be laid down in the rules or instruments of incorporation of the fund and shall ensure a sound and transparent valuation process. Moreover, the valuation procedures shall ensure that the assets are valued properly and that the asset value is calculated at least annually.[8]

14.2.2 *Special-Purpose Acquisition Companies*

Special-Purpose Acquisition Companies (SPACs) were introduced in the USA in the 1980s, but developed worldwide only in the last decade. They are listed investment vehicles whose sole purpose is to acquire an equity interest in at least one unlisted target company, which has not yet been identified at the time of the SPAC's

[8]Regulation (EU) No 345/2013, Articles 11 and 12.

incorporation. This transaction should be made within a defined time lapse, generally between 12 and 36 months.[9] The ultimate objective is to merge the SPAC with the target, thereby favouring its listing on a regulated market, after having increased its value thanks to the financial and managerial resources provided by the SPAC. The acquisition of the target and its merger with the SPAC are a joint operation, known as business combination.

Even though the target company is not known, the SPAC has to accurately communicate its investment policies, specifying the economic sector and the key features of the unlisted companies that can be acquired.

Recourse to SPACs allows a team of private equity operators, known as promoters or sponsors, to raise the funds required for an acquisition operation, addressing the stock market and therefore a broad audience of potential investors.

On incorporation of the SPAC, the sponsors underwrite the share capital at a symbolic price. At the same time, they acquire warrants against payment, i.e. call options on the SPAC's shares, which represent the true capital contributed by promoters. A capital increase is then launched and the new units, each made up of a common stock and a warrant, are placed in the market by means of an IPO. This capital increase leads to a dilution of the shareholding held by the sponsors, known as sponsor equity, which, on completion of the placement operation, is cut to about 20% of the SPAC capital. Unlike the sponsor equity, the units are subscribed by investors against payment, each one at a price usually equal to a conventional amount of 10 dollars or euros.[10]

Once the units have been listed, the warrant is traded independently and therefore the units are listed as common stock. The warrants issued by SPACs are usually issued in the money, as their strike price (commonly between 7 and 8 dollars or euros) is lower than that of the share, but cannot be exercised before the business combination and, therefore, their value will be equal to zero if the SPAC does not close the operation, for which it was incorporated, by the established date. The purpose of this financial instrument is therefore to guarantee the seriousness of the sponsors' commitment to closing the acquisition of an unlisted company, as the sponsor warrants were subscribed against payment of their premium by the SPAC's promoters, which in this case would lose 100% of what they had invested. At the same time, the warrants are an incentive for the SPAC's shareholders to vote in favour of the proposal for business combination submitted by the sponsors to the shareholders' meeting, because, if this transaction is not closed, then the warrants cannot be exercised and therefore their value will be equal to zero at the SPAC's maturity date.

Since, at the time of placement, the SPAC is yet to own any assets, whether they be real or financial, subscribers are not acquiring a portion of a specific company, as they would in the case of a common joint stock company, but are instead providing

[9]Berger (2008), Boyer and Baigent (2008), and D'Alvia (2022).

[10]If sponsors underwrite units, also during IPOs, then this investment, made under the same conditions as for the other shareholders, takes the name of sponsor coinvestment.

funds for a future acquisition. Consequently, SPACs are also known as blank-check companies on the US market, as the shareholders entrust their money to the sponsors without being able, at the time of subscription, to know the identity of the target company in which they will invest or assess the potential profitability of the operation. The decision to participate in a SPAC, as a consequence, is made only on the basis of the reputation of the private equity experts who promoted its incorporation.

In order to protect the common stock subscribers, the funds raised by the placement of shares and sponsor warrants, net of the commissions paid to the arranger bank and a minimum sum withheld for current expenditure, are immediately credited by the SPAC on an escrow account or allocated to an investment and management trust, i.e. a fiduciary relationship in which the trustee is a third party, independent from the sponsors and warrantor of the proper management of the amounts transferred to the trust. These segregated funds are allocated to assets of the money market or short-term bonds until the SPAC implements the business combination. For this reason, the shares of the SPAC should be listed on the market at a price floor of no less than the sum generated by investment in those assets, at least until the promoters of the investment vehicle announce which target company they intend to invest in and the details of the operation, i.e. the price offered for the acquisition, the share swap rate, where applicable, the merger timeframe and the business plan developed.

Additional protection for SPAC shareholders involves their right of withdrawal, which they can exercise when a meeting is called to resolve the proposal of business combination presented by the sponsors. If any non-sponsor SPAC shareholders do not agree with the combination, they have the option to withdraw from the company, receiving the share of liquidity they initially paid into the escrow account or the trust, increased by the interest accrued in the meantime. The same sum is reimbursed to the non-sponsor shareholders even if the SPAC has not completed a business combination within the set deadline, either because it has not been identified by the managers or because it has not been approved by the shareholders.

The focal point of SPAC activity is the presentation of the business combination. Sponsors can in fact use the number of withdrawals to calculate the exact total of the funds on which they can rely to conclude the transaction that they proposed. For this reason, the articles of association of SPACs provide for a maximum number of withdrawals, usually equal to 20–30% of the common stock, in excess of which the business combination cannot be implemented and therefore the company is liquidated. The informative function played by the SPAC's market price can be used to support shareholders' decisions. This price should reflect the expectations of financial agents for the business combination project and therefore it can position itself, with respect to the price floor:

- At a discount, if the operation proposed by the sponsors would destroy value.
- At a premium, if the business combination can instead provide financial benefits for the SPAC's shareholders.

Empirical analyses conducted on US operations have highlighted the information efficiency of the market with regard to the share price trends of SPACs. The predictive capacity of price deviations from the floor, in fact, has been observed on approval of the business combination by the SPAC shareholders' meeting.[11]

It is important to point out that, even with a very negative performance of SPAC's shares, the sponsors may guarantee themselves high returns in any case whenever the business combination is approved. In this case, the shareholders' right of withdrawal is no longer in force and therefore the sponsor equity, i.e. the SPAC's capital share subscribed by the promoters, takes on the same economic and administrative rights as ordinary shares. Rights which, on the other hand, did not require considerable expense by the sponsors.

14.2.3 Other Listed Vehicles of the "Private Equity" Asset Class

Investments can be made in listed financial instruments linked to private equity (listed private equity) using vehicles with quite different characteristics, depending on the national juridical system to which they are subject. They are, on the other hand, investment vehicles that, despite the adaptations to national law in the different countries of the European Union, derive from the operational and legal experience of the USA, to which, therefore, this section refers.

Investment companies were established in the USA by the 1940 "Investment Company Act" (ICA), which encompasses mutual funds, closed-end funds, and unit investment trusts in this category. These US legal forms can essentially be likened to private equity funds; please refer to Sect. 14.2.1 for an analysis of their economic and financial characteristics.

In order to promote investment in newly established companies (so-called start-ups), the US Congress introduced some amendments to the ICA in 1980, known as "Business Development Companies Provisions."[12] This legislative development led to the creation of a new investment instrument in private equity: the business development company (BDC)—a company whose purpose is to contribute capital and expertise to unlisted companies with significant growth expectations, capitals which in turn are raised from share subscribers of the BDC. The timeframe of BDCs is open-ended, thus allowing the managers greater freedom to define and implement

[11]Empirical studies (Jenkinson and Sousa 2011) have shown that in presence of a positive "announcement effect", SPAC shares increase by 25.5% their mean premium on the floor. These shares, then, attain a negative mean cumulative return, over the year following the decision, of −6.2%, thus generating a positive return of 17.7% = (1 + 25.5%) (1 − 6.2%) −1 since the time of announcement. On the contrary, SPACs whose shares have accumulated a discount of −4.5% between the announcement and the resolution of the acquisition then increase losses by another −79% over a year, with an overall loss of −80% = (1 − 4.5%) (1 − 79%) −1.

[12]Boehm and Krus (2001).

the investment policies. Consequently, unlike closed-end funds and, at least before completing the business combination, for SPACs, the shareholders are not able to know in advance the timeframe chosen by the investment company managers, who, moreover, are not obliged to share the risks taken by the company, as they are instead within SPACs.

In addition to making use of private equity funds, exposure to this asset class may be achieved by purchasing shares in companies that manage these funds, known as private equity firms.[13] The rationale underpinning this allocative choice involves the return profile of these companies. Private equity fund managers benefit from the positive trend of the portfolios they manage by collecting performance fees: if the returns exceed the hurdle rate, the fees are a considerable income for the management company. Even if indirectly, the ability to generate profits is also measured by management fees: the greater the ability to select target companies to which allocate resources, the larger the assets entrusted to the private equity firm by the market and, at the same time, the higher its value, thanks to the appreciation of the shareholdings held. Before investing in a private equity firm, however, it is necessary to verify the level of specialisation of the company, as private equity investment management is not subject to an exclusivity clause and therefore can be accompanied by that of products belonging to different asset classes.[14]

If the investors wish to employ an indexed approach to the private equity market, they can make use of exchange-traded funds specialised in replicating indices representative of the listed private equity. These indices present diverse constructive characteristics, especially in the choice of financial instruments to include in their basket. This variety originates from the qualitative features typical of the various investment vehicles, in accordance with the legal form assumed and the legal system to which they are subject. However, as a general rule, we can classify all those listed companies whose return derives, in turn, from a portfolio of unlisted shareholdings, as listed private equity companies.[15]

14.3 Management Strategies

Private equity as an asset class is characterised by a marked segmentation of the different strategies implemented by the operators. Different investment methods can be identified, which can be classified according to the stage of the business life cycle

[13] In the case of limited partnerships, the private equity firm corresponds to the general partner.

[14] For example, the Blackstone Group is often classified as a private equity firm, but its management activity is extremely diversified, with various alternative asset classes, to the extent that revenues from private equity in 2015 made up only 30.3% of the total.

[15] Of the indices of listed private equity securities, we would like to mention the following: S&P Listed Private Equity Index, Red Rocks Capital Global Listed Private Equity, LPX Composite and the Société Générale Privex.

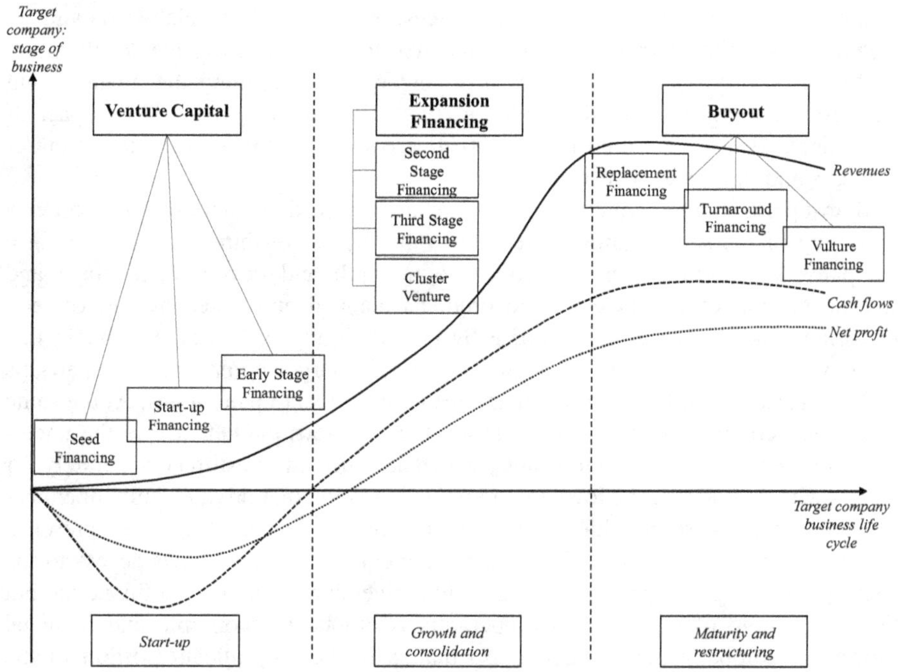

Fig. 14.1 Private equity strategies and the life cycle of the enterprise

of the target company (Fig. 14.1).[16] The needs of the companies, both in terms of funding and expertise, change over time and therefore lead to the structuring of different operations. Specifically, the contribution of private equity makes it possible to maximise value generation when the target company is in a critical stage of its life cycle in which there is a significant requirement for funds but where, at the same time, recourse to the credit and securities markets is impossible or not favourable.

14.3.1 Venture Capital

Investment strategies focused on venture capital address target companies that are in the early stages of their life cycle and share the purpose of supporting the creation and development of new business initiatives through seed financing, start-up financing, or early-stage financing operations.

Seed financing is aimed at transforming research and development projects into concrete corporate plans. The private equity dealers operating in this sector should

[16]Cumming (2010), Fraser-Sampson (2010), and Caselli et al. (2021). For an in-depth analysis of cases of private equity deals refer to Gompers et al. (2019).

be able not only to provide financial resources, but also, and especially, to evaluate the potential of the business ideas they intend to follow and favour the development of. Private equity operators also assist the enterprise in defining the financial and business planning and in initial relations with the credit system. The impact of reputational risk is very high in this investment strategy, for both the target entrepreneurs and the investment vehicle.

Start-up financing—which is closely related to seed financing and is often a follow-up of the same—aims to generate value by establishing and developing a company, successfully going beyond the research and development stage and starting the marketing stage for the products planned and tested as part of seed financing. The provision of expertise by private equity is less significant in this strategy than the purely financial dimension. The target companies are start-ups without a track record that would guarantee them access to bank loans. At the same time, these companies often cannot access the securities market, due to the uncertainty of their future economic results, which are difficult to estimate accurately for traders who do not have the level of internal information that private equity operators can benefit from as shareholders of the target company. The less significant presence of specific expertise required from the private equity investor allows access to the start-up financing for operators who are more diversified than in seed financing, but the engagement of private equity operators is crucial in economic and financial aspects of company management, also thanks to their significant portion of the shareholding in the target's capital. Whilst the risk profile of start-up financing is similar to that of seed financing, the provision of funds is more significant, therefore leading to higher potential losses if measured in absolute terms. On the other hand, just like in seed financing, there are significant opportunities for high returns, even with a low percentage of successful investments.

Early-stage financing, often also known as first-stage financing, is directed at those enterprises that have successfully gone beyond the planning and initial start-up of business and that have demonstrated the technical feasibility of the business plan that was initially proposed. The financial situation instead remains critical, with the need for additional resources, no longer in order to acquire inputs, but to satisfy the demand for working capital, which is starting its own growth phase. The know-how aspects on which private equity operators have to focus are related to market analysis and placement with respect to competitors. Although the target companies in this area already have their own structure, there is still a high risk of business failure.

14.3.2 Expansion Financing

Having successfully moved through the early stage of the company, which is typical of venture capital, private equity operators can focus on target companies committed to development and consolidation of their business. Expansion financing experts operate in this stage of the company's life cycle and their strategies can be classified as follows:

- Second stage financing: The financial resources provided by private equity investors are used to finance the internal growth strategies of the target company, by increasing or diversifying production capacity and market outlets. The management provided by private equity is limited and financing is lower than in the previous life cycle stages of the company, as the latter has begun to generate positive cash flow independently and therefore to self-finance, at least with regards to ordinary operations.
- Third stage financing: The intervention of the private equity investors contributes to financing the target company's external growth strategies, aimed at acquiring other companies or company branches. The private equity operator's know-how is relevant for evaluating the enterprises, accessing networks of knowledge and expertise useful for researching and identifying potential partners of the target companies.
- Cluster venture: No longer focused on just one target, but rather on a number of companies belonging to different economic sectors or stages of the production chain and suitable for a horizontal or vertical integration process. The result of the operation is the establishment of a holding company that controls an economic group with an overall value higher than that of its members put together. Private equity plays both a financial and a managerial role in this strategy, even if the shareholders of target companies often remain in control of the production processes.

Overall, expansion financing demonstrates a lower risk profile than that of the early stage, as it focuses on target companies that have already demonstrated their capacity to generate value.

14.3.3 Buyouts

Once the target companies have reached the maturity stage, the intervention of the private equity operators is focused on buyout transactions. The role played by private equity in this area assumes both financial and advising/managerial significance. The financial aspect takes on the form of acquisition of a shareholding, even a majority share, in the capital of the target. If the private equity specialist uses high financial leverage, then this type of transaction is known as leveraged buyout. Recourse to debt can be favoured by the financial profile of the target, which generates positive cash flows in its maturity stage. The advising/managerial dimension concerns specifically the next stage, in which the shareholdings are divested by both the private equity vehicle and, where applicable, by the original shareholders. The measurement of fair value of the shares and the search for new counterparties through the network of relations developed over time by the private equity operators are a prominent part of the value added of this kind of financing. The increased value of the target company, thanks to the positive reputation input, linked to the presence, as a shareholder, of a professional operator known to the markets, should also be considered.

Three possible types of buyouts can be identified: replacement financing, turnaround financing, and vulture financing.[17]

The main aim of the replacement financing techniques is to enable historical shareholders to dispose of their shareholdings. Private equity operators support or replace existing shareholders, by totally or partially acquiring their shareholdings and providing their own know-how resources to the company. The subsequent divestment stage can involve managers of the same target company who wish to become the new controlling shareholders: in this case, the replacement financing transaction is referred to as management buyout (MBO). If the company has instead been sold to managers outside the target company, we can speak of management buy in (MBI). Lastly, the private equity operator can place the shares of the target on the market by means of an IPO. When listing is the divestment method of choice, then the private equity strategy used is known as fourth-stage financing or bridge financing. The term "bridge" should therefore be seen as a metaphor for the support provided by the private equity operator to the target company in the transition to public equity.

Turnaround financing focuses on target companies that have reached the stage of full maturity and therefore may show the first symptoms of deterioration of their competitive position on the market, in relation to declining profits and cash inflows, with the onset of the first symptoms of stress in the financial structure. In these conditions, entry of a private equity vehicle in the company's capital has a dual purpose: to provide funds, aimed at re-negotiating debt, and management expertise, in order to restructure and renew the production processes where applicable, as well as to repositioning the target markets. Turnaround financing is very risky, as the target company could be already in a phase of irrecoverable decline.

Lastly, in vulture financing, private equity operators wind up the company of which they gained control, selling its branches or individual assets by means of a break-up transaction. This control was acquired by the private equity operator, with the aim of making a profit, at a price lower than the assumed liquidation price of the target company's assets. However, the inherent difficulty of estimating ex ante the liquidation value of a company must be underlined; this limitation makes vulture financing a risky strategy, usually implemented by highly specialised operators.

[17] Please note that, even though the maturity stage has the highest concentration of replacement financing operations, these operations can also take place at other stages, as their purpose is not closely tied to a defined economic and financial situation.

14.4 Risk and Return Measurement

The risk and return profile of private equity as an asset class is intrinsically linked to the value creation process typical of the potential strategies of the various operators. In any case, private equity is substantially different from traditional asset classes and thus performance measurement requires both innovative metrics as well as classic metrics adapted to the specificities of this category of alternative investments.[18]

14.4.1 *Performance Measurement Techniques*

The investment strategies typical of private equity are characterised by a finite time horizon, as they generally involve an established date by which the transaction should be completed through the sale of the shareholding. During the investment phase, private equity operators provide financial resources once, or often more than once, to the target company, whilst, during the following divestment phase, the operation generates cash flows originating from any profits of the target company and/or the sales of the shareholding in one or more tranches.

As maturity is predetermined, the most suitable performance measurement for a financial profile of this kind is the Internal Rate of Return (IRR). Although this measurement is usually used for financial investments, it has to be specifically adapted for the peculiarities and complexities of private equity. In addition to the presence of a number of cash disbursements, instead of just one as occurs when purchasing a bond, in private equity there are no contractual provisions indicating the level and frequency of remuneration of invested capital or guaranteeing repayment at maturity.

For these reasons, the IRR of a private equity transaction is known accurately only ex post. On the other hand, the information needs of the investors should be met periodically and well before the final liquidation of the shareholding, which usually occurs many years after the first capital contribution. This makes it necessary to calculate IRR periodically, which becomes more accurate and representative of the ex post IRR the closer the transaction is to its conclusion.

Invest Europe, an organisation dedicated to defining best practices in private equity, has proposed recourse to two forms of IRR, broadening the scope of analysis from a single deal to a portfolio managed by a fund.[19] The first measurement, known as the Gross Portfolio IRR (IRR_{GP}), is aimed at analysing the performance of investments made in a closed-end fund and at assessing the managers' ability to

[18]Baker et al. (2015).

[19]These guidelines replace previous ones issued by the International Private Equity and Venture Capital Board (IPEV). Note that the rationale underpinning the Invest Europe procedures can also apply to a single operation or to private equity investment portfolios with different legal forms, as long as they are characterised by a closed-end structure. Invest Europe (2018).

create value. The Gross Portfolio IRR, calculated at a time n equal or previous to the maturity of the fund T, is the rate that sets the algebraic sum of the values discounted at the date of set up of the fund ($t = 0$) equal to zero:

- All cash outflows (OUT_t) paid by the fund into its portfolio of target companies, both to purchase shareholding and for any lending operations.
- All cash inflows (IN_t) returned to the fund as dividends, interests, repayments of loans granted to the portfolio companies, proceeds from the sale of shares and stock options etc.
- The portfolio of existing shareholdings at a specific date n (PP_n), assessed at fair value and net of any liabilities.[20]

The Gross Portfolio IRR is, therefore, the variable in the following equation, which can be estimated by an iterative process:

$$\sum_{t=0}^{n\leq T} \frac{\text{OUT}_t}{(1+\text{IRR}_{\text{GP}})^t} = \frac{\text{PP}_n}{(1+\text{IRR}_{\text{GP}})^T} + \sum_{t=0}^{n\leq T} \frac{\text{IN}_t}{(1+\text{IRR}_{\text{GP}})^t}$$

It is useful to reflect on the present value of the portfolio of shareholdings known at a specific date n. As it is reasonable to assume, without additional information, that the portfolio of shareholdings will be liquidated by the manager at the fund's maturity date at the latest, the relative collection date is set at T as a conservative assumption. Likewise, its value in T is also unknown: as a result, it is considered to be identical to the value measured when (time n) the Gross Portfolio IRR is calculated.

The second form of IRR proposed by the IPEV is the Net Portfolio IRR (IRR_{NP}), the purpose of which is instead to measure the return obtained by participants in a private equity fund by discounting all the cash outflows, i.e. the subscriptions of fund shares ($\text{OUT}_{\text{part.},t}$), and inflows, such as the redemption of units and the distributions of income ($\text{IN}_{\text{part.},t}$), also considering the assessment of the fair value of the portfolio of shareholdings, net of any fees and expenses due on the fund itself. In this case, the net portfolio of shareholdings is therefore represented by the NAV known at time n, as it includes the effect of the periodic fees and expenses.[21] The Net Portfolio IRR is therefore the variable in the following equation:

$$\sum_{t=0}^{n\leq T} \frac{\text{OUT}_{\text{part.},t}}{(1+\text{IRR}_{\text{NP}})^t} = \frac{\text{NAV}_n}{(1+\text{IRR}_{\text{NP}})^T} + \sum_{t=0}^{n\leq T} \frac{\text{IN}_{\text{part.},t}}{(1+\text{IRR}_{\text{NP}})^t}$$

Figure 14.2 shows the trend of the IRR over time, in this case, the Net Portfolio IRR, in a private equity fund with a maturity of 10 years.[22] The IRR is recalculated

[20] In the Invest Europe terminology, this quantity is called unrealized portfolio.

[21] Harris et al. (2014).

[22] By way of an example, cash outflows are shown with the negative sign in Fig. 14.2.

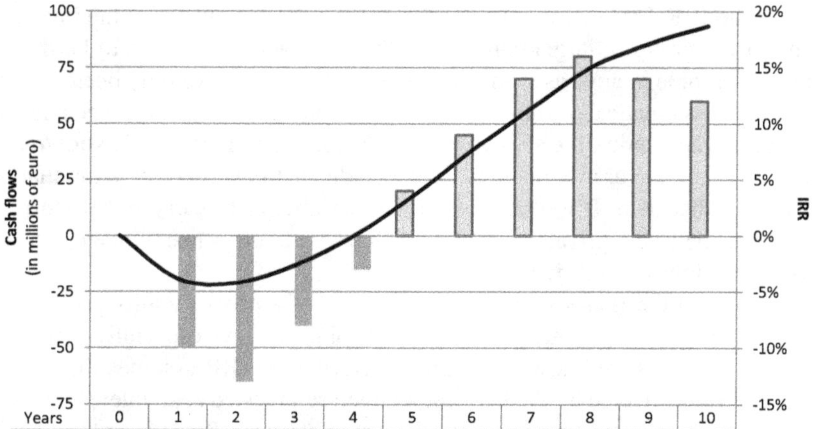

Years	0	1	2	3	4	5	6	7	8	9	10
In	0	0	0	0	0	20	45	70	80	70	60
Out	0	-50	-65	-40	-15	0	0	0	0	0	0
NAV	0	35	80	125	170	200	215	200	185	165	155
IRR	0.00%	-3.89%	-4.21%	-2.63%	0.00%	3.54%	7.63%	11.41%	15.01%	17.15%	18.69%

Fig. 14.2 The J-curve of the internal rate of return of private equity investments

every year based on the information available each time, i.e. previous cash flows and the last NAV published by the manager. Note that after it has grown, by virtue of the increase in value of the portfolio of target companies, the NAV—shown in the table of data in Fig. 14.2—subsequently decreases, mainly due to the advance reimbursements, shown by the investor's cash inflows. The last IRR shows the yearly average return actually obtained by the participant in the fund and has been estimated with increasing precision by the IRRs calculated as the fund approaches its maturity.

The "J" shape of the IRR curve (J-curve) is not a coincidence. Known in the literature as the J-effect, this phenomenon originates from the succession in time of cash flows in a private equity investment and from the different net values over time of the shareholdings in the target companies.[23] Cash flows are initially significantly negative, whilst shareholdings in the target companies have not yet generated a value increase. Also in the initial stages, we should consider the presence of transactions that failed prematurely, a problem which is encountered especially in the area of venture capital, as well as the cash outflows required to implement the management strategy. Conservative considerations also push managers not to fully revalue the shares held until they have been able to fix a market price based on one or more transactions of a significant amount. As the investment continues, the cash flows instead increase in value both thanks to the proceeds distributed by the target companies—and not reinvested—and to the first divestments, the amount of which can be credited to the participants in the private equity investment vehicle.

[23] Kaserer and Diller (2004), Meyer and Mathonet (2005), and Mathonet and Meyer (2007).

Whilst the IRR J-curve can take on different trends depending on the investment strategy or the group of target companies, the time factor continues to be the most significant variable in any case: on average, portfolios that have only been active for a few years systematically show lower IRRs than those that have been in place longer. In private equity, the year in which portfolio activities start is known as the vintage year, and the term vintage is used to indicate the number of years that have passed since that date. It is therefore preferable always to carry out performance comparisons between private equity portfolios with the same vintage, so as to prevent distortions caused by the J-effect.

In addition to the issues related to the vintage of a private equity portfolio, the IRR measurement can be criticised from other points of view, especially if applied to this asset class.[24] As we know, the logic underlying the IRR assumes that intermediate cash flows collected by the investor are then re-invested at an interest rate equal to the IRR and, at the same time, the intermediate cash flows transferred to the investment have been borrowed by the investor at a rate equal to the IRR. As the IRRs of private equity transactions often take extreme values, it seems unrealistic to assume that the same rate can be available even outside of the specific investment on which it was calculated. As a result, to overcome these excessive simplifications we can make use of the Modified Internal Rate of Return (MIRR). In private equity, this measurement can be calculated by assuming that the intermediate flows are invested, if incoming, or borrowed, if outgoing, at an interest rate fixed in advance and equal to the hurdle rate. Alternatively, and more realistically, two different rates can be applied: one for reinvestment of the inflows (r_{IN}) and the other for the debt required to finance the outflows (r_{OUT}).[25] The MIRR is calculated by the ratio of the total revenue of the private equity transaction, equal to the sum of all the cash inflows capitalised before maturity T, to the total disbursement required by that investment, equal to the sum of the outflows discounted at the date on which the first payment was made ($t = 0$):

$$\text{MIRR} = \sqrt[T]{\frac{\text{NAV}_T + \sum_{t=0}^{T} \text{IN}_t (1 + r_{\text{IN}})^{T-t}}{\sum_{t=0}^{T} \frac{\text{OUT}_t}{(1 + r_{\text{OUT}})^t}}} - 1$$

Alongside the IRR, good practice and the Invest Europe guidelines suggest recourse to the so-called money multiples. These are atemporal performance measures, as they do not consider the sequence of the various cash flows, but only their overall total by a given date n. Consequently, the multiples can be considered as informative sources for simple and instant calculation, but with limited rationality, even if they do enjoy broad use by operators in the sector. In particular, we analyse four different money multiples:

---

[24] Phalippou (2008).

[25] Kaserer and Stucke (2013).

- The paid-in to committed capital (PICC) is the ratio of resources already allocated to the fund (IN_t) to those which, in the event of multiple issues of shares, the investors originally undertook to pay in ($IN_{committed}$). It measures managers' ability to promptly identify target companies for their private equity strategies. This indicator becomes not significant upon completion of the introductory stage of the fund:

$$PICC_n = \frac{\sum_{t=0}^{n \leq T} IN_t}{IN_{committed}}$$

- The distributed to paid-in capital ratio (DPI) is calculated by the ratio of total financial resources drawn out of the investment (IN_t) to the total resources allocated by the date on which the multiple was calculated (OUT_t). This measurement is suitable for valuing funds at the end of their lives, as it does not account for the value of investments that are yet to be liquidated. It is useful for investors to identify the time at which the financial resources received fully have balanced out those previously paid into the fund:

$$DPI_n = \frac{\sum_{t=0}^{n \leq T} IN_t}{\sum_{t=0}^{n \leq T} OUT_t}$$

- The remaining value to paid-in capital ratio (RVPI) is the ratio of the value of the portfolio of shareholdings net of expenses and fees, represented therefore by the NAV of the fund, to the capital invested in the fund itself by the calculation date (OUT_t). It is a complementary indicator compared to the DPI and is therefore more suited for valuing funds in the first years of business. On the other hand, if the NAV has been object of a conservative estimate, the RVPI may underestimate the value added by the investment:

$$RVPI_n = \frac{NAV_n}{\sum_{t=0}^{n \leq T} OUT_t}$$

- The total value to paid-in capital ratio (TVPI) considers the total value of the fund as the numerator, adding up both the flows received by the investors (IN_t) and the value of the existing portfolio (NAV_n). As in the other two multiples, the denominator is the capital allocated in the fund (OUT_t). It measures the managers' ability to generate value against the resources invested by shareholders:

$$\mathrm{TVPI}_n = \frac{\mathrm{NAV}_n + \sum_{t=0}^{n \le T} \mathrm{IN}_t}{\sum_{t=0}^{n \le T} \mathrm{OUT}_t} = \mathrm{DPI}_n + \mathrm{RVPI}_n$$

The performance measurement techniques presented so far can be used for purposes of comparison within the private equity asset class, but they do not allow comparisons with financial assets belonging to different investment categories, for example with public equity. The measurement of return of listed shares is based on the calculation of price changes and, if total return, of the re-investment of payouts generated by the shares (e.g. dividends). On the contrary, the IRR applied to investments in private equity is calculated based on the trend over time and on the sum of a series of cash flows discounted at a conventional date.

To overcome this issue of non-comparability between performance measures, the scientific literature has introduced a measurement known as the public market equivalent (PME).[26] Different versions of this indicator have been proposed over the years, due to certain distortions present in the original formulation, but in this chapter we are going to focus on the most well known and used one, i.e. the Kaplan-Schoar PME.[27] The logic underlying this indicator is that of replicating the cash flows associated with the investment in private equity through a series of identical cash flows that, however, focus on an equivalent portfolio, usually a stock market index such as the S&P 500 TR. In other words, cash flows originating from investments in the private equity fund are discounted at time zero, i.e. the date on which the fund was incorporated, using the total returns of the equivalent portfolio as discount rates occurring from time zero to the time t of the actual occurrence of each flow (shown with $r_{0,t}$). Lastly, the PME is the ratio of the total discounted cash inflows, including the final one, represented by the NAV at maturity, to the total discounted outflows:[28]

$$\mathrm{PME} = \frac{\frac{\mathrm{NAV}_T}{1 + r_{0,T}} + \sum_{t=0}^{T} \frac{\mathrm{IN}_t}{1 + r_{0,t}}}{\sum_{t=0}^{T} \frac{\mathrm{OUT}_t}{1 + r_{0,t}}}$$

If the total return of the equivalent portfolio is identical to the IRR of the fund, then the PME takes on a value equal to one; it is important to recall that the IRR is the rate that makes the total present values of the inflows and outflows equal. When the fund over-performs with respect to the equivalent portfolio, the PME is instead greater than one, as the IRR of the private equity investment is higher than the total returns used as discount rates. The opposite judgement should be expressed when the PME is lower than the unit. The PME is therefore a relative performance measure with respect to the equivalent portfolio.

---

[26] The first study suggesting its use, under the name of index comparison method, was Long and Nickels (1996).

[27] Kaplan and Schoar (2005).

[28] Note that $1/(1 + r_{0,t})$ is the specific discount factor of each cash flow.

Table 14.2 Example of public market equivalent calculation

Time	Cash flows (euro/mln)		TR index (Equivalent portfolio)	Discount factors	Present value (euro/mln)	
	IN_t	OUT_t			IN_t	OUT_t
0	0	0	100			
1	0	50	102	0.98039	0.00	49.02
2	0	65	120	0.83333	0.00	54.17
3	0	40	145	0.68966	0.00	27.59
4	0	15	180	0.55556	0.00	8.33
5	20	0	210	0.47619	9.52	0.00
6	45	0	150	0.66667	30.00	0.00
7	70	0	155	0.64516	45.16	0.00
8	80	0	185	0.54054	43.24	0.00
9	70	0	200	0.50000	35.00	0.00
10	60	0	220	0.45455	27.27	0.00
	155				70.45	
Total					260.66	139.11
PME = Σ_t(PresentVal. IN_t)/Σ_t(Present Val. OUT_t)=					1.87	
$1.87 \cdot \sum_{t=0}^{10} \frac{OUT_t}{(1+IRR_{PME})^t} = \frac{NAV_{10}=155}{(1+IRR_{PME})^{10}} + \sum_{t=0}^{10} \frac{IN_t}{(1+IRR_{PME})^t}$					$=IRR_{PME} = 7.32\%$	
IRR_{PME} spread = $IRR_{fund} - IRR_{PME} = 18.69\% - 7.32\% =$					11.37%	

From a different point of view, the PME is the coefficient by which to multiply the cash flows invested in the equivalent portfolio in order to obtain the same inflows that were generated by investing in the private equity fund, keeping the frequency of all cash flows identical. For example, a PME of 1.87 (Table 14.2) shows that each outflow should be multiplied by 1.87 in order to obtain the same result achieved by the investment in private equity. Once the outflows have been corrected for the coefficient obtained from the PME, then it is possible to calculate the internal rate of return of the equivalent portfolio.

The PME method makes it feasible to estimate the internal rate of return of the equivalent portfolio. First of all, we must calculate the PME in order to achieve this result. The next step involves multiplying the total outflows by the PME, using it as a scale factor. Finally, rather than using the total returns of the equivalent portfolio to calculate the discount factors, the internal rate of return is considered as the variable in the equation. The solution to this equation, through an iterative process, enables the estimation of the public market equivalent Internal Rate of Return (IRR_{PME}):

$$\text{PME} \cdot \sum_{t=0}^{T} \frac{OUT_t}{(1 + IRR_{PME})^t} = \frac{NAV_T}{(1 + IRR_{PME})^T} + \sum_{t=0}^{T} \frac{IN_t}{(1 + IRR_{PME})^t}$$

The financial interpretation of the IRR_{PME} is the same as the comparison between two different internal rates of return: if the IRR_{PME} is greater than the IRR of the fund, then the latter performs worse than the equivalent portfolio; on the contrary, an

IRR_{PME} that is lower than the IRR of the fund indicates a better performance in comparative terms. The difference between IRR and IRR_{PME}, also known as IRR_{PME} spread, is therefore the measure of excess performance of the private equity fund against the equivalent portfolio.

Table 14.2 uses the same fund data of Fig. 14.2 and gives an example of PME and IRR_{PME} calculation by using a generic total return index as the equivalent portfolio, rescaled to 100 at time zero.

In order to compare the risk-adjusted performance of private equity with that of the other asset classes, we can resort to private equity instruments listed on regulated markets. The availability of their quoted prices makes it possible to calculate their return in the same way as with any other equity securities. On the other hand, listed private equity is only a fraction of the broader private equity asset class and does not make up a sufficiently representative sample, as the characteristics of the listed instruments are usually different from those of the other products of the same sector. Whilst private equity operators invest almost exclusively using structures that are financially similar to mutual funds, as we pointed out earlier, listed private equity is mainly made up of managing companies (known as private equity firms), which, moreover, are not always specialised only in this asset class, SPACs, BDCs, and, generically, listed private equity companies, with a very limited number of listed funds when compared to the investable universe.[29]

Indices developed on the basis of unlisted private equity are instead more representative of the asset class overall. By definition, the components are assets without a market price; as a result, these indices can only be calculated using the following approaches:[30]

- Transaction-based indices are founded on the information regarding closed transactions, supplied on a voluntary basis to the index providers by private equity managers. If, on the one hand, they are indices useful for assessing the profitability of the sector over time, on the other hand, the timings of communication of the results are different. This time imbalance, therefore, makes it impossible to accurately compare these indices with those of other asset classes.
- Appraisal-based indices are developed using the NAV valuations disclosed periodically (every 3 or 6 months, according to current legislation) by fund managers. These indices too are adversely affected by time-lag phenomena, which cause a significant level of autocorrelation of returns, because the valuations of the shareholdings are rarely updated until the exit from a specific

[29] The analysis by Jegadeesh et al. (2009) focuses on listed funds only, paying special attention to a sub-sample of 26 funds of funds which, in turn, invest in unlisted private equity funds and therefore are the only opportunity available, even if minimal, to extrapolate market returns for the latter investment vehicles.

[30] An advanced econometric technique was proposed in Cumming et al. (2013), aimed at constructing unlisted private equity indices with optimal properties. However, it should be considered that the changes made to the base indices used in the procedure are such that they could deprive them of their representativeness.

investment, involving maintenance over time of the same value attributed to each shareholding, except for showing discontinuities upon the closing of each operation. To overcome this problem, known as smoothing, we need to make use of special techniques to adjust the time series.[31]

14.4.1.1 Performance of Private Equity Deals

The returns achieved from private equity transactions, known as transaction performances, are calculated regardless of the investment vehicle used to accomplish them. As a result, they are always gross returns, which do not include compensation for private equity operators, whether in the form of fees or profit share.

The scientific literature has to face the limited availability of reliable data: as private equity transactions usually involve unlisted subjects and investment vehicles that are not addressed to the general public, the rules of information transparency aimed at protecting retail investors, typical of regulated markets, are not in force. Despite this, recourse to certain proprietary databases, where the most significant information on private equity deals is gathered, has made it possible to analyse samples of a large number of transactions that are diverse in terms of both time and the strategy implemented, thus allowing some general information to be gleaned on both the levels of return and the drivers underpinning the results obtained.[32]

In venture capital, the analysis conducted by Cochrane on 16,638 transactions in the time period of 1987–2000 highlights average returns and standard deviations close to those measured on stocks listed on the NASDAQ.[33] Unlike the companies listed on this market, however, the distribution of returns of venture capital transactions is characterised by high positive skewness: whilst the majority of deals have returns close to the mean, certain investments are characterised by extremely significant positive returns. This configuration can be compared to the payoff of a call option. This statistical comparison is also based on economic affinity, as venture capital investments concern start-ups (or even companies yet to be established, as in seed financing). In favourable market and corporate conditions, these companies are able to generate significant growth in their value against a minimum initial disbursement by a venture capitalist. The financial profile of this operation is similar to that of a call option: the premium paid can generate a capital gain if the call is deep in the money at expiry.

[31] Geltner (1993). See Sect. 14.4.2 for a more in-depth examination.

[32] We would like to highlight, specifically, the CEPRES, Preqin, Burgiss, and Cambridge Associates databases and the public informative reports that have to be provided by certain US institutional investors operating in private equity, such as pension funds for civil servants (e.g. CalPERS).

[33] Cochrane (2005). Advanced econometric techniques had to be used to calculate the returns of individual private equity transactions in order to ensure comparability with the stock market.

Similar results were achieved on the basis of a different sample, made up of 2419 transactions, mainly in the venture capital sector, in the 1971–2003 period.[34] This analysis also observed the tendency of managers to announce markedly higher prospective IRRs for ongoing transactions (more than two times) than those that were actually realised upon their closing. This discrepancy can be explained by the regulatory differences in the various markets where these transactions take place, with overestimation prevalent in systems with limited regulation about investor relations.

Legal aspects regarding the level of detail of contractual agreements reached between private equity operators and target companies were identified as explanatory factors in the performance of 834 transactions, mainly expansion financing and buyouts, concluded between 1999 and 2005.[35] In the case in point, the deals requiring especially complex and articulate contracts later showed considerably better performance compared to less binding investments, probably because target companies with better profitability expectations are inclined to accept a greater level of contractual detail, as they themselves are sure of the positive future profit. In line with previous studies,[36] this analysis has also underlined the importance of the appointment of directors in the target company by part of the private equity investor: more profitable transactions are linked to a lower level of control by the private equity operator, who tends to delegate the management of the target company to the directors who are already in office, whereas more problematic investments are followed by private equity managers or by personnel connected to them.

A further study of the performance of private equity transactions was able to use a sample of 10,328 transactions diversified for the strategy adopted, 3296 of which were buyouts.[37] The use of this sample has enabled the valuation of this asset class in its entirety for the first time, highlighting average IRRs between 10 and 50% (between 8 and 70% for buyouts only), depending on the economic sector of business of the target company, whereas the average DPIs fall between 1.8 and 4.2 (between 1.8 and 7.0 for pure buyouts) and average PMEs between 1.5 and 3.6 (between 1.4 and 5 for buyouts). Going beyond a simple measurement of the results, the analysis was extended to performance attribution, dividing the return amongst the three active levers that can be used in private equity:

- Market timing: In this area, it is defined as the ability of the private equity manager to identify trends in the valuations, more or less high, attributed to companies in a specific economic sector in different stages of the economic cycle and of the stock market.
- Financial leverage: Diversified according to the strategy adopted, minimal or absent in the context of venture capital and high for buyout operations.

[34] Cumming and Walz (2010).

[35] Caselli et al. (2013).

[36] Lerner (1995).

[37] Graf et al. (2012).

– Operational improvements: Resulting from the ability of the private equity operators to optimise the production and commercial processes of the target company.

In light of the empirical analysis conducted on a sub-sample of transactions concerning target companies operating in the financial sector, the third factor is the most significant in the creation of value by private equity, with a contribution equal to around 87% of the performance.

The previous research findings on the performance of private equity deals have been updated and revised by a subsequent study conducted on 500 buyout transactions spanning from 2007 to 2018. This study has measured a significantly superior return generated by these investments when compared to the S&P 500 and MSCI World equity indices.[38]

14.4.2 Performance of Private Equity Funds

Once the characteristics of the private equity transactions have been analysed, we can extend the evaluation to the main investment vehicles implementing these strategies, i.e. private equity funds. In this case, the perspective shifts towards performance measures net of the fees paid to the portfolio managers, thus obtaining results that are more in line with the performance profile of the investors focused on this asset class.

The difficulty in finding data on unlisted instruments, often reserved only for qualified investors, causes the measurement of performances somewhat differentiated, according to the samples available. Two recent analyses have been able to take advantage of databases that are large and diversified enough to overcome the selection bias that plagued previous works. In particular, we recall the study carried out by Higson and Stucke, which was able to analyse 1169 buyout funds with vintage years between 1980 and 2008, covering 85% of the capital raised overall by managers operating in that strategy during the period examined.[39] The research was conducted by calculating the net IRRs of the funds, in order to measure their absolute performance, and the PME measurement, in order to compare their return against the S&P 500 index, used as the equivalent portfolio. The authors observed the following phenomena:

– On average, the buyout funds exceeded the S&P 500 by over 5% per year in terms of IRR_{PME} spread. This performance is partly explained by the presence of few funds with extremely positive returns, but in any case 63% of the sample benefited from returns above the S&P 500.

---

[38] Gruener and Marburger (2022).

[39] Higson and Stucke (2012).

- A strongly cyclical trend of the returns, with marked growth for funds with vintage year in correspondence with "bear" phases of the United States stock market. The underlying causes of this cyclic nature can probably be traced to the high availability of undervalued target companies in the initial investment stage of the funds, due to more conservative valuations being made during periods of crisis, and the abundance of liquidity at low rates, following the intervention of the central banks functioning as a stimulus to the economy.
- A significant decline in average performance during the period studied. This phenomenon is not easy to understand, but it is probable that, on the one hand, the increase in the number of operators in the field of private equity has led to the entrance of less skilled managers, and on the other, the rise in competition amongst private equity companies may have increased the overall efficiency of this market, therefore reducing the range of opportunities for higher profits.

Another study that was able to benefit from a diversified, large sample was conducted by Harris, Jenkinson, and Kaplan.[40] The authors were able to analyse 1373 private equity funds concentrated in North America, 598 of which operating in buyouts and 775 in the field of venture capital strategies, with vintage years between 1984 and 2008 inclusive. The results applicable to the buyout sector are substantially consistent with Higson and Stucke (2012), whereas the conclusions on venture capital are innovative. In this sub-sample, the IRRs varied strongly over the years. Whilst a cyclical phenomenon was already evident in the context of buyouts, the vintage year takes on a fundamental role in venture capital, especially due to the increased exposure of start-up companies to speculative bubbles. Faced with an average annual IRR of 19.3% in the period examined, the funds incorporated in 1996 presented an annual IRR of more than 76%, mainly thanks to the exposure towards target companies operating in the so-called New Economy. Also in comparative terms, with respect to the S&P 500, an unsteady and downwards trend was observed, with PMEs below the unit for all vintage years after 1999.

14.4.3 Risk in Private Equity

Risk assessment in private equity concerns both the variability of returns and, ultimately, the value creation process in target companies. From a portfolio viewpoint, the former is the most significant aspect, but there are major problems surrounding its measurement.

As we have shown, the most suitable performance measurement in this asset class is the internal rate of return. On the other hand, the volatility of this rate is not comparable to the standard deviation of returns of the listed shares. As a result, advanced econometric techniques have been proposed to measure the exposure of

[40] Harris et al. (2014).

samples of private equity funds to the market risk factor, by replacing the IRR with interest rates calculated according to the CAPM.[41]

The risk profile of the listed private equity can instead be measured with the same metrics as the equity asset class, thanks to the availability of time series of market prices, but is less representative of private equity overall. Moreover, the limited liquidity typical of listed instruments belonging to this asset class causes a marked autocorrelation, which can lead to underestimation of the actual risk if not corrected with unsmoothing techniques.[42]

Risks inherent in the transactions implemented by private equity managers lie on the basis of the risk profiles of private equity instruments. These transactions are subject to two separate sources of risk: business risk and agency risk.[43] The first factor refers to the probability associated with achieving negative returns when investing in a target company and is linked to the economic sector in which it operates. As the level of business risk is not fully known beforehand, private equity operators should constantly monitor the market segment in which they invested and be able to adapt the planned strategy, also by a constant interaction with the original management of the target company. The relationship with the representatives of the target company actually poses the most perceived risk for private equity operators, known as agency risk. This factor is firstly linked to the significance of informative asymmetries, both in initial negotiations, during which the private equity manager has to implement an in-depth business due diligence, and during subsequent investment management. Private equity managers can hire internal specialists from the target's business sector, or take advantage of external consultants.

Investors in private equity instruments can diversify the risk inherent in their exposure to this asset class by diversifying their portfolio according to the following factors:[44]

- Strategy, by selecting investment vehicles diversified by the life cycle of the target enterprises.
- Vintage years. Given the different financial profiles of portfolios established in separate years, investors can contain the J-effect and ensure that there are inflows of an amount high enough to compensate for any outflows to newly established funds, thus reducing the so-called commitment risk, i.e., the need to procure resources to cope with mandatory payments for amounts higher than the income and reimbursements collected in the same period.
- Economic sector, which makes it possible to contain the exposure to overvalued sectors and reduce the impact of investments in cyclical sectors.
- Managers, by increasing the variety of the investment approaches, necessary for diversification by strategy and/or economic sector.

---

[41] Driessen et al. (2012).

[42] Lahr and Herschke (2009), Anson (2013), Ang et al. (2018), and Baz et al. (2022).

[43] Sapienza et al. (1996); Cumming and Johan (2014).

[44] Mathonet and Meyer (2007).

14.4.3.1 Discount-to-NAV

The market prices of units of closed-end private equity funds show a negative gap compared to the NAV published periodically by the manager. This phenomenon is known in the literature as closed-end fund puzzle and various causes have been identified to explain its existence and importance.

The units of private equity funds can be traded both on regulated markets, for funds admitted to listing, and over-the-counter, for unlisted funds. In the latter case, the operations only involve specialised dealers, such as institutional investors, or funds of private equity funds belonging to the category of secondaries. Given the limited availability of data on the secondary market, the discount-to-NAV phenomenon is only examined for listed funds.

It has been empirically demonstrated, based on a sample of 97 listed funds in the 1992–2009 period, that the discount of the price with respect to the average NAV of the private equity funds increases linearly for the first 2 years of activity of the fund, then staying at around −26% after that.[45] The underlying causes of this phenomenon can be linked to the following factors:

- Investors' expectations: The discount becomes less marked when there are signs of optimism amongst agents, such as the growth in the consumer confidence index and in the prices of small-cap companies or the number of new IPOs on the stock markets.
- Credit markets: Lower discount-to-NAVs correspond to lower spreads between corporate bond rates and sovereign bonds, probably due to the higher growth expectations of the target companies held in the fund portfolio.
- Illiquidity: We observe a reduction in the discount-to-NAV as the liquidity of the secondary market increases, measured by the bid-ask spread of the fund units. The buyers of fund units request a premium for the liquidity they bring to a market characterised by particularly low trading volumes.[46]
- Volatility: Increases in volatility on markets have a negative impact on the discount-to-NAV, as the increased riskiness of the investment is compensated by buyers by bidding lower prices.
- Regulatory transparency: We observe a lower discount-to-NAV in markets with sharper regulations aimed at reducing the information asymmetry between fund managers and the market.

[45] Lahr and Kaserer (2009).

[46] Nadauld et al. (2019) measured that the average difference between the PME of buyers and that of sellers of private equity funds in the secondary market stands at 0.05 on average. This discrepancy denotes the liquidity premium granted to buyers.

References

Ang A, Chen B, Goetzmann W, Phalippou L (2018) Estimating private equity returns from limited partner cash flows. J Finance 73(4):1751–1783
Anson M (2013) Performance measurement in private equity: another look at the lagged beta effect. J Private Equity 7(1):29–44
Baker HK, Filbeck G, Kiymaz H (2015) Private equity: opportunities and risks. Oxford University Press, New York
Baz J, Davis J, Han L, Stracke C (2022) The value of smoothing. J Portfolio Manage 48(9):73–85
Berger R (2008) SPACs: an alternative way to access the public markets. J Appl Corp Financ 20(3): 68–75
Boehm S, Krus C (2001) The business development company solution. Rev Securities Commodities Regulation 34(4):39–46
Boyer C, Baigent G (2008) SPACs as alternative investments: an examination of performance and factors that drive prices. J Private Equity 11(3):8–15
Caselli S, Negri G (2021) Private equity and venture capital in Europe: markets, techniques, and deals, 3rd edn. Academic Press, Burlington
Caselli S, Garcia-Appendini E, Ippolito F (2013) Contracts and returns in private equity investments. J Financ Intermed 22(2):201–217
Caselli S, Gigante G, Tortoroglio A (2021) Corporate and investment banking. A hands-on approach. Bocconi University Press
Cochrane J (2005) The risk and return of venture capital. J Financ Econ 75(1):3–52
Cumming D (2010) Venture capital: investment strategies, structures, and policies. Wiley, Hoboken
Cumming D, Johan S (2014) Venture capital and private equity contracting: an international perspective, 2nd edn. Elsevier, London
Cumming D, Walz U (2010) Private equity returns and disclosure around the world. J Int Bus Stud 41(4):727–754
Cumming D, Haß LH, Schweizer D (2013) Private equity benchmarks and portfolio optimization. J Bank Financ 37(9):3515–3528
D'Alvia D (2022) Mergers, acquisitions and international financial regulation: analysing special purpose acquisition companies. Routledge, New York
Driessen J, Lin TC, Phalippou L (2012) A new method to estimate risk and return of nontraded assets from cash flows: the case of private equity funds. J Financ Quant Anal 47(3):511–535
Fraser-Sampson G (2010) Private equity as an asset class, 2nd edn. Wiley, Chichester
Geltner D (1993) Estimating market values from appraised values without assuming an efficient market. J Real Estate Res 8(3):325–345
Gompers P, Kaplan S (2022) Advanced introduction to private equity. Edward Elgar, Cheltenham
Gompers P, Ivashina V, Ruback R (2019) Private equity. A casebook. Anthem Press
Graf C, Kaserer C, Schmidt D (2012) Private equity: value creation and performance. In: Cumming D (ed) The Oxford handbook of private equity. Oxford University Press, New York
Gruener A, Marburger L (2022) New insights into private equity: Empirical evidence from more than 500 buyouts. The Journal of Portfolio Management 48(9):103–116
Harris R, Jenkinson T, Kaplan S (2014) Private equity performance: what do we know? J Financ 69(5):1851–1882
Higson C, Stucke R (2012) The performance of private equity. London Business School and Oxford University working paper
Invest Europe (2018) Investor reporting guidelines. Brussels
Jegadeesh N, Kräussl R, Pollet J (2009) Risk and expected returns of private equity investments: evidence based on market prices. NBER Working Paper, no. 15335
Jenkinson T, Sousa M (2011) Why SPAC investors should listen to the market. J Appl Financ 21(2):38–57
Kaplan S, Schoar A (2005) Private equity performance: returns, persistence, and capital flows. J Financ 60(4):1791–1823

Kaserer C, Diller C (2004) European private equity funds: a cash flow based performance analysis, in performance measurement and asset allocation for European private equity funds. EVCA Research Paper

Kaserer C, Stucke R (2013) Performance of private equity. In: Baker HK, Filbeck G (eds) Alternative investments: instruments, performance, benchmarks, and strategies. Wiley, Hoboken

Lahr H, Herschke F (2009) Organizational forms and risk of listed private equity. The Journal of Private Equity 13(1):89–99

Lahr H, Kaserer C (2009) Net asset value discounts in listed private equity funds. CEFS Working Paper, no. 12

Lerner J (1995) Venture capitalists and the oversight of private firms. J Financ 50(1):301–318

Long A, Nickels C (1996) A method for comparing private market internal rates of return to public market index returns. The University of Texas System working paper

Manigart S, Wright M (2013) Reassessing the relationships between private equity investors and their portfolio companies. Small Bus Econ 40(3):479–492

Mathonet PY, Meyer T (2007) J curve exposure. Wiley, Chichester

Meyer T, Mathonet PY (2005) Beyond the J-curve. Wiley, Chichester

Nadauld T, Sensoy B, Vorkink K, Weisbach M (2019) The liquidity cost of private equity investments: evidence from secondary market transactions. J Financ Econ 132(3):158–181

Phalippou L (2008) The hazards of using IRR to measure performance: the case of private equity. J Perform Meas 12(4):55–66

Sapienza H, Manigart S, Vermeir W (1996) Venture capitalist governance and value added in four countries. J Bus Ventur 11(6):439–469

Chapter 15
Private Debt

Guido Abate

15.1 Private Debt Features

Private debt, sometimes referred to as private credit, is the investment in debt securities issued by companies, mainly small and medium-sized enterprises, as well as the direct provision of financing to the same borrowers. The object of the investment, even when incorporated in bonds or other debt securities, is usually not listed on regulated markets, hence the term private (in the sense of not publicly listed) debt.[1] Private debt operators aim to offer investors higher returns than those obtainable on the listed bond market and, at the same time, to meet the financial needs of companies during the introduction and development phases, on the occasion of extraordinary financial operations, or during economic downturns. Since the companies financed are often also the subject of private equity transactions and the professional requirements of private debt managers are partly overlapping with those of private equity, the two asset classes are also subsumed under the term private market.

Investment in the private debt sector is implemented through the allocation of financial resources to both non-listed bonds and credit contracts entered into between the fund and the financed company. In the first case, the bonds are issued on the primary market, normally through a private placement aimed exclusively at specialized collective investment vehicles and other institutional investors. In the second case, instead, they are credit contracts called direct lending. Investment in direct lending operations can be first lien, if the investor coincides with the lender in the

[1] Ippolito (2020).

G. Abate (✉)
Department of Economics and Management, University of Brescia, Brescia, Italy
e-mail: guido.abate@unibs.it

I. Basile, P. Ferrari (eds.), *Asset Management and Institutional Investors*, Contributions to Finance and Accounting,
https://doi.org/10.1007/978-3-031-59819-7_15

credit transaction, or second lien, if the investor is the buyer of a credit ceded by a third party.

Private debt, therefore, constitutes a true investment class, alternative to the listed bond markets, as it can provide the benefits of diversification effects and interesting opportunities in terms of risk-return combination. In fact, it:

- Allows the company to diversify its sources of financing, reducing its dependence on banks.
- Is generally more flexible than a bank loan, as it offers the possibility of tailoring the characteristics of the instruments used to the needs and preferences of the financed company and investors.
- Allows the company to access sources of financing with an average duration longer than bank financing, thus making the sources of funding more stable for companies.
- Is able to provide higher returns than the traditional bonds asset class, thanks to the presence of the liquidity premium.
- Is often characterized, although with significant differences depending on the fund considered, by lower default rates than those of listed bonds issued by debtors with the same credit rating.
- Guarantees the provision of new debt capital and is less intrusive than private equity, as the private debt fund does not interfere with the governance of the company. Private equity and private debt should therefore not be considered as alternative instruments, but rather as potentially complementary. The company can in fact make use of both solutions to raise capital simultaneously.

In relation to the last point, however, it should be noted that there is sometimes an obligation for the financed company to provide guarantees, according to the logic of asset-based lending, or to comply with certain financial constraints, to protect funds and other private debt operators from the risk of potential opportunistic behavior by the company itself. Private debt funds do not adopt a uniform policy in requiring guarantees and in resorting to covenants, i.e., the contractual obligation to comply with certain financial ratios, but the use of these controls is evaluated on a case-by-case basis and depends on the characteristics of the financed company, the type of instrument, the risk associated with the investment, as well as the approach adopted by the fund itself.

15.2 Investment Vehicles

Access to the private debt asset class is made through several alternative investment vehicles, each with its own distinctive characteristics. The private debt market is segmented in terms of the offer of investment vehicles, as the Alternative Investment Funds (AIF) in this sector are divided into two categories:

- Closed-end funds in the private debt segment.
- Credit funds.

In addition to these collective investment vehicles, in the following paragraph other investment vehicles will also be analyzed, both typical of the US market only and present in the main developed economies.

15.2.1 Private Debt Funds

Given the illiquid nature of the financial assets in which they invest, i.e., unlisted bonds, private debt funds have always been characterized by the closed-end form and a consequent fixed maturity since their appearance on the US market.

In the European Union, in line with the AIFMD directive, they fall into the category of Alternative Investment Funds (AIF). Therefore, there is no one type of mutual fund specialized in private debt investment in the European Union, and the investment strategies typical of this asset class prevent the funds of this cluster from complying with the requirements of the UCITS directives.

Private debt funds are characterized, in terms of operations, by a typical closed-end fund life cycle, characterized by an investment period, in which the manager proceeds to purchase financial instruments and manage them, and a divestment period, aimed at liquidating the assets for the purpose of proceeding, upon maturity of the fund, to the reimbursement of the shares. In this segment, the two periods tend to have a similar duration and the overall reference time horizon is usually around 10 years. The distribution of proceeds is prevalent among closed-end private debt funds.

15.2.2 Credit Funds

Credit funds, also known as direct lending funds, are characterized by the peculiarity of their asset allocation, which distinguishes them. In fact, they do not invest in securities, in fund shares or, like real estate funds, in real assets, but rather directly lend to subjects other than consumers. Given the highly illiquid nature of their activities, credit funds are always incorporated in closed form.

Credit funds almost always provide for the distribution of proceeds, which facilitates the placement of shares with institutional investors. Since the contracts signed with the financed companies provide for the payment of high commissions at the time of signature, commensurate with the significant workload required by the structuring and origination of the transactions, the proceeds tend to be higher in the investment phase, unlike what usually happens in private equity funds.

In operational terms, credit funds follow a life cycle similar to that of all closed-end funds, also in this case divided into an investment phase and a divestment phase.

The first has a relatively short duration, concentrating the disbursements within a period of usually 3 years, while the second is structured over a typical horizon of 5 years. However, the right of early termination granted to fund takers tends to make the actual duration of credit funds uncertain, which can therefore terminate their operations before the date set out in the fund regulation.

Thanks to the extreme customization that can be achieved in credit agreements, credit funds are able to offer financing terms consistent with the needs of the borrowers, almost always small and medium-sized enterprises, in terms of duration, subordination, guarantees, amortization, covenants, etc. All this translates into an effective support for fund borrowers who, due to their size and peculiarities, are not easily able to access the capital market, given the high information asymmetries and the cost necessary to overcome them borne by their potential financiers. It follows that credit funds fit a niche in the financial market previously dominated by banks, with the advantage, at the systemic level, of operating by excluding maturity transformation and collecting primarily from institutional investors and not from the public.

The specialization of credit funds in particular sectors or types of contracts is pursued to make the most of the learning economies of their management teams, which combine a deep experience in business credit with a strong focus on due diligence. The latter necessarily focuses not only on verifying the accuracy of the information provided by the financed companies and on the analysis of their financial statements, but also on the concrete feasibility of the development plans submitted by them. Thus, the gap that exists between the professional profiles that are required of managers in the private debt asset class appears evident: while the managers of private debt funds can adopt investment policies in unlisted bonds not dissimilar to those of traditional bond funds, in credit funds the process of selecting counterparties and structuring contracts brings the managers of these investment vehicles closer to the logic and procedures typical of the banking system, albeit with greater flexibility, which results also from the absence of prudential constraints applied to banks by supervisory authorities.

The labor-intensive nature of the credit fund's borrower selection and loan monitoring processes can be mitigated, however, in the case of so-called sponsored transactions, i.e., in which the financed company is itself owned or more often controlled by a private equity operator. This entails the benefit of professional management of the company, and its eventual recapitalization needs can also be covered by the private equity investor, further protecting the private debt investor. Sponsored transactions represent the majority of direct lending transactions, highlighting a fruitful collaboration and the existence of important synergies between the operators of the two segments of the private market.

15.2.3 Other Investment Vehicles in the Private Debt Asset Class

Investment in the private debt asset class can be made through additional investment vehicles, which are primarily available in the US market.[2]

Two peculiar categories of funds are active in private debt distributed in the United States. The first, called interval funds, can be defined as "semi-closed," as they are of indefinite duration but provide for the redemption of units at the Net Asset Value (NAV) on a quarterly basis, more rarely monthly or annually, within a limit indicated in the prospectus and ranging from 5% to 25% of the fund's net assets. The second, called tender funds, pose a significantly higher liquidity risk for participants, as managers have complete freedom to decide not only when to repurchase units from participants, but also their total countervalue and price, which may deviate from the NAV.

In addition to funds, Business Development Companies (BDCs) can also participate in the private debt market. As already mentioned with regard to private equity, BDCs are US companies that are dedicated to the allocation of financial resources to unlisted companies. Their activity in the private debt sector can be carried out directly, but it is mainly carried out through investment in Small Business Investment Companies (SBICs), which are companies that provide financing to small and medium-sized enterprises by raising resources from both their private investors and the Small Business Administration, a US federal agency established in 1953 for the purpose of subsidizing SMEs. Alternatively, BDCs access private debt by participating in the capital of Special Purpose Vehicles (SPVs), which are set up as joint ventures with other institutional investors for the purpose of providing financing to businesses.

In general, outside the US market, which is of primary importance globally, it is finally possible to identify the generic category of debt companies, i.e., companies that invest in unlisted bonds and corporate loans. Notably, some of these are listed on regulated markets and, therefore, fall within the scope of listed debt companies. The index provider LPX includes them in its DLX Direct Lending Index, which is currently the only benchmark available for listed private debt.

15.3 Management Strategies

The private debt market offers a wide and varied range of strategies that require a deep understanding of the sector to fully understand the specific risks and benefit from the diversification characteristics of each strategy.

Within private debt, strategies can be classified according to three key elements:

[2]Nesbitt (2023).

- Intervention in the primary or secondary market.
- Type of debt capital provided and stage of the company's life cycle.
- Possible sector specialization.

With reference to the first element, a private debt operator can intervene both in the primary market, either by directly underwriting the bonds issued by the company or by entering into a credit agreement with it, and in the secondary market for securities and loans, which, since they are not listed on regulated trading venues, impose the manager to conduct direct research and autonomous negotiation with previous holders. It is evident how the adoption of the first strategy, typical of private debt funds that carry out direct lending transactions, makes the private debt fund a kind of shadow bank, effectively performing a credit function in competition with banks. Shadow banking is in fact the provision of financing by entities operating outside the banking system.

Based on the type of debt capital provided and the stage of the company's life cycle, private debt fund strategies can be distinguished as follows:

- Senior debt.
- Subordinated and mezzanine debt.
- Distressed debt.
- Special situation.

The senior debt strategy is the most conservative for the private debt investor and can be used throughout all stages of the company's life cycle. It is based on the financing of the company, typically small and medium-sized, through the use of senior debt, that is, debt without subordination clauses. It is essential, within such financing, to distinguish between secured operations, guaranteed by assets of the debtor company and therefore less risky, and unsecured operations, without guarantees.

Subordinated and mezzanine debt strategies are typical of the expansion and development phase of the enterprise. Subordinated debt, also known as junior debt or, in the presence of certain characteristics, mezzanine debt, consists of the debts whose repayment is postponed to senior debt, i.e., in the event of a company's default, it occurs only after senior creditors have been fully satisfied. The technical forms of subordinated debt are various: medium- to long-term loans, bonds, bonds with deferred interest payments, convertible bonds, bonds with warrants on the shares of the issuer, and other hybrid debt-equity instruments. In many cases, in order to overcome the asymmetric profile of the results offered by bonds and loans, in which the return for the investor is limited above by interest and capital repayment, but the potential loss may concern the entire capital invested, the remuneration of subordinated debt also includes so-called equity kickers (or equity sweeteners). Equity kickers are, in detail, options that allow the creditor, by the maturity date of the loan, to subscribe shares of the issuing company at an exercise price calculated to meet the return targets of mezzanine debt underwriters. These instruments may take the form of a warrant on the shares of the issuing company or the right to convert bonds into shares of its common stock, rights that will be exercised by mezzanine investors only if they are in the money, thus allowing them to integrate the yield obtained in the form of interest with participation in the equity capital. The presence of these clauses accentuates the hybrid nature of the loan granted, which jointly has the characteristics of equity and debt capital, causing it to take on the status of mezzanine debt.

The distressed debt strategy is typical of the advanced maturity and decline stages of the debtor firm. In most cases, nonperforming assets are purchased at a significant discount to their par value, aiming to maximize return through capital appreciation. Investment selection is carried out through a very in-depth evaluation of the assets in question, analyzing the probability of success or failure of the corporate restructuring and the asset values in the event of a positive outcome of the restructuring or in the event of liquidation, with the aim of generating the expected returns in the different possible scenarios.

Managers of special situation funds invest in the market in an opportunistic way through a wide range of instruments. The aim is often to exploit inefficiencies or mispricing on the credit market related to the low liquidity of an instrument, the presence of forced sellers or other specific situations, looking for companies or assets that are still performing, but in financial distress and undervalued by the market, for reasons related to a specific corporate event.

As in the case of distressed debt funds, special situation funds also buy assets at a discount from their par value, albeit at lower expected return levels than distressed funds, because the companies targeted by them are not in severe financial distress. The goal of special situation funds is to target a positive future event, such as refinancing, merger, or acquisition, to liquidate the investment and realize the desired returns.

Finally, there are private debt funds on the market that focus their investment on a particular production sector and thus achieve a form of sector specialization, which offers the investor the opportunity to combine the liquidity premium typical of private debt with an exposure to a particular sector judged at the time to be more profitable than others.

15.4 Risk and Return Measurement

The risk-return profile of private debt as an asset class is a function of the value-creation process that is inherent in the strategies employed by private debt operators. The specificities of this asset class, moreover, distinguish it from traditional asset classes on the one hand and, on the other hand, make it similar to private equity, especially regarding returns measurement.

15.4.1 Performance Measurement Techniques

The strategies of the private debt sector are characterized by a finite time horizon, since the assets in which they invest always have a maturity date by which the financed entity must repay the principal received by issuing a bond or entering into a loan agreement.[3] This means that the performance measurement of private debt requires the use of the same metrics used in private equity, and therefore we refer to Sect. 15.4.1 for a more in-depth discussion of the topic.

However, a specific analysis is needed about the relative performance measure known as Public Market Equivalent (PME). The problem with the use of PME is the definition of the benchmark, i.e., the equivalent portfolio. In fact, the constructive logic of the PME assumes that the inflows and outflows of an investment in a private market vehicle are remunerated at the rate of return of the equivalent portfolio, but the replication of bond market indices is significantly more complex than in the case of the stock market.[4] This complexity is due in particular to the size and variety of fixed-income securities in terms of issuer, maturity, type of interest rate, coupon, and secondary market liquidity. It should also be considered that the finite maturity of fixed-income securities implies that, in order to maintain the maturity and average duration of the index constant over time, frequent rebalancing and reconstructions are necessary. All these aspects significantly affect the process of selecting securities to include in or exclude from index baskets, which in turn determine variable but usually much lower degrees of replicability than in the case of stock indices. If the ability to replicate the equivalent portfolio under cost-effective conditions is lost, the PME is an unrealistic measure and therefore unsuitable for properly measuring the relative performance of the private debt sector.

15.4.2 Performance of Private Debt Transactions

The performance of private debt investment vehicles is derived from the performance of the underlying transactions executed by their managers. For the sake of uniformity, it is important to note that the returns of private debt transactions must be calculated independently of the investment vehicle used to implement them. As a result, these are always gross returns, which do not take into account the remuneration of private debt managers, either in the form of fees or profit sharing.

The academic literature has had to contend with the limited availability of reliable data: since private debt transactions typically involve non-listed entities and investment vehicles not aimed at the general public, there are no requirements

[3] Demaria et al. (2021).

[4] Baker et al. (2019).

regarding information transparency aimed at protecting retail investors and typical of regulated markets.[5]

The first analysis of private debt transactions, carried out by Cumming and Fleming, evaluated 311 investments in bonds and loans made in 25 countries between 2001 and 2010, with exits from the same between 2004 and 2010.[6] The results of the study indicate that the returns of private debt transactions are influenced by both the characteristics of the manager and the debtor. Due diligence on the financed entities and loan monitoring plays a central role, as an increase of one standard deviation in the number of positions in the portfolio per manager is associated with a reduction in returns, ceteris paribus, of 8.1% on average, with the exception of the distressed debt strategy. Thus, it can be assumed that learning economies are significant and that the work carried out by the private debt manager to overcome information asymmetries is essential to ensure the successful outcome of transactions.

The study conducted by Cumming, Fleming, and Liu is based on the information collected on 443 direct lending transactions, both primary and secondary, carried out in 13 markets in the Asia-Pacific region by 15 private debt funds between 2001 and 2016.[7] The authors measure the Internal Rate of Return (IRR) and Return On Investment (ROI) of the loans, taking into account a number of variables, such as the seniority, the amount of the loan, the economic sector of the financed company, the market, the time horizon, and the type of fund manager. There is an inverse relationship between the amount of the loan and the returns of the investment itself because smaller companies are subject to greater information asymmetries and a lower ability to access capital markets. A secondary direct lending strategy outperforms strategies that focus on holding primary direct lending transactions. The median investment in a primary transaction provides investors with an IRR of 22%, while the median secondary loan offers an IRR of 26%. The higher IRR is due to the discount at which secondary loans are purchased and the relatively short holding periods compared to loans granted on a buy-and-hold basis. The authors also note that the return of a diversified portfolio of investments in direct lending, both primary and secondary, consistently outperforms listed bonds. The average monthly return for the private debt index constructed by the authors was 1.53% per month, compared to 0.58% per month for the JP Morgan Asia Credit index.

The construction of a private debt index is extremely complex, given the lack of data, and the pioneering operation conducted on the rather limited sample described so far has been subsequently surpassed by the alternative fund management company Cliffwater, which has been providing the Cliffwater Direct Lending Index (CDLI) since 2015. The time series of the CDLI has been reconstructed ex post back

[5] However, it should be noted that we are witnessing a process commercially described by asset managers as "democratization of private markets," with investment solutions and vehicles also targeting the private investor audience and no longer limited to institutional investors only.

[6] Cumming and Fleming (2013).

[7] Cumming et al. (2019).

to 2004. As of the end of 2022, it covers 11,837 direct lending transactions executed by US business development companies and is based on the reports provided to the SEC on a quarterly basis.[8] Therefore, like all indices constructed on the basis of supervisory reports or financial statements, the CDLI is characterized by a purely retrospective view and significant return smoothing. It is therefore subject to the same problems that are discussed in Sect. 15.4.2, to which the reader is therefore referred.

15.4.3 Private Debt Fund Performance

The illiquidity and credit risk characteristics of the assets in which private debt funds invest affect their returns, as these returns must also include the relative risk premiums. Consequently, this asset class performs a return enhancer function within the portfolio. Given the significant information asymmetries and management peculiarities, the fund manager plays a crucial role in achieving positive performance. This human factor implies the need, especially for institutional investors, to practice careful due diligence of potential private debt fund managers.

An example of such a procedure can be drawn from the study by Block, Jang, Kaplan, and Schulze, who conducted an extensive review of the investment policies of managers, by personally inquiring from a sample of 191 companies about the selection and evaluation of private debt transactions, the monitoring of investments and the interactions with other financial intermediaries, in particular, banks and private equity firms.[9]

The first quantitative study was carried out by Munday, Hu, True, and Zhang, who analyzed the net returns of a sample of 476 private debt funds and business development companies in the period 2004–2016.[10] The results showed positive internal rates of return for the first three quartiles of the sample distribution, albeit with strong variability depending on the vintage year, i.e., the year of the first investment. In fact, investment vehicles established in 2004 showed an IRR of 1.2%, while those established in 2011 achieved an IRR of 14.2%. The public market equivalent confirmed this temporal variability, indicating that investment vehicles established before 2010, except for 2005, have on average underperformed the listed bond market, while those active since 2011 have outperformed it.

Böni and Manigart corroborated these findings by conducting a study on a sample of 448 funds with vintage years between 1986 and 2018.[11] The analysis was based on the classic measures of closed-end funds, namely the internal rate of return, the

[8] Nesbitt et al. (2023). Also of note are the Burgiss Private Capital Indexes series of private debt indices and the Preqin Pro database.

[9] Block et al. (2023).

[10] Munday et al. (2018).

[11] Böni and Manigart (2022).

Public Market Equivalent (PME), and investment multiples. The results showed a significant performance dispersion between the best and worst managers, confirming the crucial importance of the human factor in this peculiar alternative asset class. On average, the PME of private debt funds outperforms the investment grade bond benchmark by 8%, that of high-yield bonds by 6% and the S&P 500 US equity index by a further 6%. However, these average measures are not at all representative of the typical private debt fund, as the top quartile achieved outperformance of 38%, 33%, and 42%, respectively, compared to the three traditional asset classes, while the bottom quartile underperformed traditional investments by −18%, −19%, and −21%, respectively.

15.4.4 Private Debt Risk

The risks underlying private debt transactions are similar to those found in the fixed-income segment, albeit with some specific features.[12]

The most significant risk for investors in this segment is credit risk. In the publicly traded debt segment, its measurement is typically delegated to credit rating agencies, but they do not cover direct lending transactions and rarely privately placed bonds. Consequently, as noted above, the experience of the private debt manager is essential for mitigating this risk, as its measurement cannot be delegated to third parties. Moreover, the availability of a time series of average probabilities of default or losses given default is extremely limited. In relation to the former, of note is the Proskauer Private Credit Default Index, constructed from a representative sample of US direct lending transactions, while realized credit loss data are collected on the components of the Cliffwater Direct Lending Index. In particular, the latter source shows an average loss rate of 1.07% from 2005 to 2021, compared to 1.54% for high-yield bonds and 2.38% for corporate loans from US banks. This is a confirmation of the professionalism of private debt managers, as well as of the anomaly of this asset class, which is characterized by credit spreads significantly higher than those practiced by other lenders.[13]

Interest rate risk affects private debt transaction portfolios both in terms of income, by affecting variable rates, and in terms of capital, by causing a change in the present value of future cash flows and therefore in the current value of the portfolio of loans and bonds. Exposure to these two components of interest rate risk is a function of both the contractual terms of the private debt investment vehicle's assets and its financial structure, in terms of maturity and leverage. Private debt transactions, especially in direct lending, are characterized by a predominant use of indexed rates, which are therefore strongly correlated with the trend of the money

[12] Fabozzi (2022).

[13] Carey (1998).

market. This characteristic, conversely, has resulted in lower exposure to the variability of asset values, as the duration of such transactions is extremely short.

Liquidity risk affects private debt to a greater extent than publicly traded debt. While in the latter, by definition, there is a secondary market, and therefore liquidity is measurable in terms of daily trading volume and bid-ask spread, in private debt these data are absent. However, when the private debt manager implements an optimal maturity matching between its assets and liabilities, this risk appears less significant. It is also worth noting that these transactions are usually carried out with the aim of carrying them to their natural maturity, thus making the difficulty of early liquidation of the same inessential. Moreover, the significant credit spreads applied to interest rates on transactions in this asset class lead to a positive spread between active and passive interest rates, paid on debts contracted to finance the investment portfolio. As a result, the transformation of maturities, aimed at exploiting the natural positive slope of the yield curve, is not strictly necessary for enhancing the profitability of private debt investments, simplifying their asset-liability management. The presence of significant periodic cash inflows, arising from interest and dividends received, is an additional aspect that mitigates liquidity risk, both from the point of view of the manager and, in the usual case of income distribution, of the investor in a vehicle of this asset class. However, the latter subject must also take into account the typical life cycle of closed-end funds and, therefore, to limit fluctuations over time in their cash flows, should pursue diversification between funds based on their respective vintage years.

Since private debt is a fixed-income segment, i.e., an investment in instruments with a contractually fixed return, inflation risk, arising from the loss of purchasing power of money due to rising consumer prices, must be carefully considered. The prevalence of variable-rate transactions in private debt should, apparently, provide implicit coverage from this risk, but it must be remembered that monetary policy can pursue, as has already happened in the past, financial repression aimed at maintaining real interest rates negative over a medium- to long-term time horizon.

The risk that most distinguishes private debt from publicly traded debt is specifically operational risk. In detail, in this segment it is divided into:

- Sourcing risk, which refers to the ability of managers to identify new transactions promptly and continuously and depends in particular on existing contacts and synergies with the private equity segment.
- Due diligence risk, linked to the probability that the preliminary analysis of the potential borrower is not carried out effectively.
- Structuring risk, which is mainly related to the legal sphere and depends on the ability to define the contractual clauses best suited to protect the lender.
- Valuation risk, which depends on the accuracy of the pricing of transactions and is therefore closely correlated with the estimation of credit risk, as well as the spreads applied by the capital markets to remunerate it and the liquidity premium.
- Monitoring risk, due to the failed or untimely monitoring of positions in credits or bonds after they have been opened.

As in all alternative asset classes, operational risk can be mitigated by investors through appropriate diversification across different managers.

References

Baker HK, Filbeck G, Spieler AC (2019) Debt markets and investments. Oxford University Press, New York

Block J, Jang YS, Kaplan SN, Schulze A (2023) A survey of private debt funds. NBER Working Paper Series 30868

Böni P, Manigart S (2022) Private debt fund returns, persistence, and market conditions. Financ Anal J 78(4):121–144

Carey M (1998) Credit risk in private debt portfolios. J Financ 53(4):1363–1387

Cumming D, Fleming G (2013) Debt investments in private firms: legal institutions and investment performance in 25 countries. J Fixed Income 23(1):102–123

Cumming D, Fleming G, Liu Z (2019) The returns to private debt. Primary issuances vs. secondary acquisitions. Financ Anal J 75(1):48–62

Demaria C, Pedergnana M, He R, Rissi R, Debrand S (2021) Asset allocation and private markets. A guide to investing with private equity, private debt, and private real assets. Wiley, Chichester

Fabozzi FJ (2022) Overview of investing in private corporate debt. J Portfolio Manage 48(9): 221–238

Ippolito R (2020) Private capital. Wiley, Chichester

Munday S, Hu W, True T, Zhang J (2018) Performance of private credit funds: a first look. J Alternat Invest 21(2):31–51

Nesbitt SL (2023) Private debt. Yield, safety and the emergence of alternative lending, 2nd edn. Wiley, Hoboken

Nesbitt SL, Beyers F, Zadra G, Sokolov E, Topor J (2023) 2022 Q4 report on U.S. direct lending. Cliffwater, Los Angeles

Chapter 16
Real Assets

Guido Abate

16.1 Real Asset Features

Real assets are composed of two asset classes with unique distinctive features in the asset management sector: real estate and infrastructure. In this chapter, we will analyse the real estate market in its entirety and, subsequently, the financial instruments that allow investors to go beyond the simple direct investment in real estate. The factors influencing the returns and the risks of these investment vehicles present, in turn, characteristics that are often not found in other asset classes and, therefore, allow for greater portfolio diversification. Subsequently, the theme of infrastructure investment will be explored in depth, evaluating its implementation methods and the returns and risks associated with it.

16.2 Real Estate

Historically, the real estate market has played a central role in capital investment. Nevertheless, this asset class was included in financial portfolios for the first time only in 1960.[1] Overall, the real estate market comprises two distinct, but interdependent sectors: the asset market, where ownership and actual real estate rights are traded, and the space market, i.e. the lease market.

[1] With the establishment of the real estate investment trusts in the USA by the REIT Act of 14 September 1960.

G. Abate (✉)
Department of Economics and Management, University of Brescia, Brescia, Italy
e-mail: guido.abate@unibs.it

I. Basile, P. Ferrari (eds.), *Asset Management and Institutional Investors*, Contributions to Finance and Accounting,
https://doi.org/10.1007/978-3-031-59819-7_16

We can also distinguish between direct investment in real estate, known as private property real estate, and indirect or financial investment, i.e. equity real estate.[2] In the former case, the assets may have been purchased either for investment or for direct use by an economic agent. In contrast, in the latter case, purchase of real estate by the end-investor is mediated through a company or a common fund, and exclusively aimed at investment purposes.

Before looking in detail at the aspects of both listed and unlisted equity real estate investment vehicles, we must decide if and when such an asset class presents distinctive characteristics compared to the more common financial assets.

An investor can consider this asset class with a view to improving the risk-adjusted return profile of his/her portfolio. On the one hand, the return for an investor in the equity real estate sector presents the typical and known forms of all financial assets, i.e. it derives from the change in value of the financial instrument and the proceeds that it distributes. On the other hand, what makes real estate investment particular, even more so in the form of equity real estate, is the process of value creation through the administration of the actual asset by the managers of the chosen investment vehicle. Ultimately, then, direct and indirect investments are subject to the same drivers.

The value enhancement of a real estate portfolio can follow different, albeit interdependent, paths. The first envisages the generation of income from the property units by using the so-called space market. The second aims to increase the value of the good with a view to making a capital gain from its subsequent sale on the asset market.

Market is understood here in its broadest definition, i.e. the set of transactions undertaken by a plurality of subjects operating in real estate in the absence of a specific and centralised organisational structure. It is, therefore, an example of a brokers' market, in case there is the involvement of an intermediary, or of a direct search market, in the rarer case in which offer and demand match autonomously.

The presence of high transaction costs is closely linked to the micro-structure adopted by this market. These are determined, first of all, by the remuneration of the intermediaries involved in facilitating trades. In addition, and depending on national legislation, taxes and duties are often applied to the selling price, to which fees for lawyers or public officials responsible for the compilation, verification, and subsequent transcription in public registries of the deeds of the real estate market have to be added.

This market presents strong segmentation, as it is subdivided into local, rather than regional or national, areas, and is specialised by intended use (residential, commercial, industrial etc.), size and building quality of the properties.

The complexity and variety of the characteristics of the buildings and the area in which they are located mean that buyers, in particular, need to be aware of and analyse a large amount of complex information. For these reasons, there are serious

[2] Brueggeman and Fisher (2022) and Geltner et al. (2014).

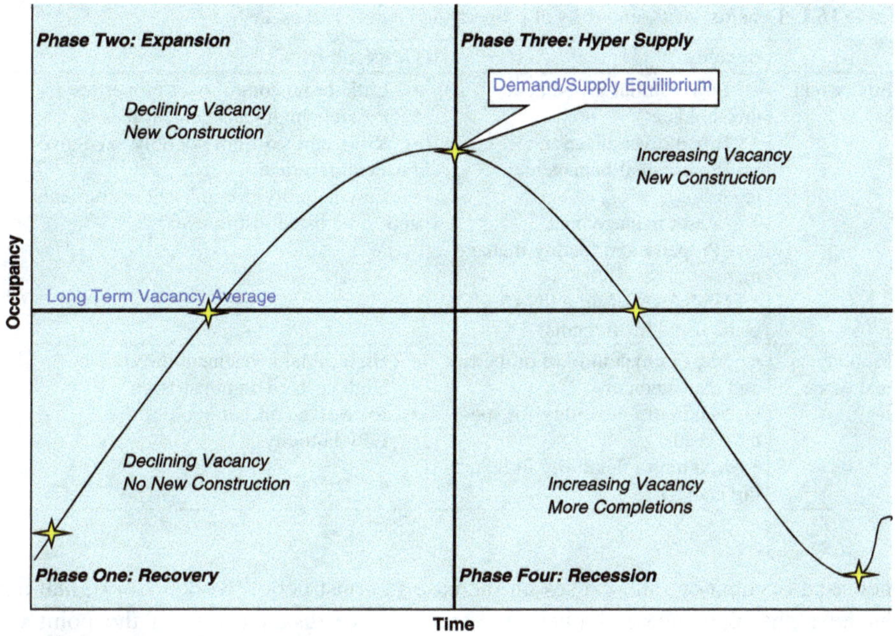

Fig. 16.1 The phases of the real estate market cycle. Source: Mueller (1999)

information asymmetries in the real estate market, which the presence of specialised brokers can only overcome in part.[3]

Buildings are immovable, scarce, and durable goods. Consequently, the supply side presents extremely limited elasticity to variations on the demand side. In other words, as demand increases, the increase in available goods is very slow, because of the time required for construction and the fact that demand in one geographic area or segment cannot be covered by goods in other areas or with different intended uses or quality characteristics. At the same time, a fall in demand is not compensated by a simultaneous reduction in the stock of available properties, as they are durable goods. For these reasons, the real estate market is rarely in a state of equilibrium, thus preventing the clearing of all positions.[4]

The economic cycle of the real estate sector can be divided into four phases as shown in Fig. 16.1. In phase 1, the demand for real estate is growing, resulting in an increased occupancy rate (i.e. the percentage of buildings used) or, in other words, a reduction in the vacancy rate, as a result of the absence of new construction. In phase 2, supply begins to respond to demand, albeit with a time lag: the marginal rate of change in the occupancy rate decreases and reaches, for a brief period, the point of equilibrium. However, the euphoria following the rise of prices, as a result of the

[3]Garmaise and Moskowitz (2004).

[4]Jowsey (2011).

Table 16.1 Benefits, costs, and risks of property and equity real estate

	Benefits	Costs and risks
Equity real estate	• Lower initial investment thresholds • Broad diversification • Professional management regarding: – Asset management – Property and facility management • High level of liquidity (for some listed instruments)	• Little or no control over properties • Possible high management fees • Risks derived from possibly excessive financial leverage • Low liquidity (for unlisted instruments and some listed instruments)
Property real estate	• Maximum control of properties and management • Maximum flexibility for specific needs • Maximum flexibility in financing choices	• High initial investment thresholds • High costs of administration • Excessive concentration of risk • Low liquidity

increased occupancy rate, causes an increase in construction works, so originating the next phase. In phase 3, there is a progressive distancing from the point of equilibrium and a consequent fall in prices, which may turn into a complete price collapse that forces real estate professionals to suspend new construction. Given the development time for construction work, however, new buildings continue to arrive on the market in phase 4, during which the vacancy rate reaches a maximum and, as a result, prices on the space and asset markets reach a minimum, prior to the beginning of a new cycle.

We can conclude, therefore, that real estate markets are subject to reduced informational efficiency derived from their structural characteristics, heterogeneity of goods involved, consequent segmentation, and disequilibrium between supply and demand.

16.2.1 Characteristics of Property and Equity Real Estate

The use of financial instruments in real estate investments can offer significant benefits compared to direct acquisition (Table 16.1). Firstly, portfolio diversification, even for smaller holdings, can be increased thanks to the lower initial investment thresholds. In the case of equity real estate, large portfolios of assets can be represented by financial instruments with a low unit price. With property real estate, on the other hand, the cost of each building is such that any asset allocation strategy aimed at dividing the investment into a diversified portfolio of properties and/or market segments would require substantial expenditure, especially for a retail investor.

In terms of liquidity, if an equity investment is listed on a regulated market, the costs linked to the bid-ask spread are markedly lower than those for property real estate. However, in some market segments, real estate financial instruments can be subject to less trading than other asset classes.

Amongst the negative factors of equity real estate, we must consider that the financial investor's level of control of the assets is substantially lower if compared to direct ownership and that the inclusion of a third party results in the payment of management fees. On the other hand, the presence of a professional operator may be only partially negative, given the greater experience of asset, facility, and property managers compared to a common investor who should, in any case, sustain administrative costs as a result of his/her recourse to specialised professionals.

Direct investment offers a high level of flexibility with respect to the specific needs of the investor, who can decide autonomously between a pure investment and a direct use of the good. This latter option is, instead, completely absent with equity real estate. With property real estate, the investor enjoys maximum freedom in the choice of the financial structure (above all, the level of leverage), unlike the situation with indirect investment. In this case, the decision is independent of the individual investor who, should the strategy not correspond to the personal risk aversion profile, could exercise, first and foremost, the right of "exit" (i.e. abandonment of the investment), unless the holding is such as to allow the right of "voice" (i.e. an attempt to influence the choices made by the real estate managers).

16.3 Real Estate Investment Vehicles

The forms that real estate investment can assume in the case of equity real estate are as follows:

- Real estate funds (or, more properly, collective investment vehicles);
- Real Estate Investment Trusts (REITs);
- Real estate Exchange-Traded Funds (ETFs);
- Property companies;
- Property derivatives.

Unlike the first three investment vehicles, property companies are not subject to particular regulations and can take different legal forms. Furthermore, they may or may not be listed on regulated markets. Given their extreme variety, as well as the freedom in the allocation of their resources and in the business objectives pursued, property companies will not be examined further here.

The following discussion focuses mainly on the institutional context in the European Union, but the economic and financial considerations have, for the main part, general, international relevance.

16.3.1 Real Estate Funds

In the European Union, real estate (or property) funds are part of the category of the Alternative Investment Funds (AIFs), as defined by the Alternative Investment Fund Managers Directive (AIFMD).[5] This legal classification is the reason for the variety in the characteristics of these investment vehicles in different countries, as local regulations are set independently by the respective supervisory authorities.[6]

A first distinction can be made regarding investors' rights to subscribe and redeem units or shares. Despite the typical illiquidity of real estate assets, the legislation in the main European countries allows for open-end real estate funds, in which the maturity of the fund is not limited and subscribers can enter and leave at any time. Of necessity, these funds have a capital structure in which the liquidity of the assets—invested in real estate—is significantly lower than the liquidity of the liabilities, given that participants can request reimbursement on-demand. For this reason, the open-end real estate funds always maintain a liquidity cushion, and national legislation may envisage reimbursement blocks if requests are such as to undermine the solvency of the fund. In order to mitigate further the liquidity risk, the funds may be semi-open-end, in which case reimbursements are possible only in given time periods.

Alternatively, real estate funds can be closed-end, with a pre-determined maturity in line with the investment policy adopted. As in all closed-end funds, where units cannot be issued or reimbursed after initial constitution of the fund, real estate funds must also be liquidated by a defined date specified in the regulations and coherent with the investment policies undertaken. Normally, the agreed duration of a fund is some decades, for example a maximum of 50 years in some European countries. Given the complexity of the liquidation of substantial real estate assets, especially if the sector is experiencing an economic crisis, these maturities can be extended for a period of some years, known as the grace period. In order to allow unit holders to sell their investments, closed-end funds may be listed on a regulated market (listed funds). Alternatively, along with a rigidly closed form, the regulations for real estate funds may allow new subscriptions to be accepted on certain pre-set dates and, simultaneously, offer the opportunity for early redemption of, at most, a comparable sum (semi-closed-end funds).

Real estate funds may have three basic legal structures in the EU: the contractual form, the trust form, and the corporate form. Some countries allow for only one legal form for collective investments, whilst many others allow for more than one form.

In the contractual form, the fund is not a separate legal entity and, therefore, is managed by an Alternative Investment Fund Manager (AIFM), which is responsible for the administration of the assets and in whose name the fund's real estate is held, albeit in the form of separate capital. This is the classical form for common funds in

---

[5] Directive 2011/61/EU of the European Parliament and of the Council of 8 June 2011 on Alternative Investment Fund Managers.

[6] Abate and Losa (2017).

many European countries (Germany, Italy, France etc.), but it is not fully suitable for a real estate fund, as the assets of these funds are non-fungible goods that are sometimes conferred to the fund by participants who might wish to maintain a certain discretion in their management.

In the trust form, a portfolio of assets is constituted and managed by the trustee, an AIFM in the EU, on behalf of the beneficiaries. Investors in the fund are, from a legal point of view, the beneficiaries of the trust and own units of the trust. This form is typical of countries of Anglo-Saxon law and, as for the contractual form, the separation between investors and management may not always be the most suitable solution for real estate funds.

By contrast, in the corporate form the fund is a legal entity and the investors are shareholders of this entity. The adopted governance model is comparable to that of a joint-stock company, and the subscribers holding ordinary shares have full rights in the definition of the investment policies applied by the administrative body, which can be internal or externalised to an AIFM. Other types of shares with fewer administrative rights might also be envisaged, which would differentiate shareholders in terms of their objectives: control of a real estate asset, favouring the possibility to hold voting rights, or merely a financial investment, limiting interferences in management policy. Moreover, the assets of a fund set as a legal entity can be assigned directly to the fund itself, so ensuring greater transparency both for shareholders and other stakeholders. The latter can include the credit sector, which can easily assess the real estate portfolio of an individual fund separately from that of other funds promoted by the same AIFM.

Real estate funds can be classified as retail, if targeted indistinctively at investors, or reserved, if intended for qualifying investors. The main difference between the two categories of funds regards the restrictions that may be placed on the asset allocation, which typically serve three purposes. The first is to guarantee a coherent asset allocation policy, by requiring that the majority of assets be in the form of real estate, with the remaining part linked to the real estate market, as holdings in real estate companies or debt securities issued following securitisation of leased properties or of mortgages. A second restriction usually regards diversification of investments, requiring that, in order to reduce exposure to idiosyncratic risk, they cannot be allocated to a small number of real estate assets. The third restriction is a limitation on leverage, in order to avoid disequilibrium resulting from excessive debt.

The last two restrictions are partially or totally waived by qualifying investor funds, for whom lawmakers have assumed that protection can be less stringent. In particular, the restriction on the diversification of portfolios is always loosened or even absent, as qualifying investors frequently use real estate funds to manage a limited number of major operations, in which the value added is sought precisely in the manager's specialisation in a specific real estate sector.

Given the frequent recourse to financial leverage of real estate funds, it is important to distinguish between their Gross Asset Value (GAV) and Net Asset Value (NAV). If the fund is indebted, its overall value, i.e. the GAV, is different from its net assets, i.e. the NAV.

Financial leverage allows the return on the fund's assets (r_{NAV}) to be increased, as long as the profitability of the GAV (r_{GAV}) is greater than the cost of debt (k_D):

$$r_{\mathrm{NAV}} = r_{\mathrm{GAV}} + (r_{\mathrm{GAV}} - k_D) \cdot \frac{D}{\mathrm{NAV}}$$

By virtue of the additional resources deriving from debt, the fund can realise greater portfolio diversification thanks to an increase in the size of its assets, which can then be divided across further market segments.

Once the restrictions on the assets of the real estate funds are known, it is important to understand the process that has led to the creation of the portfolio of assets. Indeed, real estate funds can be divided into three categories:

- Collection or ordinary funds.
- Contribution funds.
- Mixed funds, i.e. part collection and part contribution.

Initial subscription of ordinary funds involves the following procedure: the management company places the units to the subscribers and simultaneously requests them the payment of the corresponding sum. These financial resources are then allocated to investments.

With contribution funds, on the other hand, the management company can assign units to the subscribers on the basis of their contributions to real estate assets. Therefore, this is a creation-in-kind procedure for the units in the real estate funds. There are some advantages to creation by contribution. Firstly, the properties included in contribution funds are known to investors prior to subscription of units. Furthermore, if units are subsequently placed on the market, for the contributors it is a simple way to sell many assets in a single operation, so avoiding separate negotiations for each transaction. For the subscribers it may bring the advantage of a more efficient use of resources because they have been already invested in real estate without having to wait for the properties to be bought over time.

Along with these advantages, we must also note the risk of moral hazard when the value of the real estate assets is estimated by an expert chosen by the management company, because the management company itself was also chosen by the main contributors, whose primary interest is to maximise the value of their units in the fund. Should the valuation not be congruous, the other unit holders would acquire units in the real estate fund with an actual market value below that established by the estimates.

Without relinquishing its management activities, the management company may externalise certain operational aspects regarding the real estate assets. In particular, external experts can be engaged for the following functions:

- Property management: Management of the real estate assets from an administrative, accountancy, tax, and legal point of view, in order to generate income from the assets and respect compliance requirements.

- Facility management: Planning and implementation of ordinary and extraordinary maintenance, together with day-to-day management of the properties.
- Project management: Real estate development through analysis of the technical and legal aspects of the new projects in which the real estate fund may participate.

16.3.2 REITs and Real Estate ETFs

The model used by national legislations in Europe to define the characteristics of companies dedicated exclusively to investment in real estate through collection of savings from the public is the US "Real estate investment trust" (REIT).[7] Despite the name, a REIT is not always a trust, but can be a corporation or an association. It must have at least 100 shareholders (or beneficiaries in the case of a trust). Moreover, it must not be closely held, i.e. five or fewer individuals cannot hold, directly or indirectly, more than 50% of its capital.[8] REITs benefit from special tax regulations, thanks to their status as pass-through entities, and are not liable to tax on the profits distributed to shareholders, which must be at least 90% of the taxable income. REITs can be private, if they are not listed, or listed, if they are publicly tradable on a stock exchange. The latter are the main US financial instruments offering investors the opportunity to access the financial real estate.

REIT participants have the rights to intervene in management policy typical of public companies, thus if returns are unsatisfactory, management can be replaced. In addition to the performance of the NAV (the term is also used for the net assets of REITs, even though they are not collective investment vehicles from a legal point of view), the returns of a REIT depend, above all, on the distributed dividends, given the requirement of a minimum pay-out ratio. This aspect makes the REITs particularly suitable for institutional investors such as foundations or pension funds, as they must make planned and periodic payments and therefore will favour investments that do not accumulate profits.

Whilst equity REITs (eREITs), the most common form of this vehicle, mainly invest in physical real estate, mortgage REITs (mREITs) are only partly tied to the real estate sector. The latter invest in Mortgage-Backed Securities (MBSs), whose return is linked to a basket of securitised mortgages. To increase returns, REITs make extensive use of financial leverage, as there are no restrictions on levels of debt for eREITs or mREITs. Usually, in order to exploit the typically positive inclination of the term structure, mREITs make short-term debts and invest in securities representing medium/long-term loans. If the MBSs in a portfolio are predominantly at a fixed interest rate, this strategy will be particularly risky, as the repricing and duration gaps will be structurally negative, and both these values indicate that the mREIT is exposed to market risk determined by an increase in interest rates. If there

[7] Sotelo and McGreal (2013).

[8] United States Code, Title 26, § 856.

were such an increase, the repricing gap would indeed lead to an increase in the cost of debts greater than the possible increase in the return on the assets, producing a negative impact on profits. In addition, with a negative duration gap, mREITs are subject to the risk of a depreciation in assets greater than in debts, resulting in a potential imbalance.

Following experience in the USA, many European countries also have fiscal regimes comparable to the REITs. In 1969, the Netherlands was the first country in Europe to introduce tax-exempt companies known as "Fiscale beleggingsinstelling" (FBI), whilst Belgium introduced the "Sociétés d'investissement à capital fixe immobilière" (SICAFI) in 1995, which have been replaced by the "Sociétés immobilières réglementées" (SIR) in 2014. In 2003, the "Sociétés d'investissement immobilier cotées" (SIIC) was launched in France, whilst in 2007 regulations on the "Società di investimento immobiliare quotate" (SIIQ) in Italy and the "Real estate investment trusts" in the UK and Germany were approved. To distinguish the latter two investment vehicles from the US model, the denominations UK-REIT and G-REIT are commonly used.

In all cases, the national regulations state that REITs are completely exempt from income tax, making them essentially pass-through entities. On the other hand, in order to avoid an indefinite deferral of income tax, European REITs are required to distribute a significantly large proportion of their net profits, which are then taxed on the REIT shareholders. In order to be able to benefit from the REIT tax regime, the national regulations require that companies satisfy certain requirements.

The REITs usually have been conceived as an instrument to facilitate access to the real estate market for retail investors. Consequently, it is a fundamental requirement that the REITs be listed on a regulated market. To ensure that share ownership is as broad-based as possible, there may be a required minimum of floating stock, or constraints on the maximum holdings allowed for a limited number of subjects. REITs must also pass the so-called asset test, i.e. their asset allocation must respect certain limitations, usually regarding the prevalence of investment in the space market and/or diversification of the portfolio. In line with the composition of the assets, income must also derive predominantly from property leasing, in order to satisfy a further requirement, known as income test. The only aspect that significantly varies in the approach of the different European legislations to the REITs is whether or not there is a limitation on the use of financial leverage. In some countries, e.g. Italy, REITs enjoy freedom of action, essentially allowing free recourse to leverage, as is the case with standard companies. In contrast, when protection of retail investors is considered to be more important, the debt taken on by REITs cannot exceed a fixed amount, generally calculated as a ratio to specific categories of assets.

A further investment vehicle in the real estate sector is the Exchange-Traded Funds (ETFs), which seek to replicate indices composed of REITs or of other types of listed property companies. These ETFs are technically identical to those linked to equity or bond asset classes and, therefore, do not need further clarification. The performance of indices of private property real estate, which follows real estate prices, is not tracked by ETFs, as these indices do not provide for replicability. In other words, the ETFs would not be able to buy a real estate portfolio identical to that of the index and, at the same time, investors would not be able to subscribe/redeem units in the ETF by delivering/receiving the real estate assets, meaning that the mechanism of creation and redemption in kind, typical of the ETFs, could not be actuated.

16.3.3 Real Estate-Backed Securities

Real estate securitisations are carried out through the issuance of securities by a Special Purpose Vehicle (SPV), whose value is linked to a separate portfolio of real estate rights transferred from an originator to the issuer SPV.[9]

More in detail, the cash flows paid to the holders of Asset-Backed Securities (ABS) depend on the cash flows generated by the SPV's separate portfolio based on two different sources:

- Revenue from the sale of real estate rights (in particular, the right of ownership).
- Rental income from real estate leasing.

In the first case, real estate securitisation is initiated by an originator for the purpose of quickly transferring a significant portfolio of real estate assets. This transfer is completed by the sale of the properties to the SPV, which in turn finances itself through the issuance of asset-backed securities, which are subscribed by investors interested in receiving periodic payments linked to the subsequent liquidation of the SPV's separate portfolio of real estate assets.

In the second case, the originator may have operated in two distinct ways. In the first case, it may have transferred to the SPV a portfolio of already leased properties, the purchase price of which was financed by the SPV through the issuance of asset-backed securities. Alternatively, the originator may have carried out a sale and leaseback transaction, with the purpose of raising financial resources in the immediate future but, at the same time, continuing to use the properties sold. In this transaction, too, the funding is implemented by the SPV through the issuance of asset-backed securities. Investors in asset-backed securities will then receive periodic cash flows from the payment of rents by the tenants of the properties, whilst the repayment of the principal at maturity is linked to the redemption of the assets themselves by the originator or, if this is not possible or foreseen, to the sale of them to third parties.

Real estate securitisations often also include, albeit in a broad sense, transactions involving the transfer by the originator of credits secured by real estate (mortgages and real estate leases). The value of these securities is linked to the real estate sector only in terms of the probability of default of the transferred debtor. In the case of transferred credits that are performing, then it is the debtor's ability to service the

[9]Fabozzi and Dunlevy (2001).

debt that is the real driver of the profitability of these securities, which therefore do not properly fall within the real estate segment. If, instead, the transferred credits are non-performing, then the value of the securities is solely a function of the recovery rate of the credit, that is, the market value of the real estate assets provided as collateral divided by the amount of the credit exposure at the time of default (exposure at default). Therefore, in this second circumstance, the securities, which almost always derive from the transfer of mortgages to the SPV and are therefore called Mortgage-Backed Securities (MBS), are rightfully part of the real estate asset class.

16.3.4 Property Derivatives

The possibility to undertake both speculative and hedging operations on private property real estate indices is provided by property or real estate derivatives.[10] These financial instruments are listed only on some of the main derivative markets in the world, including the Chicago Mercantile Exchange (CME) and Eurex, or traded over the counter.

As these indices are non-tradable, the equilibrium between the price of the derivative and the underlying asset is not guaranteed by the implementation of arbitrage strategies between the two assets.[11] It is possible, however, to estimate the equilibrium price of the futures by recalling that it is such as to provide the investor with a rate of return proportional to the risk inherent in the investment itself. Therefore, the spot equilibrium price, S_0, of the real estate index is:

$$S_0 = \mathrm{E}[S_T]/(1 + \mathrm{E}[r_s])^T$$

where S_T is the spot price at the expiry of the futures, $\mathrm{E}[r_s]$ the expectation at time 0 of the equilibrium return given the risk of the real estate index.

Starting from the canonical no-arbitrage condition for a generic future, i.e. $F_T = S_0(1 + i)^T$, and substituting S_0, the equilibrium price of the property future is:

$$F_T = \mathrm{E}[S_T] \cdot \{(1 + i)/(1 + \mathrm{E}[r_s])\}^T$$

Amongst the OTC derivatives, an important role is played by the Property Total Return Swaps (PTRSs).[12] In these contracts, the buyer of the index pays the fixed

[10] Syz (2008).

[11] Geltner and Fisher (2007). For a survey of the numerous attempts to reach more precise and realistic pricing techniques, which have not resulted in a definitive and shared outcome, see: Fabozzi et al. (2010).

[12] Newell and Sieracki (2010).

amount, i.e. the cash flow equal to the interest matured on the notional at a pre-defined fixed rate,[13] to the seller of the index who, in turn, delivers the so-called property amount to the counterparty. This amount is calculated by applying the total return of the property index underlying the swap contract to the same notional. If the return of the index is negative, the absolute value of the percentage variation must be added to the fixed amount rate, thereby increasing the cash flow due by the buyer of the index.

16.4 Real Estate Management Strategies

So far, we have looked at the possible instruments of a financial real estate investment. Beyond this institutional view, it is useful to analyse the asset management strategies that can be implemented in this sector. For this purpose, the classification proposed by the NCREIF, the US association of real estate professionals, is the most used. It divides real estate portfolios into four categories: core, core-plus, value added, and opportunistic (Table 16.2).[14]

Table 16.2 Management strategies in real estate

	Core	Core-plus	Value added	Opportunistic
Property type	Leased or easily placed on the space market	Leased or easily placed on the space market	Suitable for valorisation	Suitable for valorisation and/or new development plans
Property sectors	Offices, industrial and commercial buildings, residential complexes	Offices, industrial and commercial buildings, residential complexes	Offices, industrial and commercial buildings, residential complexes	All (even niche) and land
Property life cycle	Completed and in a good condition	Completed and in a good condition, but with opportunities of value enhancement	Completed, with possibilities for renovation or re-allocation in new market segments	To be renovated or reconstructed completely. Building plots for new developments
Rollover	Constant and not concentrated	Constant and not concentrated	Moderate and relatively concentrated on the medium term	Limited to the time required for construction and sale. Very concentrated
Financial leverage	≤ 50%	≤ 60%	≤ 70%	> 70%

[13]Until 2007, it was common practice on the most important PTRS market (i.e. the London OTC) to use the Libor variable rates to calculate the financial indexed flow of the swap. In that case, the flow would be called "floating amount."

[14]Baczewski et al. (2003).

Managers of core portfolios invest their resources in real estate assets that are already leased or can be easily placed on the space market, and therefore are able to generate significant cash flows with little volatility. The target real estate assets of this approach are varied: offices, industrial and commercial buildings, and residential complexes. These properties are introduced into the portfolio when they have already been completed—even some time previously—but do not require any extraordinary maintenance. In terms of rollover, i.e. the rate of substitution of the properties held by the real estate investment vehicle (fund or property company), the core strategy envisages constant and not concentrated operations, with purchases and sales diversified both over time and with regard to the individual buildings involved. Despite the availability of constant cash flows, core management strategies make limited use of leverage. Although the NCREIF has not determined a restriction, the literature proposes a maximum Loan-To-Value (LTV) of 50%.[15] A variation of the core strategy is the so-called core-plus, in which the re-qualification of properties (in any case, already generating income) plays a more important role, both in terms of their value enhancement and expectations of rising prices in the area in which they are placed. The use of leverage is higher than in the core strategy.

Value-added management seeks to obtain high returns through an increase in the value of the assets. In this case, then, attention is focused on the resale value of the property, rather than on the rental income. The buildings involved in the investment are essentially the same as for the core strategy, but they must undergo renovation and, possibly, a subsequent renegotiation of the rental contracts with the aim of increasing their value and then placing them back on the market. For this reason, rollover is greater and more concentrated than with the core strategy. Given the need to raise money to fund the renovation works, recourse to leverage is greater than with core management, but usually not greater than 70% of the value of the assets.

Finally, opportunistic management focuses on the development of new construction projects or on the complete renovation or reconstruction of empty buildings or property that is not able to generate income in its given condition at the time of purchase by the fund or the property company. The target properties for the opportunistic strategy are varied, but the concentration of the investment is particularly high, as only a few projects are developed simultaneously. This asset allocation results in marked volatility in returns, compensated—at least in the eyes of investors—by significantly higher returns compared to the other strategies. It is clear that this management style is strongly influenced by both the economic climate in the market and the quality of the selected project, meaning that the management team must have both economic/financial competence and engineering expertise. Once constructed or renovated, the buildings are immediately put up for sale on the asset market, without first attempting to place them on the space market.

---

[15] Fuerst and Marcato (2009). Loan-to-value usually means the ratio between the amount of funding and the value of a good provided as collateral by the debtor. In the case of real estate portfolios, this two-way relationship between debt and collateral is not relevant. Indeed, the concept of LTV is broader and synonymous with the ratio between total debt and total assets.

Consequently, rollover lasts only the time required for construction and sale, and is particularly concentrated both in time and on individual properties. At least for as long as the credit system allowed, financial leverage was always high, usually above an LTV of 70%. This financial imbalance was possible when prices were rising, thanks to the high returns and, even more importantly, the swiftness with which supply and demand were matched on the real estate market.

The NCREIF classification allows us to divide the strategies only on the basis of qualitative indications and without guarantee of objectivity.

Other approaches, closer to those typical of equity portfolio style analysis, have been proposed, with the aim of overcoming the arbitrariness of the NCREIF taxonomy.[16] However, the use of style analysis techniques poses serious problems. We may use price indices of various real estate sectors (identified by intended use and/or geographic area) as regressors, but then we would face the omitted variable bias, i.e. a distortion related to the failure to include explicative variables, such as the effect on economic results deriving from recourse to leverage (financial value-added) or factors regarding the process of creation of value in real estate, e.g. engineering/ construction aspects (physical value-added), or commercialisation/leasing aspects (operational value-added) associated to the direct control a manager exerts over the property, which is much more incisive than is the case with investments in securities.[17] Furthermore, we can recall the well-known inefficiency of real estate markets and the presence of significant information asymmetries that can be exploited in different ways depending on the manager's experience in a specific real estate sector and geographic area. In other words, the specificities of active management make it problematic to use an analysis based on the regression on market indices.

16.5 Risk and Return Measurement in Real Estate

The risk and return profile of real estate as an asset class is intrinsically tied to the process of value creation typical of the various strategies used by managers. Investment in real estate is highly specific and therefore it is useful to examine in detail the factors underlying the valuation of real estate. Once the theoretical framework and the operational techniques used to measure the value of real estate have been defined, we can undertake an analysis of the risks and returns associated to the investment vehicles in this asset class.

[16] Myer and Webb (2000) and Fuerst and Marcato (2009).

[17] Conner and Liang (2003).

16.5.1 Real Estate Value

The market value of real estate can be defined as "the estimated amount for which the property should exchange on the date of valuation between a willing buyer and a willing seller in an arm's-length transaction after proper marketing wherein the parties had each acted knowledgeably, prudently and without being under compulsion."[18] However, this definition appears rather tautological, as it reduces the concept of value simply to the mere verification of a selling price, without any examination of the process underlying the pricing of properties.

To do this, we need to consider the close correlation existing between the space and asset markets. As for financial assets, the economic value of real estate too can be estimated by actualising the future cash flows that can be drawn from the assets. We can use, therefore, the financial criterion of discounted cash flows. This means that, on the real estate market, assets will appreciate or depreciate on the basis of the cash flows they generate. These flows will always depend ultimately on rental payments, which are received if the building is leased, and not paid (meaning no outflow of financial resources) if the building is designated for the owner's use. These flows are usually treated as potential gross income, where the adjective potential is used to recall the possibility of vacancy periods or of collection loss. Rectifying the lease income to take account of these potential deductions, we arrive at the effective gross return, from which the multiple types of operating expenses must be deducted to obtain the so-called Net Operating Income (NOI), used in real estate valuations as representative of the net cash flows in the period.[19] Once the net cash flows have been calculated, we need to estimate the discount rate for their actualisation, the so-called capitalisation rate. Many techniques can be used to calculate this rate, and there is no universally accepted standard. The alternatives can, however, be grouped into three categories:[20]

- Comparative expected Internal Rate of Return (IRR). The IRR realised in similar real estate investments is calculated, taking into account the purchasing cost, the intermediate flows, and the income derived from the sale of the property.
- Build-up approach. The risk premiums for the specific sources of risk in a real estate investment are summed to the risk-free rate.
- Weighted Average Cost of Capital (WACC). Used subjectively from the point of view of a specific purchaser, having rectified the Net Operating Income (NOI) to take account of taxes and the impact of the financial structure of the interested party.

[18] Regulation (EU) no 575/2013, art. 4, § 1, n. 76. This definition, in turn, can be found in the International Valuation Standards: Royal Institution of Chartered Surveyors (2021).

[19] Floyd and Allen (2021) and Geltner et al. (2014). Note that in operational practice of real estate valuation, the NOI (an economic result) is often used as an approximation of cash flows over a period, i.e. in place of a financial flow.

[20] Damodaran (2012).

Finally, the many techniques used to discount the NOI are those commonly employed in the valuation of financial assets, depending on whether a defined (with a terminal value) or an undefined horizon is used, or on whether or not there is a growth rate in flows (with one or more stages).

Given the known difficulties in defining the probabilistic structure of future scenarios and of the discount rates of the relative pay-offs, financial valuation techniques are often accompanied (and sometimes replaced) in operational practice by estimation methods based on different theoretical approaches, such as:

- Comparative methodology:

  (a) Comparative market criterion.
  (b) Hedonic prices criterion.

- Reconstruction cost methodology

The comparative market criterion is not based on any theory of value, i.e. market prices are taken as given rather than assuming any underlying process. Therefore, following necessary rectifications to take account of the fact that this type of good is not fungible, the comparative method estimates the value on the basis of the prices of the most recent transactions involving properties with similar characteristics to the one under valuation.

The hedonic prices criterion is an evolution of the comparative criterion and is particularly suitable for properties, especially residential, whose characteristics mean there is no sample of suitable comparables. Following numerous academic contributions since the 1920s, the hedonic prices theory was fully formulated in Lancaster's consumer theory and in Rosen's model.[21] The procedure to value the unit price, P_k (e.g. euro/m^2), of a property, k, envisages, first of all, an analysis of a sample of N sales that have occurred in the market, which is represented by a vector of unit prices, $\mathbf{P}_{1\times N}$, found for each of these. Subsequently, for each of these N properties, the M distinctive characteristics are given in the matrix, $\mathbf{Z}_{M\times N}$, often represented by dummy variables (i.e. values of 1 or 0, e.g. with or without elevator, and reception) or ordinal variables (e.g. level of finish and distance from the centre). Then, an OLS regression is performed to identify the vector of the weights, $\boldsymbol{\beta}_{1\times M}$, that the market attributes to each determinant, z.[22] Given these weights and the characteristics of the building to value, the unit price can be estimated in this way:

[21] Lancaster (1966) and Rosen (1974).

[22] The regression may or may not envisage an intercept, itself devoid of defined economic significance but indicating generically the value attributed by the market to all those variables that could not be included in the model. Given its purely residual economic significance, it is indicated by the Greek letter ε, rather than by the more frequently used α. When estimating the price of a building, k, the mean value of ε found in each of the N previous sales can be used.

$$P_k = \beta_1 z_1 + \beta_2 z_2 + \ldots + \beta_M z_M + \varepsilon$$

In contrast, the reconstruction cost method assumes that the value of a property is at least the amount necessary to re-build it from new. This heuristic approach is not suitable for valuations of assets by rational investors, even if it is used in operational practice. Indeed, this technique does not take account of supply and demand dynamics in the market and is based, therefore, on the hypothesis that a good has its own intrinsic value, irrespective of the use that can be made of it at a given time. This problem becomes extremely important in periods of speculative bubbles, when the construction process tends to exceed the level of demand, so provoking a subsequent price collapse even below the cost of construction, which therefore is no longer a prospective price, but merely a representation of the past.

16.5.2 Returns on Real Estate Investments

Listed equity real estate investment instruments present two different capital gain profiles: variation in the NAV per unit or share[23] and variation in market price. On the one hand, the ΔNAV per unit or share depends on the profit in the period (influenced not only by the returns of the real estate portfolio, but also by the level of debt) and the increase/decrease in the valuations of the real estate assets. On the other hand, the return for the investor prior to redemption on maturity is tied merely to the price movements of the units or shares in the investment vehicle traded in the securities market. Rarely these two returns match.

Real estate assets are usually evaluated with backward-looking criteria, such as comparative methodologies or the cost of reconstruction. In contrast, market prices are the result of expectations of future returns and therefore are forward looking. This, together with the time lag between the estimated valuations and the quoted prices, which are fixed in real time by supply and demand, is the reason why the listed financial real estate indices can anticipate the performance of the indices of the direct property market.[24]

Another phenomenon that distinguishes the indices of the direct property market from those of the listed investments is the smoothing of the returns of the former. In other words, the indices based on valuations—the so-called appraisal-based indices—present an autocorrelation of returns, as a result of the infrequent valuation of the underlying assets. This distortion leads to smoothing in the time series and,

---

[23] For the sake of simplicity and in agreement with common practice, the term NAV is also used for investment vehicle that are not collective investment vehicles.

[24] In the period 1978–2002 this relationship was statistically significant. The result can be demonstrated by a regression of the time series of quarterly returns of a US direct property index (e.g. the NCREIF Property Index) on the series of total returns of an aggregate index of US REITs (e.g. NAREIT Equity REIT Index) in the four previous quarters. Gyourko and Keim (1992) and Gyourko (2004).

consequently, to an underestimation of the real volatility of its returns over time. One particularly simple unsmoothing technique takes the unsmoothed return, r_t, as a function of the smoothed return, r^*, not only at t, but also at $t - 1$, given a smoothing parameter, a, between 0 and 1:[25]

$$r_t = \frac{r_t^* - (1 - a) \cdot r_{t-1}^*}{a}$$

Alternatively, it would be possible to use indices of the property market calculated only on the basis of the sales in the period, i.e. transaction-based indices, but the limited availability of information makes the use of this approach infrequent. Moreover, the fact that only a minimum percentage of overall real estate assets is traded in each year means that such an index is not fully representative of the overall market.

A typical characteristic of both direct and indirect property investment is the marked correlation of returns with inflation rates, so generating hedging against inflation risk. The reasons reside in the increase in construction costs for new buildings, which is tied to general price trends, and the indexing of rents to inflation, as a result of specific clauses or following renegotiation when the contract expires. We can also consider behavioural factors, such as investors' need to turn to real goods in the presence of inflation shocks.[26]

The first and best-known study on the topic is based on the general model of expected nominal returns, E[R], given the information, φ, available at $t - 1$ regarding the n-th asset at time t:[27]

$$\mathrm{E}[R_{n,t}|\varphi_{t-1}] = \mathrm{E}[r_{n,t}|\varphi_{t-1}] + \mathrm{E}[i_t|\varphi_{t-1}]$$

The real return of the asset n is given by $r_{n,t}$, whilst i_t is the inflation rate. Regressing the time series of $r_{n,t}$ on the series of expected inflation rates, $\mathrm{E}[i_t|\varphi_{t-1}]$, we can assess the level of hedging offered by the asset i with respect to variations in prices:

$$R_{n,t} = \alpha_n + \beta_n \mathrm{E}[i_t|\varphi_{t-1}] + \varepsilon_{n,t}$$

The coefficient β_n is the hedge ratio of asset n compared to expected inflation. However, in order to incorporate the unexpected inflation component, we have to use the following regression:

[25] Geltner (1993). We underline that there is no unanimous consensus regarding the best unsmoothing procedure, in particular, with regard to the estimate of the smoothing parameter. Note that if the series of unsmoothed returns were independently and identically distributed, a would be equal to the first order autocorrelation coefficient.

[26] Damodaran (2012).

[27] Fama and Schwert (1977).

$$R_{n,t} = \alpha_n + \beta_n \mathrm{E}[i_t|\varphi_{t-1}] + \gamma_n(i_t - \mathrm{E}[i_t|\varphi_{t-1}]) + \eta_{n,t}$$

This formulation also includes γ_n, which measures the capacity of asset n to provide an hedge on unexpected inflation (i_t − E[$i_t|\varphi_{t-1}$]), calculated as the difference between actual inflation at t and the figure expected on the basis of the information set available at t − 1. Analysing a time series of returns for US residential properties in the period 1953–1971, the study showed the coefficients β and γ to be greater than one, meaning that hedging would have been greater than both expected and unexpected inflation.

Given these first encouraging results, scientific literature has subsequently divided between supporters and opponents of inflation hedging of real estate.[28] However, more recent studies, using more innovative econometric approaches compared to the one presented here, confirm that real estate can generate returns above inflation.[29]

16.5.3 Risks of Real Estate Investments

The typical risks of direct real estate influence, in turn, the distribution of returns and so, ultimately, the risk borne by investors in financial real estate instruments.

The factors that may increase the variability of the economic results of a direct investment can be subdivided between exogenous and endogenous risks. The former refers to elements that have an impact on the entire real estate market, whilst the latter relates to the characteristics of a specific portfolio of properties.

Exogenous risks include both the main macroeconomic variables, such as, in particular, the performance of GDP and the term structure of interest rates, and those associated to the geographic and sectorial localisation of the properties. Changes that may occur in the urban environment, i.e. independently of an investor's choices, cause indeed significant variation in the commercial value of a property. This is influenced by its intended use. For example, an industrial building may rise in price following the opening of a new main road nearby, whilst the same infrastructure may depress the residential market, as a result of worries concerning its environmental impact (atmospheric and acoustic pollution). In addition, the economic environment of the geographic area in which a property is located influences its value, even more so than at a national level, given the local segmentation of the real estate market.

Endogenous risk essentially takes the form of counterparty risk associated to the creditworthiness of the tenant. The rate of collection loss, i.e. non-collected rents as a percentage of the total annual rents, is the ex post measure representing the mean reliability of real estate tenants. Another typical risk in a real estate investment is linked to the time required to place a vacant property on the space market, i.e. the

[28] For a review of the literature, see: Adrangi et al. (2004, 2015).

[29] Goetzmann and Valaitis (2006) and Case and Wachter (2011).

vacancy risk. Vacancy rate measures the vacant properties as a percentage of the total assets at a specific date: as this value refers to a particular time, it is preferable to use a mean calculated over a period covering a number of distinct market phases. Note that vacancy and collection loss are inter-dependent risks. The creditworthiness of the tenant is more important for the less fungible properties, as the vacancy periods of buildings with particular characteristics are on average longer, given the relative scarcity of interested tenants.

The dichotomy between exogenous and endogenous risks is not always evident. The economic scenario certainly influences tenant insolvency rate and vacancy period, just as the intended use of the property may or may not generate its dependence on exogenous factors.

A risk that is half way between the above two categories is the liquidity risk. The ability to sell a property on the asset market quickly and with a minimum offset, compared to the initial requested price, is tied to macroeconomic (interest rates, economic growth), geographic, and specific (condition of the property, occupied or vacant etc.) factors. Given the inefficiency of the real estate market, the liquidity risk has a significant impact on sales and would appear not to be eliminable. On the other hand, econometric studies have revealed significant premiums for liquidity, incorporated in the returns of direct property investments, proportional to the mean duration of the period between placement of a property on the market and completion of the sale.[30]

These sources of risk and the significant recourse to financial leverage influence the performance of returns from financial instruments in the real estate sector. These reveal clear skewness and kurtosis, whilst the correlation with the equity market is about 0.5, but the latter has shown strong instability over the years (Fig. 16.2).[31]

In view of these characteristics, the inclusion of financial real estate in diversified portfolio results in an improvement in the risk-adjusted performance profile, even though this benefit tends to be annulled in crises, precisely when diversification would be more useful and necessary.

16.5.4 Discount-to-NAV

The prices of listed equity real estate instruments often show a significant discount compared to the NAV, giving rise to the phenomenon known as discount-to-NAV.[32] Market price and NAV are two quantities measured at distinct times. Whilst the former may vary during every market session, the latter is published periodically in

[30] Lin and Vandell (2007).

[31] Case et al. (2012) and Yang et al. (2012).

[32] In accordance with consolidated practice in the treatment of common funds, the NAV is always taken here as NAV per unit and not as overall NAV. As already stated, NAV is also the definition used to indicate the net assets of REITs and similar companies, for which the term "book value" would be more precise.

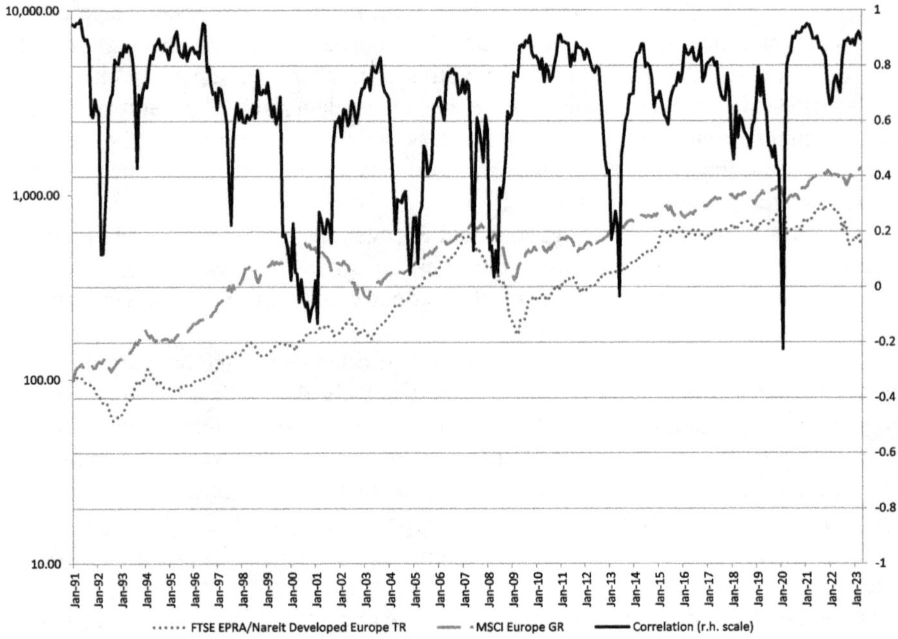

Fig. 16.2 Correlation between FTSE/EPRA NAREIT Developed Europe TR and MSCI Europe GR (1991–2023). Trailing correlation calculated on the 12 previous months. Indices: base 100 at 31/12/1990, logarithmic scale. Source: data from Morningstar Direct

accordance with national regulations and the level of disclosure adopted. Outside the publication dates, the only variations in the published NAV known to investors regard the possible distribution of proceeds.[33] In order to take account of this asynchrony between the two values, the discount must not be calculated with reference to the published NAV, but to its value net of dividends paid, also known as adjusted NAV:

$$\text{Adjusted NAV}_T = \text{NAV}_{t=0} - \sum\nolimits_{t=0}^{T} \text{Dividends}_t$$

The proceeds distributed in the period from $t = 0$ (publication date of the NAV) to T (date of calculation of the adjusted NAV) are subtracted from the NAV, as they constitute a reduction in the fund's net assets.[34] Therefore, the percentage discount of the market price compared to the NAV at time T equals:

[33]We use the expression "published NAV" to underline the fact that the daily changes in the NAV, as a result of the natural performance of the fund and the variation in the value of the real estate assets, are disregarded. This daily performance, in fact, is not calculated, in contrast to funds investing in listed assets, and therefore the only known NAV is that published periodically.

[34]Working from an ex post point of view, we can estimate the NAV at time T by linear interpolation of the NAV at the beginning and the end of the semester, as described in Biasin et al. (2010).

$$\text{Discount}_T = \frac{\text{Price}_T - \text{Adjusted NAV}_T}{\text{Adjusted NAV}_T}$$

Market prices different from the NAV are not an isolated phenomenon typical only of equity real estate but are part of the broader closed-end fund puzzle. Scientific literature has researched a number of factors that may explain this situation, amongst which the main are the following:

- Financial leverage. The increment in the level of debt increases the variability of the economic result (interest on debt is a cost unrelated to the amount of income) and may lead to the onset of so-called distress costs, so provoking an increase in discount-to-NAV.[35]
- Asset allocation. Diversification of the real estate portfolio causes an increase in discount-to-NAV. The reason may be the lower information transparency derived from the complexity of the assets managed and the increase in management costs.[36]
- Illiquidity. As the liquidity of units or shares on the secondary market increases, discount-to-NAV declines.[37] The reason for this is probably the lower transaction costs for investors trading in a liquid market, or one in which there is at least one market maker.
- Potential recourse to the grace period. The fact that unit or share redemption can be postponed forces investors to discount the cash flow to a later estimated date, so calculating a lower present value.
- Management participation. Greater management participation in the capital of a real estate investment vehicle proportionately reduces discount-to-NAV. Consequently, the market believes that agency costs are lower when investors share the same risks and returns as the managers of the real estate assets.[38]

16.6 Infrastructures

Within the real assets, the infrastructure asset class is represented by investments made in the management of complex real assets organised to provide essential services to consumers and businesses. It is, therefore, an asset class focused not only on the ownership of real assets and their eventual lease to third parties, as is the case in real estate, but specifically on their management for the purpose of profit.[39]

---

[35] Brounen and terLaak (2005).

[36] Capozza and Seguin (1999).

[37] Clayton and MacKinnon (2001).

[38] Capozza and Seguin (2003).

[39] Weber et al. (2016) and Gatti and Chiarella (2020).

Investments in infrastructure are characterised by their essentiality, as they provide essential services for the real economy, and by their high capital intensity, as their realisation requires the investment of significant initial capital. From these two aspects, some specific characteristics follow. Essentiality determines the presence of stable and predictable cash flows for the investor. In some areas, there is a limited correlation with the macroeconomic cycle, as the services offered cannot be substituted easily by users and therefore their demand is relatively inelastic with respect to price. In contrast, the high capital invested in infrastructure results in the use of high operating leverage and therefore the presence of significant barriers to entry, which make it economically unsustainable for new entrants to participate in the market. As a result, the activity in infrastructure is often characterised by an oligopolistic or monopolistic regime that, given the associated positive (services to users essential for economic development) and negative (lack of competition) externalities, requires pervasive regulation.

The authorities responsible for overseeing infrastructure have the power to set the tariffs charged by operators, calculated with the aim of remunerating risk capital, determining a return on investment that can be easily estimated ex ante and is often hedged against inflation risk, as it is indexed to inflation itself.

The allocation of financial resources in this asset class typically takes place through the interposition of two financial instruments between the investor, often institutional,[40] and the physical infrastructure. Instead of investing directly in such real assets, in fact, entities interested in this segment can buy shares of funds, both open-end and closed-end. In the first case, these are UCITSs or open-end AIFs that purchase listed financial instruments (shares and less frequently bonds, with different levels of subordination) issued by companies whose business activity consists of the management of infrastructures (listed infrastructure). In the second case, they are closed-end AIFs and the financial instruments in which they invest are not listed on a secondary market (unlisted infrastructure).

16.7 Infrastructures Management Strategies

Value creation in this asset class can occur along different strategic lines.

First, target companies can operate at three different stages of infrastructure development. In the case of new construction, the assets are called greenfield. If, instead, the assets have a significant goodwill because they have already been in the market for some time, they are called brownfield. A third niche is finally that of yellowfield assets, i.e. existing infrastructure, but in need of substantial investments in modernisation and development.

[40] Andonov et al. (2021).

Table 16.3 Management strategies in infrastructures

	Core	Core-plus	Value added	Opportunistic
Income sources	Regulated tariffs; long-term contracts with public and assimilated entities; public concessions	Long-term contracts; public concessions subject to conditions (GDP growth, volumes etc.)	Contracts of any duration, including with counterparties with reduced creditworthiness; revenues subject to conditions (GDP growth, population served etc.)	Revenues linked to volumes and market prices
Infrastructures sectors	Utilities, energy, transports	Utilities, energy, transports	Communications, energy	All
Development stage	Brownfield	Brownfield, greenfield	Yellowfield, greenfield	Greenfield
Time horizon	>7 years	>6 years	5–7 years	3–5 years

The investor must then take into consideration the economic sector of relevance of the infrastructure: utilities (public services), energy, communications, transportation, and social infrastructure.

Once the characteristics of the infrastructure to be invested in have been defined, managers can adopt one of the following approaches: core, core-plus, value added, and opportunistic (Table 16.3).

Core managers invest in natural monopoly positions, with mature infrastructure and low operational risks, subject to well-defined and consolidated regulations and tariffs. Consequently, this strategy focuses on activities that generate significant and constant cash flows.

The core-plus strategy focuses on infrastructure similar to that of the previous investment style but with a higher correlation of revenues with the macroeconomic cycle. This implies the need for more active management, aimed at ensuring higher levels of operational efficiency and growth. In the case of greenfield infrastructure, they must still be managed within a strong regulatory framework that protects the significant initial investment from the presence of potential new entrants. The core-plus strategy is therefore linked not only to the collection of relatively constant flows but also to the capital appreciation of the managed infrastructure.

Managers who adopt a value-added strategy focus on both new infrastructure (greenfield) and infrastructure in need of modernisation, more efficient management, as well as the renegotiation of existing contracts (yellowfield). The managers' goal is not to earn periodic income but to increase the value of assets in relatively short periods of time and then sell them to core investors, earning a capital gain.

In conclusion, the opportunistic strategy focuses on investments with higher risk, also because these infrastructures are not only greenfield, but also active in markets that are not fully regulated, often in emerging countries, or where the regulations are still in development and therefore subject to potential changes over time, especially in relation to the prices to be charged to users and to entry barriers. The correlations of this strategy with the economic cycle and commodity prices are very high and, therefore, the timing of the investment is crucial for achieving a positive outcome. The aim is to sell the infrastructures in a short- or medium-term time horizon, generating a strong capital gain, commensurate with the high volatility of results.

16.8 Risk and Return Measurement in Infrastructure Investments

The performance in the infrastructure asset class must be evaluated differently depending on whether the financial instruments issued by the target companies are listed or unlisted.

In the first case, performance measurement is, in itself, subject to the same issues and methods as investments in traditional asset classes. In particular, with regard to the limited investable universe of shares of companies active in the infrastructure sector, the global benchmark of reference is the S&P Global Infrastructure Index, composed of the 75 leading global companies operating in the utilities, energy, and transportation sectors.

In the second case, the valuation of the fair value of the securities plays a crucial role in performance measurement. Amenc, Blanc-Brude, Gupta, and Whittaker have proposed the following approach to the valuation of shares of the *i*-th company in the infrastructure sector:[41]

$$P_i = \sum_{t=1}^{T} \frac{D_{i,t}}{\left(1 + r_f + \gamma_i\right)^t}$$

where P is the fair value of the position, D_t is the cash flow deriving from the position at each time t, T is the investment horizon, r_f is the risk-free rate, and γ is the risk premium.

The estimation of the risk premium must be conducted using a multifactor model, in which the *j*-th independent variables are identified in the following factors: financial leverage, asset size, profitability, capital expenditures, country risk, business model, and corporate structure. Formally, therefore, the risk premium is given by the following equation:

$$\gamma_i = \sum_{j=1}^{J} \beta_{i,j} \cdot \lambda_j$$

where $\beta_{i,j}$ is the exposure of the *i*-th company to the *j*-th risk factor and λ_j is the price of the risk premium associated with the *j*-th factor.

---

[41] Amenc et al. (2022).

The model has been calibrated on a sample of over 1000 transactions of unlisted companies active in the infrastructure sector from 2000 to 2020 and has been verified on a subsample of 250 transactions over the same time horizon, denoting an estimation error of less than 5% of the observed prices. It is therefore capable of estimating the fair value of unlisted companies with considerable accuracy, providing investors in this alternative asset class with an essential reference in the absence of an active secondary market, useful not only at the time of purchase or subsequent sale of the shares but also for periodic estimates of the value of the securities portfolio. However, it is necessary to note that the absence of a market price and, at the same time, the use of estimates conducted at least with quarterly frequency lead to the emergence of returns smoothing and consequently to their autocorrelation.

The same study also analysed the risk exposure of this asset class, using the EDHEC infra300 index as a benchmark of the unlisted infrastructure sector. The returns were found to be subject to skewness and excess kurtosis. The correlations with the fixed-income asset class, although variable over time, range from 0.2 for government bonds to 0.4 for corporate bonds and tend to decline during financial crises. In contrast, the average correlation with stocks is statistically indistinguishable from zero. However, the co-skewness and co-kurtosis with traditional asset classes highlight the occurrence of extreme losses in conjunction with those recorded on the stock and bond markets, especially during the subprime mortgage and COVID-19 crises.

The underlying risk factors for these statistical estimates can be identified as follows:

- Country and regulatory risk: This risk depends on the political stability of the country in which the infrastructure is located and the complexity of the regulatory framework to which the relevant operating sector is subject. This risk affects the variability of results in relation to the timing of actual commissioning of the infrastructure, the remuneration of the services offered, the probability of expropriation or forced sale at prices not in line with the market etc.
- Macroeconomic risks: Infrastructure activity can be linked to the real economy, both in terms of volumes and prices, if these are calculated according to models defined by the authorities that take into account certain macroeconomic variables.
- Liquidity risk: Both infrastructure, as real assets, and shares of unlisted companies active in their management are illiquid. The presence of periodic cash flows is therefore of great importance for the investor, especially institutional, and depends in particular on the strategy adopted by the manager.
- Interest rate risk: The impact of this risk depends on the financial structure of the transaction, and in particular on the extent to which debt is used and the method of calculating interest (fixed, indexed etc.).
- Counterparty risk: This risk depends on the solvency of the users of the service offered. It is essential to consider the granularity of the positions, that is, the number and individual size of the counterparties. An extreme and very risky case, especially in developing countries, occurs in the presence of monopsony, that is

when the revenues from the service provided derive from an exclusive relationship with a single entity, usually public.

- Operational risk: This risk concerns the technical management of infrastructure and has a fundamental impact on the quantity and quality of the services provided. It concerns the obsolescence of assets, catastrophic events, and, in greenfield strategies, the construction risk, that is, any delays and additional costs in the construction of infrastructure.
- ESG risk: This risk arises from environmental, social, or governance events or conditions that could cause an actual or potential negative impact on the value of infrastructure assets.

References

Abate G, Losa G (2017) Real estate in Italy. Markets, investment vehicles and performance. Routledge, London

Adrangi B, Chatrath A, Raffiee K (2004) REIT investments and hedging against inflation. J Real Estate Portfolio Manage 10(2):97–11

Amenc N, Blanc-Brude F, Gupta A, Whittaker T (2022) An infrastructure investment primer. From valuation to allocation and manager selection. J Portfolio Manage 48(9):117–202

Andonov A, Kräussl R, Rauh J (2021) Institutional investors and infrastructure investing. Rev Financ Stud 34(8):3880–3934

Baczewski J, Hands K, Lathem C (2003) Real estate investment styles. NCREIF, Chicago

Biasin M, Giacomini E, Quaranta AG (2010) Public REITs' governance and regulatory structure: effects on NAV discount: evidence from the Italian market. J Eur Real Estate Res 3(3):161–181

Brounen D, terLaak M (2005) Understanding the discount: evidence from European property shares. J Real Estate Portfolio Manage 11(3):241–251

Brueggeman WB, Fisher JD (2022) Real estate finance and investments, 17th edn. McGraw-Hill, New York

Capozza D, Seguin P (1999) Focus, transparency and value: the REITs evidence. Real Estate Econ 27(4):587–619

Capozza D, Seguin P (2003) Inside ownership, risk sharing and Tobin's q-ratios: evidence from REITs. Real Estate Econ 31(3):367–404

Case B, Wachter S (2011) Inflation and real estate investments. University of Pennsylvania ILE Research Paper, n. 11-33

Case B, Yang Y, Yildirim Y (2012) Dynamic correlations among asset classes: REIT and stock returns. J Real Estate Financ Econ 44(3):298–318

Clayton J, MacKinnon G (2001) The time-varying nature of the link between REIT, real estate and financial asset returns. J Real Estate Portfolio Manage 7(1):43–54

Conner P, Liang Y (2003) Institutional real estate investment: evolution and innovation. Real Estate Finance 20(2):2–10

Damodaran A (2012) Investment valuation: tools and techniques for determining the value of any asset, 3rd edn. Wiley, Hoboken

Fabozzi FJ, Dunlevy JN (2001) Real estate-backed securities. Wiley, Hoboken

Fabozzi FJ, Shiller RJ, Tunaru RS (2010) Property derivatives for managing European real-estate risk. Eur Financ Manag 16(1):8–26

Fama E, Schwert GW (1977) Asset returns and inflation. J Financ Econ 5(2):115–146

Floyd C, Allen MT (2021) Real estate principles, 12th edn. Dearborn, La Crosse

Fuerst F, Marcato G (2009) Style analysis in real estate markets: beyond the sector and region dichotomy. J Portfolio Manage 35(5):104–117

Garmaise MJ, Moskowitz TJ (2004) Confronting information asymmetries: evidence from real estate markets. Rev Financ Stud 17(2):405–437
Gatti S, Chiarella C (eds) (2020) Disruption in the infrastructure sector. Challenges and opportunities for developers, investors and asset managers. Springer, Cham
Geltner D (1993) Estimating market values from appraised values without assuming an efficient market. J Real Estate Res 8(3):325–345
Geltner D, Fisher J (2007) Pricing and index considerations in commercial real estate derivatives. J Portfolio Manage 33(5):99–118
Geltner D, Miller N, Clayton J, Eicholtz P (2014) Commercial real estate: analysis and investments, 3rd edn. On Course Learning, Mason
Goetzmann W, Valaitis E (2006) Simulating real estate in the investment portfolio: model uncertainty and inflation hedging. Yale ICF Working Paper, n. 06-04
Gyourko J (2004) Real estate returns in the public and private markets: a reexamination following the rise of equity REITs. University of Pennsylvania Working Paper, n. 447
Gyourko J, Keim D (1992) What does the stock market tell us about real estate returns. Real Estate Econ 20(3):457–485
Jowsey E (2011) Real estate economics. Palgrave Macmillan, Basingstoke
Lancaster KJ (1966) A new approach to consumer theory. J Polit Econ 74(2):132–157
Lin Z, Vandell K (2007) Illiquidity and pricing biases in the real estate market. Real Estate Econ 35(3):291–330
Mueller GR (1999) Real estate rental growth rates at different points in the physical market cycle. J Real Estate Res 18(1):131–150
Myer F, Webb J (2000) Management styles of REIT funds. J Real Estate Portfolio Manage 6(4): 339–348
Newell G, Sieracki K (eds) (2010) Global trends in real estate finance. Wiley-Blackwell, Chichester
Rosen S (1974) Hedonic prices and implicit markets: product differentiation in pure competition. J Polit Econ 82(1):34–55
Royal Institution of Chartered Surveyors (2021) RICS valuation—Global standards 2022. Royal Institution of Chartered Surveyors, London
Sotelo R, McGreal S (eds) (2013) Real estate investment trusts in Europe: evolution, regulation, and opportunities for growth. Springer, Heidelberg
Syz JM (2008) Property derivatives: pricing, hedging and applications. Wiley, Chichester
Weber B, Staub-Bisang M, Alfen HW (2016) Infrastructure as an asset class. Investment strategy, sustainability, project finance and PPP, 2nd edn. Wiley
Yang J, Zhou Y, Leung WK (2012) Asymmetric correlation and volatility dynamics among stock, bond, and securitized real estate markets. J Real Estate Financ Econ 45(2):491–521

Chapter 17
Commodities

Guido Abate

17.1 Commodity Features

The term commodities indicates all those real goods denoted by maximum fungibility, meaning they are fully interchangeable and can be identified on the basis of a limited number of specific characteristics. As a result of their standardisation, commodities can be traded on regulated markets, eliminating the direct and unique relationship between two parties that is usually found in the trading of real assets.[1]

17.1.1 Classification of Commodities

Commodities may either be produced directly by the primary sector (in this case, we speak of raw materials) or be the result of transformation processes. They can be classified according to the production sector from which they derive. First, there are the products from the agricultural sector (soft commodities) and the mineral and energy sectors (hard commodities). Soft commodities are, in turn, divided into agricultural products and livestock. Hard commodities include three macro-categories: metals, energy and chemicals. Figure 17.1 shows in detail the subdivision of commodities into the various categories.

This taxonomy is based on the qualitative aspects of commodities, but in order to understand the price trend of these goods, other features must also be considered.

[1] Baker et al. (2018), Radetzki and Wårell (2021).

G. Abate (✉)
Department of Economics and Management, University of Brescia, Brescia, Italy
e-mail: guido.abate@unibs.it

I. Basile, P. Ferrari (eds.), *Asset Management and Institutional Investors*, Contributions to Finance and Accounting,
https://doi.org/10.1007/978-3-031-59819-7_17

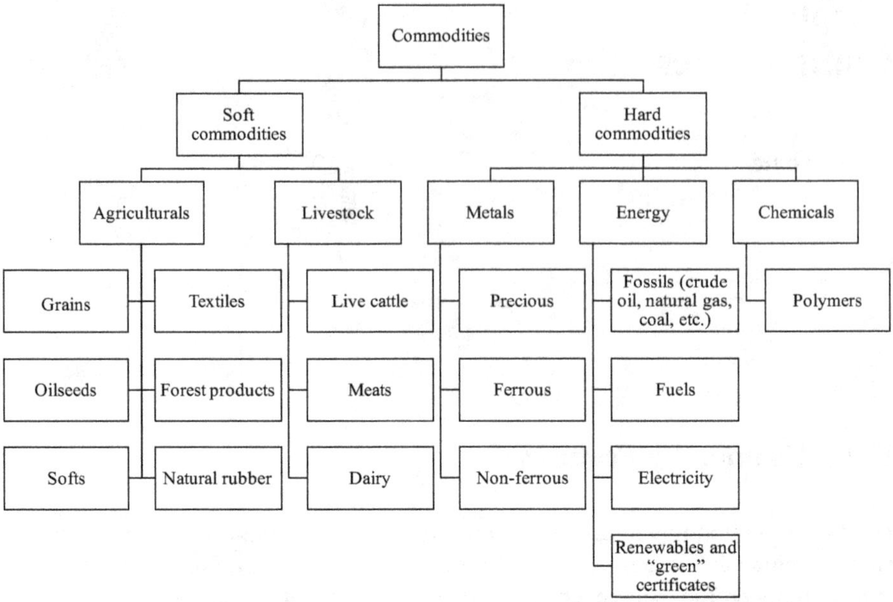

Fig. 17.1 The classification of commodities

Commodities can be distinguished by whether or not they can be stored. Only very few commodities, e.g. electrical energy, cannot be stored over time, whilst for those that can, the relative cost of storage and any seasonality in production and/or consumption are of great importance.[2] For example, industrial metals can be stored indefinitely at a limited cost and are not subject to seasonality. Fossils (like natural gas), on the other hand, are subject to varying seasonal consumption according to temperature and demand for electricity. Agricultural products are only produced at certain times of the year, but consumed constantly.[3]

17.1.2 Commodity Markets and Futures

Commodity markets have existed since antiquity, but have only been organised and regulated since the mid-1800s.[4] Commodity contracts are always settled by physical delivery and the settlement day may be set a few days after stipulation of the contract

[2]Geman (2008).

[3]Geman (2015).

[4]The first modern organised market was the Chicago Board of Trade (CBOT), opened in 1848 and still one of the main world markets. The Futures Industry Association (FIA) provides on its website the ETD Tracker monthly report, which provides data on the volume and open interest of derivatives contracts, broken down by market and by type of underlying, including in particular

(spot) or at a future date. In the latter case, the contract is called a forward or, if standardised and traded on a regulated market, a futures.

The futures with the closest expiry is called front-month or nearby, and is usually indicated on markets by the letter *M* after its name. Subsequent monthly expiries are identified progressively with $M + 1$, $M + 2$ etc. When the front-month contract expires, the names of the following contracts are reset and a new contract for the latest maturity is added.

The theoretical price of a commodity futures, F_T, can be determined by the interaction of different factors that constitute the cost of carry, i.e. the difference between the futures price and the spot price, S_t. The price of a futures on an asset that does not generate cash-flows prior to expiry, *T*, would simply be equal to the spot price of the underlying asset and the interested calculated at the risk-free rate in continuous time:

$$F_T = S_t \cdot e^{r(T-t)}$$

However, unlike financial assets, the transport and storage of commodities generate costs for the purchaser. Moreover, as commodities are not pure capital goods, but are used as production factors by part of the real economy, it is possible to measure their convenience yield, i.e. the benefit derived from the physical possession of the good. This benefit is absent, instead, when the economic agent holds a mere derivative. In addition to their industrial applications, precious metals such as gold, silver and platinum can generate income if they are loaned to operators who require these goods. In this case, the owner of the metal could earn interest from the loan.

Thus, if *c* is the cost of transport and storage and *y* the convenience yield, the formula to calculate the theoretical price of commodity futures becomes:[5]

$$F_T = S_t \cdot e^{(r+c-y)(T-t)}$$

In actual market practice, however, this equivalence is not always given. This phenomenon derives, first of all, from the asymmetry of the arbitrage positions that can be implemented. Whilst it is always possible to develop a strategy based on a long investment in the commodity, short-selling is almost impossible, except for precious metals that can be loaned. For example, let us assume that, on the London Metal Exchange, tin has a spot price of $ 20,000 per metric ton, the risk-free rate is 2% per year, the cost of storage is 0.5%, and the convenience yield is 3%. With these figures, the theoretical price of the 6-month futures would be:

commodities. This makes it possible to identify the major organized markets specializing in this asset class.

[5] Hull (2022).

$$F_{6m} = 20,000 \cdot e^{(2\% + 0.5\% - 3\%) \cdot 6/12} = \$ \ 19,950.06$$

If, on the other hand, the futures price on the LME were \$ 20,100, an arbitrageur could sell the futures and buy the assets at the spot price, borrowing the funds needed for the acquisition for 6 months at the risk-free rate. The profit on expiry of the futures would be \$ 149.94, i.e. the difference between the futures price (\$ 20,100) and the theoretical price, assuming that the cost of storage and the convenience yield for the arbitrageur were the same as the market average.

In contrast, if the opposite had occurred, i.e. $F_{6m} < S$, then the arbitrageur would have bought the futures and sold short the tin on the spot market. However, as there are no operators specialised in the loan of commodities (except for precious metals), this operation cannot be implemented and so results in asymmetrical pressure on the prices of derivatives.

The presence of futures contracts with different maturities allows us to trace the so-called commodity curve, which shows the prices at each expiry date. For those commodities with little or no seasonality, the curve generally presents a monotonous trend that may be upwards or downwards. In the first case, the futures with later expiry dates are priced higher than those with less time to expiry, and the curve is said to be in contango (Fig. 17.2), whilst in the second, the price trend is the opposite, and the curve is termed in backwardation (Fig. 17.3). In a situation of backwardation, the convenience yield, y, of a commodity is such as to compensate the storage cost, c, and the cost of financing a spot purchase, which is approximated

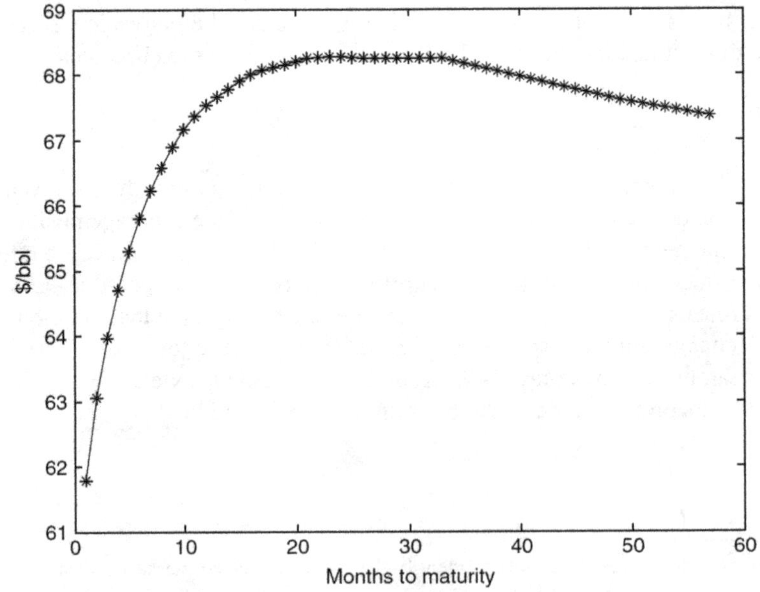

Fig. 17.2 Contango (crude oil curve). Source: Geman (2008)

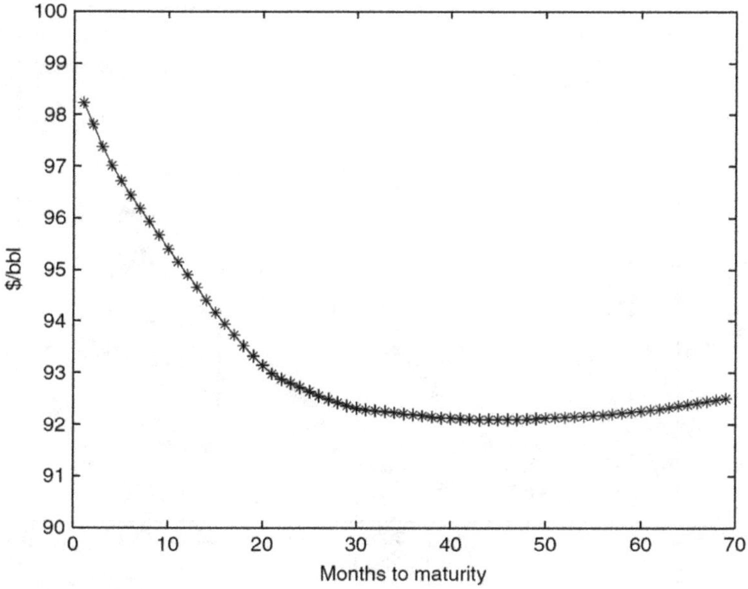

Fig. 17.3 Backwardation (crude oil curve). Source: Geman (2008)

by the risk-free rate, r. This configuration may occur, for example when the market predicts a scarcity or unstable availability of the underlying real asset.

Given the asymmetry of arbitrage operations, the backwardation phenomenon is theoretically unlimited, whilst it is possible to calculate the so-called contango limit, i.e. the maximum no-arbitrage price that generic futures with expiry T may reach assuming a convenience yield of zero:[6]

$$F_T \leq S_t \cdot e^{(r+c)(T-t)}$$

The curves for commodities with marked seasonality in either supply or demand are completely different. Transport and storage costs may be substantially higher or lower than the convenience yield depending on the month of expiry. Consequently, a sinusoidal curve will result (Fig. 17.4), which may show an upwards or downwards trend. Removing seasonality, the first case implies the presence of contango and the second that of backwardation.

[6]Roncoroni et al. (2015).

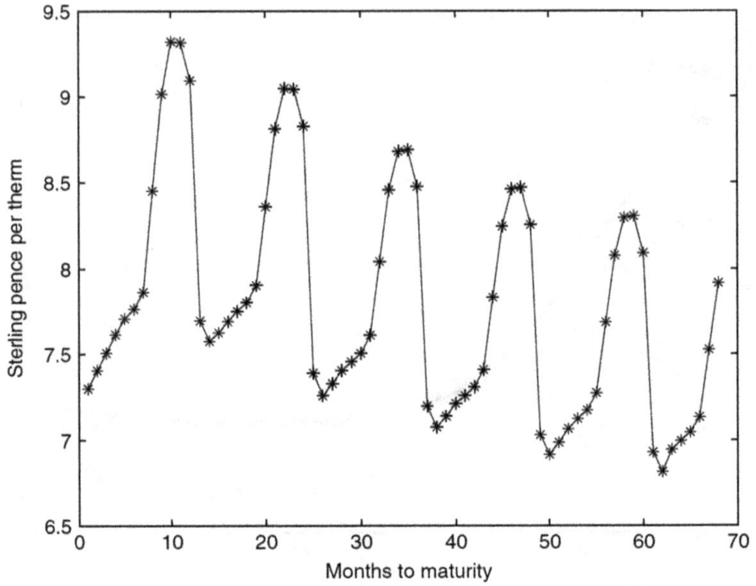

Fig. 17.4 Natural gas curve with seasonality phenomena and backwardation. Source: Geman (2008)

17.2 Risk and Return of Commodities

Whilst the risk-free rate and the cost of transport and storage can be measured relatively easily, the calculation of the convenience yield presents greater difficulty in defining the price of commodities and the shape of the term structure. These prices can be obtained using fundamental or quantitative models. The former assesses the impact of the opposing forces of supply and demand, whilst the latter involves the use of stochastic differential equations to model the price performance over time using empirically measured parameters.[7]

One of the fundamental models is the Theory of Storage, which can be applied to all storable commodities. The basis of this theoretical approach was established by Working and Kaldor in the 1930s,[8] but research has developed significantly in subsequent decades, even though the scientific community has not yet reached a generally recognised model. The common feature of the different formulations of the Theory of Storage is the hypothesis that the prices of commodities are significantly influenced by the volumes stored (which themselves depend on the temporal lags in supply and demand), and that the convenience yield can be estimated on the basis of these volumes and their trends. Scheinkman and Schectman were the first to

[7] Geman (2008).

[8] Working (1933) and Kaldor (1939).

highlight the link between this theory and the rational expectations model.[9] According to the authors' approach, economic agents allocate production of a commodity between immediate consumption at time-point t and storage (equivalent to delayed consumption at time-point T), basing their decision on information available at t on the quantities produced and stored. Given the asymmetry of stock volumes, restricted by non-negativity, if there is strong spot demand for a commodity, stockpiling would be irrational and the market would 'punish' accumulation by paying forward prices that would not compensate the cost of storage. The futures price curve would, therefore, change from contango to backwardation.

Geman and Smith have measured the relationship between prices and stock volumes empirically.[10] In order to simplify the demonstration and calculation, the yield is represented as a percentage spread, shown as ψ, in the discrete time between t and T:[11]

$$\psi = \frac{F_T - S_t e^{(r+c)(T-t)}}{S_t}$$

In other words, ψ indicates the discount between futures and spot prices: the greater the level of backwardation of the futures curve, the greater the utility derived from the physical possession of the commodity, whose spot price is indeed proportionately higher than the futures price.

The analysis was conducted on a time series of the storage volumes of the six main non-ferrous metals stored in the London Metal Exchange (LME) approved warehouses and measured not in tonnes but rather in days of world consumption, in order to standardise the differing volumes. For example, if 4620 t of aluminium are stored and the daily consumption is 110 t, the stock equals 42 days of demand and this time period is used as the unit of measure of the volumes in the LME warehouses. By applying a simple curve-fitting algorithm to the graphical representation of the sample used, we obtain the Working curves that illustrate the relationship between the spreads (ψ) of the spot and futures prices of the commodities and the volumes stored.

The authors have found a statistically significant inverse relationship between stocks and spreads, as shown in Fig. 17.5.[12] This relationship implies that spot prices rise strongly with reduced stock volumes, in particular, when volumes are below 10 days of world consumption.

Despite the correlation between stockpiled quantities and prices, the models based on the Theory of Storage may not be empirically valid under anomalous market conditions. A significant case occurred in 2005–2006 when the prices of

[9] Scheinkman and Schechtman (1983).

[10] Geman and Smith (2013).

[11] The decision to abandon a continuous time model in favour of a discrete time model is based exclusively on the greater computational simplicity of the latter.

[12] Gorton et al. (2013) also observed the same phenomenon.

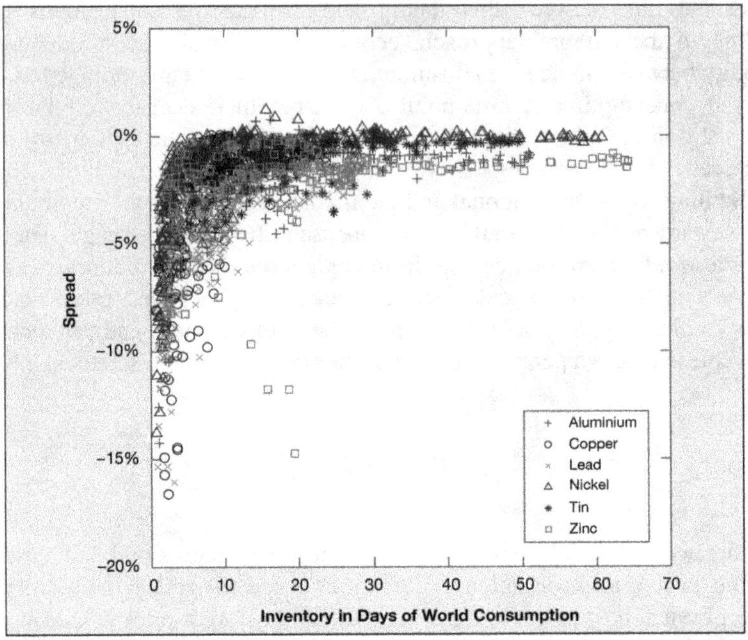

Fig. 17.5 Relationship between spread and stored volumes for the six main non-ferrous metals. Source: Geman and Smith (2013)

energy commodities and their stocks rose simultaneously. The reasons underlying this phenomenon reside in the volatility of supply and/or demand. When operators perceive a greater level of risk, they tend to increase stored volumes and accept higher prices, given the perceived scarcity of the goods. These aspects have been modelled by Pirrong in a modified version of the Theory of Storage that also takes account of stochastic shocks on the demand side.[13]

The volatility of commodity returns can also be modelled in the context of the common Theory of Storage. The stored volumes have a stabilising effect on price trends, absorbing the impact of excessive supply or demand; yet, when stocks are depleted or exhausted, prices can experience sudden fluctuations. Geman and Smith have demonstrated the validity of the theory empirically, revealing an inverse relationship between stock volumes and the standard deviation of returns, even though the statistical significance is less marked in this case than that found for prices.

Although commodities are subject to high volatility, studies undertaken over different time periods have reached conflicting conclusions. Measurements by Kat and Oomen[14] place commodities on the same level of volatility as equity returns,

---

[13] Pirrong (2011).

[14] Kat and Oomen (2007a).

whilst Geman[15] finds their standard deviation to be greater than that of traditional asset classes. However, the same studies agree that there is a significant dispersion in the volatility of different commodities, underlining that this asset class is particularly heterogeneous.

The volatility of commodity futures increases as maturity approaches. This phenomenon, known as the 'Samuelson effect', may be caused by the stronger influence of news regarding trends in supply, demand and stocks on contracts that are due to expire in the near future, whilst operators working with futures with a longer residual life might assume that the imbalances will be re-absorbed prior to expiry of the contracts.[16]

The scientific literature has proposed other models to explain trends in the prices (and thus the returns) of commodities.[17] The hedging view is linked to the Keynesian hypothesis regarding the existence of normal backwardation, according to which backwardation should be the usual conformation of the commodity forward curves.[18] According to Keynes, hedgers would systematically short-sell commodity futures in order to cover the risk of unexpected price contractions for an asset of which they hold large volumes and in which, therefore, they have a long position. Prevalent insurance-related use of futures would consequently result in a premium paid by hedgers for this protection. In a market in backwardation, this premium would consist of the increase shown by the price of the futures, close to its expiry T, converging towards the spot price S_T.

However, this assessment has not been proven empirically, as it is based on the limiting hypothesis that the operators interested in hedging are for the main part sellers of commodities.[19] To overcome this asymmetry, the hedging pressure hypothesis has been proposed. In this view, the importance of the presence of hedgers can vary based on the specific commodity.[20] Depending on whether the operators most interested in hedging are the producers/sellers or the purchasers of the commodity, the forward curve would be either in backwardation (in agreement with the hedging view) or in contango.

17.2.1 Return Decomposition of Commodity Futures

Financial investments in commodities are mainly affected by trends in futures prices rather than spot prices, as with spot contracts it is impossible to avoid the physical delivery of the underlying asset and thus they are targeted only at industrial and

[15] Geman (2005).
[16] Samuelson (1965).
[17] Erb and Harvey (2006).
[18] Keynes (1930).
[19] Kolb (1992).
[20] Cootner (1960) and Deaves and Krinsky (1995).

Table 17.1 Roll return on commodity futures

Futures positions	Contango	Backwardation
Long	Roll return < 0	Roll return > 0
Short	Roll return > 0	Roll return < 0

commercial traders. It is, therefore, useful to analyse in detail the components of futures' returns.

When futures contracts come closer to their expiry date, financial operators close the existing position and simultaneously open a new one on futures expiring at a later date. This operation is called rollover, and given the difference in price between the expiring futures and the contract expiring at a later date, it will generate a profit or a loss for the investor, which is known as roll return (or roll yield).

In particular, when the commodity forward curve shows backwardation, an operator holding long positions gains a positive return, because futures with longer residual lives have progressively lower prices. In contrast, a curve in contango indicates the opposite effect: a reduction in returns for operators holding long positions (Table 17.1).

Note that an analysis of the period 1982–2004 has shown that, historically at least, roll return has explained 91.6% of the differences in the returns of the main commodities.[21] In terms of volatility, on the other hand, the same empirical study has shown that the variation over time in spot price is a determining factor in the riskiness of each commodity futures.

In order to break down the total return of an investment in commodity futures, we need to introduce two simple assumptions:

- Rolling occurs exactly at the expiry date of the futures.
- The position in the derivative is fully collateralised, i.e. the notional of the futures is fully invested at the risk-free rate. One portion is assigned to the central counterparty as a margin, whilst the rest is set aside by the investor, thereby eliminating any financial leverage.[22]

Given the above assumptions, we can demonstrate that the total return (r_{Tot}) of an investment in commodity futures over the period from t to T is the sum of three components, i.e. the spot return, the roll return and the collateral return:[23]

[21] Erb and Harvey (2006). Subsequent studies concentrated on the period 1991–2010 have confirmed these results: Kaplan (2010).

[22] This second working hypothesis originates both from the theoretical decision to evaluate an investment position that is comparable to those in other asset classes represented by securities, rather than derivatives, and from the practical need to specify just one type of leverage (in this case equal to 1) that is identical for all commodities and regardless, therefore, of the margin rules set forth in each futures market.

[23] Kat and Oomen (2007a) and Fabozzi et al. (2008).

$$r_{Tot} = r_S + r_R + r_C$$

1. Spot return: despite the name, it is important to underline that spot return refers to the price change of the futures closest to its expiry and not of a spot commodity. Consequently, it is calculated as: $r_S = (F_T - F_t)/F_t$. Indeed, the presence of storage and transport costs, as well as the convenience yield, would make the calculation of the spot price indices unrealistic, as they do not include these factors.
2. Roll return: is the difference between the price of a futures with an expiry T and that of the following futures with expiry Θ on the rollover date: $r_R = (F_T - F_\Theta)/F_T$. However, since on expiry the price of a futures is identical to that of the underlying asset (i.e. $F_T = S_T$), the roll return is equivalent to: $r_R = (S_T - F_\Theta)/S_T$.
3. Collateral return: is the return on the equivalent of fully collateralised futures which, by assumption, is invested in a risk-free asset: $r_C = r_F \cdot (T - t)$.

The sum of the spot and roll returns is also known as the excess return and represents the return of a commodity futures position irrespective of the return of the collateral and of the leverage used.

There are indices that reproduce spot, excess and total return trends, whilst the roll return cannot be replicated independently, as it is only part of a futures strategy and, therefore, is not calculated by any index provider. In particular, we can recall the Standard & Poor's GSCI family of indices, previously known as the Goldman Sachs Commodity Indices. The weight attributed to each commodity is proportional to its yearly production: this index, therefore, is calculated using production weighting. Figure 17.6 shows the performance over time of the three versions of the S&P GSCI, which currently comprises a representative basket of futures contracts on the 24 main commodities.[24] Note that the collateral return (i.e. risk-free rate) is a very important component in the total return. Whilst this index has an average annual return of 9.6%, the excess return version has increased on average only 4.5% per year, with an implicit collateral return of 5.1%. At the same time, the roll return generates a variable contribution over time, as shown in Fig. 17.7, which shows the cumulative roll return, calculated as the differential in the return between the GSCI Excess Return and the GSCI Spot indices (on average −0.6% per year): when the average performance of the futures in the basket showed backwardation, it was positive; in periods of contango, it was negative.

[24] The number of commodities in the index has grown with time. Note, in particular, that the first energy commodity was added in December 1982, bringing the total in the basket to 12 commodities at that time.

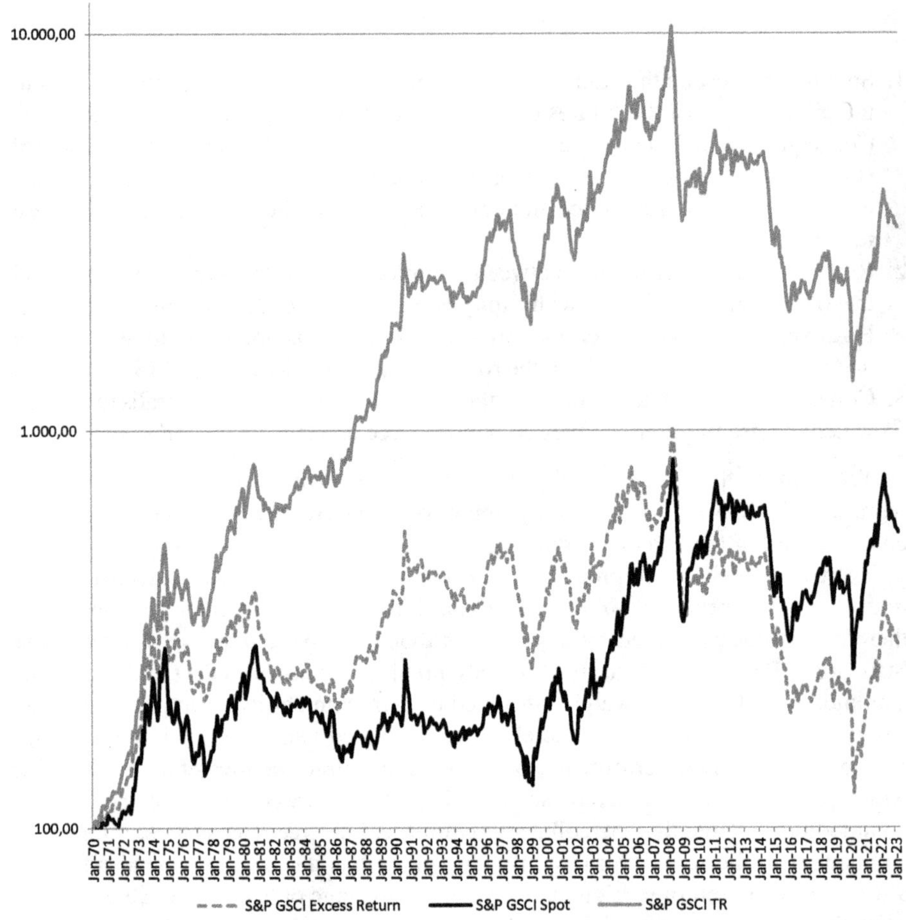

Fig. 17.6 Performance of the S&P GSCI index in the spot, excess and total return versions. Base 100 at 31/01/1970, logarithmic scale. Source: Data from Morningstar Direct

17.3 Commodities Within the Financial Assets Portfolio

The structural characteristics of the indices representing the commodity asset class have assumed considerable importance (significantly greater than in the other asset classes), and they can influence decisions regarding the weight allocated to commodities in a portfolio of financial assets. Besides the different options with respect to the type of return followed by the index, the weighting technique used is crucial. Figure 17.8 shows the position of different indices—US and EMU Equity, US Bonds and Commodities—on the mean-standard deviation plane. Note that the risk-adjusted performance profiles of traditional asset classes are the same, even if they are measured with indices provided by different sources, whilst the commodity

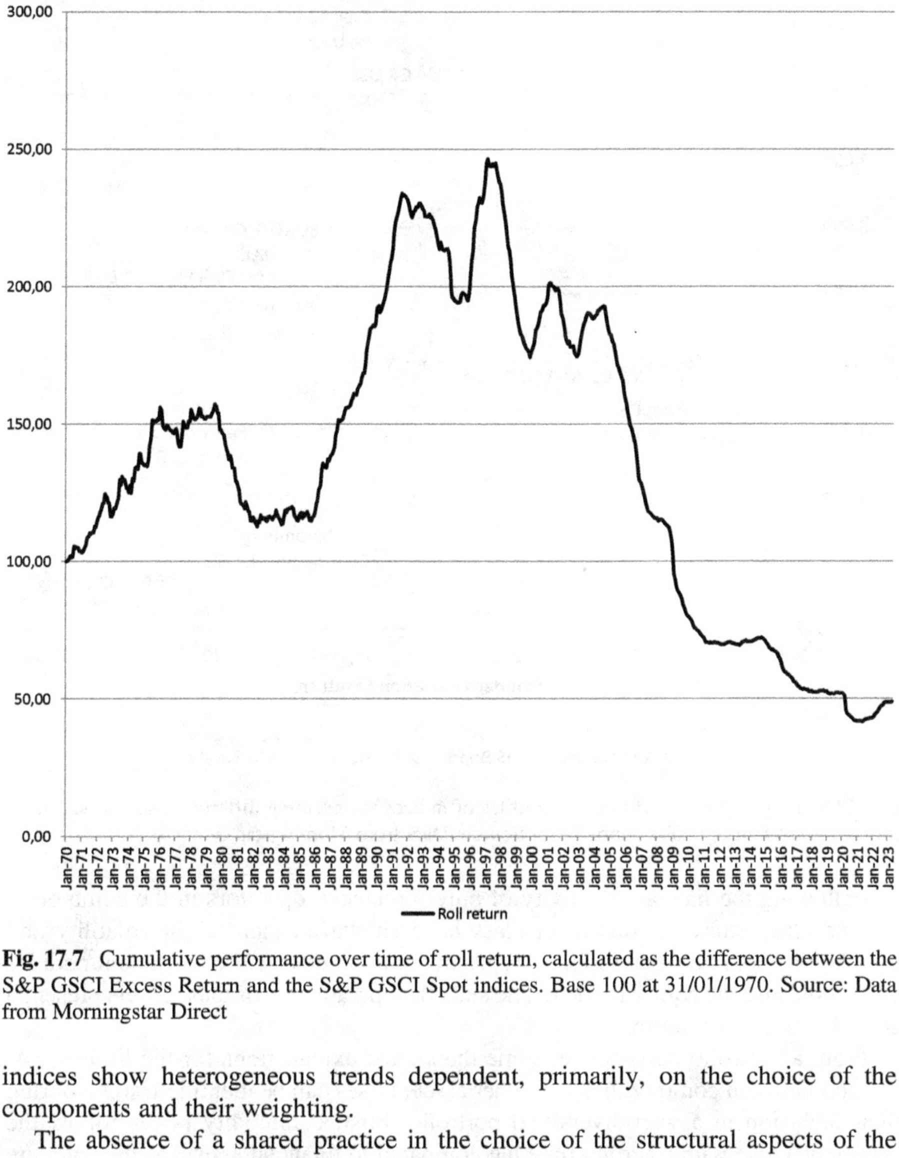

Fig. 17.7 Cumulative performance over time of roll return, calculated as the difference between the S&P GSCI Excess Return and the S&P GSCI Spot indices. Base 100 at 31/01/1970. Source: Data from Morningstar Direct

indices show heterogeneous trends dependent, primarily, on the choice of the components and their weighting.

The absence of a shared practice in the choice of the structural aspects of the commodity indices is exacerbated by the statistical properties of the commodity futures returns. Empirical analyses have highlighted a limited correlation between the different kinds of commodities.[25] The differing trends in returns result in particularly significant effects on the benchmarks representing the commodity market, as different baskets show clearly distinct results in terms of risk-adjusted performance.

[25] Erb and Harvey (2006) and Gorton and Rouwenhorst (2006).

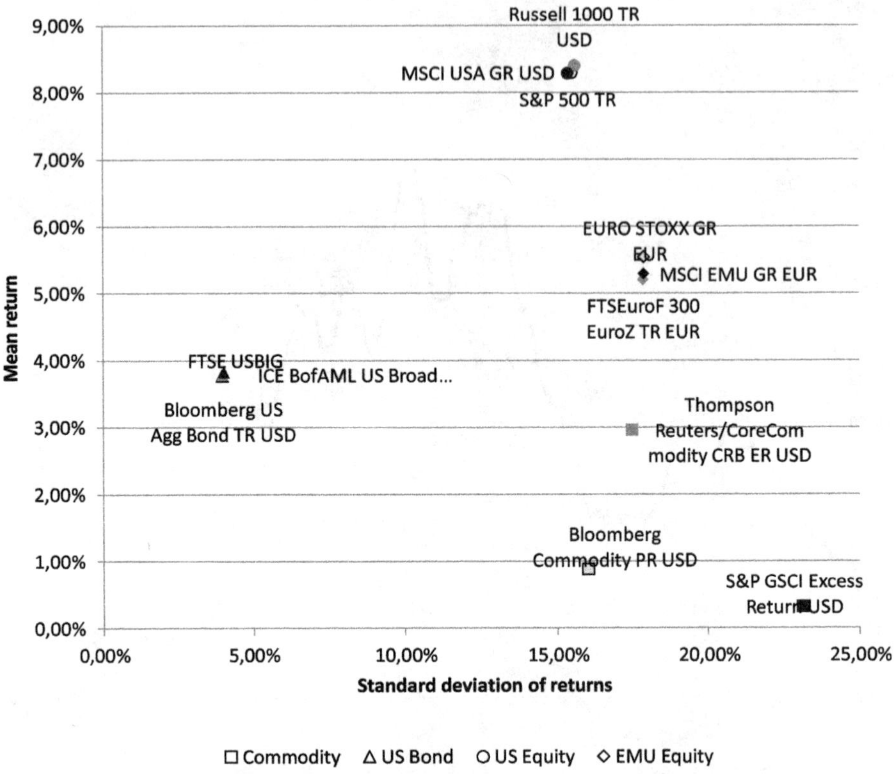

Fig. 17.8 Risk-adjusted performance profiles of indices representing different asset classes. Data for the period January 2001–April 2023. Source: Data from Morningstar Direct

Following the increased activity of purely financial operators in the commodity markets, the returns of this asset class have shown an increase in volatility and correlations both between different types of commodities and between commodities as a whole and the equity market. The statistical parameters of these assets present a so-called time-varying trend.[26]

From a portfolio perspective, some theoretical explanations for the limited correlation between commodities and other asset classes can be identified, so favouring their inclusion in a well-diversified portfolio. First, commodity prices follow the economic cycle with a certain time lag compared to financial assets, as the latter are priced by rational agents on the basis of the expectations of future income and not, as in the case of commodities, of their relative scarcity. Furthermore, shocks that are exogenous to the financial system, such as, for example geopolitical crises or adverse climatic events, have a positive impact on commodity prices, unlike their effect on the equity and bond asset classes. Finally, the profitability of capital, which is,

[26] Rouwenhorst and Tang (2012).

ultimately, the underlying driver of an increase in the value of equity and of the solvency of bond issuers, is negatively correlated to the cost of the factors of production, amongst which commodities play, directly or indirectly, a key role.

The inflation hedge property attributed to commodities differentiates this class from other asset classes and is a further justification for its inclusion in a portfolio of financial assets. In fact, a clarification is indeed necessary. The relationship between inflation and the commodity market depends directly on the weight of the commodities in the basket of goods and services measured by the consumer price index. However, given that in an efficient market the price of every asset should already discount expected inflation, we have to concentrate on unexpected inflation in order to assess whether commodities really offer protection for the investor. Assuming that the best forecast of expected inflation is current inflation, the unexpected component is equal to its variation in the time period from t to T. Empirical analyses of the US market have shown that some commodities, particularly those in the energy and industrial metals sectors, can act as an unexpected inflation hedge, therefore confirming the value of including this asset class in a diversified portfolio.[27]

17.4 Investment Vehicles in Commodities

Investors can access the commodity market by means of vehicles with different and specific features. Starting with the least distance and abstraction between the operator and the commodity, there is a direct, physical investment. However, given its administrative complexity, the entry barriers and the economies of scale involved, this technique is suitable only for professional operators. In particular, the entry barriers can be both economic and regulatory: the former case refers to the substantial investments that may be necessary to equip and manage a dedicated commodity warehouse, whilst the latter case concerns access to specialised markets and their associated storage facilities, which is allowed only to selected professional operators.

Leaving aside, then, direct investment, which plays an absolutely marginal role in commodities as an asset class, it is useful to consider indirect or financial investment. The latter can be realised with various financial instruments, including, in particular:

- Shares of companies operating in the commodity sector.
- Commodity derivatives.
- Commodity Exchange-Traded Products (ETP).

[27] Erb and Harvey (2006) and Kat and Oomen (2007b). However, the same studies reveal that precious metals do not offer any protection from unexpected inflation. On the complex relationship between commodities and inflation, see also Futerman and Sarjanovic (2022).

17.4.1 Commodity Stocks

When selecting a company as a target for investment, it is first necessary to establish which position to select in the supply chain of a specific commodity. An operator may focus on firms producing the goods, on those working in distribution, or again, on those selling to an industrial or retail end-user. In actual fact, the various phases of the commodities transformation process are often performed by the same economic agent, making it difficult to identify a company on the basis of a specific activity. Moreover, companies are subject to a series of idiosyncratic risks linked to factors such as their financial structure or management quality, so that the correlation between the prices of their shares and that of the commodities with which they work is not necessarily significant. On the contrary, empirical studies have calculated that, on average, the share prices of companies active in this asset class have a beta, measured against the stock market, not dissimilar to that of other listed companies.[28]

17.4.2 Commodity Derivatives

Commodity derivatives have developed over the course of the last decades such as to allow a very high level of customisation, coupling simple plain vanilla tools with multiple, exotic contracts traded over the counter or on highly specialised, regulated markets reserved exclusively for professional operators.[29] Note that, given their greater liquidity and the possibility to avoid physical delivery of the asset by rolling positions, the prices of main futures have assumed the role of a primary source of information regarding the respective, underlying commodities, replacing spot prices on the commodity markets and often serving as an underlying asset for further derivatives, such as options. If we exclude this particular informative function, commodity derivatives do not present any specific feature that significantly distinguishes them from the better-known financial derivatives; consequently, we deem it is unnecessary to discuss these instruments in further detail,[30] whilst the characteristics of futures have already been covered in Sect. 17.2.1.

In order to diversify a portfolio of commodity derivatives, an investor can follow an active approach using managed futures accounts administered by Commodity Trading Advisors (CTAs), i.e. professional operators registered with the US supervisory authorities.[31] Initially, the CTAs operated only in commodity futures, but now their portfolios cover different asset classes and derivatives, even though most

[28]Geman (2005).

[29]Schofield (2021).

[30]For more details see Fanelli (2020) and Schofield (2021).

[31]Specifically, they must be registered with the Commodity Futures Trading Commission and be members of the National Futures Association.

operators are still specialised in commodities. The services provided by the CTAs can be classified according to the level of customisation:[32]

- With individually managed accounts, the CTA manages the portfolio of an institutional investor.
- In private commodity pools, management is targeted at a restricted group of subjects composed of institutional investors or high net-worth individuals.
- Public commodity funds are comparable to common investment funds and are targeted at retail investors.

The quantification of the value added resulting from active CTA management compared to passive approaches has been at the centre of extensive debate in the scientific community. The strongly dynamic nature of CTA operations, which are always open to new techniques and have been influenced by the spread of ever more sophisticated automated systems, has led to varying conclusions depending on the chosen timeframe and the sample used. As a result, a consensus amongst scholars has not been reached.[33]

17.4.3 Commodity Exchange-Traded Products

The ETPs group into a single category three financial instruments with distinctly different characteristics: commodity Exchange-Traded Funds (commodity ETFs), Exchange-Traded Commodities (ETCs) and commodity Exchange-Traded Notes (commodity ETNs).

Commodity ETFs, like all ETFs, are listed investment vehicles that have the form of open-end mutual funds and aim to replicate the trend of a commodity index. Only indices that comply with the property of investibility can be used as benchmarks for ETFs. If they conform to the European UCITS directives, ETF managers cannot invest directly in physical commodities and can implement replication only through derivatives with the constraint of not exposing more than 20% of their NAV to the same financial instrument.[34] Since, however, replicable indices are, in turn, composed of diversified portfolios of futures, these legal restrictions do not pose any particular problems to the operability of ETFs on such baskets of commodities.

Unlike ETFs, ETCs are typical financial instruments of this asset class. From a legal point of view, they are zero coupons, non-subordinated, perpetual securities issued by a Special Purpose Vehicle (SPV) against an investment in the underlying commodities. The price of the ETC depends directly on the price performance of a

[32]Gregoriou et al. (2004).

[33]Schneeweis et al. (2013).

[34]Directive UCITS 85/611/EEC, Articles 19 and 22. Although the UCITS directive was issued by the European Union, UCITS funds are now regarded globally as efficiently regulated funds and are accepted for sale in many countries outside the European Union.

basket of assets held by the SPV according to the proportion ratio known as entitlement, which is reduced daily, following the charging of the management commissions. For example, at time t, an ETC on copper has an entitlement of 0.01, and the asset is worth \$ 7000 per metric ton. Therefore, the price of the ETC is $\$ 7000 \times 0.01 = \$ 70$. At $t + 1$, the management commissions have reduced the entitlement to 0.0098, whilst the value of copper is \$ 7100 per metric ton; thus, the price of the ETC is now $\$ 7100 \times 0.0098 = \$ 69.58$.

An investment using ETC can occur either directly, through the purchase and subsequent storage of the real asset—in this case, the ETC is called physically backed—or indirectly, via derivatives that have the same underlying asset as the ETC, which is termed synthetic. Direct purchase of commodities usually occurs in cases in which the raw material in question is not perishable and has a unit value such as to limit storage costs, e.g. precious metals.

As they are not mutual funds, ETCs are not included amongst the UCITS-compliant instruments and are therefore not subject to the allocation restrictions typical of the latter. This freedom of action allows the ETCs to follow the performance not only of diversified portfolios, but also of single commodities.

To ensure a correct alignment of the price with the underlying asset and so implement the principle of no-arbitrage, despite the fact that these are perpetual securities, the ETCs allow specialised operators to request early redemption. Furthermore, issuing is continuous and not limited to the initial placement on the market. The ETC creation/redemption process may envisage the possibility—exclusively for authorised participants operating on the primary market—to contribute/receive the commodities or the underlying derivatives. However, unlike with ETFs, this procedure, called creation/redemption in kind, is not the most widespread amongst ETCs, where cash creation/redemption involving only the exchange of money in return for the ETC securities is preferred.

The techniques used to issue the different ETCs on the primary market vary according to the replication strategy implemented. The following process exemplifies the most common practices used in cash creation/redemption:

1. An authorised participant, often the market maker of the ETC on the secondary market, transfers a sum of money to the SPV.
2. The SPV acts differently depending on the chosen replication strategy:
   (a) In the case of physical replication, the SPV directly purchases a quantity of commodities equivalent to the sum received and deposits it with a specialised company that ensures correct storage.
   (b) With synthetic replication, the SPV pays the sum into a time deposit with the depositary bank and then uses it as a guarantee to stipulate commodity derivatives with a specialised dealer.
3. The SPV issues ETCs for an amount equivalent to the sum received at point 1, which are delivered to the authorised participant.

Commodity Exchange-Traded Notes (ETNs) are zero coupon, medium/long-term securities (typically with a maturity between 10 and 30 years), the price of which, both on the market and at maturity, depends on the performance of the underlying assets. Opportunities for arbitrage are averted by the presence of early redemption clauses and the practice (usually not explicitly stated in the prospectus) of allowing the issue of new ETNs, in both cases by cash payment, for an amount equal to the value of the underlying asset, between the issuer and an authorised participant. On the other hand, early redemption is not immediate and may take over a week.

The distinction between ETCs and ETNs is often a source of confusion and therefore must be underlined. ETCs are securities of a SPV set up with the sole aim of issuing these financial instruments, and whose capital is exposed only to the performance of the asset via a physical or synthetic investment. In contrast, issuers of ETNs, generally investment banks, do not allocate any of their own assets to guarantee the value of these securities, unless in presence of specific clauses in the issue prospectus. Furthermore, the issuer has no obligation to invest the financial resources collected through the issue of the ETNs either physically or synthetically in commodities.

Having described the main characteristics of ETPs, now it is useful to look at the limits and risks of this form of investment in commodities.

ETPs cannot replicate exactly the price performance of a commodity, as they have the extra burden of management commissions. In addition, with synthetic replication (i.e. for all ETFs and the large majority of ETCs), tracking is carried out using futures contracts, and so the return of these ETPs does not reflect the trend of spot prices. As illustrated in Sect. 17.2.1, a futures position in commodities is subject to roll return and collateral return, which vary according to the level of financial leverage used. On the other hand, rolling of futures positions can be optimised to minimise costs (in contango) or maximise benefits (in backwardation). For example, implementation of this tactic may envisage using rolling at times when the implicit convenience yield is higher than its recent average or may take advantage of the Samuelson effect. In this second case, the greater volatility of futures with an earlier expiry compared to that of those with a later expiry means that the prices of the former will fluctuate more around the equilibrium value, so offering opportunities to exploit arbitrages.

With physical replication, instead, an ETC would not, in any case, be in a position to follow the spot prices of the commodity precisely, as spot prices do not include costs of storage and, depending on contractual agreements, transport. However, the possibility, where provided, to loan physically held commodities may allow the ETC to earn a commission comparable to a form of convenience yield.

If the commodities are not loaned, physical replication annuls all counterparty risks, presenting just a limited operational risk derived from possible fraud or theft of stored material. On the other hand, every synthetic replication exposes the investor to the counterparty risk present between the issuer of the ETF or the ETC and its derivatives dealers, unless these contracts have been stipulated through a central counterparty.[35]

[35] Amenc et al. (2012).

ETCs can, however, present a systemic risk regarding the commodity markets in which they operate. According to the Theory of Storage, the trends in price and volumes stored for sale are inversely proportional. Hoarding of commodities by physically backed ETCs results in their relative scarcity, so causing, all other conditions being equal, an increase in prices. This phenomenon has been measured empirically for precious metals, whilst similar conclusions have not yet been reached for other commodities.[36]

Apparently, the capacity of an ETN to track the trend of the underlying asset should be appreciably greater than that of an ETF or an ETC, given that the price of an ETN is contractually linked to that of the commodities and is therefore not subject to possible transaction costs related to the underlying asset. The only explicit source of tracking error with respect to the commodity is the periodic management fee, deducted from the market price. As ETNs are securities without collateral guarantees, however, their price can be influenced by the issuer's creditworthiness: they are thus subject to counterparty risk. Nevertheless, empirical studies have not revealed any discount, with respect to the underlying asset, in the market price of these financial instruments, probably as a result of the market-making activity undertaken by the issuing banks.[37]

References

Amenc N, Ducoulombier F, Goltz F, Tang L (2012) What are the risks of European ETFs? EDHEC-Risk Institute, Nice

Baker HK, Filbeck G, Harris JH (eds) (2018) Commodities: markets, performance, and strategies. Oxford University Press, Oxford

Cootner P (1960) Returns to speculators: Telser versus Keynes. J Polit Econ 68(4):396–404

Cserna B, Levy A, Wiener Z (2013) Counterparty risk in exchange-traded notes (ETNs). J Fixed Income 22(5):76–101

Deaves R, Krinsky I (1995) Do futures prices for commodities embody risk premiums? J Futur Mark 15(6):637–648

Erb C, Harvey C (2006) The strategic and tactical value of commodities futures. Financ Anal J 62(2):69–97

Fabozzi FJ, Füss R, Kaiser D (2008) The handbook of commodities investing. Wiley, Hoboken

Fanelli V (2020) Financial modelling in commodity markets. Chapman and Hall/CRC, Boca Raton

Fassas A (2012) Exchange-traded products investing and precious metal prices. J Derivatives Hedge Funds 18(2):127–140

Futerman AG, Sarjanovic IA (2022) Commodities as an asset class: essays on inflation, the paradox of gold and the impact of crypto. Palgrave Macmillan, Cham

Geman H (2005) Commodities and commodities derivatives. Wiley, Chichester

Geman H (ed) (2008) Risk management in commodities markets: from shipping to agriculturals and energy. Wiley, Chichester

---

[36] Fassas (2012).

[37] Cserna et al. (2013). Note that holders of ETNs issued by Lehman Brothers sustained a loss of 91% of the value of these instruments following the sudden insolvency of the investment bank in September 2008.

Geman H (2015) Agricultural finance: from crops to land, water and infrastructure. Wiley, Chichester
Geman H, Smith W (2013) Theory of storage, inventory and volatility in the LME base metals. Res Policy 38(1):18–28
Gorton GB, Rouwenhorst KG (2006) Facts and fantasies about commodity futures. Financ Anal J 62(2):47–68
Gorton GB, Hayashi F, Rouwenhorst KG (2013) The fundamentals of commodity futures returns. Eur Finance Rev 17(1):35–105
Gregoriou G, Karavas V, Lhabitant FS, Rouah F (2004) Commodities trading advisors: risk, performance analysis, and selection. Wiley, Hoboken
Hull JC (2022) Options, futures, and other derivatives, 11th edn. Pearson, Hoboken
Kaldor N (1939) Speculation and economic stability. Rev Econ Stud 7(1):1–27
Kaplan PD (2010) The long and short of commodity indexes. J Index Investing 1(1):61–71
Kat H, Oomen R (2007a) What every investor should know about commodities. Part I: Univariate return analysis. J Invest Manage 5(1):4–28
Kat H, Oomen R (2007b) What every investor should know about commodities. Part II: Multivariate return analysis. J Invest Manage 5(3):40–64
Keynes JM (1930) Treatise on money. Macmillan, London
Kolb RW (1992) Is normal backwardation normal? J Futur Mark 12(1):75–91
Pirrong C (2011) Commodities price dynamics: a structural approach. Cambridge University Press, Cambridge
Radetzki M, Wårell L (2021) A handbook of primary commodities in the global economy, 3rd edn. Cambridge University Press, Cambridge
Roncoroni A, Fusai G, Cummins M (2015) Handbook of multi-commodity markets and products: structuring, trading and risk management. Wiley, Chichester
Rouwenhorst KG, Tang K (2012) Commodities investing. Annu Rev Financ Econ 4:447–467
Samuelson P (1965) Proof that properly anticipated prices fluctuate randomly. Indus Manage Rev 6(2):41–49
Scheinkman J, Schechtman J (1983) A simple competitive model with production and storage. Rev Econ Stud 50(3):427–441
Schneeweis T, Spurgin R, Szado E (2013) Managed futures: a composite CTA performance review. J Altern Invest 15(3):32–61
Schofield NC (2021) Commodity derivatives: markets and applications, 2nd edn. Wiley, Chichester
Working H (1933) Price relations between July and September wheat futures at Chicago since 1885. Wheat Stud 9(6):187–238

Chapter 18
Currency Overlay

Guido Abate

18.1 Passive Versus Active Currency Overlay

International diversification of investments exposes portfolios to exchange rate risk, which is a typical speculative risk. In this case, the overall performance of a portfolio is tied not only to the returns of individual investments, but also to variations over time of the exchange rates of the different currencies against the so-called home currency, taken as the base currency by the investor. Given that this is a speculative risk, it can have a positive or negative impact depending on whether the foreign currency strengthens or weakens against the home currency. In the first case, the return is higher than the local return, i.e., the financial return of investments calculated in the foreign currency in which they are denominated, while in the second case, the return is lower than the local return.

As exchange rates are subject to specific factors that are distinct from those of traditional financial assets, currencies can be seen as an autonomous alternative asset class.[1] In its passive version, currency overlay seeks to manage currency risk, in order to limit its potentially negative impact on investments denominated in foreign currencies. In contrast, active currency overlay combines the management both of risk and of the currency asset class, with the aim of increasing overall portfolio performance.

In asset management, a currency overlay strategy occurs when there is a process of risk separation. Decisions regarding the management of exchange rate risk are made independently of other investment choices by a specialist (overlay manager),

[1] Pojarliev and Levich (2014).

G. Abate (✉)
Department of Economics and Management, University of Brescia, Brescia, Italy
e-mail: guido.abate@unibs.it

I. Basile, P. Ferrari (eds.), *Asset Management and Institutional Investors*, Contributions to Finance and Accounting,
https://doi.org/10.1007/978-3-031-59819-7_18

who may be part of or outside the financial asset management team.[2] It is important to note that this separation in decision-making does not necessarily imply independence from the asset allocation policies. The currency overlay is undertaken autonomously, but in its most developed forms is part of portfolio optimization. All the financial tools used to manage currency exchange risk overlap with the positions denominated in foreign currency as if they were an additional level of investment that may be managed passively or actively.

In formal terms, currency return, i.e., the return dependent on changes in the spot exchange rates in the period from t to T, can be defined as follows, where S is the spot rate:[3]

$$s = (S_T - S_t)/S_t$$

If the forward exchange rate fixed today with a trade date of T is added to or subtracted from the numerator of the right-hand side, we obtain:

$$s = (S_T - F_T + F_T - S_t)/S_t$$
$$s = (S_T - F_T)/S_t + (F_T - S_t)/S_t$$

This allows us to break down the currency return into its two components: the currency surprise (given as e) and the forward premium/discount (f):[4]

$$e = (S_T - F_T)/S_t$$
$$f = (F_T - S_t)/S_t$$

Therefore, exchange rate risk derives only from the currency surprise, i.e., the difference between the spot exchange rate at T and the forward exchange rate F_T. On the other hand, the forward premium and the forward exchange rate are known from time t and are calculated on the basis of the theory of covered interest rate parity:[5]

[2]For example, there are companies specialised in currency overlay that offer their services to managers of financial assets. Xin (2011).

[3]The values at times t or T are given in upper case, while their percentage changes from t to T are given in lower case. In line with the literature on currency overlay, but not with operational practice, here exchange rates are calculated according to the direct or price quotation, i.e. indicating the amount of base currency required to purchase a unit of the quoted currency.

[4]Below, in agreement with Ankrim and Hensel (1994) and the main literature, we refer to forward premium, irrespective of whether its value is positive or negative.

[5]Note that, irrespective of the convention used for the exchange rate, in the calculation of the forward premium the numerator must be the interest rate of the quoted currency and the denominator that of the base currency. In this case indirect quotation is used, and therefore the home currency is the variable quantity in the exchange rate.

$$f = \frac{1 + r_{\substack{\text{quoted} \\ \text{currency}}} \cdot (T - t)}{1 + r_{\substack{\text{base} \\ \text{currency}}} \cdot (T - t)} - 1$$

and so:

$$F_T = S_t(1 + f)$$

This analysis considers exchange rates as a risk factor for portfolios and, therefore, part of risk management, albeit applied to asset management. However, there are strategies that can be used to extract value from currency management, as in the case of active currency overlay. Scientific literature shows an almost total absence of any correlation between the returns generated by active currency management and traditional asset classes.[6] While this result was initially attributed entirely to the managers' ability to choose the most remunerative positions on exchange rates, more recent studies have highlighted market factors, specific to the currency asset class and independent of those present in traditional sectors, which can be exploited by active currency overlay techniques.[7] Consequently, currencies constitute an alternative asset class: the performance of currency managers is explained in the main part by the exposure to this class of investment and is no longer wholly attributable to their skills in timing and selection regarding trends in exchange rates.

18.2 Passive Currency Overlay

Passive currency overlay techniques seek to reduce exchange rate risk using predefined strategies. Once these have been set and developed by the overlay manager, they are implemented without any further discretional measures. There are several tools that can be used for this purpose, but all have the same primary objective: establish a short position with regard to currency surprise by borrowing in foreign currency or using financial derivatives.[8]

[6] Note, in particular, the study by Burnside et al. (2011).

[7] Pojarliev and Levich (2008) and Pojarliev and Levich (2010).

[8] Wystup (2017).

18.2.1 Implementing Passive Currency Overlay

The easiest passive currency overlay procedure is a full hedging approach, i.e., 100% coverage of the expected value of the portfolio at a specified future date, T, representing the time horizon of the procedure.[9]

For a portfolio P, denominated in dollars and held by an investor in the Eurozone, the returns without hedging and then those with currency risk hedging are calculated below. The unhedged (p_U) return is generated by P in the home currency without currency overlay and is calculated from the product of the future values of the local return (p_L) and the currency return (s):

$$p_U = (1 + p_L) \cdot (1 + s) - 1 = p_L + s + p_L \cdot s$$

A numerical example will help to illustrate the mechanism more clearly:

- EUR/USD exchange rate at t:[10] $1/S_t = 1.2500$, therefore $S_t = 1/1.25 = 0.8000$
- EUR/USD exchange rate at T: $1/S_T = 1.3000$, therefore $S_T = 1/1.30 = 0.7692$
- Portfolio value in foreign currency at t: $P_t = \$100$
- Portfolio value in foreign currency at T: $P_T = \$120$
- Currency return from t to T: $s = (0.7692 - 0.8000)/0.8000 = -3.85\%$
- Local return from t to T: $p_L = (\$120 - \$100)/\$100 = +20.00\%$
- Unhedged return from t to T: $p_U = 20\% - 3.85\% + 20\% \cdot (-3.85\%) = +15.38\%$

A full hedge strategy is possible, for example, by selling forward the foreign currency in which the investment is denominated at time T at a notional (N) equal to the expected value of P_T in the foreign currency:

$$N = \mathrm{E}_t[P_T] = P_t \cdot (1 + \mathrm{E}_t[p_L])$$

With a full hedge, the hedge ratio (H_T), i.e., the proportion of portfolio P covered by the forward contract, is 100%: $H_T = N/\mathrm{E}[P_T] = 1$. However, as the expected local return of the portfolio $\mathrm{E}[p_L]$ is not known ex ante, but can only be estimated, in operational practice it is assumed to be 0, so $\mathrm{E}[P_T] = P_t$, and consequently the notional N is equal to P_t.[11]

[9] Schmittmann (2010). The study by Chen et al. (2015) focuses on identifying the optimal hedge ratio.

[10] Conventionally, financial markets use indirect or quantity quotation for the euro exchange rate, which gives the amount of foreign currency needed to purchase a unit of home currency. For this reason, it is indicated here by the symbol $1/S$ (see also note 2). Moreover, for the sake of simplicity, the bid-ask spread is not considered. The rates given here must be interpreted as mid-spread and assumed to be identical both for the buyer and the seller of the currency.

[11] Eun and Resnick (1988).

This hypothesis is useful to simplify the calculation of the hedged return p_H, which is the price change of P expressed in the investor's home currency when using a currency overlay strategy. If the expected local return $E[p_L]$ were not 0, then p_H would be calculated with the following three elements:

- The hedged portion of the portfolio, without exchange rate risk: $H_T(1 + E[p_L])(1 + f)$.
- The unhedged portion of the portfolio (different from zero only when the hedge ratio H_T is not equal to 1): $(1 - H_T)(1 + E[p_L])(1 + s)$.
- The unexpected local return, which cannot be hedged as it is not known when the strategy is implemented: $(p_L - E[p_L])(1 + s)$.

Therefore, the hedged return derives from the sum of these three components:

$$p_H = \underbrace{H_T(1 + E[p_L])(1 + f)}_{\text{Hedged share of } P} + \underbrace{(1 - H_T)(1 + E[p_L])(1 + s)}_{\text{Unhedged share of } P \text{ (only if } H_T \neq 1)} +$$

$$+ \underbrace{\overbrace{(p_L - E[p_L])}^{\text{Unexpected return of } P} \quad (1 + s)}_{\text{Currency impact of the unexpected return of } P} - 1$$

On the other hand, if $E[p_L] = 0$, the hedged return of the portfolio can be calculated more simply as the unhedged return minus the effect of the hedging of the currency surprise (e):

$$\begin{aligned} p_H &= H_T(1 + f) + (1 - H_T)(1 + s) + p_L(1 + s) - 1 = \\ &= \underbrace{p_L + s + p_L \times s}_{\text{Unhedged return } (p_U)} - H_T \underbrace{(s - f)}_{\substack{\text{Currency} \\ \text{surprise}(e)}} = p_U - H_T(s - f) \end{aligned}$$

If the hedge ratio (H_T) equals 1 (full hedge), the calculation of the hedged return is further simplified:

$$p_H = p_L + s + p_L \cdot s - s + f = p_L + p_L \cdot s + f = p_L(1 + s) + f$$

Continuing the previous example, with a full hedge we must take into account the following terms of the forward contract:

- Time to maturity: $T - t = 1$ year
- Notional: $N = \$100$
- Interest rate on home currency (EUR) deposits from t to T: $r_D = 2.50\%$
- Interest rate on foreign currency (USD) deposits from t to T: $r_X = 4.00\%$

Now, we can calculate the forward exchange rate F_T using the theory of covered interest rate parity:

- Forward premium from t to T: $f = (1 + 2.50\% \cdot 1)/(1 + 4.00\% \cdot 1) - 1 = -1.44\%$;
- Forward exchange rate fixed at t for the period T: $F_T = 0.8000 \cdot (1 - 1.44\%) = 0.7885$.

Therefore, the hedged return of the portfolio in the investor's home currency is:

$$p_H = p_L(1 + s) + f = 20.00\% \cdot (1 - 3.85\%) - 1.44\% = 17.79\%.$$

The result can also be verified by dividing the hedging strategy into its components:

- Cost in EUR of the $100 purchased on the market at time T and transferred to the buyer of the forward:[12] $N \cdot S_T = \$100 \cdot 0.7692 = €76.92$.
- Income in EUR of the sale of dollars to the buyer of the forward: $N \cdot F_T = \$100 \cdot 0.7885 = €78.85$.
- Profit from the forward hedging: €78.85 − €76.92 = €1.93.
- Hedged return of portfolio P:

$$p_H = \frac{\overbrace{(\$120 \times 0.7692)}^{\text{Portfolio in } T\ (P_T)} - \overbrace{(\$100 \times 0.7692) + (\$100 \times 0.7885)}^{\text{Gain/Loss of the forward}}}{\underbrace{(\$100 \times 0.8000)}_{\text{Portfolio in } t\ (P_t)}} - 1 =$$

$$= \frac{€92.30 - €76.92 + €78.85}{€80.00} - 1 = 17.79\%$$

The results of the example may seem to suggest that this technique does not completely cancel the exchange rate risk. While the local return of the portfolio is 20%, in fact, the hedged return in EUR is only 17.79%. Actually, this conclusion would be wrong, as the exchange rate risk only refers to the currency surprise, and it is this component that is subtracted from the unhedged return by the full hedging. The currency surprise e is:

$$e = s - f = -3.85\% - (-1.44\%) = -2.41\%$$

Consequently, the hedged return is:[13]

[12] It is important to note that a forward contract does not require payment of an initial margin. It is sufficient to have a line of credit in foreign currency. Alternatively, for the main non-convertible currencies (i.e. currencies of those nations that impose restrictions on the movement of capital), a non-deliverable forward can be contracted, which does not require physical delivery of the future sum, but a net cash settlement.

[13] Note that the currency surprise is zero only in the case of full hedging.

$$p_H = p_U - H_T \cdot e = 15.38\% - 100\% \cdot (-2.41\%) = 17.79\%$$

Having analyzed the impact of passive currency overlay on performance, we can see that the volatility of the hedged returns of P is less than that of the unhedged returns.[14] As the product $p_L \cdot s_L$ is small in magnitude, at least for sufficiently short horizons, the variance of p_{H} is approximately equal to:

$$\text{V}[p_{\text{H}}] = \text{V}[p_L + p_L \cdot s + f] \approx \text{V}[p_L + f] = \text{V}[p_L] + \text{V}[f] + 2 \cdot \text{Cov}[p_L, f]$$

In contrast, the volatility of the unhedged portfolio is:

$$\text{V}[p_U] = \text{V}[p_L] + \text{V}[s] + 2 \cdot \text{Cov}[p_L, s]$$

Empirical studies have shown that the volatility of the currency return, V[s], is systematically greater than that of the forward premium, V[f], and, consequently, the hedging of the exchange rate risk leads to a reduction in portfolio variance.[15] Given this risk mitigation effect, currency hedging has been called a free lunch, as it enhances the risk-adjusted performance apparently at no extra cost. However, recent studies have questioned the effective benefit of this technique, as the reduction in volatility would be more than compensated by a fall in returns and by their greater non-normality.[16]

The importance of passive currency overlay in portfolio management depends, in particular, on the ratio of the volatilities of the exchange rates and of the hedged financial assets. When, as in the case of the equity asset classes, the standard deviation of returns dominates the usually smaller standard deviation of the exchange rates, the benefit of a passive currency overlay is more limited. This can be seen, for example, in the case of the MSCI EAFE index, used as a benchmark of foreign investments held by US subjects, as it reflects the equity markets of all developed countries in Europe, Asia, and the Pacific.[17] The trend over time of the US dollar versions of the index with and without currency hedge is shown in Fig. 18.1.

---

[14] James et al. (2012).

[15] For the sake of completeness, we should also consider the correlations between the portfolio of financial assets and the spot and forward exchange rates. On the other hand, the difference between the volatility of these two types of exchange rates is so marked as to make almost negligible the impact of any greater correlation between the portfolio and the forward rates compared to the spot rates. Indeed, Schmittmann (2010) calculates the volatility of f as between 7% and 16% of the volatility of s.

[16] The first article to propose the 'free lunch' theory—albeit uncommon in the financial field—was Perold and Schulman (1988). The theory has been re-stated in various forms by numerous academic studies in the course of the subsequent two decades. For a review of the literature and an empirical assessment (with a negative outcome), see De Roon et al. (2012).

[17] Therefore, in the so-called developed markets, the MSCI EAFE index excludes only the United States and Canada.

Fig. 18.1 Impact of passive currency overlay on the Equity asset class. Source: Data from Morningstar Direct

The graph clearly shows that the two-time series are strictly correlated. The standard deviation of the hedged version of the index is 86.7% of that of the common MSCI EAFE PR USD, with a risk mitigation of only 13.3%. Since the average return of the hedged index is 113.1% of the unhedged index, the return per unit of risk following application of the currency overlay is 1.3 times that generated without hedging.

The impact of the passive currency overlay is appreciably greater if it is applied to traditional asset classes with limited volatility, as, for example, bonds. In this case, the currency effect dominates the overall portfolio risk, as illustrated in Fig. 18.2, showing the hedged and unhedged versions of the Bloomberg Global Treasury ex US TR index, which comprises fixed-rate bonds issued by major world governments with the exception of the United States. The absence of the United States is useful to show the effect of exchange rate hedging on investors who take the US dollar as their base currency. The standard deviation of the hedged version of the index is just 37% of the unhedged index. The average return after hedging is instead slightly higher at 117.9% of the unhedged index. The performance of the hedged basket in terms of return per unit of risk is thus approximately 3.2 times that of the unhedged basket, a substantial improvement in terms of risk-adjusted performance, clearly higher than the one recorded with the equity indexes of the previous example.

Fig. 18.2 Impact of passive currency overlay on the Bond asset class. Source: Data from Morningstar Direct

18.2.2 Advanced Techniques of Passive Currency Overlay

Another passive currency overlay technique, which unlike full hedging is integrated with portfolio theory, is minimum-variance or regression hedging. The amount of hedging H_T is not arbitrarily fixed, but calculated so as to minimize portfolio variance.[18] Note that the variance of the returns of a hedged portfolio, in the general case in which H_T can be a value other than 1, is:

$$\mathrm{V}[p_H] = \mathrm{V}[p_L + s + p_L \cdot s - H_T(s - f)] = \mathrm{V}[p_U - H_T(s - f)]$$

Given the statistical properties of the ordinary least squares regression, this variance can be minimized simply by setting H_T as equal to the coefficient β estimated using the regression of the series of unhedged returns $\mathbf{p}_U$ on that of the currency surprises $\mathbf{e} = (\mathbf{s} - \mathbf{f})$:[19]

[18] Campbell et al. (2010).

[19] Vectors and matrices are indicated in bold type. In this case, each vector is composed of a time series of returns in periods prior to t, with an identical length of $T - t$.

$$\mathbf{p}_{\mathrm{U}} = \alpha + \beta \mathbf{e} + \boldsymbol{\varepsilon}$$

For a portfolio exposed to more than one currency, instead, a separate hedge ratio must be estimated for each one using, in this case, a multivariate regression. If there is a particularly large number of currencies, we can use currency basketing, i.e., the grouping of different currencies into baskets, the value of which is given by the most relevant currency included. The criterion used may be statistical, i.e., based on the levels of correlation between the currencies, or macroeconomic, i.e., combining the currencies of countries whose economic and financial systems are particularly integrated.

The passive currency overlay technique known as downside or option-based hedging involves the use of currency options,[20] which limit the downside risk of an investment. Their premium is, therefore, the cost of this strategy.[21] Given the variety of options available, from plain vanilla to the exotic, the strategy can be applied on a single currency, on currency baskets, or on the entire portfolio. In the latter case, the notional of the option must be calculated as the weighted average of the various currencies, each with a weight equal to its share of the investor's portfolio.[22]

In option-based strategies, the overlay manager defines the hedge ratio on the basis of the sensitivity of the option's position to the trend in the underlying currencies. This sensitivity is approximated by the Greeks delta and gamma, but in operational practice normally only the former (which assumes a linear relation between the underlying asset and the option price) is considered, and managers implement the so-called delta hedging. Since the delta of an option varies over time, this form of hedging requires repeated re-balancing, even if it is undertaken mechanically. This case, therefore, is known as dynamic hedging.[23]

The use of this technique that, albeit passive, can also adapt to the current market situation, may result in a significant increase in transaction costs derived from the recurring changes to the overlay portfolio. A measure of the distance of the hedge from the underlying portfolio is the tracking error, in this case calculated as the difference between the hedged and the local returns: $\mathrm{TE} = p_H - p_L$. This difference reduces, *coeteris paribus*, as the reference time horizon shortens, i.e., the more frequently the overlay portfolio is rebalanced.[24] On the other hand, the more

[20] A European put option grants the holder the right to sell a specific currency at a given future date in exchange for a sum expressed in another currency calculated on the basis of a specified exchange rate. European currency options are priced using a modified version of the Black-Scholes formula called Garman-Kohlhagen, which uses two risk-free rates, one for each currency.

[21] Celebuski et al. (1990).

[22] The topic of currency options is broad and complex. For details of applied aspects and an exhaustive empirical analysis, see: James et al. (2015).

[23] Record (2003).

[24] Note that investors make decisions based on a single-period horizon, so the returns path in the home currency between t and T is not relevant. If instead this discrete-time model is replaced by a

frequently the hedging is rebalanced in order to follow changes in P more closely, the higher the transaction costs. With forward, for example, we must consider:

- The bid-ask spread applied to the forward exchange rates.
- The forward roll tenor, i.e., the remaining life of the contracts used for the hedging.
- The cost of maintaining a liquid deposit or, alternatively, the effect of the sale of part of the portfolio to cover possible losses arising from the hedging strategy.

The overlay manager, therefore, must assess the trade-off between the tracking error and the increase in the transaction costs of the rebalancing strategy of the currency hedge.[25]

18.3 Active Currency Overlay

The techniques of active currency overlay aim to create value for the investor, so as not only to compensate the related transaction costs, but also to increase the risk-adjusted return of the portfolio by acting on the return rather than intervening only in terms of risk mitigation.[26] However, it is important to note that the distinction between active and passive management is less clear-cut than it may appear. Indeed, deciding whether or not to use hedging is itself a gamble on future exchange rates and therefore constitutes the implementation of the managers' view.

The enhancement of portfolio returns, which is the target of active currency overlay, aligns these techniques with active currency management. However, it must be underlined that they are two distinct sectors of asset management. Currency management is a so-called funded technique, in which managers operate in foreign exchange markets through their own portfolio of bonds and monetary market instruments. In contrast, currency overlay—even in its active form—is an unfunded technique, in which initially overlay managers have not a net active or passive position and use the financial assets of the portfolio to be hedged only as a guarantee for the credit lines needed for their strategies.[27]

While the tools used in active currency overlay are substantially the same as in passive approaches, the implementation is different. The overlay managers modify the hedging on the basis of discretional or systematic strategies. Discretional strategies allow management absolute freedom of action with the consequent advantages of adaptability to the contingent market situation, but with the typical disadvantages of operations subject to behavioral bias. Systematic strategies, instead, follow

continuous-time one, a constant rebalancing of the hedge ratio should be assumed. This assumption is theoretically acceptable, but impossible in operational reality.

[25] James (2004).

[26] Huttman and Harris (2006).

[27] James et al. (2012). The only exception is the possible use of currency options.

pre-defined rules, but these may be changed by managers over time or if the market situation requires.

18.3.1 Currencies as an Asset Class

The scientific literature has identified four risk factors specific to the main active strategies used by overlay managers.[28]

The first strategy makes use of the momentum factor, i.e., the exploitation of persistence in exchange rate trends. At time point t, the manager buys (sells) a foreign currency if the long (short) position has generated a profit in the period from $t - 1$ to t. In theoretical terms, this is a controversial question, as even though the analysis of currency managers' returns shows that it is effective, the possibility of exploiting exchange rate trends is not shared by all authors.[29] Practical implementation of this strategy involves the use of directional trades, in which managers modify the hedge ratio of the currency overlay in function of the expected changes in exchange rates based on previous trends.

Another strategy applied by managers uses the so-called value factor.[30] A currency is considered over- or undervalued on the basis of the Purchasing Power Parity (PPP) theory, according to which an exchange rate in equilibrium should make the prices of equivalent baskets of goods denominated in different currencies identical when they are converted into the same currency. Therefore, managers concentrate on exchange rates that show the greatest offset from the PPP, thus exploiting a subsequent return to equilibrium. An inherent risk in this strategy is the time taken for realignment, which may be so long as to make the exploitation of the PPP uneconomical. Furthermore, the exchange rate in equilibrium varies over time as local purchasing power varies. Consequently, the strategy must take a long-term view and not merely assess the PPP at a given moment in time. As for the momentum factor, managers again make use of directional trades by modifying the hedge ratio in line with the offsets from the PPP.

The third factor exploited by active overlay managers is the carry factor, underlying the so-called carry trade.[31] This technique, which is also known as the forward-rate bias strategy, derives from the non-applicability in the real world of the uncovered interest rate parity theory, according to which the value of the spot exchange rate at time point T can be identified ex ante and is the same as the future exchange rate fixed at time point t for the time point T. As, instead, the presence of a so-called forward discount anomaly has been demonstrated empirically, currencies do not follow a deterministic trend. In particular, those with relatively low (high)

---

[28] Pojarliev and Levich (2008) and Kroencke et al. (2014).

[29] Menkhoff et al. (2012a).

[30] Menkhoff et al. (2017).

[31] Menkhoff et al. (2012b).

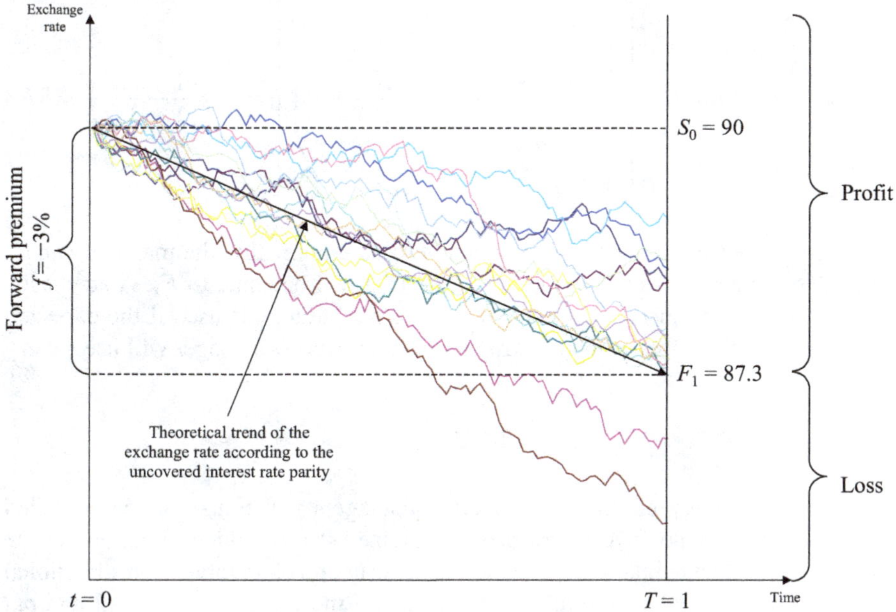

Fig. 18.3 Example of the profits and losses of carry trade operations

interest rates tend not to appreciate (depreciate) as much as would be required by the uncovered interest rate parity theory.[32] This is the phenomenon underlying the carry trade, which involves making investments in currencies with relatively high interest rates using financial resources borrowed in currencies with lower interest rates.

Figure 18.3 shows an example of carry trade applied to the exchange rate of the Australian dollar and the Japanese yen (AUD/JPY). Let the spot exchange rate be 90 yen to the dollar (S_0), the interest rate on 1-year deposits in AUD 3.5%, and on those in JPY 0.4%. The overlay manager contracts a 1-year loan in JPY at an interest rate of 0.4% and converts the proceeds into AUD at an exchange rate of 90, investing the sum in a 1-year deposit in AUD at an interest rate of 3.5%. The manager can generate a profit if the future value of the deposit in AUD, when it is converted into JPY at the spot exchange rate in 1 year, S_1, is greater than the amount of the loan in JPY including interest. This is the case when S_1 is greater than the forward exchange rate 1 year from today, F_1. Therefore, this forward exchange rate is the break-even point of a carry trade strategy and is equal to:[33]

[32] Engel (1996).

[33] Note that as the yen is the quoted currency in the exchange rate, the interest rate on deposits denominated in yen is the numerator in the formula to calculate the forward premium (indicated as f).

$$F_1 = S_0 \times (1+f) = S_0 \times \left[1 + \underbrace{\left(\frac{1 + r_{\mathrm{JPY}} \times \overbrace{(1-0)}^{T-t}}{1 + r_{\mathrm{AUD}} \times (1-0)} - 1\right)}_{\text{forward premium}}\right] = 90 \times \frac{1+0.4\%}{1+3.5\%} = 87.3$$

As uncovered interest rate parity is not verified empirically, the manager cannot assume that the spot exchange rate in 1 year (S_1) will be equal to F_1. A stochastic view must be taken instead, in which a carry trade strategy is used if the expected value today of S_1 is greater than F_1. In practice, an overlay manager will use a carry trade strategy if:

$$\mathrm{E}[S_1] > F_1$$

A further strategy exploited by overlay managers makes use of the so-called volatility factor. A positive return can be obtained if the sudden variations in the volatility of exchange rates can be exploited. This approach involves non-directional trades, such as strangles or straddles, by means of the combination of call and put options in strategies correlated to implicit volatility. Market timing, i.e. the choice of the point in time and time horizon in which to implement the operations, is extremely important in exploiting the volatility factor.

18.3.2 Implementing Active Currency Overlay

The views guiding active overlay managers in the choice of risk factors in the foreign exchange market can be subject to three distinct theoretical approaches:

- Macroeconomic models, which assume that exchange rates follow variables of the real economies.[34] The models that market forecasters follow most frequently are PPP and the Balassa-Samuelson effect, i.e., those linked to variables such as inflation and GDP growth rates.
- Technical analysis, which uses graphical methods and measures of the trends (known as oscillators) in exchange rates.
- Quantitative models, based on computer programs that analyze every statistical aspect of exchange rates time series in order to identify market inefficiencies (possible arbitrage positions on currency derivatives) or patterns that can be exploited by managers.

[34] Hauner et al. (2011).

Although the quantitative models are a development of the macroeconomic models and technical analysis, they cannot ignore a particularly serious risk in currency markets, commonly known as the "peso event." This definition was coined following the collapse of the Mexican peso in 1994 and indicates a sudden change in exchange rates of a magnitude not previously recorded in the available time series. The potential occurrence of such an extreme event—a sort of currency "black swan"—must necessarily be taken into account in any currency overlay strategy by using macroeconomic models, both to limit its negative impact on returns and, from the point of view of return enhancement, as a possible source of profits.

References

Ankrim E, Hensel C (1994) Multicurrency performance attribution. Financ Anal J 50(2):29–35

Burnside C, Eichenbaum M, Rebelo S (2011) Carry trade and momentum in currency markets. Annu Rev Financ Econ 3:511–535

Campbell JY, Serfaty-De Medeiros K, Viceira LM (2010) Global currency hedging. J Financ 65(1): 87–121

Celebuski MJ, Hill JM, Kilgannon JJ (1990) Managing currency exposures in international portfolios. Financ Anal J 46(1):16–23

Chen W, Kritzman M, Turkington D (2015) Alternative currency hedging strategies with known covariances. J Invest Manage 13(2):6–24

De Roon F, Eiling E, Gerard B, Hillion P (2012) Currency risk hedging: no free lunch. INSEAD Faculty & Research Working Paper, 2012/87/FIN

Engel C (1996) The forward discount anomaly and the risk premium: a survey of recent evidence. J Empir Financ 3(2):123–192

Eun CS, Resnick BG (1988) Exchange rate uncertainty, forward contracts, and international portfolio selection. J Financ 43(1):197–215

Hauner D, Lee J, Takizawa H (2011) In which exchange rate models do forecasters trust? IMF Working Paper, WP/11/116

Huttman M, Harris L (2006) Generating alpha from currency programs. Financ Anal J 8(4):48–54

James J (ed) (2004) Currency management: overlay and alpha trading. Risk Books, London

James J, Marsh I, Sarno L (2012) Handbook of exchange rates. Wiley, Hoboken

James J, Fullwood J, Billington P (2015) FX option performance. An analysis of the value delivered by FX options since the start of the market. Wiley, Chichester

Kroencke TA, Schindler F, Schrimpf A (2014) International diversification benefits with foreign exchange investment styles. Eur Finance Rev 18(5):1847–1883

Menkhoff L, Sarno L, Schmeling M, Schrimpf A (2012a) Currency momentum strategies. J Financ Econ 106(3):660–684

Menkhoff L, Sarno L, Schmeling M, Schrimpf A (2012b) Carry trades and global foreign exchange volatility. J Financ 67(2):681–718

Menkhoff L, Sarno L, Schmeling M, Schrimpf A (2017) Currency value. Rev Financial Stud 30(2): 416–441

Perold A, Schulman E (1988) The free lunch in currency hedging: implications for investment policy and performance standards. Financ Anal J 44(3):45–50

Pojarliev M, Levich R (2008) Do professional currency managers beat the benchmark? Financ Anal J 64(5):18–32

Pojarliev M, Levich R (2010) Trades of the living dead: style differences, style persistence and performance of currency fund managers. J Int Money Financ 29(8):1752–1775

Pojarliev M, Levich R (eds) (2014) The role of currency in institutional portfolios. Risk Books, London
Record N (2003) Currency overlay. Wiley, Chichester
Schmittmann JM (2010) Currency hedging for international portfolios. IMF Working Paper, WP/10/151
Wystup U (2017) FX options and structured products, 2nd edn. Wiley, Chichester
Xin H (2011) Currency overlay: a practical guide, 2nd edn. Risk Books, London

www.ingramcontent.com/pod-product-compliance
Lightning Source LLC
Chambersburg PA
CBHW050820120825
30947CB00044B/47
* 9 7 8 3 0 3 1 5 9 8 1 8 0 *